ANNUAL REVIEW OF NUTRITION

EDITORIAL COMMITTEE (1992)

ANNUAL REVIEW OF NUTRITION

VOLUME 12, 1992

ROBERT E. OLSON, *Editor*

State University of New York, Stony Brook

DENNIS M. BIER, *Associate Editor*

Washington University

DONALD B. McCORMICK, *Associate Editor*

Emory University

ANNUAL REVIEWS INC.   4139 EL CAMINO WAY   P.O. BOX 10139   PALO ALTO, CALIFORNIA 94303-0897

ANNUAL REVIEWS INC.
Palo Alto, California, USA

International Standard Serial Number: 0199–9885
International Standard Book Number: 0–8243–2812-4

Annual Review and publication titles are registered trademarks of Annual Reviews Inc.

⊗ The paper used in this publication meets the minimum requirements of
American National Standard for Information Sciences—Permanence of Paper
for Printed Library Materials, ANSI Z39.48-1984.

Annual Reviews Inc. and the Editors of its publications assume no responsibility for the
statements expressed by the contributors to this *Review*.

Typesetting by Kachina Typesetting Inc., Tempe, Arizona; John Olson, President;
Janis Hoffman, Typesetting Coordinator; and by the Annual Reviews Inc. Editorial Staff

PRINTED AND BOUND IN THE UNITED STATES OF AMERICA

PREFACE

Over fifty years ago nutrition was defined by C. Glen King as the science of food and its relationship to health. This quite accurate definition implies that nutrition is not a single science but a cluster of sciences relating to the production and utilization of food. Furthermore, nutrition, like medicine, is a field for both scientists and practitioners. Nutrition scientists are as diverse as molecular biologists who study nutrient-related gene expression and epidemiologists who track the movement of nutrient-related diseases in populations. The thread that unites them is the study of various aspects of food. The nutritional sciences, in fact, include essentially all biological sciences that can be applied to the study of nutritional problems.

This observation is well illustrated by Volume 12 of the Annual Review of Nutrition. Approximately 45% of the reviews are concerned with the basic sciences, 40% with clinical science, and 15% with epidemiology and public health. Of the ten basic science reviews, three deal with various aspects of the retinoids, including molecular biology of signal transduction, physiological and biochemical processing of vitamin A, and cancer prevention. The other reviews are devoted to the regulation of iron homeostasis, the regulation of enzymes of the urea cycle, dietary effects on biliary lipids, cellular and molecular aspects of adipose tissue development, multisite regulation of cellular energy metabolism, nutritional aspects of collagen metabolism, and the dietary impact of food processing. Seven of the ten use the techniques of molecular biology to elucidate nutrient effects on gene expression.

Of the nine clinical reviews, one deals with the physiological effects of dietary fiber, two are directed toward human milk and lactation, one discusses the role of homocysteine in the pathogenesis of atherosclerosis, and another unravels the pathogenesis of the eosinophilic-myalgia syndrome in patients taking tryptophan as a health food. Two reviews are devoted to vitamins: one focuses on the effects of cobalamin on the nervous system and the other on carotenoids, vitamin E, and vitamin C as anti-cancer agents. Completing the clinical group are a review of the physiology of placental nutrition and a discussion of the nutritional requirements of humans in space.

Three reviews concern epidemiology and public health. The preface by Dr. Gopalan, president of the Nutrition Foundation of India and former president of the International Union of Nutritional Sciences, describes the contribution of nutrition research to the control of undernutrition in India. He recounts the important research discoveries that have aided in the reduction of protein-energy malnutrition, xerophthalmia, pellagra, goiter, lathyrism, and iron deficiency in India during the past three decades. The second review is an epidemiologic study of hypocholesterolemia and cancer, and the third paper discusses the safety and efficacy of fat substitutes and the role of the Food and Drug Administration in monitoring new foods.

The breadth and the variety of technologies used in the nutritional sciences, as revealed by any thorough study of nutritional research (including the range of subjects considered in Volume 12), tend to create varying standards for reaching conclusions in different scientific areas. The precision of measurement is high and the average error of conclusions is fairly low in the basic science investigation of nutritional events. As one examines clinical investigations, generally, and specifically those dealing with nutrition, one finds somewhat less precision in measurements and a larger mean error in the derived conclusions. This finding is partly due to genetic heterogeneity in the human studies and less control over the experimental variables in the investigation.

Finally, as one surveys the results of epidemiologic investigations of nutritional disease, the precision of measurement is lowest, often because of the need to measure life style events that are difficult to document like diet and drug (including alcohol) intake, smoking, exercise, and work environment. Finally, the potential errors in conclusions reached are the sum of the errors in input factors, like diet, and output factors, like disease incidence, that require long-term follow-up for accuracy.

Although epidemiologists freely admit in seminars that *associations* of input and output variables do not constitute evidence of cause and effect, when an epidemiologic study is viewed in the light of public health practice the tendency is for epidemiologists to use their studies as a basis for public health action. For example, there is currently no acceptable scientific evidence that a diet containing 30% of calories from fat, 10% of calories from saturated fat, and 300 mg of cholesterol per day will extend life by preventing coronary heart disease (CHD) and cancer. Furthermore, the hypothesis that low-fat diets will prevent CHD has been tested mainly in middle-aged males. Yet most cardiologic and health societies the world over, including the US Committee on Diet and Health of the National Academy of Sciences and various expert panels of the National Heart, Lung, and Blood Institute, as well as a WHO study group, have recommended that this low-fat diet become a health standard for *all populations over two years of age worldwide*. This recommendation is made in the *belief* that this dietary regimen will improve health and reduce the incidence of noncommunicable diseases even though scientific data are lacking.

This recommendation is particularly erroneous when applied to children who, as a group, are not at risk for the chronic degenerative diseases, who need more fat calories to supply their energy requirement, and who, when placed on low-fat diets, have been shown to be undernourished with respect to other essential nutrients. Within medical practice, however, one can identify some adults at high risk for CHD who have benefited from such a dietary regimen, usually coupled with other measures to reduce risk.

The bottom line is that when data from epidemiologic studies and clinical trials are in conflict, and they often are, a consensus of experts in epidemiology usually recommends a course of public health practice that is scientifically radical rather than conservative. Although the nutritional sciences vary in terms of their study materials and research objectives, there should be a uniform standard for the evaluation of data, the development of conclusions, and the extrapolation of findings to generate nutrition policy.

I thank my associate editors, Dennis M. Bier and Donald B. McCormick, for reviewing manuscripts, all members of the Editorial Committee for their help in assembling the list of topics and authors, and the authors who contributed the excellent reviews that appear in Volume 12. Production editor Joan Cohen in Palo Alto, California, deserves our thanks for her diligence in producing this volume.

ROBERT E. OLSON
EDITOR

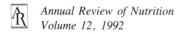

Annual Review of Nutrition
Volume 12, 1992

CONTENTS

x   CONTENTS   (*continued*)

SOME RELATED ARTICLES IN OTHER ANNUAL REVIEWS

From the *Annual Review of Biochemistry,* Volume 61 (1992):

Inositol Phosphate Biochemistry, *Philip W. Majerus*
The Ubiquitin System for Protein Degradation, *Avram Hershko and Aaron Ciechanover*

From the *Annual Review of Medicine,* Volume 43 (1992):

The Molecular Basis of Colon Cancer, *Anil K. Rustgi and Daniel K. Podolsky*
Prospects for the Prevention of Breast Cancer, *I. S. Fentiman*
The Role of Oxidized Low-Density Lipoproteins in the Pathogenesis of Atheroscler-
osis, *Sampath Parthasarathy, Daniel Steinberg, and Joseph L. Witztum*
Glucose Transporters, *Louis J. Elsas and Nicola Longo*

From the *Annual Review of Physiology,* Volume 54 (1992):

Arachidonic Acid Metabolism in Airway Epithelial Cells, *M. J. Holtzman*

From the *Annual Review of Public Health,* Volume 13 (1992):

How Much Physical Activity Is Good for Health, *S. N. Blair, H. W. Kohl, N. F.
Gordon, and R. S. Paffenbarger, Jr.*

For the convenience of readers, a detachable order form/envelope is bound into the back of this volume.

Annu. Rev. Nutr. 1992. 12:1–17

THE CONTRIBUTION OF NUTRITION RESEARCH TO THE CONTROL OF UNDERNUTRITION: The Indian Experience

C. Gopalan

Nutrition Foundation of India, B-37 Gulmohar Park, New Delhi 110 049, India

KEY WORDS: nutritional deficiency diseases of India, protein-energy malnutrition, hypovitaminosis A, pellagra, fluorosis, lathyrism

CONTENTS

1

0199-9885/92/0715-0001$02.00

INTRODUCTION

The eradication of undernutrition in developing countries can only be achieved through socioeconomic development and the elimination of poverty. But this consideration should in no way obscure the urgent need for scientific research in human nutrition and the important practical contributions that such research can offer. In India, as in all developing countries, a wide spectrum of specific nutritional deficiency diseases is found; the precise pathogenesis and approach to the prevention of each of these diseases needs careful elucidation.

Scientific research in India on problems of undernutrition during the last three decades has helped generate several practical methods of action. These contributions have also enriched nutrition science in general. Some of the recent developments are briefly discussed below. What follows is by no means a comprehensive catalogue of notable nutrition research contributions from India but only selected examples that have a direct bearing on major nutritional diseases encountered in the country today.

MAJOR NUTRITIONAL PROBLEMS

The outstanding nutritional problems in the country that currently account for significant impairment of the country's human resources are protein-energy malnutrition (PEM), which leads to physical and mental retardation and underdevelopment of several thousands of children; vitamin A deficiency, which results in keratomalacia and nutritional blindness; widespread iron-deficiency anemia, which impairs productivity and increases the vulnerability of poor populations to infections; and endemic goiter and other iodine-deficiency manifestations, which affect physical growth and mental development.

Other nutritional problems are somewhat more limited in distribution but, nevertheless, present fascinating challenges to biologists and health scientists: pellagra in the Deccan plateau, fluorosis in some parts of the country, lathyrism in Central India, and "lactose intolerance."

PROTEIN-ENERGY MALNUTRITION

Protein energy malnutrition (PEM) has held the central stage in global nutrition research for nearly three decades. Numerous studies have been published on this subject in India, but we highlight two important Indian contributions that have helped to bring about a major change in prevailing perceptions regarding its pathogenesis and approaches to its prevention.

The first was the clear demonstration that the primary dietary problem underlying PEM in India was *not* a deficiency of protein (as was hitherto widely claimed) but rather a deficiency of calories (21). Careful surveys of diets of children under five years of age in different parts of the country in communities where PEM was common showed that the daily protein intake ranged from 2.8 g/kg body weight to 1.7 g/kg—levels that on the basis of widely accepted international and national recommendations could be considered adequate. The daily calorie intakes, however, were of the order of 70 to 75 kcal/kg versus the figure of 100 kcal/kg, which is generally considered adequate. While the diets of over 90% of children were deficient in calories, the diets of only 35% of the children were deficient in protein. And if food intake had been raised to meet the calorie requirements of the latter group of children, their protein needs would have been met. Practically no situation was found in which the children's diets were adequate in calories but deficient in protein alone (30).

That the calorie gap in the prevailing diets of children of poor communities was the crucial factor was further demonstrated in yet another longitudinal study (of 14 months) (33) of a community of poor children whose daily diets provided no more than 700 kcal (with 18 g protein). This study showed that when the calorie gap in these diets was bridged by supplementation with 300 additional kcal daily, derived from ("empty calories") carbohydrate and fat sources (wheat flour, sugar, and edible oil) with little additional protein (no more than 3 g), growth performance was significantly improved and clinical manifestations of PEM were averted.

These findings indicated that the prevailing emphasis on "the protein gap" and "protein concentrates" was wholly misplaced; and that the solution to the problem of PEM fortunately need not depend on imports of expensive protein-rich concentrates but rather could be achieved through proper use of inexpensive traditional cereal-legume-based diets within the economic reach of poor families and within the country's resources.

The Practical Challenges

The real challenge in the prevention of PEM is to ensure that children under five years of age (especially under three years) get their habitual cereal-legume-vegetable foods in amounts adequate to meet their calorie needs. A hurdle in feeding cereal-legume-based diets to very young children stems from the low calorie-density of these diets, "the bulk factor." Research in India has also attempted to address this issue and to identify traditional home-based techniques through which this bulk factor can be overcome, the viscosity of cooked cereal foods can be reduced, and their calorie density increased (15, 34).

We now proceed to the second major contribution to the understanding of the PEM problem. It is well recognized that two distinct clinical syndromes are associated with PEM, namely kwashiorkor and marasmus. (Marasmus in early infancy associated with highly inadequate intakes of milk has to be considered as a separate category. Here we address marasmus in the preschool child.) At any given point of time throughout the 1960s and 1970s one could calculate on the basis of available survey data that while roughly about 1% of children under three years of age in poor communities may exhibit kwashiorkor, nearly 2 to 3% may show marasmus. Thus, at a given point of time, in the 1960s and 1970s in poor communities we could expect to see several thousands of poor children suffering from kwashiorkor and several thousands more suffering from marasmus, both of these conditions existing almost side by side in the same villages.

The earlier widely held postulate was that these two manifestations were different diseases with entirely different dietary etiologies: kwashiorkor was thought to be due primarily to "protein deficiency and calorie excess" while marasmus was thought to be due to calorie deficiency. If this was really the case, poor developing countries like India would have had on their hands two major public health problems requiring two entirely different approaches to their prevention and control. Studies carried out in India led to the conclusion that this fortunately was not the case.

A major clarification of immense practical significance was that kwashiorkor and marasmus are *not* two different diseases but just two facets (clinical manifestations) of one and the same central problem of PEM, with a common dietary etiology, and therefore requiring identical approaches for their solution.

On the basis of intensive studies of the actual diets and hormonal profiles of children suffering from these two syndromes, it was postulated that marasmus represents the stage of attempted "adaptation" to the nutritional stress. Thus hormonal mechanisms are invoked to ensure that the integrity of highly vulnerable tissues with a high protein-turnover, like the liver, pancreas, and viscera, is maintained at the expense of the muscle (21). Kwashiorkor represents the stage when this adaptation breaks down. Further studies helped to elucidate the probable nature of the hormonal changes that may be involved in such an adaptation mechanism leading to marasmus at one stage of the disease and to a breakdown of adaptation leading to kwashiorkor at a later stage (35, 36). Thus in marasmus, elevation of plasma cortisol levels was found to be of a higher order than in kwashiorkor; the adrenal cortical response to injection of corticotrophin was exaggerated. Plasma growth hormone levels and their response to stimuli that were found to be raised in kwashiorkor were not altered in marasmus. Plasma somatomedin activity was found to be low in kwashiorkor but not in marasmus.

These hormonal changes may help to ensure that in the face of stress posed by nutritional deprivation, muscle tissue is preferentially broken down so that the structural and functional integrity of more vital tissues like the liver, pancreas, and viscera is maintained. Marasmus may thus be looked upon as an extreme stage of adaptation—the farthest limit of what was described as a "contraction of the metabolic frontiers." When adaptation eventually breaks down because of continued stress or its aggravation by additional factors like fresh infections, etc, the fatty infiltration of liver, a fall in serum albumin, a reduction in serum enzymes, and edema ensue with the resultant picture of kwashiorkor. The fact that marasmus and kwashiorkor exist side by side in the same community subsisting on the same diet, as well as the fact that marasmus and kwashiorkor can exist *in the same child* at different points of time, lends support to the postulate that the two syndromes are but two facets of one and the same disease.

This clarification has helped place the entire problem of marasmus and kwashiokor in proper perspective as far as the public health approach to these diseases is concerned. It became clear that we were dealing with two manifestations of a *single* problem and that we did not need two divergent strategies for their control. It is hardly necessary to emphasize here the far-reaching practical implications of this conclusion.

The extensive work done on the foregoing and other aspects of the problem of PEM in India in the 1950s and 1960s has been reviewed in an earlier publication (62).

Misuse of the Term Adaptation

In the above discussion the term *adaptation* has been used to refer to the organism's response to stress. It is, however, important to emphasize that "adaptation" as used here is not synonymous with normalcy and therefore something that is "acceptable." Even a severely marasmic child with extreme emaciation but a normal liver is "adapted"! I emphasize this point because of the loose manner in which the term *adaptation* is now being misused to propagate the view that stunting and "moderate malnutrition" (which are not of such severity as to be life threatening) arising from PEM in Third World children may be viewed as an acceptable adaptation, consistent with their "culture" and environment.

In several recent publications (23–25, 27, 28) I have cautioned against the danger of the misuse of the concept of adaptation in a manner likely to promote social and political indifference to (and acquiescence in) "moderate malnutrition" in children.

Outstanding Indian contributions bearing on the question of "adaptation" to chronic energy deficit, and on the functional significance of small body size, are those of Shetty (59) and Satyanarayana et al (55–58). These authors emphasize the point that while small body size and behavioral alterations in work patterns arising from chronic energy deficit may facilitate survival under marginal living conditions, they cannot be viewed as beneficial satisfactory adaptive responses consistent with optimal levels of productivity and quality of life. Physical activity, especially with respect to moderate and strenuous work, is compromised in subjects with low body weight and poor muscle mass (55, 56), and there is no convincing evidence of increased metabolic efficiency with respect to energy handling by the residual active tissues of the body in chronically energy deficient small-sized individuals (59).

KERATOMALACIA AND NUTRITIONAL BLINDNESS

The current global approach to the prevention of keratomalacia arising primarily from vitamin A deficiency is to distribute two massive annual oral doses of synthetic vitamin A, one each at six-month intervals, to children under three years of age. This approach was developed and pioneered in India on the basis of experimental, clinical, and field studies. That vitamin A can be stored in the liver for prolonged periods and is released gradually to meet tissue needs has long been well-established. It was important, however, (a) to establish an optimal dosage of vitamin A that while not being toxic would adequately protect children against keratomalacia for fairly long durations; (b) to identify the most effective and feasible route and form of administration of the vitamin, (c) to demonstrate that under real-life conditions in the field, the administration of vitamin A in the dosage, frequency, and form identified above helps raise and maintain serum vitamin A levels in children over several months and thus does in fact offer protection against keratomalacia, and finally (d) to develop a practical procedure for the routine evaluation of the program by the public health agency.

Development of the Massive Dose Prophylaxis Approach

An indication of the extensive amount of work that was involved in the development of this prophylaxis program can be obtained from the following brief account of the study that preceded the introduction of this program in the National Health System.

CHOICE OF PREPARATION   In a preliminary trial in which a single dose of 300,000 IU of vitamin A was given orally as a water-immiscible preparation to a group of preschool children, 25% developed signs of acute though transient vitamin A toxicity characterized by raised intracranial tension (bulging fontanelles), restlessness, and fever (63). When the same amount was given as an oil-soluble preparation, the incidence of toxic signs was 4%. Moreover, animal studies had earlier shown that the best hepatic storage was achieved with oral administration of oil-soluble vitamin A (52). Therefore an oil-soluble preparation was chosen. Oral administration of 100,000 IU of oily vitamin A produced significant increases in serum vitamin A levels, but the same dose given intramuscularly had no such effect because much of the vitamin continued to remain at the site of infection (52). Oral dosage was therefore preferred; in addition, it was a more convenient method of administration.

OPTIMAL DOSAGE LEVEL   Longitudinal studies on groups of children showed that a single oral dose of 300,000 IU was able to sustain normal levels of serum vitamin A in children for a period of 6 months. Studies in which 200,000 IU of vitamin A were given along with labelled retinyl acetate (urinary excretion of the label was monitored) indicated that 70% of the dose was absorbed and somewhat less than 50% of the total dose was retained (51). The fact that urinary excretion of lysosomal enzymes arylsulphatase and acid phosphatase showed either no increase or an insignificant transient increase following the administration of such a massive dose demonstrated that lysosomal damage was not significant (49).

The real test of the efficacy of this prophylaxis approach was demonstrated by a prolonged 5-year field trial in which 2500 children under five years of age and drawn from several villages received 300,000 IU of vitamin A administered orally once a year for a period of 5 years. This study showed (a) a 75% reduction in the overall incidence of vitamin A deficiency in the community, (b) that no new cases of keratomalacia occurred during this period, and (c) serum vitamin A levels were consistently higher in children who received the dose than in children who had not (64).

After these extensive, time-consuming tests were completed, scientists advised the Government of India to include this program as part of routine primary health care in at least nine states of the Indian Union where evidence indicated that vitamin A deficiency was widely prevalent.

As a precautionary measure in order to reduce the risk of toxicity to the absolute minimum, it was also recommended that the dose be reduced to 200,000 IU and that it be given twice a year at six-month intervals. A practical simple method that was feasible under field conditions for evaluation of the implementation of the program was also developed (65).

This may seem to be an unqualified success story. However, looking back on these efforts that were initiated a quarter of a century ago, and now looking at the results, we may legitimately ask whether all the expectations that prompted these efforts by Indian nutrition scientists have in fact been fulfilled.

The control of nutritional blindness through the "short-cut" of administering synthetic vitamin A had been envisaged as a short-term approach—not as the permanent solution to the problem. It was always recognized that the ultimate solution lay in the promotion of the optimal use of β-carotene-rich foods, green leafy vegetables, in the diets of poor children. Unfortunately, the euphoria and complacency created by the introduction of the prophylaxis through massive dosage of synthetic vitamin A have to a considerable extent retarded research designed to develop and promote the better use of inexpensive β-carotene-rich foods in the country. If such research has not altogether come to a standstill, it is proceeding, at best, at a snail's pace as a program of low priority.

Secondly, implementation of the prophylaxis program is obviously slow, especially in states like Bihar. Most disconcerting is the fact that we do not have any authentic indication of what real impact the prophylaxis program has had on nutritonal blindness. The official figures of the annual incidence of cases of nutritional blindness will not stand scientific scrutiny. We do not even seem to have reliable data on changes in the annual incidence of keratomalacia in our leading ophthalmic and pediatric hospitals since the introduction of the program. In the absence of such data, we are in no position to counter or confirm the claims that are frequently made.

IRON DEFICIENCY ANEMIA

A major contribution of immense practical value has been the development of a technique for the fortification of common salt with iron. This research was not just a simple exercise in food technology but included studies of bioavailability and field trials to determine acceptability and efficacy.

Contrary to the general belief that iron deficiency anemia is mostly a disease of women in the reproductive age group, studies carried out under the auspices of the Indian Council of Medical Research showed that it is also very much a disease of preschool children and indeed even of adult men. A more recent study by the National Institute of Nutrition (47) showed that 65% of adult women, 75% of pregnant women, 77% of preschool children, and nearly 45% of adult men in poor rural communities were anemic. Anemia is probably the most extensive nutritional deficiency disorder in the country. Recent research indicates that apart from impairing productivity, the disease also has other functional implications. Though Indian diets generally provide 20 to 30 mg of iron daily, because of their high phytate content owing to the predominance of cereals, the bioavailability of dietary iron as determined by radioisotope technique is only 1 to 5%.

Iron Fortification

The rational ultimate answer to the problem would, of course, consist in the diversification and improvement of diets—a goal likely to take many years to achieve. The program of distribution of iron-folate tablets though the health system can reach only a small proportion of the population. Under the circumstances, a sensible practical approach would be to increase intake of iron through fortification of a suitable dietary item. Since common salt is a food commodity in universal use and since the poor take it in almost the same amounts as the rich, common salt was the obvious suitable candidate for iron fortification.

THE FORMULA   The challenge was to identify a formula for fortification that would satisfy the conflicting requirements of stability, acceptability, and bioavailability. The formula identified by the National Institute of Nutrition as satisfying these requirements consisted of ferro orthophosphate (3.5 g per kg) and sodium acid sulphate (5 g per kg) as an absorption promoter and provided 1 mg iron per gram of salt. Later this formula was further improved by replacing ferric phosphate with much less expensive ferrous sulphate (3500 ppm) and orthophosphoric acid or sodium orthophosphate (2800 ppm). With an estimated intake of 15 g of common salt per adult per day, common salt fortified as noted above will provide an additional 15 mg of iron.

FIELD STUDIES   The acceptability and efficacy of salt fortified as above was investigated through a field trial lasting for 18 months among 1600 (boys and girls) school children between 5 and 15 years of age who were divided into two matched groups, one receiving fortified salt and the other unfortified salt. The culinary acceptability and physiological efficacy of the fortification procedure were clearly demonstrated (47). Subsequently, a multicentric study coordinated by the National Institute of Nutrition and covering a population of 6000 was also undertaken. In this study the salt was made available to the population through the regular food distribution system. Analysis of data on hemoglobin levels in the experimental and control group again helped to confirm the significant impact of the procedure on the anemia problem (53). The government of India has now been persuaded to undertake this program initially in some parts of the country.

A major hurdle was the need to ensure that the fortification of common salt with iron was compatible with the government's decision to support the universal iodation of common salt intended for human consumption in the country as a method of prevention and control of endemic goiter. Scientists of the National Institute of Nutrition recently developed a feasible procedure for the simultaneous fortification of common salt with iron and iodine.

ENDEMIC GOITER

According to some estimates more than 40 million people in the country suffer from goiter. The National Goiter Program based on iodation of common salt was initiated in the latter half of the 1950s, but after an initial promising start had languished because of poor implementation and inept supervision (22). The emergence of new goiter-endemic areas has added fresh dimensions to the problem (2).

Recent studies from India have provided important indications of hitherto unsuspected serious dimensions of the problem of neonatal chemical hypothyroidism (NCH) in endemic goiter zones (37, 39). As high as 13% of neonates in endemic goiter areas have been shown to be functionally decompensated on the basis of T4 and TSH levels in their cord blood as determined by radioimmunoassays techniques. This observation corresponds closely to the finding of a study under the auspices of the Nutrition Foundation of India that nearly 15% of school children investigated in endemic goiter districts showed evidence of varying degrees of mental underdevelopment. These findings have lent urgency and importance to our National Goiter Control Program, which has yet to achieve its full potential. A somewhat complacent view of the role of iodine deficiency in mental underdevelopment had been taken earlier because of the very low incidence of cretinism and deaf mutism in the endemic goiter zone.

Parenteral administration of iodized oil to pregnant women is now being promoted in some quarters as a suitable prophylactic approach in relatively inaccessible areas until such time as the salt iodation program gathers full momentum. Recent Indian studies (38), however, sound a note of caution against resorting to this approach. According to these studies, iodized oil injections, when given to mothers particularly in the last trimester of pregnancy, do *not* help to reduce the incidence of neonatal chemical hypothyroidism; the relevance or even the safety of administering iodized oils to pregnant mothers has been seriously questioned. These views have been challenged, and apparently some controversy exists. Clearly, however, it would not be prudent to push ahead with any procedure regarding whose safety serious doubts have been expressed, especially when a time-tested, safe, and inexpensive alternative (salt iodation) is already available.

The foregoing account deals with major nutritional deficiency disorders affecting vast numbers of the country's population. A brief account now follows of scientific contributions from India towards the better understanding of four other nutritional disorders that, although not as extensive as those described earlier, are of considerable interest to health and nutrition scientists all over the world.

PELLAGRA

Pellagra is a classical nutritional deficiency disorder traditionally associated with poor populations whose staple is maize (corn). The low content in maize of the essential amino acid tryptophan, the precursor of nicotinic acid, has been generally held responsible. The important finding from India, which ran clearly counter to this well-accepted view, was that endemic pellagra in the Deccan plateau of India occurred in populations subsisting not on maize but on the millet sorghum (jowar), which is *not* poor in tryptophan. A feature common to both maize and sorghum, however, is the high content of the amino acid leucine. This finding sparked a new series of studies on pellagra, starting with a paper (32) in which we proposed that the high level of leucine in sorghum may play a positive role in the pathogenesis of the disease. Subsequent studies showed that excess leucine in otherwise poor diets could induce disturbances in the tryptophan—niacin pathway, which were reflected in increased urinary excretion of quinolinic acid on leucine feeding (11), a decreased rate of synthesis of nicotinamide nucleotides by erythrocytes (48), decreased activity of quinolinate phosphoribosyl transferase (QPRT), a key enzyme in NAD synthesis in liver (10), and a fall in platelet 5-hydroxy-tryptamine levels (43).

These studies showed that excess leucine in poor sorghum diets could bring about significant changes in key enzymes in the tryptophan-niacin pathway that ultimately resulted in decreased nicotinamide nucleotide formation from dietary tryptophan and thus led to conditioned deficiency of nicotinic acid.

Further studies showed that these effects of excess leucine could be countered by pyridoxine. Posttryptophan load excretion of xanthurenic acid, kynurenic acid, and quinolinic acid, which were initially raised in pellagrins, were reduced after pyridoxin treatment (42).

Thus the Indian studies indicate that in the pathogenesis of pellagra (which is by no means exclusively confined to maize eaters but can also occur in sorghum eaters), apart from tryptophan deficiency (in maize eaters), an excess of leucine (in sorghum eaters) and deficiencies of pyridoxin and nicotinic acid (in both maize eaters and sorghum eaters) may all play a part. The above observations on the possible role of leucine in pellagra have been contested and challenged by some scientists from Europe and the United States. Some recent reports from England, however, have lent support to the observations from India. Magboul & Bender (44) showed that diets that provide excess leucine brought about a "significant reduction in the concentrations of nicotinamide nucleotides in liver and blood." The effect was only apparent when the diets provided less than adequate amounts of nicotinamide. The addition of leucine was also shown to bring about "significant activation of tryptophan oxygenase and inhibition of kynurenase." In a subsequent communication Bender (12) reported that "dietary excess of leucine led to inhibition of kyureninase and increased the activity of piconilate carboxylase—which could be expected to explain decreased synthesis of nicotinamide nucleotides."

LATHYRISM

Neurolathyrism characterized by spastic paraplegia affecting the lower extremities is an ancient disease and is endemic in areas in which diets are predominantly based on the pulse *Lathyrus sativus*. Though the association of lathyrism with the consumption of the pulse has been known for over a century, the toxic factor in the pulse responsible for the disease could not be identified, mainly because the disease could not be reproduced in experimental animals.

A major breakthrough was achieved at the National Institute of Nutrition (54) when it was demonstrated that alcoholic extracts of lathyrus sativus could produce neurotoxic manifestations when injected into baby chickens. The toxic factor was subsequently isolated and identified as BOAA (B-oxalyl aminoalanine) (1, 46). A simple household method by which the toxin can be completely removed from the seed by steeping the seeds in hot water for about an hour, or by parboiling the seed in a process similar to the parboiling of rice, was also developed (45). Simultaneously, attempts were also made by agricultural scientists in India to identify and selectively propagate genetic strains of lathyrus sativus low in BOAA. These attempts were not successful, but recently new attempts have been made in other parts of the world (Canada).

Thus far, the scientific research efforts that have gone into the elucidation of the problem of lathyrism have not directly resulted in the eradication of the disease. Attempts to ban the cultivation of the offending crop failed because the crop is hardy and able to grow on unirrigated land; it has been the staple of the poor and there was no easy substitute. Recently, however, following the relative decline in production of pulses in the wake of the green revolution, *Lathyrus sativus* has found a flourishing market as an adulterant of other more expensive pulses like Bengal gram and reportedly is being widely exported out of the endemic zone for this purpose (26). To the extent to which these new developments dictated by commercial considerations reduce sole reliance by the poor of the endemic regions on *Lathyrus sativus* as their staple food, they may be of some benefit, but if the profitability of adulteration should act as an incentive for intensive cultivation of *Lathyrus sativus,* the problem would be disseminated well beyond the present "endemic" zones.

FLUOROSIS

While in other parts of the world, there is active support for the fluoridation of water as a method for the prevention of dental caries, in India the problem in some parts of the country (especially Panjab and Andhra Pradesh) is the presence of excess fluoride in drinking water that leads to skeletal changes—sometimes so severe as to be incapacitating.

Endemic fluorosis was in fact first identified in the country in some areas of the present Andhra Pradesh that were then parts of the erstwhile Madras Presidency (60). Subsequently, endemic fluorosis belts were also identified in Punjab (61). The disease affects the rural poor in areas where the drinking water may contain as high as 15 ppm of fluoride. While attempts to defluoridate water using inexpensive adsorbents like paddy-husk carbon have failed to make any significant dent in the problem, recent studies in India have shown that the disease has acquired new serious dimensions. In parts of Andhra Pradesh where the disease has been known to be endemic, it was noticed that large numbers of adolescents and young adults had started to develop serious bone deformities generally characterized by marked genu-valgum (40, 41) manifestations that had never been seen in those areas in earlier years. The prevalence of these deformities ranged from about 2% in some areas to as high as 17% in others and was found to be higher in sorghum eaters than in those not subsisting on sorghum.

A series of studies revealed that this new aggravation of an ancient disease was related to the construction of the large Nagarjunasagar Dam, which had impounded large amounts of water. The sequence of events leading to these new manifestations was described as follows: Construction of dam and impounding of water → elevation of subsoil water in wide areas in the vicinity of the dam → soil alkalinity, changes in the concentration of trace elements in food grains grown in the area, and, in particular, an increase in concentration of molybdenum → increased urinary excretion of copper, osteoporosis (superadded to fluorosis) → genu-valgum. Positive evidence in support of this hypothesis has been forthcoming from several studies at the National Institute of Nutrition.

As a preventive measure, it was suggested that the rural poor should be advised against drawing water for drinking purposes from the wells in the area because of high fluoride concentrations. Instead the Government was advised that part of the impounded water, which was being diverted almost entirely for irrigation purposes through canals, should be made available for drinking purposes.

Here is an instance of an unexpected ecological repercussion of a developmental program that was envisaged as an unmixed blessing and would help irrigate vast tracts of land and grow more food.

LACTOSE INTOLERANCE

Chronic diarrhea arising as a result of intolerance to disaccharides owing to a deficiency of disaccharidases is now being reported from some parts of the world. The incidence of lactose intolerance is reported to be high among Asians and Africans and rare among Caucasians (5, 13, 14, 16). On the basis of these findings it was postulated that inclusion of milk in the diets of undernourished populations of developing countries might lead to undesirable sequelae such as abdominal discomfort and diarrhea.

Indian studies (50) showed that there was no correlation between signs of lactose intolerance, as determined by lactose-overloading tests, and the levels of the enzyme. They pointed out that lactose intolerance demonstrated under the artificial conditions of the tolerance tests did not necessarily imply milk intolerance. Thus there was no reason to withhold milk from undernourished Asian children nor to provide them with lactase tablets every time they had a milk drink as was being suggested by some commercial interests. These observations helped to dispel doubts about a traditionally highly valued item of Indian diets.

OTHER CONTRIBUTIONS

The foregoing account does not do full justice to all the important recent Indian work in the field of human nutrition. The emphasis here has been only on studies related to major public health nutrition problems. Significant contributions to nutrition science such as those of Ganguly (17) on vitamin A metabolism, of Bamji et al (3, 4) on riboflavin nutrition, and of Bhavani Belavady (6–9) and Gopalan (18–20, 29) on aspects of human lactation, to mention only a few, have not been discussed.

Perhaps the most important contributions, although the least spectacular, are the continuing investigations, compilations, and updating of information on the nutritive value of Indian foods (31)—the work that provides the basic foundation for all dietary recommendations and programs for dietary improvement. However, as mentioned at the outset, I have sought not to present a comprehensive catalogue but rather to give a few select examples of Indian scientific contributions to the amelioration of undernutrition in developing countries.

SUMMARY

Since diseases directly related to undernutrition are the major public health problems of India, nutrition research in the country has been largely directed towards elucidating their causes and identifying the most feasible methods for their prevention and control. This effort is an interdisciplinary exercise carried out in the laboratory, the clinic, and the field, with close interaction among biochemists, clinicians, and epidemiologists. Some of the identified solutions have found practical application; but, as in other areas of scientific endeavor, a gap exists between the acquisition of knowledge in the laboratories and its application in the field. Today, thanks to research efforts of the last few decades, we have the knowledge with which most diseases related to undernutrition can be prevented. Unfortunately, however, we do not always have the means of applying this knowledge under real-life conditions in the field. Even so, nutrition research during the last few decades has contributed significantly to the amelioration of undernutrition among poor communities in India.

Literature Cited

1. Adiga, P. R., Rao, S. L. N., Sharma, P. S. 1963. Some structural features and neurotoxic action of a compound from *L. sativus* seeds. *Curr. Sci.* 32:153
2. Agarwal, K. N., Agarwal, D. K., Srivastava, S. 1984. Emergence of goitre in Delhi. *Nutr. Found. India Bull.* 5(4):6–7
3. Bamji, M. S. 1969. Glutathione reductase activity in red blood cells and riboflavin nutritional status in humans. *Clin. Chem. Acta* 26:263–69
4. Bamji, M. S., Bhaskaram, P., Jacob, C. M. 1987. Urinary riboflavin excretion and erythrocyte glutathione reductase activity in preschool children suffering from upper respiratory infections and measles. *Ann. Nutr. Metab.* 31:191–96
5. Bayless, T. M., Rosenweig, N. S. 1966. A racial difference in incidence of lactase deficiency. A survey of milk intolerance and lactase deficiency in healthy adult males. *J. Am. Med. Assoc.* 197(12):968–72
6. Belavady, B. 1969. Nutrition in pregnancy and lactation. *Indian J. Med. Res.* 57:63–74
7. Belavady, B. 1980. Dietary supplementation and improvement in lactation performance of Indian women. In *Maternal Nutrition During Pregnancy and Lactation,* ed. H. Aebi, R. Whitehead, pp. 264–73. Bern: Hans Huber
8. Belavady, B., Gopalan, C. 1959. Chemical composition of human milk in poor Indian women. *Indian J. Med. Res.* 47:234–45
9. Belavady, B., Gopalan, C. 1960. Effect of dietary supplementation on the composition of breast milk. *Indian J. Med. Res.* 48:518–23
10. Belavady, B., Gopalan, C. 1965. Production of black tongue in dogs by feeding diets containing jowar (Sorghum vulgare). *Lancet* 2(424):1220–21
11. Belavady, B. L., Srikantia, S. G., Gopalan, C. 1963. The effect of the oral administration of leucine on the metabolism of tryptophan. *Biochem. J.* 87:652–55
12. Bender, D. A. 1983. Effects of a dietary excess of leucine on the metabolism of tryptophan in the rats: a mechanism for the pellagragenic action of leucine. *Br. J. Nutr.* 50:25–32
13. Bolin, T. D., Davis, A. E. 1969. Asian lactose intolerance and its relation to intake of lactose. *Nature* 222:382–83
14. Cook, G. C., Kajubi, S. K. 1966. Tribal incidence of lactose deficiency in Uganda. *Lancet* 1(440):725–29
15. Desikachar, H. S. R. 1981. Malting as an aid in reduction of viscosity of cereal and legume based diets. *Nutr. Found. India Bull.,* April, p. 6
16. Flatz, G., Saengudom, C., Sanguanbhokhai, T. 1969. Lactose intolerance in Thailand. *Nature* 221:758–59
17. Ganguly, J. 1989. *Biochemistry of Vitamin A.* Boca Raton, Fla: CRC Press. 221 pp.
18. Gopalan, C. 1956. Protein intake of breastfed poor Indian infants. *J. Trop. Pediatr.* 2:89–92
19. Gopalan, C. 1958. Studies on lactation in poor Indian communities. *J. Trop. Pediatr.* 4:87–97
20. Gopalan, C. 1962. Effect of nutrition on pregnancy and lactation. *Bull. WHO* 26:203–11
21. Gopalan, C. 1968. Kwashiorkor and marasmus. Evolution and distinguishing features. In *Calorie Deficiencies and Protein Deficiencies,* ed. R. A. McCance, E. M. Widdowson, pp. 49–58. London: Churchill
22. Gopalan, C. 1981. The National Goitre Control programme—a sad story. *Nutr. Found. India Bull.,* July, pp. 1–2
23. Gopalan, C. 1982. The nutrition policy of brinkmanship. *Nutr. Found. India Bull.,* October, pp. 5–6
24. Gopalan, C. 1983. Small is healthy? For the poor, not for the rich? *Nutr. Found. India Bull.,* October, pp. 1–5
25. Gopalan, C. 1984. Child survival and child nutrition. *Natl. Found. India Bull.* 5(1):1–3
26. Gopalan, C. 1984. The lathyrism problem—current status and new dimensions. *Sci. Rep. 2, Nutr. Found. India,* pp. 54–55
27. Gopalan, C. 1988. Stunting: Significance and implications for public health policy in linear growth retardation in less developed countries. *Nestle Nutrition Workshop Series, Nestec Ltd., New York,* ed. J. C. Waterlow, 14:266–69. Vevey: Raven
28. Gopalan, C. 1989. Undernutrition: measurement and implications. In *Nutrition, Health and National Development, Nutr. Found. India Spec. Publ. Ser.* 4:69–100
29. Gopalan, C., Belavady, B. 1961. Nutrition and lactation. *Fed. Proc.* 20(Suppl. 7):177–84
30. Gopalan, C., Narasinga Rao, B. S. 1971. Nutritional constraints on growth and development in current Indian dietaries. *Proc. Nutr. Soc. India* 10:111
31. Gopalan, C., Rama Sastri, B. V., Balasubramanian, S. C. 1971. *Nutritive Value of Indian Foods.* Revised and updated by B. S. Narasinga Rao, Y. G.

Deosthale, K. C. Pant. 1989. Hyderabad: Natl. Inst. Nutr., Indian Counc. Med. Res. 156 pp.

32. Gopalan, C., Srikantia, S. G. 1960. Leucine and pellagra. *Lancet* 1:954–57

33. Gopalan, C., Swaminathan, M. C., Krishnakumari, V. K., Rao, D. H., Vijayaraghavan, K. 1973. Effect of calorie supplementation on growth of undernourished children. *Am. J. Clin. Nutr.* 26:563–566

34. Gopaldas, T., Mehta, D., Chinnama, J. 1988. Reducing the bulk of cereal weaning gruels—a simple technology for rural homes. *Natl. Found. India Bull.* 9(1):5–8

35. Jaya Rao, K. S. 1974. Evolution of kwashiorkor and marasmus. *Lancet* 1(860):709–11

36. Jaya Rao, K. S., Srikantia, S. G., Gopalan, C. 1968. Plasma cortisol levels in protein calorie malnutrition. *Arch. Dis. Child.* 43:365

37. Kochupillai, N. 1984. Neonatal thyroid status in iodine deficient environment of sub Himalayas region. *Indian J. Med. Res.* 80:293–99

38. Kochupillai, N., Godbole, N. M. 1986. Iodised oil injections in goitre prophylaxis: possible impact on the newborn. *Nutr. Found. India Bull.* 7(4):1–3

39. Kochupillai, N., Yalow, R. S. 1978. Preparation, purification and stability of high specific activity I 125-labelled thyronines. *Endocrinology* 102(1):128–35

40. Krishnamachari, K. A. V. R., Krishnaswamy, K. 1973. Genu valgum and osteoporosis in an area of endemic fluorosis. *Lancet* 2(834):877–79

41. Krishnamachari, K. A. V. R., Krishnaswamy, K. 1974. An epidemiological study of the syndrome of genu valgum among residents of endemic areas for fluorosis in Andhra Pradesh. *Indian J. Med. Res.* 62:1415

42. Krishnaswamy, K. 1979. Role of pyridoxine in pellagra—V. N. Patwardhan Prize Oration, *Indian Counc. Med. Res. Bull.* 9(1):1–3

43. Krishnaswamy, K., Rao, S. B., Raghuram, T. C., Srikantia, S. G. 1976. *Am. J. Clin. Nutr.* 29:177–81

44. Magboul, B. I., Bender, D. A. 1983. The effect of dietary excess of leucine on the synthesis of nicotinamide nucleotides in the rat. *Br. J. Nutr.* 49:321–29

45. Mohan, V. S., Nagarajan, V., Gopalan, C. 1966. Simple practical procedure for the removal of toxic factors in *L. Sativus. Indian J. Med. Res.* 54:410

46. Nagarajan, V., Roy, D. N., Mohan, V.

S., Gopalan, C. 1963. *Ann. Rep., Nutr. Res. Lab., Indian Counc. Med. Res., Hyderabad,* pp. 39–40

47. Narasinga Rao, B. S. 1981. Control of anaemia by fortification of common salt with iron. *Nutr. Found. India Bull.,* April, pp. 7–8

48. Raghuramulu, N., Srikantia, S. G., Narasinga Rao, B. S., Gopalan, C. 1965. Nicotinamide nucleotides in the erythrocytes of patients suffering from pellagra. *Biochem. J.* 96:837–39

49. Reddy, V., Mohanram, M. 1971. Urinary excretion of lysosomal enzymes in hypovitaminosis and hypervitaminosis A in children. *Int. J. Vitam. Nutr. Res.* 41:321–26

50. Reddy, V., Pershad, J. 1972. Lactose intolerance in Indians. *Am. J. Clin. Nutr.* 25:114

51. Reddy, V., Sivakumar, B. 1972. Studies on vitamin A absorption. *Indian Pediatr.* 9:307–10

52. Reddy, V., Srikantia, S. G. 1966. Serum vitamin A in kwashiorkor. *Am. J. Clin. Nutr.* 18:34–37

53. Report of the Working Group on Fortification of Salt with Iron. 1972. *Am. J. Clin. Nutr.* 35:1442–51

54. Roy, D. N., Nagarajan, V., Gopalan, C. 1963. Production of neurolathyrism in chicks by the injection of *Lathyrus sativus* concentrates. *Curr. Sci.* 32:116–18

55. Satyanarayana, K. 1989. Body mass index, nutritional status and productivity. *Nestle Found., Lausanne, Ann. Rep.*

56. Satyanaranayana, K., Naidu, A. N., Chatterjee, B., Rao, B. S. N. 1977. Body size and work output. *Am. J. Clin. Nutr.* 30:322–25

57. Satyanarayana, K., Naidu, A. N., Rao, B. S. N. 1979. Nutritional deprivation in childhood and the body size, activity and physical work capacity of young boys. *Am. J. Clin. Nutr.* 32:1769–75

58. Satyanarayana, K., Venkataramana, Y., Someswara Rao, M. 1989. Nutrition and work performance: studies carried out in India. *Proc. Int. Congr. Nutr., 14th, Seoul,* 1:302–5

59. Shetty, P. S. 1990. Energy metabolism in chronic energy deficiency. Dr. S. G. Srikantia memorial lecture. *Proc. Nutr. Soc. India* 36:89–98

60. Short, H. M., McRobert, G. R., Bernard, T. W., Nayar, A. S. M. 1937. Endemic fluorosis in Madras Presidency. *Indian Med. Gaz.* 72:369–98

61. Singh, A., Jolly, S. S. 1961. Endemic fluorosis. *Q. J. Med.* 30:357

62. Srikantia, S. G. 1969. Protein calorie

malnutrition in Indian children. *Indian J. Med. Res.* 57(8):36–53 (Suppl.)

63. Swaminathan, M. C. 1971. Prevention of vitamin A deficiency by administration of massive doses of vitamin A. *Proc. Asian Congr. Nutr., 1st, Hyderabad,* pp. 696–701

64. Swaminathan, M. C., Susheela, T. P., Thimmayamma, B. V. S. 1970. Field prophylactic trial with a single annual oral massive dose of vitamin A. *Am. J. Clin. Nutr.* 23:119–22

65. Vijayaraghavan, K., Naidu, A. N., Rao, N. P., Srikantia, S. G. 1975. A simple method to evaluate the massive dose vitamin A prophylaxis programme in pre-school children. *Am. J. Clin. Nutr.* 28(10):1189–93

Annu. Rev. Nutr. 1992. 12:19–35

THE PHYSIOLOGICAL EFFECT OF DIETARY FIBER: AN UPDATE

Martin A. Eastwood

Department of Medicine, University of Edinburgh, Western General Hospital, Edinburgh EH4 2XU, United Kingdom

KEY WORDS: structure, chemistry, function

CONTENTS

INTRODUCTION

Dietary fiber is a member of the family of dietary complex carbohydrates. These complex carbohydrates have individual and diverse actions. At present we cannot chemically identify or predict the biological action in the gastrointestinal tract of individual polymers. Each polymer is peculiar in its biological action and is modified by its physical format and processing (10). Dietary polysaccharide polymers that exceed 20 sugar residues can be classified as dietary complex carbohydrates (43). An alternative description is non-starch

0199/9885/92/0715-0019$02.00

polysaccharides (NSP) and starch (21). Both of these are chemical descriptions, but they may not satisfy physiologists and consumers who want to know what is the best source of edible fiber, what is the nutritional benefit of a particular fiber, and how will cooking and processing modify the desired effect.

PLANT STRUCTURE AND CHEMISTRY

Considerable anatomical differences exist between and within economically important plant groups (8). The variety of changes in the cell wall could be important in determining the diversity of actions of sundry dietary fibers in the gastrointestinal tract.

Cell wall structure not only differs among plant species but also during normal development within one species or even within a single cell. The composition of the cell wall is dependent not only on the plant species but also on the tissue type, the maturity of the plant organ at harvesting, and to some extent on the post-harvest storage conditions. Noncarbohydrate components of plant cell walls may influence physiology in the plant and the nutritional potential of the plant fibers, e.g. lignin, phenolic esters, cutin and waxy materials, and suberin. As the plant cell wall develops, it is constantly altered to confirm to a specific developmental pattern that results in the unique shape of any given species (3, 26, 47).

Parenchymatous tissues are the most important source of vegetable fiber. The vascular bundles and parchment layers of cabbage leaves, runner beans, pods, asparagus stems, and carrot roots are relatively immature and only slightly lignified on harvesting and digestion. The soft fruits such as strawberries contain very little dietary fiber but luxious amounts of water. Lignified tissues are of greater importance in cereal sources such as wheat bran and oat products. Cereals contain very little pectic substances, but there is substantial arabinoxylan in wheat or β-glucan in barley and oats. The distribution of polysaccharides within the plant tissue also varies. Much of the β-glucan in oats is concentrated in the cells of the outermost layer of the seeds, whereas the β-glucans in barley are more evenly distributed (3, 26, 47).

An important aspect of the plant cell wall is the interlocking of water-soluble polysaccharides to form biological barriers that are water resistant. Many of the constituents of the plant cell wall, hemicelluloses and pectins, are soluble in water after extraction. This solubility, unmasked by the extraction process, contrasts with the insolubility of the complex polysaccharide of the intact cell wall. The debate of the quantitative measurement of soluble and insoluble fiber is an analytical contrivance rather than an argument that has real significance either in regard to the physiology of the plant or to the recipient of the food fiber.

The backbone of the plant cell wall, cellulose, is a polymer of linear β-(1→4)-linked glucose molecules, several thousand molecules in length. Cellulose occurs largely in a crystalline form in microfibrils, coated with a monolayer of more complex hemicellulosic polymers held tightly by hydrogen bonds. These are embedded in a gel of pectin polysaccharides. The cellulose microfibrils are coated with a layer of xyloglucans bound by hydrogen bonds, and this enables the insoluble cellulose to be dispersed within the wall matrix. Substitution of the hydroxyl group at C6 with xylose, as in xyloglucans, renders cellulose more soluble in alkali and water. Hemicelluloses provide part of the true rigidity of the cell wall. The important hemicelluloses are xyloglucans, xylans, and β-glucans. Xyloglucan is a linear (1→4)β-D-glucan chain substituted with xylosyl units that may be further substituted to form galactosyl-(1→2)β-D-xylosyl or fucosyl-(1→2)α-D-galactosyl-(1→2)β-D-xylosyl units.

The pectins may act as biological glue, cementing cells together through ionic bonds. The precise function of pectins within the cell wall is unclear, but they are closely associated with calcium. Most pectins are probably derived from the primary cell wall and appear to be soluble only after calcium ions are removed. The principal cross-linkage is provided by the helical (1→4)α-D-galactosyluronic groups from adjacent polysaccharides, and condensation with calcium converts soluble pectin into rigid "egg-box" structures. The extent of calcium cross-bridging or esterification through aromatic linkages, and even degree of branching and size of neutral sugar side chains, influence gel flexibility, cell wall porosity, and interaction with hemicellulosic polymers.

The glycoproteins within the cell wall can provide extensive cross-linkages across the different polysaccharide components of the cell wall and may act to form a network with the cellulose microfibrils within a hemicellulose pectin gel (47).

Some plant polysaccharides are harvested from the plant and purified before inclusion as processed food components or therapeutic use. These are more often soluble polysaccharides and include the following.

Guar gum ($M_r = 0.25 \times 10^6$), linear nonionic galactomannan.

Gum karaya ($M_r = 4.7 \times 10^6$), a cylindrical complex polysaccharide partially acetylated and highly branched with interior galacturonorhamnose chains to which are attached galactose and rhamnose end groups. Glucuronic acid is also present.

Gum arabic ($M_r = 0.5–1.5 \times 10^6$) with a complex acidic heteropolysaccharide based on a highly branched array of galactose, arabinose, rhamnose, and glucuronic acid. Uronic acid residues tend to occur on the periphery of an essentially globular structure.

Gum tragacanth (M_r = 0.5–1 × 10^6), a complex gum with two major components, bassorin and tragacanthin, is composed of arabinose, fucose, galactose, glucose, xylose, and galacturonic acid.

These gums and mucilages are used as additives by food manufacturers and therefore may contribute less than 2% of a food (16).

Starch, a ubiquitous storage polysaccharide, is an α-linked glucan and is the major carbohydrate of foods such as cereal grains and potatoes. The majority of dietary starches are susceptible to hydrolysis by salivary and pancreatic α-amylases, but, a proportion of dietary starch may escape digestion by human α-amylase. This may be due either to physical inaccessibility or to the inhibition of amylase activity by a rigid stereochemical structure caused during food processing. This starch can pass undigested into the colon where it may act as a substrate for the bacteria and therefore have some of the biological properties of the undigestible plant cell wall (22).

In the plant, starch is contained within the granular structures in a closely packed, partly crystalline form. The actual crystalline structure of the starch granule is believed to depend on the chain length of amylopectin (6, 9).

PHYSICAL PROPERTIES OF DIETARY FIBER

When considering the action of cooking on cell wall structure and comparing cooked and raw plant foods, the different solubility characteristics of cell wall polysaccharides should be considered. Cell wall structures are degradable to varying degrees, depending on the structure and the conditions used. An important function of insoluble fibers is to increase lumenal viscosity in the intestine. It is not yet clear whether the soluble fibers in food have the same effect. Other polymeric components of the diet (proteins, gelatinized starch) and mucus glycoproteins liberated from the epithelia contribute to viscosity in the same way. Particulate material present in chyme, such as insoluble fiber or hydrated plant tissues, also contribute to a lesser extent to overall viscosity. Digesta viscosity is highly sensitive to changes in ionic concentration that are due to intestinal secretion or absorption of aqueous fluids. Consequently, prediction of physiological action from viscosity measurements in vitro is difficult (23, 38, 39).

Raw apples undergo little sloughing of cells upon ingestion and mastication. Gastric hydrochloric acid only solubilizes a small proportion of the pectins. Cooking the apples results in cell sloughing, and hence significant proportions of the middle lamellae pectic polysaccharides are solubilized. These make the digesta more viscous.

Vegetables undergo structural change during cooking and mastication, e.g. cellular disintegration. The cells in the intact carrot are each bounded by an intact cell wall; after cooking most, if not all, the cell walls have been ruptured and the cell contents lost. The grinding of foods before cooking and ingestion may also have pronounced effects on fiber action. Cell walls may be disrupted, and the reduced particle size of some fiber preparations such as wheat bran may be less biologically effective (27). The effect of other cooking processes, e.g. Maillard reactions, are not known.

Controlled drying of a heated starch gel can produce any of the different X-ray diffraction patterns, depending on the temperature. On cooling, gelatinized starchy foods will retrograde. During retrogradation, solubility of the starch molecule decreases and so does its susceptibility to hydrolysis by acid and enzymes. Chain length and linearity are important factors affecting retrogradation. The longer the starch chains, the greater the number of interchain hydrogen bonds formed (9).

Studies in man suggest that the mainly retrograded amylose fraction virtually resists digestion in the small intestine (22).

QUANTITATIVE MEASUREMENT OF FIBER

Chemical analysis or quantitative measurement of the fiber content of specific foods does not allow prediction of their biological action, since the physiological effects of dietary fiber depend predominantly on physical properties that do not relate in any simple or direct way to chemical composition (10).

Gravimetric and gas liquid chromatography (GLC) can be used to analyze dietary fiber. Gravimetric methods weigh an insoluble residue after chemical and enzymic solubilization of non-fiber constituents. The remaining protein is assayed and subtracted from the weight. These methods include the AOAC accepted method (41). Gas liquid chromatography involves the enzymatic breakdown of starch and the separation of the low molecular weight sugars, acid hydrolysis to free sugars, derivatization to alditol acetates, and finally separation and quantitation of neutral monomers with GLC, together with determination of uronic acid and lignin. The GLC methods enable the nature of the carbohydrate to be determined in more detail (21, 52).

ACTION OF FIBER ALONG THE GASTROINTESTINAL TRACT

Dietary fiber has major effects on (a) the rate of gastrointestinal absorption, (b) sterol metabolism, (c) cecal fermentation, and (d) stool weight (13).

Rate of Intestinal Absorption

In the upper gastrointestinal tract dietary fiber prolongs gastric emptying time and retards the absorption of nutrients. Both of these processes are dependent on the physical form of the fiber, and particularly on viscosity.

The inclusion of viscous polysaccharides in carbohydrate meals reduces the postprandial blood glucose level concentrations in humans. No correlation between the rate of gastric emptying and postprandial concentrations of blood glucose has been observed (19).

Diets that contain a substantial amount of complex carbohydrate content tend to be bulky and require longer times for ingestion. The consumption of whole apple takes longer (17 min) than that required for puree (6 min) or apple juice (1.5 min) in equicaloric amounts (25).

Gastric emptying studies are bedeviled by problems of methodology. The physical nature of the gastric contents is as important as the chemistry of the components (20). Isolated viscous fibers tend to slow the gastric emptying rate of liquids and disruptible solids. It is almost certain, however, that the gastric emptying rate for a fiber ingested alone will differ from that of a fiber ingested along with other dietary constituents such as fat and protein. Likewise, it has been shown that the gastric emptying time for different fiber sources is variable (13).

Rates of release of nutrients from dietary fiber in the intestine are influenced by factors such as the intactness of tissue histology, degree of ripeness, and the effects of processing and cooking (25, 27).

There is no evidence to suggest that viscous polysaccharides inhibit transport across the small intestinal epithelium. More likely, their viscous properties inhibit the access of nutrients to the epithelium. Two mechanisms bring nutrients into contact with the epithelium. Intestinal contractions create turbulence and convection currents that mix the luminal contents and bring material from the center of the lumen close to the epithelium. Nutrients have to diffuse across the thin, relatively unstirred layer of fluid lying adjacent to the epithelium. Increasing the viscosity of the luminal contents may impair both convection and diffusion of the nutrients across the unstirred layer. In the case of isolated polysaccharides such as guar gum, the slowing of nutrient absorption appears to be a function of viscosity (20). The reduction in absorption caused by guar gum is probably due to resistance by viscous solutions of the convective effects of intestinal contractions (20).

In the case of whole plant material, the influence on absorption appears to be due to the inaccessibility of nutrients within the cellular matrix of the plant. The effects on absorption can be decreased by grinding the food before cooking or by thorough chewing; both processes open the cellular structure (20).

Inadequate mixing of luminal contents due to increased viscosity by soluble polysaccharides may also slow the movement of digestive enzymes to their substrates (46).

Viscous polysaccharides tend to delay small bowel transit, possibly due to resistance to the propulsive contractions of the intestine (31). In rats most of this delay is secondary to alterations in ileal motility; transit through the upper small intestine is little affected (42).

Complex carbohydrates, particularly those that possess uronic and phenolic acid groups, or sulphated residues such as pectins and alginates may bind magnesium, calcium, zinc, and iron. However, other constituents of plant cells, e.g. phytates, silicates, and oxalates, also chelate divalent cations. The binding of minerals may be reduced by acid, protein, ascorbate, and citrate (13, 30).

The reduction in absorption of minerals and vitamins could, in theory, have adverse nutritional consequences, particularly in populations eating diets inherently deficient in these nutrients, i.e. in developing countries or fastidious, health food conscious communities where diets may be marginal in micronutrients but high in fiber. Children are particularly vulnerable to such conditions. Customary Western diets contain levels of minerals and vitamins in excess of daily requirements. Mineral balance studies have indicated that for people on nutritionally adequate diets, the ingestion of mixed high fiber diets or dietary supplementation with viscous polysaccharides is unlikely to cause mineral deficiencies (30).

The ingestion of dietary fiber may affect drug absorption in two ways: by reducing gastric emptying or inhibiting mixing in the small intestine. Viscous polysaccharides delay the absorption of paracetamol. Quite separately if a drug enters the enterohepatic circulation, any bacterial metabolism of the drug may be altered by coincidental fermentation of fiber and thus the half life of a drug may be increased or decreased. An example of this effect is digoxin. Digoxin has a narrow therapeutic range and is passively absorbed in the small intestine and so is affected by gastric emptying or decreased small intestinal absorption. Digoxin is also reduced to an inactive metabolite in the colon, so an inactive metabolite will be absorbed (33).

Alteration of Sterol Metabolism

Dietary fiber has been shown to have an effect on sterol metabolism (12). This effect is not simple: Possibly dietary fiber displaces fat from the diet (49); polyunsaturated fats frequently eaten in conjunction with the fiber may also be important (51). The direct effect of fiber on sterol metabolism may be through one of several mechanisms: altered lipid absorption; altered bile acid metabolism in the cecum; reduced bile acid absorption in the cecum; indirectly via short chain fatty acids, especially propionic acid, resulting from fiber fermentation.

An important action of some fibers is to reduce the reabsorption of bile acids in the ileum and hence the amount and type of bile acid and fats reaching the colon. Bile acids may be trapped within the lumen of the ileum either because of a high luminal viscosity or because they bind to the polysaccharide structure. A reduction in the ileal reabsorption of bile acid has several direct effects. The enterohepatic circulation of bile acids may be affected. In the cecum, bile acids are deconjugated and 7α-dehydroxylated. In this less water-soluble form, bile acids are adsorbed to dietary fiber in a way that is affected by pH and is mediated through hydrophobic bonds, thereby increasing the loss of bile acid in the feces (17, 32).

Consequently, the enterohepatic pool is initially reduced. It may be renewed by increased synthesis of bile acids from cholesterol, thereby reducing body cholesterol. Other fibers, e.g. gum arabic, are associated with a significant decrease in serum cholesterol without increasing fecal bile acid excretion. The fibers that are most effective in influencing sterol metabolism (e.g. pectin) are fermented in the colon, as shown by increased breath hydrogen production. That the physiological effect is due entirely to adsorption to fiber in the colon is unlikely (14). In contrast, fiber has an important sequestrating effect in the ileum. Possibly, an alteration occurs in the end product of bile acid bacterial metabolism; these bile acids are absorbed from the colon and returned to the liver in the portal vein, thus modulating either the synthesis of cholesterol or its catabolism to bile acids (29). The precise relationship between the proportions of ileal and cecal absorption of bile acids is difficult to estimate. Clearly it is a variable and very diet dependent, especially in regard to the amount and type of fiber. In particular, some bacterial colonization of the ileum may simulate cecal bacterial activity. Approximately 25% of the body pool of cholic acid and 50% of chenodeoxycholic acid pass into the cecum either to be absorbed or excreted in feces (7, 37).

An alternative mechanism for the effect of fiber on serum cholesterol is the action of propionic acid, derived from fiber fermentation, on liver cholesterol synthesis. Initial in vitro experiments have indicated that cholesterol synthesis by isolated hepatocytes is inhibited by propionic acid. Whether this inhibition occurs in vivo at physiological concentrations is not clear (4).

Substrate for Cecal Fermentation

The colon may be regarded as two organs: The right side is a fermenter, the left side affects continence. The right side of the colon is involved in nutrient salvage so that dietary fiber, resistant starch, fat, and protein are utilized by bacteria and the end products absorbed and used by the body (11).

The colonic flora is a complex ecosystem largely consisting of anaerobic bacteria, which outnumber the facultative organisms at least 100:1. The colonic flora of a single individual consist of more than 400 bacterial species. The total bacterial count in feces is 10^{10} to 10^{12} colony-forming units per milliliter. Despite the complexity of the ecosystem, the microflora population is remarkably stable. Although wide variations in the microflora are found between individuals, studies in a single subject show that the microflora are stable over prolonged periods of time. The identification of individual bacteria is desirable. It is more profitable, however, for physiological and nutritional studies to regard the cecal bacterial complex as an important organ in its own right that is complementary to the liver in the enterohepatic circulation (11).

The cecal bacterial flora are dependent upon dietary and endogenous sources for nutrition. The amounts of substances passing through the intestine from the ileum vary with an inverse relationship between cecal bacterial metabolism and upper intestinal nutrient absorption. Dietary fiber has an influence on bacterial mass and enzyme activity. The consensus view is that while the cecal bacterial mass may increase as a result of an increased fiber content in the diet, the types of bacteria do not alter (11).

The process whereby a compound is bacterially dissimilated in the cecum under anaerobic conditions is complex and varied, leading to partial or complete decomposition. The end products are absorbed from the colon and utilized as nutrients, absorbed and reexcreted in the enterohepatic circulation, and excreted in stool.

In addition, the colon is part of the excretion system provided by the liver and biliary tree, i.e. the enterohepatic circulation. Poorly water-soluble chemicals of a molecular weight or more than 300–400 are excreted in the bile with enhanced water-soluble properties through chemical conjugation with glucuronide, sulphate, acetate, etc, or are made physically soluble by the detergent properties of bile acid. These chemicals may be endogenous, e.g. bile acids, bilirubin, hormones, or exogenous, e.g. drugs, food additives, pesticides. They pass unabsorbed through the small intestine. In the cecum these biliary excretion products and also unabsorbed dietary constituents, e.g. resistant starch, fat and proteins, and mucopolysaccharides secreted by the intestinal mucosa, are fermented by the bacterial enzymes. The fermentation process of biliary excretion products removes substitutions that have enhanced water solubility and enabled biliary excretion to occur. The bacterial metabolic products are less water soluble. Some of the end products of the fermentation of biliary excretion compounds are reabsorbed, metabolically altered and reconjugated in the liver, and excreted in bile; an enterohepatic circulation is established (11).

The effects of dietary fiber in the colon may be summarized in terms of (a) susceptibility to bacterial fermentation, (b) ability to increase bacterial mass, (c) ability to increase bacterial saccharolytic enzyme activity, and (d) water-holding capacity of the fiber residue after fermentation.

Enlargement of the cecum is a common finding when some dietary fibers are fed, and this is now believed to be part of a normal physiological adjustment. Such an increase may be due to a number of factors such as prolonged cecal residence of the fiber, increased bacterial mass, or increased bacterial end products (11).

The fermentation of fiber yields hydrogen, methane, and short-chain fatty acids. Hydrogen is readily measured in the breath; it has a diurnal variation with its nadir at midday, and it increases in the afternoon. Diverse sources of fiber influence the evolution of hydrogen in different ways. Disaccharides generate hydrogen more rapidly than trisaccharides, which in turn evolve hydrogen more quickly than oligosaccharides. More complex carbohydrates may not be fermented as rapidly and may require induction of specific enzymes before they can be utilized (13).

Methane-producing organisms are said to be strict anaerobes. The proportion of breath methane exhalers in different healthy adult populations vary widely, ranging from 33 to 80%. The breath methane status of an individual remains stable throughout the day and over prolonged periods. Yet feces from healthy individuals regardless of breath methane excretion status will always produce methane. This suggests that all individuals produce methane, but a certain amount must be produced to spill over into the breath (35).

The fermentation of feces from herbivorous animals produces methane, whereas the fermentation of feces from carnivores does not. Differences also exist between the two feeding groups in the production of short-chain fatty acids. This suggests that methanogenic animals require a dietary fibrous residue (34).

It has been suggested that considerable methane excretion only takes place when sulphate-reducing bacteria are not active. The metabolic end product of dissimilatory sulphate reduction is thought to be toxic to methanogenic bacteria. When sulphate is present, sulphate-reducing bacteria have a higher substrate affinity for hydrogen than do methanogenic bacteria (24).

Some nonabsorbed carbohydrates, e.g. pectin, gum arabic, oligosaccharides, and resistant starch, are fermented to short-chain fatty acids (chiefly acetic, propionic and n-butyric), carbon dioxide, hydrogen, and methane. The molar ratio of acetate, propionate, and butyrate is of the order of $60:20:15$, though small amounts of isobutyrate, valerate, and isovalerate are present; the latter originate from the breakdown of protein and in particular from the breakdown of branch-chained fatty acids. The production of short-chain fatty acids has several possible actions on the gut mucosa. All of the short-chain fatty acids are readily absorbed by the colonic mucosa, but only acetic acid reaches the systemic circulation in appreciable amounts. Butyric acid is metabolized before it reaches the portal blood; propionic acid is metabolized in the liver. Butyric acid appears to be used as a fuel by the colonic mucosa, and in vitro studies of isolated cells have indicated that the short-chain fatty acids and butyric acid in particular are the preferred energy sources of colonic cells. Short-chain fatty acids are potent stimulants of cellular proliferation not only in the colon but also in the small intestine (5).

Short-chain fatty acids are the predominant anions in the human feces (5). If the daily content of the diet is increased from 63 g of protein and 23 g of dietary fiber, type unspecified, to an isocaloric 136 g of of protein and 53 g of fiber, though the amount of ammonia in fecal water (1 mmol per liter) is doubled, the short-chain fatty acids remain unchanged. Unless the dietary intake of fermentable carbohydrate is severely restricted or antibiotics are given, fecal short-chain fatty acid concentrations and molar ratios remain relatively constant in man. Cummings & Branch have estimated that 40–50 g of carbohydrate will yield 400–500 mmol total short-chain fatty acids, 240–300 mmol acetate, and 80–100 mmol of both propionate and butyrate (5). Almost all of these short-chain fatty acids will be absorbed from the colon. This means that fecal short-chain fatty acid estimations do not reflect cecal and colonic fermentation but rather reflect only the efficiency of absorption, the ability of the fiber residue to sequestrate short-chain fatty acids, and the continued fermentation of fiber around the colon, which presumably will continue until the substrate is exhausted. The absorption of short-chain fatty acids from the colon in man is concentration dependent and is associated with bicarbonate secretion. Bicarbonate appears consistently in the colonic lumen during short-chain fatty acid absorption, a process independent of the chloride-bicarbonate exchange. Possibly, an acetate-bicarbonate exchange takes place at the cell surface, but the precise mechanism is not understood. Short-chain fatty acids have a stimulatory effect on sodium absorption from the colonic lumen. This may be related to the recycling of hydrogen ions. The unionized short-chain fatty acid crosses into the cell where it dissociates and hydrogen ion is moved back into the lumen in exchange for sodium. Thus short-chain fatty acids provide a powerful stimulant to sodium and water absorption (5).

The bacteria in the colon produce an "organ" of intense metabolic activity. This activity in the colon is mainly reductive, in contrast to the activity in the liver, which is oxidative. The intestinal flora perform a wide range of metabolic transformations on ingested compounds. The major enzymes involved in these activities include azoreductase, nitrate reductase, nitroreductase, β-glucosidase, β-glucuronidase, and methylmercury-demethylase. The action of fiber on the activity of these enzymes may be species-dependent, and animal studies do not always indicate what happens in man (45).

Stool Weight

Feces are complex and consist of 75% water. Bacteria make a large contribution to the dry weight; the residue is unfermented fiber and excreted compounds. There is a wide range of individual and mean stool weights. In a study in Edinburgh the variation in stool weight was between 19 and 280 g during a 24-hr period. The amount of stool excreted by an individual varies quite markedly from individual to individual and by that individual over a period of time. Of the dietary constituents, only dietary fiber influenced stool weight (15). It is not known why there is such individual variation in stool weight.

The most important mechanism whereby dietary fiber increases stool weight is through the water-holding capacity of unfermented fiber, e.g. wheat bran. The greater the water-holding capacity of the bran, the greater the effect on stool weight (44, 48).

Fiber may influence fecal output by another mechanism. Colonic microbial growth may be stimulated by ingestion of fermentable fiber sources such as apple, guar, or pectin. However, an increase in stool weight does not always result from eating these fibers (50). An osmotic effect of products of bacterial fermentation on stool mass may also occur, though this as yet is not a well-defined contribution (19).

One of the major functions of the colon is to absorb water and produce a plasticine-type of stool that can be readily voided at will from the rectum. The ileum contains a viscous fluid; the viscosity is created by mucus and water-soluble fibers whose molecular weight, degree of cross-linkages, and aggregation will affect the viscosity. If the viscosity increases to a certain point, peculiar to the constituent macromolecules, then a sol or hydrated carbohydrate complex will result. The sol will be coherent and homogeneous.

The concentration of ileal effluent in the cecum and colon is the result of the absorption of water. This might be expected to create a gel. Feces are not a gel, however, but a plasticine-like material, heterogeneous without viscosity, and made up of water, bacteria, lipids, sterols, mucus, and fiber. In the cecum, therefore, a marked physical change occurs, in part as a result of bacterial activity, in part by the presence of bacteria themselves. Such a plasticine structure is lost in watery diarrhea. The mechanism of this change, physiological or pathological, is unknown but some of the steps involved are described below (13).

In the colon water is distributed in three ways: (*a*) free water, which can be absorbed from the colon; (*b*) water that is incorporated into bacterial mass; (*c*) water that is bound by fiber. Stool weight is dictated by (*a*) the time available for water absorption to take place through the colonic mucosa, (*b*) the incorporation of water into the residue of fiber after the fermentation of fiber, and (*c*) the bacterial mass.

Wheat bran added to the diet increases stool weight in a predictable linear manner and decreases intestinal transit time. The increment in stool weight is independent of the initial stool weight. Wholemeal bread, unless it is of a very coarse nature, has little or no effect on stool weight. The particle size of the fiber is all important. Coarse wheat bran is more effective than fine wheat bran (48). The greater the water-holding capacity of the bran, the greater the effect on stool weight. The effect of the water-binding by wheat bran is such that in addition to an increase in stool weight, other fecal constituents such as bile acids, which in absolute amounts do not increase, are diluted by fecal water and hence their concentration decreases (13). The increment in stool weight per gram of wheat bran varies in different populations. For control subjects, an increase in stool weight, while being dependent on the particle size of the bran, is generally of the order of 3 to 5 g wet stool weight per gram fiber. However in individuals with the irritable bowel syndrome and symptomatic diverticulosis the increment is of the order of 1 to 2 g wet weight per gram fiber. This suggests a difference in the handling of the fiber in the intestine in these situations (40).

Bacteria are an important component of the fecal mass (50). What percentage are living and what percentage are dead and as such are being voided is not known. The fermentation of some fibers results in an increase in the bacterial content and hence in the weight of stool. Other fibers (e.g. pectin) are fermented without any effect on stool weight. Possibly some fibers that increase stool weight in association with an increased bacterial mass do so because of an increase in excreted bacteria adherent to unfermented fiber (50).

The degree to which free water is absorbed from the colon is affected by a number of factors that are poorly understood (2). In the rat a comparison of cecal and fecal contents has shown that the fermentation of some complex carbohydrates, e.g. ispaghula and gellan, has a significant effect on the content of luminal short-chain fatty acids in the more distal colon. This effect appears to be related to continued fermentation along the colon (19). An increase in the short-chain fatty acid concentration of feces appears to be related to an increased output of fecal water. Thus under some circumstances short-chain fatty acid absorption may be less efficient, which may play a role in determining fecal output. This observation supports the view of Hellendoorn who suggested that fiber fermentation products play an important role in determining stool weight and transit time (28). The demonstration that short-chain fatty acids were rapidly absorbed in the colon suggested that short-chain fatty acids play no part in determining fecal output (36). However, it would appear that there is continued fermentation of some complex carbohydrates, e.g. ispaghula, in the distal colon. Under these circumstances the fecal short-chain fatty acids may influence fecal water osmolality, absorption, and stool weight (2).

MATHEMATICAL EQUATIONS TO DESCRIBE THE INTRALUMENAL EFFECTS OF FIBER

In the gastrointestinal tract, fiber will interact with (*i*) one-phase miscible mixtures or (*ii*) binary mixtures (multiphase system). These include (*a*) two-phase systems with one continuous phase and one dispersed phase, and (*b*) two-phase systems with two continuous phases. There are equations for defining each of these classes of mixtures.

Fiber and other constituents of the intestinal luminal and mural phases may be involved in single-phase systems in which the components are completely miscible or soluble in each other. The properties that then become important are the density of the liquid mixtures and the dielectric thermodynamic properties.

Alternatively, there can be two-phase or multiphase systems in which the components are insoluble or only partially soluble in each other. For two-phase systems, the shape and physical characteristics of the particles, the rate of diffusion through mixtures, the viscosity of the suspension, and the character and physical properties of the two phases become important.

Soluble fiber components can be regarded as forming a continuous sol phase within which insoluble and hydrated components are dispersed as a discontinuous, particulate phase. Other dietary components that do not form part of the homogeneous continuous phase (e.g. unmicellized fat) can be regarded as separate phases. In such two-phase or multiphase systems, certain physical properties, such as density, obey a simple rule of mixing:

$$P = P_1\phi_1 + P_2\phi_2 \ldots + Pn\phi n$$

where P is the overall physical property of the entire system, $P_1, P_2, \ldots Pn$ represent the corresponding physical property in the individual phases, and $\phi_1 + \phi_2 + \ldots + \phi n = 1$.

Other physical properties, such as viscosity, however, combine in a much more complex way and may be difficult or impossible to predict from the behavior of the individual phases in isolation.

The principal physiological effect of dietary fiber in the small intestine is to reduce the rate or the extent of release of nutrients.

The rate of release of nutrients from fibrous particles into the surrounding intestinal fluid will be inversely proportional to particle size and directly proportional to solute gradient. It will also be affected, for example, by the physical state of the solute (e.g. whether it is present in solid form or is already dissolved in water trapped within the particle), so that dissolved solids can be squeezed out by peristaltic contractions or diffuse out. The surface properties of the particle, e.g. the suface tension effects, are also important. The concentration of nutrients within the continuous aqueous phase will be constantly depleted by enteric absorption, and will be replenished, as outlined above, by release of material from food particles. The progress of these sequential release processes will, of course, also be influenced by transit time (i.e. the duration of exposure to a particular absorptive surface or digestive environment).

These processes may be expressed as

$$R = k'\text{cf}/w.$$

The rate of release of nutrients from a polymeric system in the intestine is in direct proportion to the concentration within the particle (cf) and decreases with increasing particle size (w).

The effect of fiber in the colon may be summarized as

$$\text{stool weight} = W_f (1 + H_f) + W_b(1 + H_b) + W_m(1 + H_m)$$

where W_f, W_b, and W_m are, respectively, the dry weights of fiber remaining after fermentation in the colon, bacteria present in the feces, and osmotically active metabolites and other substances in the colonic contents that could reduce the amount of free water absorbed, and H_f, H_b, and H_m denote their respective "water-holding capacities" (i.e. the weight of water resistant to absorption from the colon, per unit dry weight of each fecal constituent) (18).

Simple methods are being developed to anticipate the four major biological functions of dietary fiber whereby fiber affects absorption and metabolism in the intestine. This approach has been successful in comparing dietary fibers used in human experiments with those studied in vitro (1). The importance of measuring fiber in absolute amounts cannot be underestimated. Nevertheless, the measurement of fiber in terms that reflect function is of primary importance. The understanding of enzyme activity grew immeasurably after the establishment of Michaelis-Menten equations. Such functional classifications are an important model for the development of an understanding of the action of dietary complex carbohydrates and dietary fiber.

New methods need to be developed to determine the precise fate and disposition of the products of fiber fermentation. This is an unexplored and important field.

ACKNOWLEDGMENTS

Conversations with John Cummings, Jenny Eastwood, Christine Edwards, Hans Englyst, Ken Heaton, Dave Kritchevsky, Ed Morris, and Nick Read were most helpful in the preparation of this manuscript.

Literature Cited

1. Adiotomre, J., Eastwood, M. A., Edwards, C. A., Brydon, W. G. 1990. Dietary fiber in vitro methods that anticipate nutrition and metabolic activity in humans. *Am. J. Clin. Nutr.* 50:128–34
2. Armstrong, E. F., Brydon, W. G., Eastwood, M. 1990. Fiber metabolism and colonic water. See Ref. 31a, pp. 179–86
3. Carpita, N. C. 1990. The chemical structure of the cell walls of higher plants. See Ref. 31a, pp. 15–30
4. Chen, W. J. L., Anderson, J. W., Gould, M. R. 1981. The effects of oat bran, oat gum and pectin on lipid metabolism in cholesterol-fed rats. *Nutr. Rep. Int.* 24:1093–98
5. Cummings, J. H., Branch, W. J. 1986. Fermentation and the production of short chain fatty acids in the human large intestine. See Ref 53, pp. 131 50
6. Cummings, J. H., Englyst, H. N. 1987. Fermentation in the human large intestine and the available substrates. *Am. J. Clin. Nutr.* 45:1243–55
7. Danielsson, H. 1973. Mechanisms of bile acid biosynthesis. See Ref. 40a, pp. 1–32
8. Delmer, D. P., Stone, B. A. 1988. Biosynthesis of plant cell walls. In *The Biochemistry of Plants*, ed. J. Preiss, pp. 373–420. Orlando: Academic. 537 pp.
9. Dobbing, J., Ed. 1989. *Dietary Starches and Sugars in Man: A Comparison.* London: Springer-Verlag. 256 pp.

10. Eastwood, M. A. 1986. What does the measurement of dietary fiber mean? *Lancet* 1:1487–88

11. Eastwood, M. A. 1988. Food, drugs, bile acid enterobacterial interactions. In *Diseases of the Colon, Rectum and the Anal Canal,* ed. J. B. Kirsner, R. G. Shorter, pp. 133–57. Baltimore: Williams & Wilkins. 724 pp.

12. Eastwood, M. A. 1990. Fibres alimentaires et lipoproteines. *Cah. Nutr. Diet.* XXV:1–7

13. Eastwood, M., Brydon, W. G. 1985. Physiological effects of dietary fiber on the alimentary tract. In *Dietary Fibre, Fibre-Depleted Foods and Disease,* ed. H. Trowell, K. Heaton, pp. 105–32. London: Academic. 433 pp.

14. Eastwood, M. A., Brydon, W. G., Anderson, D. M. W. 1986. The effect of the polysaccharide composition and structure of dietary fibers on caecal fermentation and fecal excretion. *Am. J. Clin. Nutr.* 44:51–55

15. Eastwood, M. A., Brydon, W. G., Baird, J. D., Elton, R. A., Helliwell, S., et al. 1984. Fecal weight and composition of serum lipids and diet amongst subjects aged 18–80 years not seeking health care. *Am. J. Clin. Nutr.* 46:628–34

16. Eastwood, M. A., Edwards, C. A. 1992. Fibrous polysaccharide. In *Toxic Substances and Crop Plants,* ed. J. P. F. D'Mello, pp. 258–84. London: R. Soc. Chem. 339 pp.

17. Eastwood, M. A., Hamilton, D. 1986. Studies on the adsorption of bile salts to non-absorbed components of diet. *Biochim. Biophys. Acta* 152:165–73

18. Eastwood, M. A., Morris, E. R. 1992. Physical properties of dietary fibre that influence physiological function: a model for polymers along the gastrointestinal tract. *Am. J. Clin. Nutr.* In press

19. Edwards, C. A., Bowen, J., Eastwood, M. A. 1990. Effect of isolated, complex carbohydrates on cecal and faecal short chain fatty acids in stool output in the rat. See Ref. 48a, pp. 273–76

20. Edwards, C. A., Johnson, I. T., Read, N. W. 1988. Do viscous polysaccharides reduce absorption by inhibiting diffusion or convection? *Eur. J. Clin. Nutr.* 42:307–12

21. Englyst, H. N., Cummings, J. H. 1986. Measurement of dietary fiber as non-starch polysaccharides. See Ref. 53, pp. 17–34

22. Englyst, H. N., Kingman, S. M. 1990. Dietary fiber and resistant starch: a nutritional classification of plant polysaccharides. See Ref. 31a, pp. 49–66

23. Everett, D. H. 1988. *Basic Principles of Colloid Science.* London: R. Soc. Chem. 288 pp.

24. Gibson, G. R., Cummings, J. H., Macfarlane, G. T., Allison, C., Segal, I., et al. 1990. Alternative pathways for hydrogen disposal during fermentation in the human colon. *Gut* 31:679–83

25. Haber, G. B., Heaton, K. W., Murphy, D., Burroughs, L. F. 1977. Depletion and disruption of dietary fibre. Effects on satiety, plasma-glucose and serum insulin. *Lancet* 2:679–82

26. Hayward, H. E. 1983. *The Structure of Economic Plants.* New York: Macmillan. 674 pp.

27. Heaton, K. W., Marcus, S. N., Emmett, P. H., Bolton, C. H. 1988. Particle size of wheat, maize, oat test meals; effects on plasma glucose and insulin responses and rate of starch digestion in vitro. *Am. J. Clin. Nutr.* 47:675–82

28. Hellendoorn, E. W. 1978. Fermentation as the principal cause of the physiological activity of indigestible food residue. In *Topics in Dietary Fiber Research,* ed. G. A. Spiller, pp. 127–68. New York: Plenum. 223 pp.

29. Hofmann, A. F. 1976. Enterohepatic circulation of bile acids in man. *Adv. Intern. Med.* 21:501–34

30. James, W. P. T. 1980. Dietary fiber and mineral absorption. In *Medical Aspects of Dietary Fiber,* ed. G. A. Spiller, R. M. Kay, pp. 239–60. New York: Plenum. 299 pp.

31. Jenkins, D. J. A., Wolever, T. M. S., Leeds, A. R., Gassull, M. A., Haisman, P., et al. 1978. Dietary fibres, fibre analogues and glucose tolerance: importance of viscosity. *Br. Med. J.* 1:1392–94

31a. Kritchevsky, D., Bonfield, C., Anderson, J. W., eds. 1990. *Dietary Fiber.* New York: Plenum. 499 pp.

32. Kritchevsky, D., Story, J. A. 1974. Binding of bile salts in vitro by nonnutritive fiber. *J. Nutr.* 104:458–62

33. Lindenbaum, J., Rund, D. G., Butler, V. P., Tseeng, D., Saha, J. R. 1981. Inactivation of digoxin by the gut flora: reversal by antibiotic therapy. *New Engl. J. Med.* 305:789–94

34. McKay, L. F., Eastwood, M. A. 1984. A comparison of bacterial fermentation end products in carnivores, herbivores and primates including man. *Proc. Nutr. Soc.* 43:35A

35. McKay, L. F., Eastwood, M.A., Brydon, W. G. 1985. Methane excretion in man a study of breath, flatus and faeces. *Gut* 26:69–74

36. McNeil, N. I., Cummings, J. H.,

James, W. P. T. 1978. Short chain fatty acids absorption by the human large intestine. *Gut* 19:819–22

37. Miettinen, T. A. 1973. Clinical implications of bile acid metabolism in man. See Ref. 40a, pp. 191–248

38. Morris, E. R. 1984. Rheology of hydrocolloids. In *Gums and Stabilisers for the Food Industry*, ed. G. O. Phillip, D. J. Wedlock, P. A. William, 2:57–78. Oxford: Pergamon. 378 pp.

39. Morris, E. R. 1990. Physical properties of dietary fibre in relation to biological function. See Ref. 48a, pp. 91–102

40. Muller-Lissner, S. A. 1988. Effect of wheat bran on weight of stool and gastrointestinal transit time: a meta analysis. *Br. Med. J.* 296:615–17

40a. Nair, P. P., Kritchevsky, D., eds. 1973. *Chemistry, Physiology and Metabolism*, Vol. 2. New York: Plenum. 329 pp.

41. Prosky, L., Asp, N-G., Furda, I., DeVries, J. W., Schweizer, T. F., Harland, B. F. 1984. Determination of total dietary fiber in foods, food products and total diets: Interlaboratory study. *J. Assoc. Off. Anal. Chem.* 67:1044–52

42. Read, N. W., MacFarlane, A., Kinsman, R. I., Bates, T. E., Blackhall, N. W., et al. 1984. Effect of infusion of nutrient solutions into the ileum on gastrointestinal transit and plasma levels of neurotensin and enteroglucagon in man. *Gastroenterology* 86:274–80

43. Report of the British Nutrition Foundations Task Force. 1990. *Complex Carbohydrates in Foods*. London: Chapman & Hall. 164 pp.

44. Robertson, J. A., Eastwood, M. A. 1981. An investigation of the experimental conditions which could affect water-holding capacity of dietary fiber. *J. Sci. Food. Agric.* 32:819–25

45. Rowland, I. R., Mallett, A. K., Wise, A. 1985. The effect of diet on the mammalian gut flora and its metabolic activities. *Crit. Rev. Toxicol.* 16:31–103

46. Schneeman, B. O., Gallacher, D. 1985. Effects of dietary fibre on digestive enzyme activity and bile acids in the small intestine. *Proc. Soc. Exp. Biol. Med.* 180:409–14

47. Selvendran, R. R., Verne, A. V. F. V. 1990. Chemistry and properties of plant cell walls. See Ref. 31a, pp. 1–14

48. Smith, A. N., Drummond, E., Eastwood, M. A. 1981. The effect of coarse and fine Canadian red spring wheat and French soft wheat on colonic motility in patients with diverticular disease. *Am. J. Clin. Nutr.* 34:2460–63

48a. Southgate, D. A. T., Waldron, K., Johnston, I. T., Fenwick, G. R., eds. 1990. *Dietary Fibre: Chemical and Biological Aspects, Spec. Publ. No. 83.* London: R. Soc. Chem. 386 pp.

49. Stasse-Wolthuis, M., Hautvast, J. G., Hermus, R. J. J., Katan, M. B., Bausch, J. E., et al. 1979. The effect of a natural high fiber diet on serum lipids, fecal lipids and colonic function. *Am. J. Clin. Nutr.* 32:1881–88

50. Stephens, A. M., Cummings, J. H. 1980. Mechanism of action of dietary fibre in the human colon. *Nature* 283–84

51. Swain, J. F., Rouse, I. L., Curley, C. B., Sacks, F. M. 1990. Comparison of the effects of oat bran and low fibre wheat on serum lipoprotein levels and blood pressure. *New Engl. J. Med.* 322:147–52

52. Theander, O., Westerlund, E. 1986. Studies on dietary fiber. *J. Agric. Food Chem.* 34:330–36

53. Vahouny, G., Kritchevsky, D., eds. 1986. *Dietary Fiber*. New York: Plenum. 566 pp.

Annu. Rev. Nutr. 1992. 12:37 57

VITAMIN A: Physiological and Biochemical Processing

Rune Blomhoff,[1] Michael H. Green,[2] and Kaare R. Norum[1]

[1]Institute for Nutrition Research, School of Medicine, University of Oslo, 0316 Oslo 3, Norway; [2]The Pennsylvania State University, Nutrition Department, University Park, Pennsylvania 16802

KEY WORDS: retinol metabolism, retinoids, vitamin A absorption, stellate cells, retinoid-binding proteins

CONTENTS

37

0199-9885/92/0715-0037$02.00

INTRODUCTION

Knowledge about vitamin A (retinol) has advanced dramatically during the last few years. A major breakthrough came with the discovery of nuclear retinoid receptors that regulate gene expression by binding to short DNA sequences (retinoic acid–responsive elements) in the vicinity of target genes. Most of the extravisual functions of vitamin A seem to be mediated via these newly discovered receptors.

How does the body homeostatically regulate vitamin A, which plays such a critical role in vision and cellular growth and differentiation? The answer involves the following: (*a*) The liver, with its processing and storage of retinol, and several extracellular retinoid-binding proteins provide tissues with optimal amounts of retinol in spite of normal fluctuations in daily vitamin A intake. (*b*) A group of cellular retinoid-binding proteins and enzymes regulate intracellular metabolism and transport. And (*c*), the family of nuclear receptors mediates the ultimate actions of vitamin A in gene expression.

The aim of this article is to review the physiology and biochemistry of vitamin A with emphasis on the absorption, transport, cellular uptake, storage, and intracellular metabolism of vitamin A. Since several reviews of vitamin A metabolism have been published recently (5, 11, 12, 18, 99), here we provide an overview of the metabolism of vitamin A and highlight recent advances in the field. Another chapter in this volume discusses functional aspects of vitamin A, including the role of the nuclear retinoid receptors.

VITAMIN A INTAKE AND INTESTINAL ABSORPTION

Vitamin A is a term used for all compounds that exhibit the biological activity of retinol, whereas *retinoids* is a term that includes the natural forms of vitamin A as well as the many synthetic analogs of retinol, whether or not they have biological activity. The main dietary sources of vitamin A are provitamin A carotenoids from vegetables, preformed retinyl esters, and, to a lesser extent, retinol from animal sources.

Absorption of Carotenoids

Little quantitative data are available on the efficiency of intestinal absorption of provitamin A carotenoids (α-, β-, and γ-carotene and cryptoxanthin) and their conversion to retinol. Carotene absorption is by passive diffusion, and in humans, between 5 and 50% is absorbed (reviewed in Ref. 12). Absorption efficiency appears to be dependent on an adequate quantity of dietary fat. It is generally assumed that, in humans consuming a "normal" diet, one-sixth (on a weight basis) of dietary β-carotene and one-twelfth of other provitamin carotenoids are absorbed and converted to retinol in the enterocytes (12). Notably in humans and some other species such as the ferret and preruminant calf, a significant fraction of the carotenoids is absorbed intact and transported via the lymph to other cells in the body.

The enzymatic mechanisms responsible for intestinal conversion of β-carotene to retinol have been a subject of recent controversy (23, 43, 59). Work by Dmitrovski (23) indicates that carotene may be cleaved both centrally, as originally reported by Goodman, Olson, and collaborators (37, 38, 73), and peripherally by two separate enzymes. The apo-carotinals formed by peripheral cleavage may subsequently be further processed to retinoic acid or to retinol. Retinal generated by central cleavage is presumably reduced to retinol by a reductase. In contrast to the earlier reports that this was accomplished by a cytosolic enzyme (73), recent work by Kakkad & Ong (58) suggests that retinal bound to the intestinal cellular retinol-binding protein type II [CRBP(II)] may be reduced by a membrane-bound microsomal enzyme. Clearly, the conversion of carotene to retinol in vivo should be further investigated.

Absorption of Retinyl Esters

Retinyl esters from the diet are hydrolyzed in the intestinal lumen by a pancreatic enzyme, and mixed micelles containing retinol deliver retinol to the enterocytes. Retinol in physiological concentrations is apparently absorbed by facilitated diffusion, whereas at pharmacological levels it can be absorbed by passive diffusion. Published data suggest that absorption of retinol is < 75%, and it is apparently dependent on both quantity and quality of fat in the diet (reviewed in Ref. 12). More research is needed to determine what factors influence retinol absorption.

Esterification of Retinol in Enterocytes

Most of the retinol absorbed into the enterocytes leaves via the lymphatics as retinyl esters in chylomicrons (Figure 1). Two enzymes have been identified as being important for the esterification of retinol in enterocytes: acyl CoA:retinol acyltransferase (ARAT) (47, 48) and lecithin:retinol acyltransferase (LRAT) (66, 75). MacDonald & Ong found that retinol complexed to CRBP(II) was esterified by LRAT (66). In contrast, uncomplexed retinol in membranes may be esterified by ARAT. Blomhoff and co-workers suggested (12) that LRAT esterifies retinol during absorption of a "normal" load of retinol, and ARAT esterifies excess retinol when large doses are absorbed and CRBP(II) becomes saturated. Thus, CRBP(II) may play a critical role in the normal carrier-mediated absorption of retinol.

Recently, a retinoic acid responsive element (RARE) was detected in the promoter for CRBP(II) (69), thus indicating that CRBP(II) transcription is positively regulated by retinoic acid via the nuclear retinoid receptors. We

Major pathways for retinoid transport in the body.

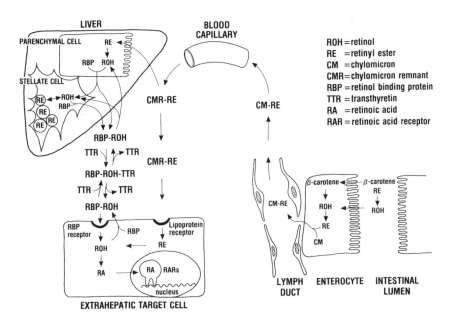

ROH = retinol
RE = retinyl ester
CM = chylomicron
CMR = chylomicron remnant
RBP = retinol binding protein
TTR = transthyretin
RA = retinoic acid
RAR = retinoic acid receptor

Figure 1   Schematic diagram of vitamin A absorption, transport, storage, and metabolism.

speculate that high doses of retinol in the diet may lead to an increased concentration of retinoic acid in enterocytes and an increased expression of CRBP(II).

TISSUE UPTAKE OF CHYLOMICRON REMNANT RETINYL ESTERS

In the circulation, hydrolysis of chylomicron triacylglycerols and several other processes are involved in formation of chylomicron remnants. Most retinyl esters present in the chylomicrons remain with the particle during conversion to chylomicron remnants (12).

Although chylomicron remnants are mainly cleared by the liver (see next section), extrahepatic uptake of remnants may be important in the delivery of retinol and carotenoids to extrahepatic tissues (Figure 1) such as adipose tissue, skeletal muscle, kidney, and carcass (13, 39). Recently, Hussain and coworkers reported that, although liver was the main site of chylomicron retinyl ester removal in all species examined, the bone marrow in rabbits and marmoset monkeys and the spleen in rats, guinea pigs, and dogs were also important (56, 57). In light of the importance of retinoids for cell differentiation, chylomicrons may be an important transport complex for delivering retinol (and carotenoids) to tissues with intensive cell proliferation and differentiation such as bone marrow and spleen. We have recently shown that human myeloid leukemia cells take up retinyl esters from chylomicron remnants and that this uptake leads to both differentiation and decreased cell proliferation (17, 97).

HANDLING OF CHYLOMICRON REMNANT RETINYL ESTERS BY THE LIVER

Constituents of chylomicron remnants including retinyl esters are taken up by liver parenchymal cells (hepatocytes) via a process that appears to involve chylomicron sequestration in the space of Disse, further lipolytic processes, and then receptor-mediated uptake into the cells (67). Both the low density lipoprotein (LDL) receptor and the LDL receptor-related protein (LRP) can bind chylomicron remnants and may be involved in uptake (67). The quantitative role of these two receptors is being debated.

Once chylomicron remnant retinyl esters have been taken up by liver parenchymal cells, their observed rapid hydrolysis may be catalyzed by a plasma membrane retinyl ester hydrolase recently described by Harrison & Gad (46). Subcellular fractionation (10) and autoradiographic studies (52) suggest that chylomicron vitamin A is rapidly transferred to endosomes and then to the endoplasmic reticulum. Retinol binds to plasma retinol-binding protein (RBP), presumably in the endoplasmic reticulum, and RBP-retinol is then transferred to Golgi for secretion (5). Secretion is influenced by vitamin A status: in vitamin A–deficient animals, secretion is reduced such that plasma levels of RBP and retinol are decreased. Vitamin A repletion leads to an immediate increase in secretion of RBP-retinol from the liver (reviewed in Ref. 5).

PARACRINE TRANSFER OF RETINOL TO STELLATE CELLS IN THE LIVER

In vitamin A–sufficient rats, most of the chylomicron remnant retinyl esters taken up by hepatocytes appear to be rapidly (within 2–4 hours) transferred as retinol to perisinusoidal stellate cells in the liver (13, 14) (Figure 1). A plausible candidate for a protein carrier that mediates the transfer is RBP, since stellate cells can take up the RBP-retinol complex (15, 36) and hepatocytes secrete retinol bound to RBP (5). The transfer of retinol from hepatocytes to stellate cells is too rapid to be accounted for by a secretion of RBP-retinol from hepatocytes to the general circulation, followed by retinol being taken up by liver stellate cells. During an in situ perfusion of rat livers, we observed that labeled retinol was transferred from hepatocytes to stellate cells. Furthermore, addition of antibodies against RBP blocked the transfer, indicating that RBP was the transport protein mediating the transfer of retinol from hepatocytes to stellate cells (9).

STORAGE OF RETINYL ESTERS IN STELLATE CELLS

In mammals, 50–80% of the body's total retinol (retinol plus retinyl esters) is normally present in the liver (18). Under most conditions, stellate cells contain about 90% of the liver total retinol (4, 6, 16, 53); the rest is in hepatocytes. Ninety-eight percent of the stellate cell vitamin A is present in the form of retinyl esters. The normal reserve of vitamin A in stellate cells is adequate to last for several months. Only when retinol concentration in the liver is very low (< 1–5 nmol/g tissue) does the proportion of liver vitamin A in hepatocytes become appreciable (3, 8).

Older morphological studies by Wake (95, 96) were important in showing that stellate cells in the liver have the ability to store large amounts of retinyl esters in lipid droplets. The size and number of these droplets is dependent on the vitamin A status of the animal. Chemical analysis of lipid droplets isolated from rat liver stellate cells revealed that between 12 and 65% of the total lipid mass is retinyl esters, depending on the vitamin A status of the animal. Triacylglycerols compose between 35 and 50% of the lipid mass (72).

A small, acute load of retinol does not accumulate in liver stellate cells of vitamin A–depleted rats (13). This observation may be related to the amount of cellular retinol-binding protein type I [CRBP(I)] in these cells. Rats fed normal levels of vitamin A have large amounts of CRBP(I) in stellate cells (16). Because CRBP(I), like intestinal CRBP(II), is an effective donor of retinol for esterification by LRAT (76, 101), reduced levels of CRBP(I) in stellate cells of vitamin A–deficient rats may account for the reduced accumulation of retinyl esters.

Several reports have shown that addition of retinol or retinoic acid to different cell types results in induction of CRBP(I) (44, 98). In addition, Smith et al (88) recently showed that a retinoic acid responsive element is present in the CRBP(I) promoter, which suggests that control of CRBP(I) gene transcription by retinoic acid and retinoic acid receptors may represent a positive feedback mechanism that is important in regulating cellular uptake of retinol and retinyl ester storage.

RETINOL MOBILIZATION FROM STELLATE CELLS

Theoretically, retinol could be secreted by stellate cells directly to the general circulation or it might first be transferred to hepatocytes. Since cultured hepatocytes have been shown to synthesize and secrete RBP, it has been assumed that hepatocytes are the exclusive site of retinol mobilization from the liver (5). As discussed below, data now available suggest that a direct mobilization of retinol from stellate cells to the general circulation occurs.

Although there is some disagreement in the literature about how much RBP stellate cells contain and whether they synthesize RBP, these cells apparently do contain the RBP necessary for exporting the vitamin (6, 51, 70). We (1) recently demonstrated that stellate cells contain mRNA for RBP, whereas Yamada et al (100) were unable to detect mRNA for RBP in stellate cells. We have also found that when stellate cells are cultured in a serum-free medium they secrete RBP-retinol. An alternative route of mobilization may also take place: Recent data suggest that apoRBP can bind to stellate cells, make a complex with stellate cell retinol, and then be released as RBP-retinol to the circulation (1). Since most of the liver RBP is found in parenchymal cells (5), one can speculate that apoRBP secreted by these cells may bind to stellate cells and mobilize retinol to the circulation (Figure 1).

An independent line of evidence supporting the possibility that stellate cells secrete RBP-retinol directly to the blood comes from a whole-body multi-compartmental model of retinol dynamics in rats (M. H. Green et al, unpublished). The model suggests that stellate cells are an important site of retinol secretion into blood. The model predicted that the stellate cell retinol pool that is responsible for the secretion was small and rapidly turning over; this prediction is compatible with the relatively small amounts of RBP observed in stellate cells.

The ability of stellate cells to control storage and mobilization of retinol ensures that the blood plasma retinol concentration is close to 2 μM in spite of normal fluctuations in daily vitamin A intake. It is likely that the retinoid-regulated CRBP(I) expression, in addition to saturation of CRBP(I) and RBP by retinol, controls retinol uptake, storage, and mobilization by stellate cells.

TURNOVER AND RECYCLING OF PLASMA RETINOL

Until a decade ago, it was assumed that, once retinol left the blood plasma, it was taken up by tissues and irreversibly utilized. However, earlier work by Vahlquist (94) indicated that retinol may be cycled to the blood. This idea has recently been verified in rats (40–42, 62, 63). In fact, it is now thought that the majority of retinol that leaves the plasma is recycled, since the plasma retinol turnover rate is more than an order of magnitude greater than the utilization rate (40–42, 62). Thus, in the rat, an average retinol molecule recycles to the plasma 7–13 times before irreversible utilization, and the whole-body vitamin A utilization rate is only about 10% of the plasma retinol turnover rate. Data from tracer kinetic experiments in the rat (42) indicate that the vehicle for retinol recycling is RBP. In rats with normal versus marginal versus nearly depleted liver vitamin A levels, kinetic studies indicate that the recycling time for retinol averages 8.4 day (41), 1.7–2.0 day (41, 42), and 0.6–0.7 day (41, 62), respectively. That is, once retinol leaves the plasma, it may take a week or more to recycle to the plasma in a normal rat. Most of this time would presumably be spent in retinyl ester pools.

In the rat, it has been estimated that ~50% of plasma retinol turnover is to kidneys (40, 62), ~20% is to liver, and the remaining 30% is to extrahepatic/extrarenal tissues. The source of RBP for retinol recycling from the kidneys and other extrahepatic tissues is not yet known. It is interesting in this regard that many extrahepatic tissues, including the kidneys, contain mRNA for RBP (68, 89). Notably, Makover et al (68) reported that adipose tissue contains relatively high concentrations of RBP mRNA. The function of adipose tissue in retinol recycling remains to be determined.

Plasma retinol turnover time, or mean transit time, is defined as the mean of the distribution of times that retinol molecules remain in plasma before leaving plasma reversibly or irreversibly. It is thus the time it takes for a mass equivalent to the plasma pool to turn over. For rats, kinetic data have been used to estimate a mean transit time of 1–3.5 hr (41, 42, 62). That is, an average retinol molecule spends 1–3.5 hr in the blood plasma before leaving the circulation. This retinol has several possible fates once it enters the interstitial spaces: Either it can be taken up by cells reversibly or irreversibly, or it can cycle to the blood from the interstitium. Based on equilibrium distributions and the mass distribution of unbound retinol, RBP-retinol, and TTR-RBP-retinol (27, 72), ~95.5% of plasma retinol is present as TTR-RBP-retinol, ~4.4% as RBP-retinol, and ~0.14% as unbound retinol. Using these distributions and considering molecular weights (286 for retinol, 21,286 for RBP-retinol, and 76,266 for TTR-RBP-retinol) and capillary permeabilities, we estimate that the vast majority of retinol leaves the circulation as RBP-retinol. TTR-RBP-retinol would leave the circulation too slowly (transit time, >18 hr), and there is too little unbound retinol (in spite of a high filtration fraction into tissues) to account for the high rates of plasma retinol turnover observed in kinetic studies (41, 42, 62). This conclusion was also drawn by Fex & Felding (27).

PLASMA RETINOL HOMEOSTASIS

In normal humans and experimental animals such as the rat, plasma retinol concentrations are maintained within a fairly narrow range despite wide fluctuations in dietary vitamin A intake. Thus it appears that plasma retinol levels are homeostatically regulated, unlike plasma retinyl esters. The control of plasma retinol is not yet well understood. It may be that the controlled variable is the pool of retinol that equilibrates between plasma RBP-retinol (free or bound to TTR), interstitial fluid RBP-retinol, and intracellular CRBP(I)-retinol. Thus, since liver stellate cells represent the main storage site of vitamin A in the body, it is likely that regulation of stellate cell CRBP(I) expression is one of many factors involved in regulation of plasma retinol homeostasis (reviewed in Ref. 12).

Control can also be mediated in many tissues by the enzymes that esterify retinol and hydrolyze retinyl esters. See Figure 2 for a schematic representation of retinol metabolism in stellate cells (or other cells which store retinol). Note the numerous transport proteins (e.g. RBP, RBP receptor, and CRBP) and enzymes [e.g. LRAT and retinyl ester hydrolase(s)] that are potentially involved in retinol homeostasis.

Control of plasma retinol is presumably also mediated by factors that affect the balance between retinol input to plasma and retinol output from plasma. Under normal conditions, output seems to be determined mainly by physical considerations (i.e. the equilibrium distribution of free retinol, RBP-retinol, TTR-RBP-retinol, and capillary permeabilities) (reviewed in Ref. 12). On the other hand, input determinants appear to be more complex. Both apoRBP and retinol (either free or more likely transferred through membranes via RBP receptors from CRBP-retinol) must be present for RBP to bind retinol and then enter the circulation.

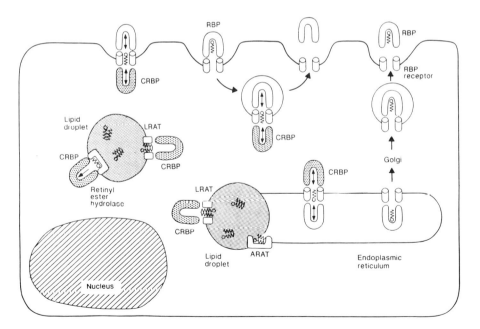

Figure 2   Hypothetical schematic diagram of retinol transport and storage in liver stellate cells.

To date, two organs have been assumed to play key roles in plasma retinol homeostasis: the liver and the kidneys. Following experiments in which retinoic acid was fed to vitamin A–depleted rats in order to spare retinol utilization, Underwood et al (93) hypothesized that vitamin A utilization resulted in the return of a "metabolite signal" to the liver and that this metabolite decreased hepatic secretion of retinol and thus plasma retinol concentrations. Recently, Gerlach & Zile (33–35) used rats with experimentally induced acute renal failure to investigate the role of the kidneys in plasma retinol homeostasis. Following nephrectomy or bilateral ligation of the renal arteries, these authors found that the plasma levels of retinol and RBP increased significantly compared with sham-operated controls. These findings are similar to observations in humans with chronic renal failure (87, 92). Gerlach & Zile concluded that the kidneys either remove or provide a specific regulatory signal to the liver that alters secretion of RBP-retinol.

Using a whole-body compartmental model of retinol kinetics in vitamin A–depleted rats (62), one can show that similar increases in plasma retinol levels would occur if the model is simulated with no output of plasma retinol to the kidneys. According to this model simulation, plasma retinol levels would increase to 41% above normal by 5 hr.

Above we noted that the controlled variable for plasma retinol homeostasis may be the pool of retinol that equilibrates between plasma TTR-RBP-retinol, RBP-retinol, interstitial fluid RBP-retinol, and intracellular CRBP-retinol (mainly stellate cell CRBP-retinol). The observation by Gerlach & Zile (34), that the injection of apoRBP into rats with bilateral ligation of the renal arteries resulted in an increase in plasma retinol, is consistent with this idea. Further studies are needed to advance our understanding of the role of various organs in plasma retinol homeostasis.

INTERCELLULAR TRANSPORT OF VITAMIN A IN VARIOUS TISSUES

Retinol-binding protein is the only retinoid-specific binding protein that has been found in plasma. However, many retinoid-binding proteins that are not found in plasma have been isolated from interstitial or other body fluids; these probably function locally to transport retinoids. An inter-photoreceptor retinoid-binding protein (IRBP) (99) has been identified in the extracellular space between the retinal pigment epithelial cells and the photoreceptor cells, where it constitutes 70% of the soluble protein. IRBP not only binds retinol and retinal but also vitamin E, fatty acids, and cholesterol. It may participate in the intercellular transport of retinoids during the visual cycle (30). Interestingly, the pineal gland, a primitive photosensitive organ, also contains IRBP (99).

Tear fluid contains retinol bound to a protein related to or identical to RBP (54). This RBP-retinol is probably a source of vitamin A for the ocular epithelia, which has an absolute requirement for vitamin A.

Several retinol-binding proteins related to but distinct from plasma RBP have been found to be secreted by the pig uterus (21). It has been suggested that RBP may carry retinol to the placenta. It has also been demonstrated that placental membranes in many species produce RBP-like proteins (64) and that the pig, sheep, and cat conceptus secrete RBP (45). The conceptus RBP is probably identical to plasma RBP. Since vitamin A is known to play a critical role in embryo development, the many RBP-like molecules produced by the conceptus and the surrounding tissues likely ensure that an optimal amount of retinol is delivered to the embryo.

Recently, Ong & Chytil observed that two proteins called epididymal-binding protein 1 and 2 (EBP1 and 2), which are secreted into the lumen of the first portion of the epididymis, were carrier proteins for retinoic acid (74). They suggested that the proteins deliver retinoic acid to sperm where it functions in sperm maturation.

CELLULAR UPTAKE OF RETINOL

The mechanism by which retinol is transferred from RBP to the plasma membrane of cells is not yet fully understood. Since a small amount of unbound retinol appears to be in equilibrium with RBP-retinol, this retinol would be available for cellular uptake by a nonspecific mechanism without involvement of membrane RBP receptors. Recent findings (28, 29, 72) suggest some nonspecific transfer of retinol from RBP-retinol to cell membranes in vivo. However, a number of other observations discussed below indicate that nonspecific partitioning is not likely to be the primary mechanism by which cells obtain retinol.

Retinal Pigment Epithelial Cells

In an autoradiographic study, Heller & Bok (50) reported that ^{125}I-RBP bound only to the choroidal surface of retinal pigment epithelial (RPE) cells. The binding was greatly inhibited by the presence of a 600-fold molar excess of unlabeled RBP. Binding to photoreceptor cells or other retinal cells was not observed. In another paper, Heller (49) presented evidence for the saturable binding of RBP to isolated RPE cells. The binding of ^{125}I-RBP was rapid (complete in about 1 min), and the level of cell-associated radioactivity was 7-fold higher at 22° than at 0°C. Since he found that unlabeled RBP-retinol could displace the cell-associated ^{125}I-RBP-retinol within 3 min, Heller concluded that RBP was not internalized by these cells.

Ottonello et al (77) more recently have shown that retinol released from

RBP after binding to RPE cells became associated with a protein of molecular weight 16,000, and they suggested that the protein might be CRBP(I). Their data indicate that a "functional link" exists between the uptake of retinol and retinol esterification, since they observed a fourfold reduction in total vitamin A incorporated when retinyl ester formation was inhibited. Since CRBP(I) seems to be involved, the enzyme responsible for the observed esterification is likely to be lecithin:retinol acyltransferase (LRAT).

Researchers in Peterson's laboratory have attempted to characterize the RBP receptor in bovine RPE cells (4a). Radiolabeled RBP was cross-linked with RPE membranes, and complexes were analyzed by electrophoresis and autoradiography. From such studies, the receptor was assumed to have a molecular weight of 63,000 (4a).

Hepatocytes and Liver Stellate Cells

That hepatocytes and stellate cells can take up both retinol and RBP from the plasma RBP-retinol transport complex in vivo has been demonstrated (15, 36). We have also applied model-based compartmental analysis to data on the kinetics of plasma [^{3}H]retinol-RBP-TTR in rats (M. H. Green et al, unpublished results). This model predicted that about half of the plasma retinol that cycled to the liver was taken up by hepatocytes and half by stellate cells.

A selective uptake of RBP by liver parenchymal and stellate cells was observed in a study by Senoo et al (82). Iodinated RBP and other ligands were injected intravenously into rats, and the uptake by liver cells was determined. They found, for example, that acetyl-LDL was recovered primarily in endothelial cells and asialoorosomucoid was recovered mainly in hepatocytes. RBP was the only protein that had appreciable recovery in both parenchymal and stellate cells.

These data agree with immunocytochemical results obtained by Senoo et al (82), in which human RBP was injected intravenously into rats and traced in cryosections of liver with gold-labeled antibodies. RBP was shown to be taken up by liver parenchymal and stellate cells but not by Kupffer cells or endothelial cells. At early times after injection, RBP was localized on the surface of liver parenchymal and stellate cells, whereas at later times it was also located in vesicles deeper in the cytoplasm.

Blood-Testis and Blood-Brain Barriers

Sertoli cells form the blood-testis barrier surrounding maturing spermatocytes. Since spermatogenesis is dependent on retinol, one would expect that Sertoli cells express RBP receptors. A saturable and specific uptake of [^{3}H]retinol-RBP was in fact recently reported by Shingleton et al (84). Since iodinated RBP was not taken up by the cells, the authors concluded that Sertoli cells take up retinol from RBP-retinol via an RBP receptor, but that

RBP is not internalized. The data also suggest that retinol taken up by the cells combines with CRBP(I) and is further processed to retinyl esters (83).

Related results were reported by MacDonald et al, who used autoradiography to study the uptake of intravenously injected ^{125}I-RBP by rat brain (65). ^{125}I-RBP was localized along the basolateral surface and inside the choroidal epithelial cells. Interestingly, high concentrations of CRBP(I) were also found in these cells. The data suggest RBP receptor-mediated uptake of retinol by the cells that constitute the blood-brain barrier and apparent further processing of retinol by CRBP(I).

Placental Brush Border Membranes

Retinol that is necessary for the developing fetus has to cross the placenta, which suggests that an RBP receptor may also be present in this tissue. Recently, Sivaprasadarao & Findlay studied binding of RBP-retinol to human placental brush border membranes (85, 86). The data are compatible with the presence of a specific receptor for RBP. Scatchard analysis of the equilibrium binding of ^{125}I-RBP revealed both high (3×10^{-9} M) and low (9×10^{-8} M) affinity binding components. The authors speculated that the higher affinity form might be converted to the lower affinity state after RBP bound and transferred its retinol or that apoRBP may have a lower affinity than RBP-retinol for the receptor.

Keratinocytes

Some investigators have not found binding of RBP to these cells. For example, in two studies (22, 55) no binding of ^{125}I-RBP to keratinocytes could be demonstrated. This is interesting in light of the fact that this cell is a classical target for retinol. Possibly some cell types express the RBP receptor while others do not.

INTRACELLULAR METABOLISM OF RETINOL

Until a few years ago, the role of the many cellular retinoid-binding proteins in intracellular retinol metabolism was neglected. Data now being presented, mainly from the laboratory of Ong and colleagues, indicate that the binding proteins direct retinoids to specific enzymes. That is, when retinoids complexed with binding proteins are added to cell homogenates, the retinoids are metabolized by other enzymes than if they are added when dissolved in an organic solvent. Three main processes are involved in the intracellular metabolism of retinol. (a) Retinol may be temporarily stored after conversion to retinyl esters. (b) Retinol may be converted to an active metabolite such as retinoic acid or retinal. (c) The retinol (or retinoic acid) molecule may be catabolized to a form that is excreted from the body.

Esterification of Retinol

LRAT seems to be the main intestinal enzyme esterifying retinol under normal conditions. LRAT was recently identified in liver stellate cells (7), and the high level of CRBP(I) in stellate cells points to an important role of LRAT in stellate cell retinol esterification. The product of the LRAT reaction in enterocytes is incorporated into chylomicrons while the product in liver stellate cells is instead diverted to intracellular lipid droplets. The reason for this difference is not understood, but it may be related to the different carrier proteins (CRBP(II) versus CRBP(I)).

LRAT activity (per mg protein) in the retinal pigment epithelial (RPE) cells is apparently about 1000 times higher than the activity in intestine and liver (81). Interestingly, the esterification of all-*trans* retinol by LRAT in RPE is linked to the direct conversion of all-*trans* retinyl esters to 11-*cis* retinol by an isomerase-like enzyme (20). The latter reaction couples the free energy of hydrolysis of an ester to the thermodynamically uphill trans to cis isomerization, thus providing the energy to drive the process. It will be important to determine whether LRAT in ocular cells is identical to LRAT in other cells.

Retinol is esterified in lactating mammary gland and transported as retinyl esters in milk lipid droplets to the developing organism. In this process, ARAT seems to be the most important enzyme for esterification based on the following reasoning. Randolph et al (80) showed that mammary gland contains lower levels of LRAT activity and CRBP(I) than does liver. In contrast, ARAT displayed a similar Vmax but a lower Km in mammary gland than in liver. Thus at physiological concentrations of retinol, esterification in mammary gland appears to occur predominantly by ARAT.

A gradient of retinyl esters exists in the human epidermis, with the highest concentration in the upper epidermal layer. Also in human epidermis, ARAT seems to be the predominant enzyme involved in retinol esterification (91). ARAT in keratinocytes has a more acidic pH optimum than ARAT in other tissues. The pH gradient that exists in the epidermis may thus be important for facilitating retinol esterification in the upper epidermis (91).

Activation of Retinol

It is generally assumed that 11-*cis* retinal covalently bound to opsin proteins and all-*trans* retinoic acid noncovalently bound to nuclear retinoic acid receptors (RARs) are the active retinoids in vision and in regulation of transcription, respectively. Recently, however, it was demonstrated that 9-*cis* retinoic acid binds and activates the three nuclear retinoid X receptors (RXRs), which also are ligand-dependent transcription factors (54a, 61a). Thus, retinoid isomerization may be important not only in vision but also in regulation of transcription.

Furthermore, it has been suggested that retinoic acid cannot substitute for all effects of retinol in growth regulation. Retinol is metabolized by many cells to 14-hydroxy-4,14-*retro*-retinol, and this compound may be the mediator of these effects (19). Many other retinoids, such as 13-*cis* retinoic acid, the retinoyl glucuronides, 4-oxo and 4-hydroxy retinol and retinoic acid, may also be important for retinoid function or as intermediate products in retinoid catabolism (32). In human epidermis and in the developing chick wing bud, 3,4-didehydro-retinol, 3,4-didehydro-retinyl esters, and 3,4-didehydro-retinoic acid comprise a large proportion of the endogenous retinoids normally present (90, 91). Retinoic acid and 3,4-didehydro-retinoic acid are equipotent in evoking digit duplication in the developing wing bud (90). Much more work is needed to clarify the possible role of these metabolites in vitamin A function.

The metabolic pathway whereby retinoic acid is synthesized in situ is poorly understood. There has been a tendency to assume that retinoic acid synthesis occurs by two distinct steps, including dehydrogenation of retinol to retinal by alcohol dehydrogenases, and a further oxidation of retinal to retinoic acid. Recently, retinoic acid was shown to regulate the human alcohol dehydrogenase gene for ADH3 directly via a retinoic acid responsive element (24). This observation was thought to indicate that ADH3 may play a regulatory role in retinoic acid synthesis. Duester et al suggested that retinoic acid activation of ADH3 constitutes a positive feedback loop regulating retinoic acid synthesis (24). Physiologically, however, such a positive regulatory loop for retinoic acid synthesis may be unlikely, since large concentrations of retinoic acid are very toxic to cells.

Although several types of dehydrogenases convert retinol to retinoic acid in vitro, it is unlikely that all are physiologically important. Since it was demonstrated that retinoic acid could be synthesized using cytosol from an alcohol dehydrogenase-negative strain of deermouse, some researchers suggest that retinol and ethanol oxidation are mediated by different enzymes (60, 78).

In vivo, most intracellular retinol is bound to cellular binding proteins, and these binding proteins may be directly involved in delivering retinol to the proper enzyme in a manner similar to that discussed above for retinol esterification. In fact, recent data of Posch et al (79) suggest that retinal synthesis is supported by retinol-CRBP(I) directly rather than by unbound retinol. Furthermore, neither ethanol nor ketoconazole inhibited retinal formation from retinol-CRBP(I), which suggests that ethanol-oxidizing enzymes and cytochrome P-450 isozymes were not involved in retinoic acid biogenesis.

Other recent data indicate that retinoic acid may also be synthesized from β-carotene in organs such as intestine, liver, kidney, and lung (71). Thus, β-carotene may be a source of retinoic acid in retinoid target cells, particularly in species such as humans that are capable of accumulating high concentrations of tissue carotenoids. Also, the recent demonstration that β-carotene dosing increases retinoic acid levels in rabbit serum is consistent with this idea (31).

Catabolism of Retinol

Several investigators have studied the catabolism of retinol by analyzing radioactive urinary, biliary, and fecal metabolites of retinol. A number of more polar metabolites are formed, and some of them have been identified (32). Rat liver microsomes may oxidize retinol to 4-hydroxy retinol and 4-oxo retinol, metabolites that are found in vivo. The cytochrome P-450 system seems to be involved in this conversion (61). Also glucuronides may be formed from retinol that is probably destined for excretion in bile and urine (2). Most of the catabolism of retinol, however, probably involves the production of retinoic acid as an intermediate. Once formed, retinoic acid cannot be converted to retinal or retinol. The catabolism of retinoic acid probably involves conjugation to retinoyl β-glucuronide and taurine, decarboxylation, oxidation at the 4 position of the cyclohexenyl ring, epoxidation, isomerization, and esterification (32).

A number of retinoids in addition to retinol and retinyl esters are present in plasma in nanomolar concentrations. These include all-*trans* retinoic acid, 13-*cis* retinoic acid, 13-*cis*-4-oxoretinoic acid, and all-*trans* retinoyl β-glucuronide (2, 25, 26, 32). The level of most of these retinoids is dependent on the intake of vitamin A and will typically increase 2–4 times after ingestion of a large amount of vitamin A (25, 26). Whether these retinoids simply reflect retinoid catabolism or whether they have a physiological role in vitamin A action is not known. However, since most of these retinoids are active in many in vitro systems in nanomolar concentrations, one should not exclude the possibility of a functional role for these plasma retinoids.

CONCLUDING REMARKS

Although advances have been made recently in our knowledge of vitamin A metabolism and function, many important topics are left for future study. For example, the absorption and metabolism of carotenoids, the mechanisms for and control of uptake of retinoids by cells, and the in situ synthesis and regulation of retinoic acid need to be studied more extensively. Although a number of proteins involved in retinol metabolism and function have been identified and their genes cloned during the last few years, many others that may be involved have not yet been cloned or characterized. Preliminary observations suggest that many other RBP-, CRBP-, and CRABP-like proteins will be discovered. Also, none of the enzymes involved in retinol metabolism, such as LRAT and ARAT, have been well characterized or cloned. In a few years, when the genes for a larger number of the proteins involved in vitamin A metabolism have been cloned and the proteins well-characterized, an important scientific challenge will be to try to understand how these proteins work and how they are regulated. Unraveling the secrets of vitamin A transport, storage, metabolism, and action will likely continue to provide us with worthwhile and stimulating research topics for many years to come.

ACKNOWLEDGMENTS

Preparation of this review was supported in part by a grant from the Norwegian Research Council for Science and the Humanities, the Norwegian Cancer Society, and the Anders Jahres Foundation. Research conducted in M. H. G's laboratory and discussed in this review was supported by USDA Competitive Research Grants 81-CRCR-1-0702 and 88-37200-3537. We thank Joanne Balmer Green for editorial assistance.

Literature Cited

1. Andersen, K. B., Kvam, L., Nilsson, A., Norum, K. R., Blomhoff, R. 1992. Mobilization of retinol from stellate cells. *J. Biol. Chem.* 267:In press
2. Barua, A. B., Olson, J. A. 1986. Retinoyl β-glucuronide: an endogenous compound of human blood. *Am. J. Clin. Nutr.* 43:481–85
3. Batres, R. O., Olson, J. A. 1987. A marginal vitamin A status alters the distribution of vitamin A among parenchymal and stellate cells in rat liver. *J. Nutr.* 117:874–79
4. Batres, R. O., Olson, J. A. 1987. Relative amount and ester composition of vitamin A in rat hepatocytes as a function of the method of cell preparation and of total liver stores. *J. Nutr.* 117:77–82
4a. Båvik, C. O., Eriksson, U., Allen, R. A., Peterson, P. A. 1991. Identification and partial characterization of a retinal pigment epithelial membrane receptor for plasma retinol-binding protein. *J. Biol. Chem.* 266:14978–85
5. Blaner, W. S. 1989. Retinol-binding protein: the serum transport protein for vitamin A. *Endocr. Rev.* 10:308–16
6. Blaner, W. S., Hendriks, H. F. J., Brouwer, A., De Leeuw, A. M., Knook, D. L., Goodman, D. S. 1985. Retinoids, retinoid-binding proteins, and retinyl palmitate hydrolase distributions in different types of rat liver cells. *J. Lipid Res.* 26:1241–51
7. Blaner, W. S., van Bennekum, A. M., Brouwer, A., Hendriks, H. F. J. 1990. Distribution of lecithin-retinol acyltransferase activity in different types of rat liver cells and subcellular fractions. *FEBS Lett.* 274:89–92

8. Blomhoff, R., Berg, T., Norum, K. R. 1988. Distribution of retinol in rat liver cells: effect of age, sex and nutritional status. *Br. J. Nutr.* 60:233–39
9. Blomhoff, R., Berg, T., Norum, K. R. 1988. Transfer of retinol from parenchymal to stellate cells in liver is mediated by retinol-binding protein. *Proc. Natl. Acad. Sci. USA* 85:3455–58
10. Blomhoff, R., Eskild, W., Kindberg, G. M., Prydz, K., Berg, T. 1985. Intracellular transport of endocytosed chylomicron [³H]retinyl ester in rat liver parenchymal cells. Evidence for translocation of a [³H]retinoid from endosomes to endoplasmic reticulum. *J. Biol. Chem.* 260:13566–70
11. Blomhoff, R., Green, M. H., Berg, T., Norum, K. R. 1990. Transport and storage of vitamin A. *Science* 250:399–404
12. Blomhoff, R., Green, M. H., Green, J. B., Berg, T., Norum, K. R. 1991. Vitamin A metabolism: New perspectives on absorption, transport and storage. *Physiol. Rev.* 71:951–90
13. Blomhoff, R., Helgerud, P., Rasmussen, M., Berg, T., Norum, K. R. 1982. In vivo uptake of chylomicron [³H]retinyl ester by rat liver: evidence for retinol transfer from parenchymal to nonparenchymal cells. *Proc. Natl. Acad. Sci. USA* 79:7326–30
14. Blomhoff, R., Holte, K., Naess, L., Berg, T. 1984. Newly administered [³H]retinol is transferred from hepatocytes to stellate cells in liver for storage. *Exp. Cell Res.* 150:186–93
15. Blomhoff, R., Norum, K. R., Berg, T. 1985. Hepatic uptake of [³H]retinol bound to the serum retinol-binding protein involves both parenchymal and peri-

sinusoidal stellate cells. *J. Biol. Chem.* 260:13571–75

16. Blomhoff, R., Rasmussen, M., Nilsson, A., Norum, K. R., Berg, T., et al. 1985. Hepatic retinol metabolism: distribution of retinoids, enzymes, and binding proteins in isolated rat liver cells. *J. Biol. Chem.* 260:13560–65

17. Blomhoff, R., Skrede, B., Norum, K. R. 1990. Uptake of chylomicron remnant retinyl ester via the low density lipoprotein receptor: implications for the role of vitamin A as a possible preventive for some forms of cancer. *J. Intern. Med.* 228:207–10

18. Blomhoff, R., Wake, K. 1991. Perisinusoidal stellate cells of the liver: important roles in retinol metabolism and fibrosis. *FASEB J.* 5:271–77

19. Buck, J., Derguini, F., Levi, E., Nakanishi, K., Hämmerling, U. 1991. Intracellular signaling by 14-hydroxy-4,14-retro-retinol. *Science* 254:1654–56

20. Canada, F. J., Law, W. C., Rando, R. R., Yamamoto, T., Derguini, F., Nakanishi, K. 1990. Substrate specificities and mechanism in the enzymatic processing of vitamin A into 11-cis-retinol. *Biochemistry* 29:9690–97

21. Clawitter, J., Trout, W. E., Burke, M. C., Araghi, S., Roberts, R. M. 1990. A novel family of progesterone-induced, retinol-binding proteins from uterine secretions of the pig. *J. Biol. Chem.* 265:3248–55

22. Creek, K. E., Silverman-Jones, C. S., De Luca, L. M. 1989. Comparison of the uptake and metabolism of retinol delivered to primary mouse keratinocytes either free or bound to rat serum retinol-binding protein. *J. Invest. Dermatol.* 92:283–89

23. Dmitrovski, A. A. 1991. Metabolism of natural retinoids and their functions. In *New Trends in Biological Chemistry*, ed. T. Ozawa, pp. 297–308. Tokyo: Japan Sci. Soc.

24. Duester, G., Shean, M. L., McBride, M. S., Stewart, M. J. 1991. Retinoic acid response element in the human alcohol dehydrogenase gene ADH3: Implications for regulation of retinoic acid synthesis. *Mol. Cell. Biol.* 11:1638–46

25. Eckhoff, C., Collins, M. D., Nau, H. 1991. Human plasma all-trans-4-oxoretinoic and 13-cis-4-oxoretinoic acid profiles during subchronic vitamin A supplementation—Comparison to retinol and retinyl ester plasma levels. *J. Nutr.* 121:1016–25

26. Eckhoff, C., Nau, H. 1990. Identification and quantitation of all-trans- and 13-cis-retinoic acid and 13-cis-4-

oxoretinoic acid in human plasma. *J. Lipid Res.* 31:1445–54

27. Fex, G., Felding, P. 1984. Factors affecting the concentration of free holo retinol-binding protein in human plasma. *Eur. J. Clin. Invest.* 14:146–49

28. Fex, G., Johannesson, G. 1987. Studies of the spontaneous transfer of retinol from the retinol:retinol-binding protein complex to unilamellar liposomes. *Biochim. Biophys. Acta* 901:255–64

29. Fex, G., Johannesson, G. 1988. Retinol transfer across and between phospholipid bilayer membranes. *Biochim. Biophys. Acta* 944:249–55

30. Flannery, J. G., O'Day, W., Pfeffer, B. A., Horowitz, J., Bok, D. 1990. Uptake, processing and release of retinoids by cultured human retinal pigment epithelium. *Exp. Eye Res.* 51:717–28

31. Folman, Y., Russell, R. M., Tang, W., Wolf, G. 1989. Rabbits fed on beta-carotene have higher serum levels of all-trans retinoic acid than those receiving no beta-carotene. *Br. J. Nutr.* 62:195–201

32. Frolik, C. A. 1984. Metabolism of retinoids. In *The Retinoids*, ed. M. B. Sporn, A. B. Roberts, D. S. Goodman, 2:177–208. Orlando: Academic

33. Gerlach, T. H., Zile, M. H. 1990. Upregulation of serum retinol in experimental acute renal failure. *FASEB J.* 4:2511–17

34. Gerlach, T. H., Zile, M. H. 1991. Effect of retinoic acid and apo-RBP on serum retinol concentration in acute renal failure. *FASEB J.* 5:86–92

35. Gerlach, T. H., Zile, M. H. 1991. Metabolism and secretion of retinol transport complex in acute renal failure. *J. Lipid Res.* 32:515–20

36. Gjøen, T., Bjerkelund, T., Blomhoff, H. K., Norum, K. R., Berg, T., Blomhoff, R. 1987. Liver takes up retinol-binding protein from plasma. *J. Biol. Chem.* 262:10926–30

37. Goodman, D. S., Huang, H. S. 1965. Biosynthesis of vitamin A with rat intestinal enzymes. *Science* 149:879–80

38. Goodman, D. S., Huang, H. S., Kanai, M., Shiratori, T. 1967. The enzymatic conversion of all-trans beta-carotene into retinal. *J. Biol. Chem.* 242:3543–54

39. Goodman, D. S., Huang, H. S., Shiratori, T. 1965. Tissue distribution and metabolism of newly absorbed vitamin A in the rat. *J. Lipid Res.* 6:390–96

40. Green, M. H., Green, J. B. 1987. Multicompartmental analysis of whole body retinol dynamics in vitamin A-sufficient rats. *Fed. Proc.* 46:1011

41. Green, M. H., Green, J. B., Lewis, K. C. 1987. Variation in retinol utilization rate with vitamin A status in the rat. *J. Nutr.* 117:694–703

42. Green, M. H., Uhl, L., Green, J. B. 1985. A multicompartmental model of vitamin A kinetics in rats with marginal liver vitamin A stores. *J. Lipid Res.* 26:806–18

43. Hansen, S., Maret, H. 1988. Retinal is not formed in vitro by enzymatic central cleavage of beta-carotene. *Biochemistry* 27:200–6

44. Haq, R. ul, Chytil, F. 1988. Retinoic acid rapidly induces lung cellular retinol-binding protein mRNA levels in retinol-deficient rats. *Biochem. Biophys. Res. Commun.* 156:712–16

45. Harney, J. P., Mirando, M. A., Smith, L. C., Bazer, F. W. 1990. Retinol-binding protein: A major secretory product of the pig conceptus. *Biol. Reprod.* 42:523–32

46. Harrison, E. H., Gad, M. Z. 1989. Hydrolysis of retinyl palmitate by enzymes of rat pancreas and liver. Differentiation of bile salt-dependent and bile salt-independent, neutral retinyl ester hydrolases in rat liver. *J. Biol. Chem.* 264:17142–47

47. Helgerud, P., Petersen, L. B., Norum, K. R. 1982. Acyl CoA:retinyl acyltransferase in rat small intestine: its activity and some properties of the enzymic reaction. *J. Lipid Res.* 23:609–18

48. Helgerud, P., Petersen, L. B., Norum, K. R. 1983. Retinol esterification by microsomes from the mucosa of human small intestine. *J. Clin. Invest.* 71:747–53

49. Heller, J. 1975. Interactions of plasma retinol-binding protein with its receptor. Specific binding of bovine and human retinol-binding protein to pigment epithelium cells from bovine eyes. *J. Biol. Chem.* 250:3613–19

50. Heller, J., Bok, D. 1976. A specific receptor for retinol-binding protein as detected by the binding of human and bovine retinol binding protein to pigment epithelial cells. *Am. J. Ophthalmol.* 81:93–97

51. Hendriks, H. F. J., Blaner, W. S., Wennekers, H. M., Piantedosi, R., Brouwer, A., et al. 1988. Distributions of retinoids, retinoid-binding proteins and related parameters in different types of liver cells isolated from young and old rats. *Eur. J. Biochem.* 171:237–44

52. Hendriks, H. F. J., Elhanany, E., Brouwer, A., De Leeuw, A. M., Knook, D. L. 1988. Uptake and processing of [³H]retinoids in rat liver studied by electron microscopic autoradiography. *Hepatology* 8:276–85

53. Hendriks, H. F. J., Verhoofstad, W. A. M. M., Brouwer, A., De Leeuw, A. M., Knook, D. L. 1985. Perisinusoidal fat-storing cells are the main vitamin A storage sites in rat liver. *Exp. Cell Res.* 160:138–49

54. Herbert, J., Cavallaro, T., Martone, R. 1991. The distribution of retinol-binding protein and its mRNA in the rat eye. *Invest. Ophthalmol. Visual Sci.* 32:302–9

54a. Heyman, R. A., Mangelsdorf, D. J., Dyck, J. A., Stein, R. B., Eichele, G., et al. 1992. 9-cis retinoic acid is a high affinity ligand for the retinoid X receptor. *Cell* 68:1–20

55. Hodam, J. R., Hilaire, P. S., Creek, K. E. 1991. Comparison of the rate of uptake and biologic effects of retinol added to human keratinocytes either directly to the culture medium or bound to serum retinol-binding protein. *J. Invest. Dermatol.* 97:298–304

56. Hussain, M. M., Mahley, R. W., Boyles, J. K., Fainaru, M., Brecht, W. J., Lindquist, P. A. 1989. Chylomicron-chylomicron remnant clearance by liver and bone marrow in rabbits. Factors that modify tissue-specific uptake. *J. Biol. Chem.* 264:9571–82

57. Hussain, M. M., Mahley, R. W., Boyles, J. K., Lindquist, P. A., Brecht, W. J., Innerarity, T. 1989. Chylomicron metabolism. Chylomicron uptake by bone marrow in different animal species. *J. Biol. Chem.* 264:17931–38

58. Kakkad, B. P., Ong, D. E. 1988. Reduction of retinaldehyde bound to cellular retinol-binding protein (type II) by microsomes from rat small intestine. *J. Biol. Chem.* 263:12916–19

59. Lakshman, M. R., Mychkovsky, I., Attlesey, M. 1989. Enzymatic conversion of all-trans-β-carotene to retinal by a cytosolic enzyme from rabbit and rat intestinal mucosa. *Proc. Natl. Acad. Sci. USA* 86:9124–28

60. Leo, M. A., Kim, C., Lieber, C. S. 1987. NAD⁺-dependent retinol dehydrogenase in liver microsomes. *Arch. Biochem. Biophys.* 259:241–49

61. Leo, M. A., Lieber, C. S. 1985. New pathway for retinol metabolism in liver microsomes. *J. Biol. Chem.* 260:5228–31

61a. Levin, A. A., Sturzenbecker, L. J., Kazmer, S., Bosakowski, T., Huselton, C., et al. 1992. 9-cis retinoic acid stereoisomer binds and activates the nuclear receptor RXRa. *Nature* 355:359–61

56   BLOMHOFF, GREEN & NORUM

62. Lewis, K. C., Green, M. H., Green, J. B., Zech, L. A. 1990. Retinol metabolism in rats with low vitamin A status: a compartmental model. *J. Lipid Res.* 31:1535–48
63. Lewis, K. C., Green, M. H., Underwood, B. A. 1981. Vitamin A turnover in rats as influenced by vitamin A status. *J. Nutr.* 111:1135–44
64. Liu, K. H., Baumbach, G. A., Gillevet, P. M., Godkin, J. D. 1990. Purification and characterization of bovine placental retinol-binding protein. *Endocrinology* 127:2696–2704
65. MacDonald, P. N., Bok, D., Ong, D. E. 1990. Localization of cellular retinol-binding protein and retinol-binding protein in cells comprising the blood-brain barrier of rat and human. *Proc. Natl. Acad. Sci. USA* 87:4265–69
66. MacDonald, P. N., Ong, D. E. 1988. Evidence for a lecithin-retinol acyltransferase activity in the rat small intestine. *J. Biol. Chem.* 263:12478–82
67. Mahley, R. W., Hussain, M. M. 1991. Chylomicron and chylomicron remnant catabolism. *Curr. Opin. Lipidol.* 2:170–76
68. Makover, A., Soprano, D. R., Wyatt, M. L., Goodman, D. S. 1989. Localization of retinol-binding protein messenger RNA in the rat kidney and in perinephric fat tissue. *J. Lipid Res.* 30:171–80
69. Mangelsdorf, D. J., Umesono, K., Kliewer, S. A., Borgmeyer, U., Ong, E. S., Evans, R. M. 1991. A direct repeat in the cellular retinol-binding protein type II gene confers differential regulation by RXR and RAR. *Cell* 66:555–61
70. Moriwaki, H., Blaner, W. S., Piantedosi, R., Goodman, D. S. 1988. Effects of dietary retinoid and triglyceride on the lipid composition of rat liver stellate cells and stellate cell lipid droplets. *J. Lipid Res.* 29:1523–34
71. Napoli, J. L., Race, K. R. 1988. Biogenesis of retinoic acid from beta-carotene. Differences between the metabolism of beta-carotene and retinal. *J. Biol. Chem.* 263:17372–77
72. Noy, N., Blaner, W. S. 1991. Interactions of retinol with binding proteins—studies with rat cellular retinol-binding protein and with rat retinol-binding protein. *Biochemistry* 30:6380–86
73. Olson, J. A., Hayaishi, O. 1965. The enzymatic cleavage of β-carotene into vitamin A by soluble enzymes of rat liver and intestine. *Proc. Natl. Acad. Sci. USA* 54:1364–69
74. Ong, D. E., Chytil, F. 1988. Presence of novel retinoic acid-binding proteins in the lumen of rat epididymis. *Arch. Biochem. Biophys.* 267:474–78
75. Ong, D. E., Kakkad, B., MacDonald, P. 1987. Acyl-CoA-independent esterification of retinol bound to cellular retinol-binding protein (Type II) by microsomes from rat small intestine. *J. Biol. Chem.* 262:2729–36
76. Ong, D. E., MacDonald, P. N., Gubitosi, A. M. 1988. Esterification of retinol in rat liver. Possible participation by cellular retinol-binding protein and cellular retinol-binding protein II. *J. Biol. Chem.* 263:5789–96
77. Ottonello, S., Petrucco, S., Maraini, G. 1987. Vitamin A uptake from retinol-binding protein in a cell-free system from pigment epithelial cells of bovine retina. *J. Biol. Chem.* 262:3975–81
78. Posch, K. C., Boerman, M. H. E. M., Burns, R. D., Napoli, J. L. 1991. Holocellular retinol binding protein as a substrate for microsomal retinal synthesis. *Biochemistry* 30:6224–30
79. Posch, K. C., Enright, W. J., Napoli, J. L. 1989. Retinoic acid synthesis by cytosol from the alcohol dehydrogenase negative deermouse. *Arch. Biochem. Biophys.* 274:171–78
80. Randolph, R. K., Winkler, K. E., Ross, A. C. 1991. Fatty acyl CoA-dependent and -independent retinol esterification by rat liver and lactating mammary gland microsomes. *Arch. Biochem. Biophys.* 288:500–8
81. Saari, J. C., Bredberg, D. L. 1989. Lecithin:retinol acyltransferase in retinal pigment epithelial microsomes. *J. Biol. Chem.* 264:8636–40
82. Senoo, H., Stang, E., Nilsson, A., Kindberg, G. M., Berg, T., et al. 1990. Internalization of retinol-binding protein in parenchymal and stellate cells of the rat liver. *J. Lipid Res.* 31:1229–39
83. Shingleton, J. L., Skinner, M. K., Ong, D. E. 1989. Retinol esterification in Sertoli cells by lecithin-retinol acyltransferase. *Biochemistry* 28:9647–53
84. Shingleton, J. L., Skinner, M. K., Ong, D. E. 1989. Characteristics of retinol accumulation from serum retinol-binding protein by cultured sertoli cells. *Biochemistry* 28:9641–47
85. Sivaprasadarao, A., Findlay, J. B. C. 1988. The mechanism of uptake of retinol by plasma-membrane vesicles. *Biochem. J.* 255:571–79
86. Sivaprasadarao, A., Findlay, J. B. C. 1988. The interaction of retinol-binding protein with its plasma-membrane receptor. *Biochem. J.* 255:561–69
87. Smith, F. R., Goodman, D. S. 1971.

The effects of diseases of the liver, thyroid, and kidneys on the transport of vitamin A in human plasma. *J. Clin. Invest.* 50:2426–36

88. Smith, W. C., Nakshatri, H., Leroy, P., Rees, J., Chambon, P. 1991. A retinoic acid response element is present in the mouse cellular retinol binding protein I (mCRBPI) promoter. *EMBO J.* 10:2223–30

89. Soprano, D. R., Soprano, K. J., Goodman, D. S. 1986. Retinol-binding protein messenger RNA levels in the liver and in extrahepatic tissues of the rat. *J. Lipid Res.* 27:166–71

90. Thaller, C., Eichele, G. 1990. Isolation of 3,4-didehydroretinoic acid, a novel morphogenetic signal in the wing bud. *Nature* 345:815–20

91. Törmä, H., Vahlquist, A. 1990. Vitamin A esterification in human epidermis: A relation to keratinocyte differentiation. *J. Invest. Dermatol.* 94:132–38

92. Underwood, B. A. 1984. Vitamin A in animal and human nutrition. In *The Retinoids*, ed. M. B. Sporn, A. B. Roberts, D. S. Goodman, 1:281–92. Orlando: Academic

93. Underwood, B. A., Loerch, J. D., Lewis, K. C. 1979. Effects of dietary vitamin A deficiency, retinoic acid and protein quantity and quality on serially obtained plasma and liver levels of vitamin A in rats. *J. Nutr.* 109:796–806

94. Vahlquist, A. 1972. Metabolism of the vitamin A-transporting protein complex: turnover of retinol-binding protein, prealbumin and vitamin A in a primate (Macaca Irus). *Scand. J. Clin. Lab. Invest.* 30:349–60

95. Wake, K. 1980. Sternzellen in the liver: perisinusoidal cells with special reference to storage of vitamin A. *Am. J. Anat.* 132:429–62

96. Wake, K. 1980. Perisinusoidal stellate cells (fat-storing cells, interstitial cells, lipocytes), their related structure in and around the liver sinusoids, and vitamin A-storing cells in extrahepatic organs. *Int. Rev. Cytol.* 66:303–53

97. Wathne, K.-O., Norum, K. R., Smeland, E., Blomhoff, R. 1988. Retinol bound to physiological carrier molecules regulates growth and differentiation of myeloid leukemic cells. *J. Biol. Chem.* 263:8691–95

98. Wei, L.-N., Blaner, W. S., Goodman, D. S., Nguyen-Huu, M. C. 1989. Regulation of the cellular retinoid-binding proteins and their messenger ribonucleic acids during P19 embryonal carcinoma cell differentiation induced by retinoic acid. *Mol. Endocrinol.* 3:454–63

99. Wolf, G. 1991. The intracellular vitamin A-binding proteins: an overview of their functions. *Nutr. Rev.* 49:1–12

100. Yamada, M., Blaner, W. S., Soprano, D. R., Dixon, J. L., Kjeldbye, H. M., Goodman, D. S. 1987. Biochemical characteristics of isolated rat liver stellate cells. *Hepatology* 7:1224–29

101. Yost, R. W., Harrison, E. H., Ross, A. C. 1988. Esterification by rat liver microsomes of retinol bound to cellular retinol-binding protein. *J. Biol. Chem.* 263:18693–18701

Annu. Rev. Nutr. 1992. 12:59–79

COBALAMIN DEFICIENCY AND THE PATHOGENESIS OF NERVOUS SYSTEM DISEASE

Jack Metz

Department of Hematology, South African Institute for Medical Research and University of the Witwatersrand, Johannesburg 2000, South Africa

KEY WORDS: Demyelination, nitrous oxide, methionine, methylation

CONTENTS

59

0199-9885/92/0715-0059$02.00

INTRODUCTION

Severe untreated cobalamin (Cbl) deficiency in humans is associated with crippling neurological disease, ataxia, and death. According to Chanarin (12), the earliest suggestion of neuropathy associated with cobalamin (Cbl) deficiency is contained in the account of Osler & Gardner (49) in 1877 of anemic patients with numbness of the fingers, hands, and forearms. The association between Cbl deficiency and spinal cord lesions was recognized in the same year (41), and the features of the characteristic subacute combined degeneration of the spinal cord were documented in 1900 (53). The demyelination lesion begins as swelling of the myelin sheaths, is followed by breakdown of myelin and disruption of the axon, and results in a spongy vacuolated appearance of the cord. The process is most severe in the posterior columns of the thoracic region, but the lateral columns are also affected in due course. Changes in the brain consist of foci of demyelination in the white matter. In the peripheral nerves, the number of myelin sheaths may be reduced.

Although the features of the Cbl neuropathy were described more than a century ago, the underlying pathogenic mechanism of the lesions remains poorly understood. This long period with relatively little progress can be largely attributed to the lack of availability of suitable animal models of the Cbl neuropathy. Only in the last decade, with the use of nitrous oxide (N_2O) to induce Cbl deficiency, have suitable models been introduced, and most of the recent advances in knowledge in this area have been derived from studies with these models. This review focuses on the data that have become available in the last decade.

The induction of Cbl deficiency by dietary manipulation alone is difficult. The small daily requirement for Cbl, coupled with the relatively large body stores, necessitates prolonged dietary deprivation, often longer than a year, before severe deficiency can occur. In addition, as feces are a rich source of the vitamin, rigid exclusion of coprophagy is necessary for deficiency to develop.

Before 1975, the only known experimental animal in which Cbl neuropathy could be induced was the monkey (34). The long period (2 to 3 years) required to induce dietary deficiency, and the cost of maintaining these relatively large laboratory animals, limited their use to one or two laboratories. The observation that in the fruit bat (*Rousettus aegyptiacus*) severe Cbl deficiency accompanied by neuropathy similar to that of Cbl-deficient humans could be induced by minimal manipulation of the diet provided the first small animal model of the Cbl neuropathy (30, 47). However, a prolonged period (9–12 months) of dietary restriction was still necessary to produce the neuropathy, which limited the usefulness of the model. The introduction of the anesthetic gas N_2O to induce Cbl deficiency represented a major advance in that its use shortened to weeks the time period required to render animals Cbl deficient. Repeated exposure of monkeys (18), fruit bats (61), and pigs to N_2O (70) leads to the development of typical Cbl neuropathy in these animals.

BIOCHEMICAL REACTIONS REQUIRING COBALAMIN, AND THE ACTION OF NITROUS OXIDE

To date, Cbl is known to be required in two biochemical reactions in humans (Figure 1). The coenzyme forms of Cbl differ in the two reactions. (a) Cbl in the form of adenosylcobalamin (AdoCbl) is required as cofactor for methyl-malonyl CoA mutase (MMCoM) (EC 5.4.99.2), which is essential for the isomerization of methylmalonyl CoA (MMCoA) to succinyl CoA, the final step in the metabolism of propionate to succinyl CoA. (b) Cbl in the form of the methylcobalamin (MeCbl) is required as cofactor for the enzyme methionine synthetase (MS) (5-methyltetrahydrofolate homocysteine methyl-transferase: EC 2.1.1.13), which catalyzes the recycling of homocysteine to methionine. Folate, in the form of methyltetrahydrofolate, is also required in this reaction (Figure 1).

Nitrous oxide oxidizes active reduced cob(I)alamin to inactive cob(III)ala-min (4). As Cbl in the form of reduced MeCbl is required as cofactor for MS, exposure to N_2O causes rapid inactivation of this enzyme (15). The inactive Cbl is excreted, so that repeated exposure to N_2O results in depletion of body Cbl stores, with reduced AdoCbl, so that the activity of the AdoCbl-dependent enzyme MMCoAM is also compromised (15).

ANIMAL MODELS OF THE COBALAMIN NEUROPATHY

To be acceptable, experimental animal models of the Cbl neuropathy must show pathological changes in the spinal cord similar to those described in humans with Cbl deficiency. This type of neuropathy can be induced in the monkey (18, 28, 57), the fruit bat (20, 30), and the pig (70) by dietary deprivation of Cbl, or by repeated exposure to N_2O, or by a combination of the two.

The Monkey

Prolonged (33–45 months) dietary deprivation of Cbl leads to severe clinical neuropathy with spastic paralysis of the hind limbs, accompanied by neuro-

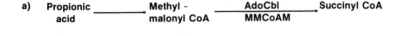

a) Propionic ————→ Methyl - ————→ AdoCbl ————→ Succinyl CoA
   acid   malonyl CoA   MMCoAM

b) 5-Methyl- ———— + Homocysteine ————→ CH₃Cbl ————→ Tetrahydrofolate + Methionine
   tetrahydrofolate   MS

Figure 1 The two cobalamin-dependent biochemical reactions in humans. Abbreviations: AdoCbl, adenosylcobalamin; MMCoAM, methylmalonyl CoA mutase; CH₃Cbl, methylcobalamin; MS, methionine synthetase.

pathological changes in the spinal cord indistinguishable from those of human subacute combined degeneration (1, 2). These changes include separation and vacuolation of myelin lamellae, with eventual complete destruction of myelin sheaths. Repeated exposure to N₂O leads to the onset of clinical neuropathy within 9–12 weeks, followed by ataxia 2–3 weeks later (18). The neuropathy is accompanied by degeneration of myelin sheaths and axis cylinders in the posterior columns of the spinal cord, and accumulation of lipid-laden macrophages. Although the degree of clinical neuropathy in N₂O- and dietary-induced models is similar, the neuropathologic changes are less severe in the N₂O-induced model (18, 57).

The Fruit Bat

As fruit bats normally subsist on an all fruit diet, and fruit contains no Cbl, these animals presumably obtain their Cbl from contaminated water or from pests infesting the fruit. Thus, dietary Cbl deficiency can be induced by simply feeding pest-free, washed fruit and uncontaminated water. Such a diet leads to severe Cbl deficiency within about nine months (30). Neuropathy is manifested clinically by loss of ability to fly, progressing to ataxia, spastic paralysis, and death. When exposure to N₂O is combined with dietary deprivation, the onset of neuropathy is after about nine weeks. The neuropathological changes in the posterior columns of the spinal cord include vacuolation of the myelin, which leads to marked distention and separation of the lamellae, with thinning or frank loss of myelin sheaths (20).

The Pig

Repeated exposure to N₂O leads to progressive ataxia within nine weeks. Neuropathological changes in the spinal cord comprise multiple foci of myelin degeneration. The lesions are less severe than those accompanying N_2O-induced Cbl neuropathy in the monkey.

Other small animals such as the rabbit, the mouse, and the rat do not develop neuropathy when Cbl deficiency is induced. The rat has been used extensively as an animal model for Cbl deficiency, and neuropathy has not been induced by dietary means or with N_2O. However, a recent report describes spongy degeneration throughout the white matter of all the spinal cord segments in rats rendered Cbl deficient by total gastrectomy (55). The changes occurred as rapidly as two months after the operation. It is difficult to assess the significance of this report, as the Cbl-deficient rats apparently did not show clinical signs of neuropathy, and histological changes were found in sham-operated rats, but to a lesser degree. Why Cbl deficiency in the rat after total gastrectomy is followed by neuropathological changes, when similar changes do not occur with deficiency induced by dietary means or exposure to N_2O, is not clear. Confirmation of these findings would provide an additional small animal model of the Cbl neuropathy.

THE ADENOSYLCOBALAMIN-DEPENDENT METHYLMALONYL CoA MUTASE AND THE COBALAMIN NEUROPATHY

Cbl is required for the normal functioning of two enzymes, MMCoAm and MS. Consequently, the lesion underlying the Cbl neuropathy has been sought in the reactions catalyzed by these enzymes. Neuropathy is a frequent complication of severe, prolonged Cbl deficiency in humans, but it is not usually associated with folate deficiency. For years, therefore, attention was focused on the AdoCbl-requiring MMCoAM reaction in propionic acid catabolism, where folate plays no role, rather than on the MeCbl-dependent MS reaction in which both Cbl and folate are involved.

Changes in Odd-Chain and Branched-Chain Fatty Acids

An understanding of the mechanism whereby deficiency of AdoCbl leads to the Cbl neuropathy is based on the hypothesis that impairment of the AdoCbl-dependent MMCoAM reaction with subsequent intracellular accumulation of the intermediates of propionic acid metabolism, propionyl CoA and methylmalonyl CoA (MMCoA), leads to the formation of abnormal fatty acids (5, 8, 9, 23–25, 35) (Figure 2). Substitution of propionyl CoA for acetyl CoA and of MMCoA for malonyl CoA would lead respectively to the formation of

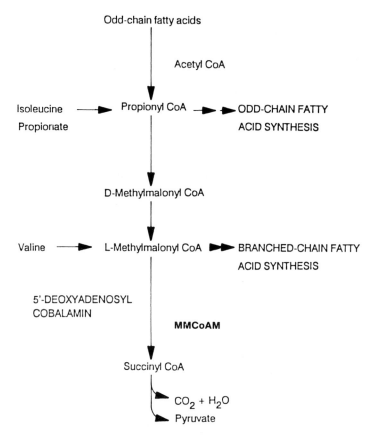

Figure 2   The propionic acid pathway in cobalamin deficiency. (Reproduced with permission from J. Metz and J. van der Westhuyzen, 1987, *Comp. Biochem. Physiol.* 88A:171–77.)

increased odd-chain and branched-chain fatty acids (9). Excessive odd-chain and branched-chain fatty acids would accumulate in membrane lipids of nervous tissue. As the synthesis of normal myelin is dependent on the availability of specific fatty acids, synthesis of abnormal in place of normal fatty acids could result in altered myelin integrity and in demyelination, thus leading to impaired nervous system functioning (25).

The evidence in support of this hypothesis has been derived from earlier in vitro studies and observations on humans with Cbl deficiency and on animals with experimentally induced deficiency. Patients with Cbl neuropathy were shown to excrete in the urine excessive amounts of methylmalonic acid (MMA), presumably derived from the accumulation of MMCoA (71). The intermediates of propionic acid metabolism, propionyl CoA and MMCoA, have been shown to inhibit in vitro fatty acid synthesis (8, 23, 25), and accumulation of propionyl CoA may result in a relative increased biosynthesis of odd-chain fatty acids (5, 24, 35).

An increase in odd-chain and branched-chain fatty acids in nervous tissues has been reported from earlier studies on humans with Cbl deficiency and on experimental animals with dietary-induced deficiency. In a young child who died with MMA accumulation owing to an inherited inability to convert Cbl to the AdoCbl coenzyme form, odd-chain fatty acids (15:0, 17:0) were significantly increased in the glycerolipids of the spinal cord (35, 40). Branched-chain fatty acids of C17 were detected in the brain and spinal cord while control tissue contained only trace amounts or none. A patient with an inherited deficiency of both the AdoCbl and MeCbl coenzymes showed a relative increase in odd-chain fatty acids (15:0, 15:1, 17:0, 17:1) in myelin lipids, when compared with controls (52). Methyl branched C17 fatty acids were also identified. Humans with Cbl-deficient pernicious anemia showed a decrease in net synthesis of fatty acids but an increase in the amounts of odd-chain fatty acids (15:0, 17:0) present in biopsies of peripheral nerve (24). Labelled propionate was shown to be incorporated into these odd-chain fatty acids.

Accumulation of odd-chain fatty acids 15:0 and 17:0 in the cerebral lipids of rats with dietary-induced Cbl deficiency has been reported from a number of studies (3, 21, 50, 51). In addition, diminished amounts of fatty acids 20:4 and 22:5 were found in one study (50); as malonyl CoA is required for the elongation of 18:2 to form 20:4 and 22:5, it was suggested that the accumulation of MMCoA in Cbl deficiency interfered with such elongation reactions. Some studies failed to demonstrate increased synthesis of odd-chain fatty acids in the Cbl-deficient rats. In one study in which propionate was administered, contrary to expectation, C15 and C17 fatty acids were present in liver triglycerides in smaller amounts than in control animals (22).

These earlier studies do not make clear what relationship, if any, the changes demonstrated in nervous system fatty acids bear to the development of the Cbl neuropathy, and whether small changes in membrane fatty acid composition could affect membrane stability and renewal to a degree that could cause demyelination. There was in fact no evidence of a correlation between the relative differences in odd chain and branched-chain fatty acids in the lipids of neural tissues and the functional changes in these tissues of humans or rats. Evidence of an etiological relationship would be the consistent finding of excess odd-chain and branched-chain fatty acids in the neural tissue of experimental animals with Cbl neuropathy and a correlation between the severity of the neuropathy and the abundance of these fatty acids. Acceleration or retardation of the neuropathy by dietary manipulation should be reflected by corresponding changes in the levels of fatty acid accumulation.

A series of studies of neural tissue of the fruit bat with Cbl neuropathy has failed to provide this evidence. The Cbl neuropathy in this animal is accompanied by decreased activity of MMCoAM and accumulation of MMA in serum and urine (67). The brains of bats deprived of dietary Cbl have a marginally higher percentage of 18:3 fatty acid in phosphotidylcholine and a slightly higher percentage of 15:0 fatty acid in sphingomyelin, with no other significant changes in odd-chain fatty acids (59). In spinal cord myelin the concentrations of odd-chain fatty acids 15:0, 15:1, 17:1, and 19:0 are all higher in Cbl-deficient bats (60). In fruit bats with Cbl neuropathy induced by a combination of dietary deprivation and exposure to N_2O, no significant differences in the concentration of odd-chain fatty acids in brain lipids were demonstrable, compared with levels in control animals, and branched-chain fatty acids were not present in detectable amounts in either group (63).

In the most recently reported study of Cbl neuropathy in the fruit bat (67), MMA metabolism was examined in relation to fatty acid concentrations in the nervous system. The Cbl-deficient bats had low hepatic MMCoAM activity, with elevated plasma and urinary MMA levels, yet no significant changes in the concentration of odd-chain or branched-chain fatty acids in the spinal cord or brain could be demonstrated. Branched-chain and odd-chain fatty acids were detectable in only trace amounts in both spinal cord and brain, and no significant increase in branched-chain or odd-chain fatty acids in phosphatylethanolamine, phosphatylcholine, or sphingomyelin was observed.

Thus the studies of fatty acids in spinal cord and brain in fruit bats with severe Cbl neuropathy have produced variable results with little or no changes in individual branched-chain or odd chain fatty acids, despite impaired MMCoAM activity and accumulation of MMA. It seems unlikely that the essentially small and inconsistent changes in fatty acids play a significant role in the pathogenesis of the fatal neuropathy in the Cbl-deficient fruit bat.

Effect of Supplementation with Amino Acids

If the Cbl neuropathy is related to the block in propionic acid catabolism and accumulation of MMA, one would expect a significant reduction in the MMA levels in animals where dietary supplementation delayed the onset of neuropathy. Supplementation of the diet with methionine delays the onset of the N_2O-induced Cbl neuropathy in the pig, but does not lead to a concomitant reduction in plasma MMA levels (70). Methionine supplementation has a similar ameliorating effect on the Cbl neuropathy in the fruit bat, but again fails to maintain MMCoAM activity, and there is persistent methylmalonyl acidemia (67). Thus it is unlikely that the effect of methionine in delaying the onset of the Cbl neuropathy is mediated by an effect on the propionic acid pathway.

Studies of dietary supplementation with valine and isoleucine have likewise failed to provide evidence that a defect in propionic acid metabolism is the

cause of the Cbl neuropathy. These two amino acids are catabolized exclusively via the propionic acid pathway. If the underlying defect is in the
propionic acid pathway, increasing the dietary intake of these amino acids
would be expected to aggravate the Cbl neuropathy by providing additional
substrate to be metabolized by the already compromised MMCoAM. However, no such effect of these amino acids was noted on the Cbl neuropathy in the
fruit bat (69). In fact, the opposite effect was noted, whereby supplementation
with valine and isoleucine delayed the onset of neuropathy. Failure of valine
and isoleucine loading to aggravate the Cbl neuropathy is not consistent with
the hypothesis that a defect in propionic acid metabolism is the underlying
basis for the Cbl neuropathy.

Additional Evidence Against the Adenosylcobalamin Hypothesis

Further evidence against the AdoCbl hypothesis comes from observations on
children with inherited disorders associated with accumulation of MMA.
Inherited mutations of the apomutase or of AdoCbl synthesis (Cbl A and Cbl
B mutants) are accompanied by marked methylmalonylacidemia and methyl-
malonylaciduria (14, 43). Children with these defects may show muscular
hypotonia and mental retardation, but they do not develop the Cbl neuropathy. Thus, accumulation of large amounts of MMA from defects other than
Cbl deficiency does not produce neuropathy.

Chronic intermittent exposure to N_2O in humans leads to a syndrome
similar to Cbl neuropathy (6, 39, 54). That the inhibitory effect of N_2O on
cobalamin metabolism is primarily on the MeCbl-dependent MS reaction
rather than on the AdoCbl MMCoAM reaction, suggests that the Cbl neuropathy is not related to AdoCbl deficiency. This interpretation is probably
correct, but during long-term exposure to N_2O, MMCoAM is affected as
body deficiency of Cbl develops, so that the eventual N_2O effect is not limited
to the MS reaction.

Thus a large body of evidence can now be marshalled against the hypothesis that decreased function of AdoCbl is the mechanism of the Cbl neuropathy. Although this hypothesis was dominant for about two decades, it seems
no longer tenable when considered in conjunction with the results derived
from more recent studies of experimental Cbl neuropathy.

DEFECTIVE METHYLATION AND THE COBALAMIN NEUROPATHY

Adenosylmethionine Concentration in the Nervous System

The first suggestion that the Cbl neuropathy might be related to defective
methylation in the nervous system came from observations on a patient with
an inherited mutation in Cbl (52) and from studies of cycloleucine-treated

mice (26, 33). In the patient with inherited deficiency of both AdoCbl and MeCbl, the amount of basic protein in myelin was found to be markedly reduced. This prompted the suggestion that lack of normal methylation of basic protein, due to the impairment of the MS reaction in Cbl deficiency, may have rendered the myelin basic protein unstable (52). Cycloleucine, an analogue of methionine, is a competitive inhibitor of the enzyme methionine adenosyltransferase (EC 2.5.1.6), which converts methionine (Met) to adenosylmethionine (AdoMet). Cycloleucine administered to mice caused a neurological condition similar to the Cbl neuropathy (26, 33). AdoMet is an important donor of methyl groups for transmethylation reactions, and possibly cycloleucine produced the neurologic lesion through defective methylation of myelin lipid.

Scott and co-workers (19, 57) produced the first experimental evidence that the Cbl neuropathy was related to the Cbl-dependent MS reaction by demonstrating that monkeys exposed to N_2O developed neurological changes resembling subacute combined degeneration found in humans. They hypothesized that the impaired MS activity attendant upon N_2O exposure resulted in decreased Met synthesis, which in turn led to decreased amounts of AdoMet for methylation reactions in myelin (Figure 3). In this way, methyl group deficiency would result in demyelination and the clinical neuropathy. In a subsequent study, supplementation of the diet with Met prevented the onset of clinical neuropathy and partially prevented demyelination in the spinal cord. Measurement of Met and AdoMet concentration in neural tissue was not performed as part of these studies. The induction of Cbl neuropathy by exposure to N_2O, and the action of dietary Met supplementation in delaying the onset of neurological signs and neuropathological changes in the spinal cord, was confirmed in the fruit bat (61) and in the pig (70).

The postulate that Cbl neuropathy was related to deficient synthesis of AdoMet was tested by direct measurement of this compound in the tissues of fruit bats with N_2O-induced Cbl neuropathy (27, 65). A small reduction of about 10% in the mean concentration of AdoMet was observed when livers of test animals were compared with those of controls. In the brain, however, contrary to expectation, the mean AdoMet concentration of animals with the Cbl neuropathy was significantly greater than that of controls. In the same experiment, groups of animals received dietary supplements of either Met or pteroylglutamic acid (PGA). As had been demonstrated previously, supplementation of the diet with Met delayed the onset of the neuropathy. Supplementation of the diet with PGA accelerated the onset of the neuropathy. The AdoMet concentrations in brains of animals receiving dietary supplements of Met or PGA were not significantly different from those of animals exposed to N_2O without supplementation. This study thus failed to reveal any reduction in AdoMet levels in brains of animals with N_2O neuro-

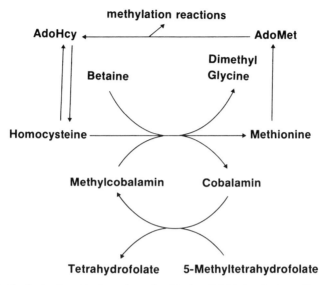

Figure 3   Synthesis of methionine, adenosylmethionine (AdoMet) and adenosylhomocysteine (AdoHcy).

pathy; furthermore, the effects of PGA in aggravating and Met in delaying the onset of progression of the neuropathy were not reflected by changes in the AdoMet concentration in the brain. Thus in the fruit bat model, these findings did not provide support for the hypothesis that the Cbl neuropathy was related to defective methylation caused by reduced concentration of AdoMet in the brain.

Similar findings were reported in the rat (42) and in the pig (70). In rats rendered Cbl deficient by repeated exposure to N_2O, the Met and AdoMet levels in brain were not reduced, compared with levels found in control animals, despite a fall in MS activity in the brain. It was suggested that the brain was able to maintain its Met levels in the absence of normal Cbl-dependent MS activity owing to increased uptake of Met from plasma or via a yet undiscovered other pathway of Met synthesis in the brain. In the pig with N_2O-induced Cbl neuropathy, the concentration of AdoMet in the nervous system was not reduced (70).

Further evidence that the Cbl neuropathy was not related to nervous system Met levels came from studies of the effect of dietary supplementation with betaine on the Cbl neuropathy in the fruit bat (66). Met is synthesized from homocysteine not only through the MS reaction but also through the reaction catalyzed by the enzyme betaine-homocysteine methyltransferase (EC 2.1.1.5) (58). This reaction is independent of Cbl or folate, and the methyl donor is betaine. Supplementation of the diet with betaine delayed the onset of the Cbl neuropathy in fruit bats exposed to N_2O, but betaine was less effective than Met. The Met levels in brains of betaine-supplemented animals were lower than those of the N_2O-exposed animals without betaine supplementation, despite the protective effect of betaine on the development of neurological impairment. The results of the betaine experiments confirm the importance of adequate Met synthesis in the prevention of the Cbl neuropathy, but indicate that the neuropathy is not related directly to the concentration of Met in neural tissue.

Although Met and AdoMet levels in the nervous system are not reduced in the fruit bat or pig models of Cbl neuropathy, evidence from one study in humans suggests that reduced Met and AdoMet levels may be associated with neurological changes similar to those of Cbl deficiency. In patients with deficiency of 5,10-methylenetetrahydrofolate (CH_2THF) reductase, formation of methyltetrahydrofolate (CH_3THF) from CH_2THF is reduced. CH_3THF is the donor of the methyl group for the conversion of homocysteine to methionine via the MS reaction. Patients with 5,10-CH_2THF reductase deficiency have reduced concentration of both Met and AdoMet in the cerebrospinal fluid (CSF) and show demyelination in the brain and subacute combined degeneration of the spinal cord (32). Methionine is effective in the treatment of only some of these patients; betaine, however, prevented the progress of the neurologic symptoms in all patients in whom it was tried, and betaine restored CSF AdoMet concentration to normal.

Although reduced levels of Met and AdoMet in neural tissue have not been demonstrated in patients with CH_2THF reductase deficiency, the finding of low levels in CSF does indicate that nervous system deficiency of Met and Adomet may occur in association with a defect in the MS reaction. Furthermore, the progression of the neurological disease in some of the children was halted by activation of the alternative betaine-dependent methionine synthesis reaction, and this was associated with restoration of the CSF AdoMet levels to normal. The relevance of these findings, in which the MS reaction is impaired owing to lack of the CH_3THF cofactor, to the Cbl neuropathy in which the impairment is due to lack of MeCbl is uncertain. In particular, the relative effects of Met and betaine supplementation are different. In experimental Cbl-deficient neuropathy, Met produces a greater and more certain protective effect on the neuropathy than does betaine (66), while the opposite is the case in children with neuropathy associated with CH_2THF reductase deficiency.

Adenosylhomocysteine Toxicity

Inability to demonstrate reduced AdoMet concentrations in the neural tissue of the Cbl-deficient fruit bat (27, 65), rat (42), and pig (70) suggested that undermethylation was not the underlying cause for the Cbl neuropathy.

Attention was then focused on the AdoMet/Adenosylhomocysteine (AdoHcy) ratio as an index of methylation. An altered methylation ratio, due to either a decrease in AdoMet or an increase in AdoHcy, would be expected to inhibit all transmethylation reactions (56). In pigs with N_2O-induced Cbl neuropathy, AdoHcy levels in the spinal cord were significantly increased while the AdoMet concentration was not elevated, thus resulting in an inversion of the AdoMet/AdoHcy ratio from a control value of 15.0 to 0.8. (70) The accumulation of AdoHcy was presumably the result of a sequence of impaired reactions in Cbl deficiency (Figure 3). AdoMet, in transferring its methyl group in transmethylation reactions, is converted to AdoHcy, which is subsequently degraded to adenine and homocysteine. As the activity of the MS reaction is impaired in Cbl deficiency, the homocysteine formed from AdoHcy cannot be remethylated to Met. Accumulation of homocysteine leads to accumulation of AdoHcy and the altered AdoMet/AdoHcy ratio. Liver behaved differently from neural tissue, in that the values for AdoMet, AdoHcy, and the AdoMet/AdoHcy ratio were not significantly different in animals with N_2O-induced neuropathy compared with control animals.

In pigs whose diet was supplemented with Met, the severity of the Cbl neuropathy was greatly reduced. In these animals, the AdoHcy levels in spinal cord were elevated, but the AdoMet levels were significantly higher, so that there was not the same degree of fall in the AdoMet/AdoHcy ratio as in the animals exposed to N_2O but without dietary supplementation with Met. The added Met was presumed to exert its protective action by increasing the tissue levels of AdoMet in neural tissue, thereby compensating for the elevated AdoHcy levels and thus maintaining a near normal methylation ratio. In the same study, exposure of rats to N_2O produced only a moderate fall in AdoMet/AdoHcy ratio, despite a comparable degree of inhibition of MS in neural tissue. This relatively modest fall in the methylation ratio might explain the absence of neuropathy in the Cbl-deficient rat.

In a further experiment, weanling pigs exposed to N_2O for seven days showed a marked increase in AdoHcy concentrations in all tissues except liver (48). The elevated AdoHcy levels resulted in a fall in the AdoMet/AdoHcy ratio, particularly in neural tissues.

The hypothesis that the Cbl neuropathy is the result of AdoHcy toxicity could not be confirmed in the fruit bat model (68). In bats with severe Cbl neuropathy induced by a combination of dietary deprivation and exposure to N_2O, no significant differences in AdoMet and AdoHcy levels in spinal cord and brain tissue were found, compared with levels in control animals, nor were there any differences in the AdoMet/AdoHcy ratios. In the liver there was a small but significant rise in AdoHcy levels but no change in AdoMet levels. Supplementation of the diet with Met resulted in a rise in AdoMet concentration in the liver, a much smaller increase in the spinal cord and brain cortex, and an insignificant fall in the AdoMet/AdoHcy ratio.

Direct Measurement of Methylation Reactions

Studies of methylation reactions (17, 46) and myelin protein (44, 45) in the tissues of Cbl-deficient fruit bats failed to provide evidence of defective methylation. When bats rendered Cbl deficient by dietary deprivation and exposure to N_2O were compared to control animals, no differences in [^{14}C]ethanolamine incorporation into liver and brain phospholipids could be detected (46). Using synaptosomes and myelin as substrates for the incorporation of methyl groups into membrane lipids, no abnormality in lipid methylation could be demonstrated in Cbl-deficient bats (64). The rate of synaptosomal and myelin protein methylation was similar in Cbl-deficient and normal bats. Thus the Cbl-deficient neuropathy in the fruit bat is not associated with changes in the rate of protein or lipid methylation. Furthermore, no differences in protein profile of the myelin membrane in Cbl-deficient bats compared to that of control bats could be demonstrated (7).

Cobalamin Mutants Affecting Methylcobalamin

Observations on patients with inherited disorders of Cbl metabolism affecting MetCbl have provided further evidence focusing on the MS reaction rather than on the mutase reaction in the pathogenesis of the Cbl neuropathy. The Cbl mutants CblC and CblD are associated with failure to form both AdoCbl and MetCbl. In contrast to those mutants (CblA and CblB) in which only AdoCbl is affected, patients with the CblC and CblD mutants may show prominent neurological disorders with spasticity and cerebral atrophy in older patients (14, 31).

Patients with the CblE and CblG mutations, which are associated with MetCbl deficiency only, have been the subject of a recent review (31). In these infants reduced activity of the MS enzyme, elevated levels of homocysteine in plasma and urine, and reduced plasma methionine have been observed. Examination of the brain by CT scans and magnetic resonance imaging revealed atrophy and hypoplasia of the brain, with delayed myelination. Although these findings provide significant additional evidence of the importance of MetCbl in the Cbl neuropathy, data derived from infants with an inherited disorder in Cbl metabolism are not necessarily valid for Cbl deficiency in general (31).

One patient with the CblG mutant presented at age 21 with clinical neuropathy closely resembling subacute combined degeneration, thus providing further evidence implicating the MetCbl-dependent MS in the pathogenesis of the Cbl neuropathy (11).

The Methionine Synthetase Reaction and the Cobalamin Neuropathy

A large body of data now identifies the impairment of the MeCbl-dependent MS reaction, rather than the AdoCbl-dependent mutase reaction, as the underlying biochemical lesion in the Cbl neuropathy, and indicates that Met plays a central role in the development of the lesion. It is uncertain how the impairment in MS activity, with the attendent changes in homocysteine and Met metabolism, results in neuropathy. The hypomethylation hypothesis, which is based on the the the accumulation of homocysteine and AdoHcy, with a fall in the AdoMet/AdoHcy ratio and resultant inhibition of methyltrans-ferases involved in myelin and other brain proteins, can be applied to the Cbl neuropathy in the pig only. Cbl neuropathy in the pig is associated with a fall in the AdoMet/AdoHcy ratio. A lesser, but still significant fall in the ratio occurs in the Cbl-deficient rat, but without the development of neuropathy. By contrast the fruit bat develops severe neuropathy without a significant change in the ratio in neural tissue. In view of the lack of consistent correla-tion between the development of Cbl neuropathy and the changes in the methylation ratio in neural tissue in various experimental animals, defective methylation cannot be accepted at present as the universal basis for the Cbl neuropathy.

These variant findings in the experimental animals could be ascribed to species differences. However, in all four species in which Cbl neuropathy occurs, namely humans, monkeys, fruit bats, and pigs, the neurological and neuropathological changes are quite similar. Furthermore, the effect of supplementation with Met in ameliorating the neuropathy is consistent for the monkey, the fruit bat, and the pig. In humans, for obvious ethical reasons, the possible effect of Met on the Cbl neuropathy will probably never be known. In view of these similarities, it is tempting to postulate a common pathogene-sis for the Cbl neuropathy that is applicable to all four species in which the neuropathy occurs.

The results of a number of studies have confirmed the central role of Met in the pathogenesis of the Cbl neuropathy in the fruit bat. In contrast with the pig, it has not been possible to show a defect in methylation reactions associated with the Cbl deficiency. This would suggest that the Met effect could be mediated via its role in formate metabolism (13, 16) or by some as yet unidentified pathway. Metabolic pathways showing the metabolism of Met and AdoMet are illustrated in Figure 4.

COBALAMIN ANALOGUE TOXICITY

In addition to the biologically active forms of Cbl, some Cbl analogues are apparently physiologically inactive. A number of such analogues have been identified in mammalian tissues and in human plasma (36, 37). We do not know to what extent, if any, Cbl analogues, when present, could substitute for active Cbl in humans and in this way exacerbate the effect of true Cbl deficiency and possibly play a role in the pathogenesis of the Cbl neuropathy.

Cbl analogues have been demonstrated in the serum of human patients with

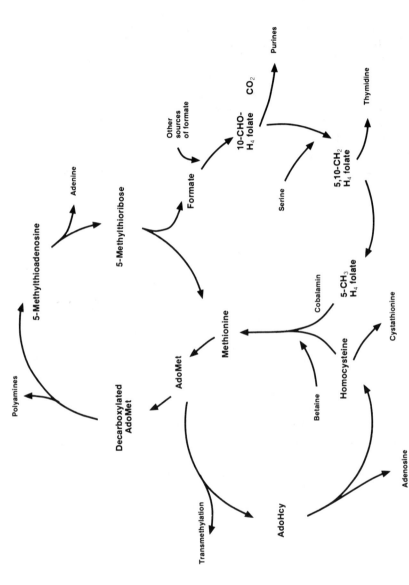

Figure 4 The central role of methionine in the metabolism of cobalamin, folate, and adenosylmethionine (AdoMet). (Reproduced with permission from E. Vieira-Makings, J. Metz, J. van der Westhuyzen, T. Bottiglieri, and I. Chanarin, 1990, *Biochem. J.* 266:707–11.)

Cbl deficiency (36). In one study, patients with Cbl deficiency and primarily neurological symptoms had significantly higher analogue levels than Cbl-deficient patients with primarily hematological abnormalities (10). On the basis of these findings, it was suggested that the retention or acquisition of analogues in Cbl deficiency may affect the nervous system preferentially.

The results of studies of Cbl analogues in animals with experimental Cbl deficiency have failed to provide evidence of a role for Cbl analogues in the development of the Cbl neuropathy. Exposure of rats to N_2O is associated with the conversion of Cbl to Cbl analogues in the liver, brain, and serum, but the animals do not develop neurological changes (38). No accumulation of Cbl analogues could be demonstrated in fruit bats with Cbl neuropathy induced by dietary deprivation (29) or by exposure to N_2O (62).

The evidence for a role of Cbl analogues in the pathogenesis of the Cbl neuropathy is thus limited. If the proposed toxicity of the analogues is due to substitution for active Cbl and aggravation of the degree of deficiency, it is difficult to see why this effect is exerted preferentially on the nervous system complications of Cbl deficiency. Furthermore, any hypothesis of the pathogenesis of the Cbl neuropathy should include an explanation of the mode of action of Met in ameliorating the neuropathy. It is not clear how Met might retard the formation of analogue or attenuate any effect analogues may have in compromising nervous system function.

SUMMARY

Neuropathy commonly complicates cobalamin (Cbl) deficiency in humans, monkeys, fruit bats, and pigs. The neuropathy is characterized by demyelination of the posterolateral columns of the spinal cord (subacute combined degeneration). The lesion was thought to arise primarily from impairment of the adenosylcobalamin-dependent methylmalonyl CoA mutase reaction, leading to the formation of abnormal odd-chain and branched-chain fatty acids and their incorporation into myelin with resultant demyelination. Data from recently developed animal models of the Cbl neuropathy induced by exposure to nitrous oxide do not substantiate this hypothesis, but rather identify impairment of the methylcobalamin-dependent methionine synthetase reaction as the more important basic defect. The key evidence for this hypothesis is the ability of methionine to delay the onset of Cbl neuropathy in experimental Cbl deficiency. In the Cbl-deficient pig, adenosylhomocysteine accumulates in neural tissue, presumably owing to the inability to recycle homocysteine via the defective methionine synthetase reaction. Accumulation of adenosylhomocysteine results in a fall in the adenosylmethionine:adenosylhomocysteine methylation ratio, and this change is believed to cause defective methylation and demyelination in the nervous system. However, in the Cbl neuropathy in the fruit bat, adenosylhomocysteine does not accumulate in the nervous system, the methylation ratio does not change, and no defect can be demonstrated in the methylation of myelin lipid or basic protein. Although a central role for methionine in the pathogenesis of the Cbl neuropathy has been established, defective methylation attendant upon impairment of the methionine synthetase reaction may not be the universal defect underlying the Cbl neuropathy. This would suggest that the methionine effect could be mediated via its role in formate metabolism or polyamine synthesis, or by some as yet unidentified pathway.

Literature Cited

1. Agamanolis, D. P., Chester, E. M., Victor, M., Kark, J. A., Hines, J. D., et al. 1976. Neuropathology of experimental B_{12} deficiency in monkeys. *Neurology* 26:905–14
2. Agamanolis, D. P., Victor, M., Harris, J. W., Hines, J. D., Chester, E. M., et al. 1978. An ultrastructural study of subacute combined degeneration of the spinal cord in vitamin B_{12} deficient rhesus monkeys. *J. Neuropathol. Exp. Neurol.* 37:273–99
3. Åkesson, B., Fehling, C., Jägerstad, M. 1979. Lipid composition and metabolism in liver and brain of vitamin B_{12} deficient rat sucklings. *Br. J. Nutr.* 41:263–74
4. Banks, R. G. S., Henderson, R. J., Pratt, J. M. 1968. Reactions of gases in solution. Part III. Some reactions of nitrous oxide with transition-metal complexes. *J. Chem. Soc. Sect. A* 2886–89
5. Barley, F. W., Sato, H. G., Abeles, R. H. 1972. An effect of vitamin B_{12} deficiency in tissue culture. *J. Biol. Chem.* 247:4270–76
6. Blanco, G., Peters, H. A. 1983. Myeloneuropathy and macrocytosis associated with nitrous oxide abuse. *Arch. Neurol.* 40:416–18
7. Cantrill, R. C., Oldfield, M., van der Westhuyzen, J., McLoughlin, J. 1983. Protein profile of the myelin membrane of the fruit bat *Rousettus aegyptiacus*. *Comp. Biochem. Physiol.* 76:881–84
8. Cardinale, G. J., Carty, T. J., Abeles, R. H. 1970. Effect of methylmalonyl coenzyme A, a metabolite which accumulates in vitamin B_{12} deficiency, on fatty acid synthesis. *J. Biol. Chem.* 247:4270–76
9. Cardinale, G. J., Dreyfus, P. M., Auld, P., Aeles, R. H. 1969. Experimental vitamin B_{12} deficiency. Its effect on tissue vitamin B_{12} coenzyme levels and on the metabolism of methylmalonyl CoA. *Arch. Biochem. Biophys.* 131:92–99
10. Carmel, R., Karnaze, D. S., Weiner, J. M. 1988. Neurologic abnormalities in cobalamin deficiency are associated with higher cobalamin "analogue" values than are hematological abnormalities. *J. Lab. Clin. Med.* 111:57–62
11. Carmel, R., Watkins, D., Goodman, S. L., Rosenblatt, D. S. 1988. Hereditary defect of cobalamin metabolism (CblG mutation) presenting as a neurological disorder in adulthood. *New Engl. J. Med.* 318:1738–41
12. Chanarin, I. 1979. *The Megaloblastic Anaemias.* Oxford: Blackwell Sci. 2nd ed.
13. Chanarin, I., Deacon, R., Lumb, M., Perry, J. 1985. Cobalamin-folate interrelations. A critical review. *Blood* 66:479–89
14. Cooper, B. A., Rosenblatt, D. S. 1987. Inherited defects of vitamin B_{12} metabolism. *Annu. Rev. Nutr.* 7:291–320
15. Deacon, R., Lumb, M., Perry, J. 1978. Selective inactivation of vitamin B_{12} in rats by nitrous oxide. *Lancet* 2:1023–24
16. Deacon, R., Perry, J., Lumb, M., Chanarin, I. 1990. Formate metabolism in the cobalamin-inactivated rat. *Br. J. Haematol.* 74:354–59
17. Deacon, R., Purkiss, P., Green, R., Lumb, M., Perry, J., et al. 1986. Vitamin B_{12} neuropathy is not due to failure to methylate myelin basic protein. *J. Neurol. Sci.* 72:113–17
18. Dinn, J. J., McCann, S., Wilson, P., Reed, B., Weir, D. G., et al. 1978. Animal model for subacute combined degeneration. *Lancet* 2:1154
19. Dinn, J. J., Weir, D. G., McCann, S., Reed, B., Wilson, P., et al. 1980. Methyl group deficiency in nerve tissue:

a hypothesis to explain the lesion of subacute combined degeneration. *Isr. J. Med. Sci.* 149:1–4

20. Duffield, M. S., Philips, J. I., Vieira-Makings, E., van der Westhuyzen, J., Metz, J. 1990. Demyelinisation in the spinal cord of vitamin B_{12}-deficient fruit bats. *Comp. Biochem. Physiol. C.* 96:291–97

21. Fehling, C., Jägerstad, M., Äkesson, B., Axelsson, J., Brun, A. 1978. Effects of vitamin B_{12} deficiency on lipid metabolism of the rat liver and nervous system. *Br. J. Nutr.* 39:501–3

22. Fehling, C., Jägerstad, M., Arvidson, G. 1978. Lipid metabolism in the vitamin B_{12} deprived rat. *Nutr. Metab.* 22:82–89

23. Forward, S. A., Gompertz, D. 1970. The effects of methylmalonyl CoA on the enzymes of fatty acid biosynthesis. *Enzymologia* 39:379–90

24. Frenkel, E. P. 1973. Abnormal fatty acid metabolism in peripheral nerves of patients with pernicious anemia. *J. Clin. Invest.* 52:1237–45

25. Frenkel, E. P., Kitchens, R. L., Johnson, J. M. 1973. The effect of vitamin B_{12} deprivation on the enzymes of fatty acid synthesis. *J. Biol. Chem.* 248: 7450–56

26. Gandy, G., Jacobson, W., Sidman, R. 1973. Inhibition of transmethylation reaction in the central nervous system—an experimental model for subacute combined degeneration of the cord. *J. Physiol.* 233:P1–P3

27. Gibson, J., van der Westhuyzen, J. 1984. Effect of L-dihydroxyphenalanine (L-dopa) and methionine on tissue S-adenosylmethionine concentrations in cobalamin-inactivated fruit bats. *Int. J. Vitam. Nutr. Res.* 54:329–32

28. Goodman, A. M., Harris, J. W. 1980. Studies in B_{12} deficient monkeys with combined system disease. *J. Lab. Clin. Med.* 96:722–33

29. Green, R., Jacobsen, D. W. 1980. No discrepancy between vitamin B_{12} radioassays using purified intrinsic factor or R binder in bats developing cobalamin deficient neuropathy. *Clin. Res.* 28:A74

30. Green, R., van Tonder, S. V., Oettle, G. J., Cole, G., Metz, J. 1975. Neurological changes in fruit bats deficient in vitamin B_{12}. *Nature* 254:148–50

31. Hall, C. A. 1990. Function of vitamin B_{12} in the central nervous system as revealed by congenital defects. *Am. J. Hematol.* 34:121–27

32. Hyland, K., Smith, I., Bottiglieri, T., Perry, J., Wendel, U., et al. 1988. De-

myelination and decreased S-adenosylmethionine in 5,10-methylenetetrahydrofolate reductase deficiency. *Neurology* 38:459–62

33. Jacobson, W., Gandy, G., Sidman, R. L. 1973. Experimental subacute combined degeneration of the cord in mice. *J. Pathol.* 109:R13–R14

34. Kark, J. A., Victor, M., Hines, J. D., Harris, J. W. 1974. Nutritional vitamin B_{12} deficiency in rhesus monkeys. *Am. J. Clin. Nutr.* 27:470–78

35. Kishimoto, Y., Williams, M., Moser, H. W., Hignite, C., Biemann, K. 1973. Branched chain and odd-numbered fatty acids and aldehydes in the nervous system of a patient with deranged vitamin B_{12} metabolism. *J. Lipid Res.* 14:69–77

36. Kolhouse, J. F., Kondo, H., Allen, N. C., Podell, E., Allen, R. H. 1977. Cobalamin analogues are present in human plasma and can mask cobalamin deficiency because current radioisotopic dilution assays are not specific for true cobalamin. *New Engl. J. Med.* 299:785–92

37. Kondo, H., Kolhouse, J. F., Allen, R. H. 1980. Presence of cobalamin analogues in animal tissue. *Proc. Natl. Acad. Sci. USA* 77:817–21

38. Kondo, H., Osborne, M. L., Kolhouse, J. F., Binder, M. J., Podell, E. R., et al. 1981. Nitrous oxide has multiple deleterious effects on cobalamin metabolism and causes decreases in activities of both mammalian cobalamin dependent enzymes in bats. *J. Clin. Invest.* 67:1270–83

39. Layzer, R. B. 1978. Neuropathy after prolonged exposure to nitrous oxide. *Lancet* 2:1227–30

40. Levy, H. L., Mudd, S. H., Schulman, J. D., Dreyfus, P. M., Abeles, R. H. 1970. A derangement in B_{12} metabolism associated with homocystinemia, cystathionemia, hypomethioninemia and methylmalonic aciduria. *Am. J. Med.* 48:390–97

41. Lichtheim, H. 1887. Zur kenntnis der perniciösen anämie. *Muench. Med. Wochenschr.* 34:300

42. Lumb, M., Sharer, N., Deacon, R., Jennings, P., Purkis, P., et al. 1983. Effect of nitrous oxide-induced inactivation of cobalamin on methionine and S-adenosylmethionine metabolism in the rat. *Biochim. Biophys. Acta* 756:354–59

43. Mahoney, M. J., Bick, D. 1987. Recent advances in the inherited methylmalonic acidemias. *Acta Paediatr. Scand.* 76: 689–96

44. McLoughlin, J. C., Cantrill, R. C.

1983. Nitrous oxide alters the pattern of myelin proteins in the nervous system of the fruit bat *Rousettus aegyptiacus*. *Neurosci. Lett.* 44:99–104

45. McLoughlin, J. C., Cantrill, R. C. 1984. Vitamin B_{12} deficiency alters the distribution of membrane proteins on linear sucrose gradients in the fruit bat brain. *Neurosci. Lett.* 49:175–80

46. McLoughlin, J. C., Cantrill, R. C. 1986. Nitrous oxide induced vitamin B_{12} deficiency: Measurement of methylation reactions in the fruit bat *(Rousettus aegyptiacus)*. *Int. J. Biochem.* 18:199–202

47. Metz, J., van der Westhuyzen, J. 1987. The fruit bat as an experimental model of the neuropathy of cobalamin deficiency. *Comp. Biochem. Physiol. A.* 88:171–77

48. Molloy, A. M., Weir, D. G., Kennedy, G., Kennedy, S., Scott, J. M. 1990. A new high performance liquid chromatographic method for the simultaneous measurement of S-adenosylmethionine and S-Adenosylhomocysteine. *Biomed. Chromatogr.* 4:257–60

49. Osler, W., Gardner, W. 1877. On the changes in marrow in progressive pernicious anaemia. *Can. Med. Surg. J.* 5:385

50. Peifer, J. J., Lewis, R. D. 1979. Effects of vitamin B_{12} deprivation on phospholipid fatty acid patterns in liver and brain of rats fed high and low levels of linoleate in low methionine diets. *J. Nutr.* 109:2160–72

51. Peifer, J. J., Lewis, R. D. 1981. Odd-numbered fatty acids in phosphatidyl choline versus phosphatidyl ethanolamine of vitamin B_{12} deprived rats. *Proc. Soc. Exp. Biol. Med.* 167:212–17

52. Ramsay, R. B., Scott, T., Banik, N. L. 1977. Fatty acid composition of myelin isolated from the brain of a patient with cellular deficiency of co-enzyme forms of vitamin B_{12}. *J. Neurol. Sci.* 34:221–32

53. Russell, J. S. R., Batten, F. E., Collier, J. 1900. Subacute combined degeneration of the spinal cord. *Brain* 23:39–110

54. Sahenk, Z., Mendel, J. R., Couvi, D., Nachtman, J. 1978. Polyneuropathy from inhalation of N_2O cartridges through a whipped cream dispenser. *Neurology* 28:485–87

55. Scalabrino, G., Monzio-Compagnoni, B., Ferioli, M. E., Lorenzini, E. C., Chiodini, E., et al. 1990. Subacute combined degeneration and induction of ornithine decarboxylase in spinal cords of totally gastrectomised rats. *Lab. Invest.* 62:297–304

56. Schatz, R. A., Wilens, T. E., Sellinger, O. Z. 1981. Decreased transmethylation of biogenic amines after in vivo elevation of brain S-adenosyl-L-homocysteine. *J. Neurochem.* 36:1739–48

57. Scott, J. M., Dinn, J. J., Wilson, P., Weir, D. G. 1981. The pathogenesis of subacute combined degeneration: A result of methyl group deficiency. *Lancet* 2:334–37

58. Sturman, J. A., Gaull, G. E., Nieman, W. H. 1976. Activities of some enzymes involved in homocysteine methylation in brain, liver and kidney of the developing rhesus monkey. *J. Neurochem.* 27:425–31

59. van der Westhuyzen, J., Cantrill, R. C., Fernandes-Costa, F., Metz, J. 1981. Lipid composition of the brain in the vitamin B_{12} deficient fruit bat *(Rousettus aegyptiacus)* with neurological impairment. *J. Neurochem.* 37:543–49

60. van der Westhuyzen, J., Cantrill, R. C., Fernandes-Costa, F., Metz, J. 1983. Effect of a vitamin B_{12} deficient diet on lipid and fatty acid composition of spinal cord myelin in the fruit bat. *J. Nutr.* 113:531–37

61. van der Westhuyzen, J., Fernandes-Costa, F., Metz, J. 1982. Cobalamin inactivation by nitrous oxide produces severe neurological impairment in fruit bats: Protection by methionine and aggravation by folates. *Life Sci.* 31:2001–10

62. van der Westhuyzen, J., Fernandes-Costa, F., Metz, J., Drivas, G., Herbert, V. 1982. Cobalamin (vitamin B_{12}) analogues are absent in plasma of fruit bats exposed to nitrous oxide. *Proc. Soc. Exp. Biol. Med.* 171:88–91

63. van der Westhuyzen, J., Gibson, J. 1984. Lipid composition of the brain in the cobalamin-inactivated fruit bat, *Rousettus aegyptiacus*. *Int. J. Vitam. Nutr. Res.* 54:208–10

64. van der Westhuyzen, J., Lashansky, G., Cantrill, R. C. 1982. Fatty acid composition of synaptosomes from normal and cobalamin-deficient bat brain. *Comp. Biochem. Physiol.* B73:2:297–99

65. van der Westhuyzen, J., Metz, J. 1983. Tissue S-adenosylmethionine levels in fruit bats with N_2O-induced neuropathy. *Br. J. Nutr.* 50:325–30

66. van der Westhuyzen, J., Metz, J. 1984. Betaine delays the onset of neurological impairment in nitrous oxide induced vitamin B_{12} deficiency in fruit bats. *J. Nutr.* 114:1106–11

67. Vieira-Makings, E., Chetty, N., Reavis, S. C., Metz, J. 1991. Methylmalonic

acid metabolism and nervous-system fatty acids in cobalamin-deficient fruit bats receiving supplements of methionine, valine and isoleucine. *Biochem. J.* 275:585–90

68. Vieira-Makings, E., Metz, J., van der Westhuyzen, J., Bottiglieri, T., Chanarin, I. 1990. Cobalamin neuropathy. Is S-adenosylhomocysteine toxicity a factor? *Biochem. J.* 266:707–11

69. Vieira-Makings, E., van der Westhuyzen, J., Metz, J. 1990. Both valine and isoleucine supplementation delay the development of neurological impairment in vitamin B_{12} deficient bats. *Int. J. Vitam. Nutr. Res.* 60:41–46

70. Weir, D. G., Keating, S., Molloy, A., McPartlin, J., Kennedy, S., et al. 1988. Methylation deficiency causes vitamin B_{12}-associated neuropathy in the pig. *J. Neurochem.* 51:1949–52

71. White, A. M., Cox, E. V. 1964. Methylmalonic acid excretion and vitamin B_{12} deficiency in the human. *Ann. NY Acad. Sci.* 112:915–21

Annu. Rev. Nutr. 1992. 12:81–101

REGULATION OF ENZYMES OF UREA AND ARGININE SYNTHESIS

Sidney M. Morris, Jr.

Department of Molecular Genetics and Biochemistry, University of Pittsburgh School of Medicine, Biomedical Science Tower, Pittsburgh, Pennsylvania 15261–2072

KEY WORDS: gene transcription, liver, kidney, intestine, hormone action

CONTENTS

INTRODUCTION

Sixty years ago Krebs & Henseleit first described the ornithine-urea cycle (67), a paradigm for the elucidation of other metabolic cycles. The urea cycle is an essential metabolic pathway for disposal of the toxic metabolite ammonia in most terrestrial vertebrates, whereas in marine elasmobranchs (sharks, skates, and rays) the urea synthesized by this pathway is used for osmoregulation (3). The urea cycle is catalyzed by five enzymes—carbamyl phosphate synthetase I [CPS-I; carbamoyl-phosphate synthetase (ammonia), EC 6.3.4.16], ornithine transcarbamylase (OTC; carbamoylphosphate: L-ornithine carbamoyltransferase, EC 2.1.3.3), argininosuccinate synthetase

0199-9885/92/0715-0081$02.00

[AS; L-citrulline: L-aspartate ligase (AMP-forming), EC 6.3.4.5], argininosuccinate lyase (AL; L-argininosuccinate arginine-lyase, EC 4.3.2.1), and arginase (L-arginine ureohydrolase, EC 3.5.3.1)—which effect the net conversion of two molecules of ammonia and one of bicarbonate into urea at the expense of four high energy phosphate bonds. Each enzyme is composed of a single type of polypeptide chain encoded by a single-copy nuclear gene (15, 59). Kinetic and structural features of these enzymes and clinical aspects of inherited defects in members of the urea cycle have been reviewed recently (5, 15, 59, 103).

The first two enzymes of the urea cycle are located within the mitochondrial matrix, and the remaining three enzymes are cytosolic. These enzymes appear to be associated with one another so that substrates and products move from one enzyme to another by channeling rather than by simple diffusion (145). Within the liver, the urea cycle enzymes are localized predominantly within periportal, rather than perivenous, hepatocytes (79). This spatial distribution, which can be modulated by diet and hormones (82), has functional significance for ammonia and glutamine metabolism (45, 79).

Although the urea cycle is generally thought of as a detoxification pathway for disposal of ammonium, the fact that one mole of bicarbonate also is disposed of for each two moles of ammonium has led to the notion that the urea cycle also is a mechanism for maintaining pH homeostasis via regulation of bicarbonate levels (4), a proposal that has not been accepted universally (42, 143). More recently, the urea cycle has been viewed also as a component of an intercellular glutamine cycle (45). Thus, even sixty years after its initial description the physiologic role of the urea cycle continues to be the subject of new ideas and controversies.

Although the complete urea cycle is expressed only in liver, some enzymes of this pathway are expressed at significant levels also in small intestine and kidney, thereby constituting an independent arginine biosynthetic pathway (Figure 1). Vertebrates require arginine for protein synthesis and for the synthesis of compounds such as creatine, polyamines, and the novel signalling molecule nitric oxide (47, 81). The relative contribution of endogenous arginine synthesis in meeting this requirement varies according to age, physiologic state, and species of animal (141). Young animals require dietary arginine for optimal growth whereas endogenous arginine synthesis largely meets the arginine requirement of most adult omnivores (16, 51, 120, 141). Both immature and adult carnivores require dietary arginine. An extreme example is the cat for whom an arginine-free diet may be fatal within hours (84). Although AS and AL are present also in brain (104), their physiologic role in this organ is unclear and thus is not considered here.

The hepatic capacity for urea synthesis is determined by regulating the abundance and catalytic efficiency of the enzyme catalyzing the rate-limiting

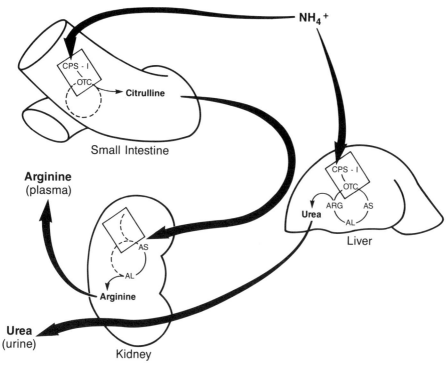

Figure 1 Interorgan relationships in urea and arginine synthesis. *Small rectangles* indicate the mitochondrial compartment. *Dotted lines* indicate missing parts of the urea cycle. Abbreviations: CPS-I, carbamyl phosphate synthetase-I; OTC, ornithine transcarbamylase; AS, argininosuccinate synthetase; AL, argininosuccinate lyase; ARG, arginase.

step. Based on enzyme activities measured under optimal conditions in the test tube, the potential ureagenic capacity of the liver is limited by AS abundance (103). However, as the urea cycle under normal physiologic conditions does not function at its full potential, rates of ureagenesis usually are controlled by substrate availability or catalytic efficiencies of the enzymes. Activities of the urea cycle enzymes are not known to be modulated by posttranslational modifications; regulation of catalytic efficiency is effected instead by varying concentrations of activators or inhibitors of the enzymes (15, 103). Thus, regulation of CPS-I by varying the concentration of its essential cofactor *N*-acetylglutamate is generally thought to be the principal mechanism for dynamically modulating ureagenic capacity in vivo (15, 30, 103). Recently, however, it has been proposed that arginase, which is at a branch point for utilizing arginine for ureagenesis or for other purposes such as protein or polyamine synthesis, may become rate-limiting under conditions that alter its affinity for the cofactor Mn^{2+} (68). As regulation of rates of ureagenesis by varying substrate availability or catalytic efficiency of enzymes has been discussed elsewhere (79, 103), this review emphasizes the regulation of enzyme and mRNA abundance. Regulation of the urea cycle enzymes in fetal and perinatal periods also has been reviewed recently (79) and thus is not considered here.

UREA CYCLE ENZYMES IN LIVER

Responses to Diet

Since the beginning of this century it has been known that urea production in adult humans varies as a function of dietary protein intake (31). Changes in liver arginase activity according to dietary protein intake were reported in 1939 (71), representing one of the earliest examples of metabolic adaptation in mammals at the level of enzyme activity. Thirty years ago it was demonstrated that activities of all five urea cycle enzymes in rat liver varied as a function of dietary protein intake (115–117). Note, however, that metabolic regulation of urea cycle enzyme activities is species-specific, perhaps depending on the type of diet to which an animal is normally adapted. For example, in a strict carnivore such as the cat, adaptation of urea cycle enzyme activities in response to varying levels of dietary protein intake does not occur (107). Whether this lack of adaptation in cats involves reduced responses of the urea cycle enzymes to agents such as cAMP, insulin, and glucocorticoid is not known.

Activities of the urea cycle enzymes are highest in response to starvation and high-protein diets and are reduced to response to low-protein or protein-free diets (12, 14, 19, 22, 30, 76, 99, 114, 116). Variations in amino acid composition between different animal or vegetable dietary protein sources have little or no effect on the enzyme activities (22, 46, 117). However, feeding rats very high doses of some, but not all, individual amino acids can lead to increases in urea cycle enzyme activity (122). Because the latter study was conducted in vivo, the mechanism of these effects—such as possible selective effects on endocrine status—is unknown, and the authors noted that they had been unable to duplicate these results using cultured rat hepatocytes. Feeding rats ammonium acetate or ammonium citrate has not produced consistent effects on the urea cycle enzymes; when noted, effects were relatively small (54, 121, 122).

Diet-dependent changes in urea cycle enzyme activity are primarily the consequence of changes in enzyme mass (83, 114, 115, 122), which in turn largely reflect altered enzyme synthesis rates (119, 137). Abundances of urea cycle enzyme mRNAs as a function of dietary protein intake generally correlate with the activities of these enzymes (83, 89, 112), thus indicating that dietary regulation of enzyme levels occurs primarily at a pretranslational step. Furthermore, this regulation is largely coordinate in degree for all five mRNAs (89). As these mRNAs are regulated also by agents such as cAMP, insulin, and glucocorticoids, it is likely that dietary responses are mediated primarily by hormones. These general assertions are probably accurate even though rigorous studies to document their validity for all five urea cycle enzymes have not been conducted, and some minor exceptions to these statements have been noted. For example, arginase activity may be regulated in vivo by varying the concentration of manganese (6, 108).

Disruption of nitrogen metabolism by conditions such as sepsis, trauma, uremia, or cancer also may result in altered urea synthesis and levels of urea cycle enzymes (12, 14, 17, 80, 101). However, as ureagenesis in these pathophysiologic states has not yet been studied extensively, information on regulatory signals and mechanisms is limited.

Hormonal Regulation

Treatment with glucagon or cAMP analogs results in increased activities and mRNA abundance for the urea cycle enzymes in rat liver, cultured hepatocytes, and hepatoma cell lines (13, 23, 35, 66, 72, 78, 94, 97, 109, 113, 124). To directly compare changes in enzyme and mRNA abundances is difficult owing to the large time differences in reaching new steady state values. The enzymes have half-lives of 3–9 days (40, 98, 119, 137, 142), whereas the mRNAs apparently have half-lives of several hours (97). Nevertheless, increases in enzyme activity apparently can be accounted for by increases in mRNA abundance (89, 97). In contrast to the other urea cycle enzymes, OTC is not responsive to cAMP analogs in cultured hepatocytes, in hepatoma cell lines, or in short-term treatment of rats (36, 89, 97). However, OTC is induced in rats treated with glucagon for several days (13, 124), suggesting that in vivo induction may represent an indirect response to glucagon or that response to cAMP is elicited only in the presence of some other agent, which has not been identified. Transcription rates of CPS-I, AS, and AL are rapidly stimulated by cAMP analogs (89, 97), which is indicative of a direct response to cAMP.

Activities of the urea cycle enzymes are reduced following adrenalectomy (19, 31, 77) and raised by glucocorticoid treatment of rats or of hepatoma cell lines (12, 19, 36, 40, 41, 77, 91, 94). As with cAMP, glucocorticoids also increase mRNA abundance in rats, cultured hepatocytes, and hepatoma lines, although precise correlations between enzyme activity and mRNA levels have not been determined in most instances (23, 26, 65, 97, 113). OTC is again unusual in that its activity or mRNA exhibits little or no response to glucocorticoid (36, 89, 97, 113, 117). Transcription rates of CPS-I and AL are stimulated by glucocorticoid in cultured hepatocytes (94).

Responses to glucagon or cAMP analogs plus glucocorticoid are greater than to either agent alone (64, 66, 97). In cultured hepatocytes, the combination of cAMP and glucocorticoid was additive for AL mRNA and synergistic for CPS-I, AS, and arginase mRNAs, whereas OTC mRNA was unaffected (97). These results conflicted with those of other studies that found that glucocorticoid alone had little or no effect on enzyme activities in cultured hepatocytes but was permissive for the response to glucagon (35). However, these latter studies may not have observed increased enzyme activities owing to the lag in response to glucocorticoid (97) and the long half-lives of the enzymes. As enzyme activities and mRNA levels were not measured simultaneously in these studies, possibly the conflicting results represent differences in culture conditions among these laboratories or differences in glucocorticoid response at pre- and posttranslational steps. A combination of cAMP analog and glucocorticoid produced large increases in transcription rates of all urea cycle enzymes genes except OTC (94).

Urea excretion increases in diabetes, in part due to increased protein catabolism and increased food consumption. Activities of CPS-I, AS, AL, and arginase are elevated in diabetic rats, whereas OTC activity remains unchanged (6, 63, 125). The increased arginase activity reflects an increased concentration of its cofactor Mn^{2+} rather than any change in enzyme protein (6, 127). In cultured hepatocytes or hepatoma cell lines, insulin alone has little effect on urea cycle enzyme mRNAs but reduces the induction of CPS-I mRNA or synthesis rate by glucocorticoid, glucagon, or cAMP (63–66, 94) and also reduces the induction of arginase mRNA by cAMP (94); effects of cAMP and glucocorticoid on the urea cycle enzyme mRNAs are otherwise unchanged. Although these effects of insulin likely involve transcriptional control, experiments addressing this possibility have not been reported.

Inasmuch as thyroid status affects both synthesis and degradation of protein, thus influencing metabolism of nitrogenous metabolites, one might reasonably anticipate that thyroid hormone would prove to be a major regulator of urea cycle enzyme activity. However, relatively few studies of thyroid hormone action on this system have been performed, and the results obtained have not always been consistent (19, 31, 32, 38, 39, 74, 125). Because different diets were used in these studies, the extent to which nutritional factors influenced the variability in results is unknown. Overall, in vivo effects of thyroid hormone on urea cycle enzyme activities are modest. Although direct effects of thyroid hormone would best be elucidated by use of cultured hepatocytes or hepatoma cell lines, hepatocytes rapidly lose responsiveness to thyroid hormone in culture and thyroxine-responsive rat hepatoma cell lines have not been described. The well-known effects of thyroid hormone on activities and mRNAs of the urea cycle enzymes in liver of the metamorphosing tadpole have been described previously (20, 85).

Effects of growth hormone on urea cycle enzymes are unclear. Treatment of normal rats with growth hormone resulted in decreased activities of AS, AL, and arginase but little or no change in CPS-I and OTC activities, which suggests that only the three cytosolic enzymes are responsive to growth hormone (77). A later study concluded that hypophysectomy decreased arginine synthetase and arginase activities whereas OTC activity remained unchanged (31); however, when enzyme activities are calculated as units per gram liver, the opposite conclusion is reached: OTC activity increases whereas activities of the other two enzymes decline slightly, if at all. Hypophysectomy of rats also increased activities of CPS-I and OTC in a more recent study, an effect that was reversed by administration of growth hormone; AS activity remained unchanged throughout (102). Differences in purity of the growth hormone preparations used in these studies could have contributed to the differing results. In any event, the activity changes reported are not great. Although hepatocytes have growth hormone receptors, we do not know whether any changes in urea cycle enzymes are the consequence of a direct action of growth hormone on these cells.

ARGININE BIOSYNTHETIC ENZYMES

Small Intestine

The small intestine is the principal source of circulating citrulline in adult mammals (146). Intestinal citrulline synthesis is catalyzed by CPS-I and OTC, which are localized in mucosal epithelial cells (43, 111). The regulation of intestinal CPS-I and OTC expression differs from that in liver. CPS-I activity in jejunum of adult mice was unaffected by altered dietary protein content (53), whereas OTC activity of rat small intestine decreased slightly when dietary protein content increased (147). As neither group assayed both enzymes, it is unclear whether these results reflect intrinsic differences in regulation of CPS-I and OTC in small intestine, species-specificity in dietary response, or differences in methodology. Moreover, assays of enzymes in intestinal homogenates are problematic owing to the presence of active proteases, and neither group apparently included protease inhibitors in the homogenization buffers.

Levels of CPS-I and OTC mRNAs in small intestine of adult rats were unaffected by treatment with glucagon or dexamethasone (113). However, levels of these mRNAs were only modestly affected in liver of the same animals, so it is unclear whether the apparent lack of an intestinal response reflects an intrinsic insensitivity to the hormones or reflects masking of a hormone response owing to high basal expression. With regard to the latter possibility, CPS-I mRNA induction by cAMP or glucocorticoids is much higher in cultured rat hepatocytes than in rat liver owing to the much lower basal expression in the cultured hepatocytes; the final induced levels are very similar (97). Moreover, hormonal responsiveness may be a function of age, as a number of intestinal enzymes in rat lose glucocorticoid responsiveness during the third postnatal week (49).

Developmental expression of CPS-I and OTC has been examined by several groups. CPS-I activity in mouse jejunum was highest at birth, gradually declining to about a third of the highest value in adults (53). A similar profile was found for CPS-I mRNA in rat intestine (113). Reported developmental profiles for OTC activity are variable. OTC activity during development of mouse intestine was similar to that reported for CPS-I (27), whereas another group found little difference in intestinal OTC activity in mice from birth to 21 days, with slightly higher levels in the adult (73). OTC mRNA abundance in developing rat intestine paralleled that of CPS-I mRNA, declining between birth and maturity (113), and matched the OTC activity profile in developing mouse intestine (27). In contrast, OTC activity in small intestine of rat increased 5- to 6-fold from birth to adulthood (50). Developmental profiles of CPS-I and OTC mRNAs in small intestine differ significantly from those in liver (86, 113). As no group measured both mRNA abundance and enzyme activity, it is unclear whether this discrepancy reflects methodologic differences between laboratories, differences in expression between mouse and rat, or regulation at the translational or posttranslational level. As with the dietary studies, no group measured activities of both enzymes nor included protease inhibitors in the homogenization buffers, although Western blotting analysis showed intestinal OTC to be intact in one of these reports (27).

The roles of small intestine and kidney with regard to arginine biosynthesis change during development. Whereas levels of AS and AL are relatively high in liver and kidney but very low in small intestine of adults, leading to the relationship diagrammed in Figure 1, activities of these two enzymes are relatively high in small intestine but low in kidney for the first week or two after birth (53). This finding is consistent with the virtual absence of arginase activity in small intestine for two weeks after birth, followed by a rapid increase in activity to adult levels (50, 53, 73). Thus, the small intestine appears to be the principal arginine biosynthetic organ at and shortly after birth, whereas this function is divided between small intestine and kidney in adults.

Kidney

That kidney has a significant capacity for converting citrulline to arginine has been known for 50 years (7, 21). This conversion is carried out by AS and AL

(105), the third and fourth enzymes, respectively, of the urea cycle (Figure 1). These enzymes are localized in the proximal tubules of the kidney (70, 90). Physiologic studies have demonstrated that renal uptake of citrulline from the blood is closely matched by renal release of arginine into the circulation with little loss to the urine (48, 135, 146) and that the kidney is a major site of arginine biosynthesis in the rat (29). Renal production of arginine in the rat appears to be limited primarily by the availability of citrulline rather than by the capacity for conversion of citrulline to arginine (25). Although arginase activity is observed in kidney, it is much lower than in liver, it includes the activity of an isozyme distinct from liver arginase (126), and renal arginase activity is segregated from the major site of renal arginine biosynthesis (70).

Renal AS and AL activities and mRNA abundances increase as dietary protein intake increases, although the magnitude of this dietary response for these enzymes is less than in liver (87, 106). The effects of cAMP and glucocorticoid on abundance of these mRNAs differ for liver and kidney (87). Renal AS mRNA is induced by dibutyryl cAMP but not by dexamethasone, whereas this mRNA is induced by both agents in liver. The absence of a renal mRNA induction by dexamethasone is in apparent contrast to increased arginine synthetase (AS + AL) activity in kidney of rats receiving repeated injections of hydrocortisone (130). However, the duration of the treatment in the latter study raises the possibility that the increase was an indirect response to the hormone. Renal AL mRNA was unaffected by dibutyryl cAMP, dexamethasone, or a combination of these agents, whereas it was induced by dibutyryl cAMP in liver (87). Thus, the hormonal responses of these mRNAs in kidney differ qualitatively and quantitatively from their hormonal responses in liver of the same animals. These organ-specific hormonal responses may account, at least in part, for the organ-specific responses of mRNA abundance to dietary protein changes.

Messenger RNAs for renal AS and AL are measurable by 15 days of gestation in the mouse, corresponding to the time of appearance of immunoreactivity for these enzymes (90). These mRNAs increase coordinately during fetal and neonatal kidney development, whereas they exhibit clearly different developmental profiles in liver (86, 90). Developmental increases in AL activity and mRNA abundance are very similar in magnitude (53). Abundances of these mRNAs in kidney are unaffected by the hormonal changes that occur at parturition.

Cultured Cells

AS and AL are among the relatively few mammalian enzymes whose levels are regulated directly by metabolite concentration. Schimke reported nearly 30 years ago that activities of AS and AL in HeLa, KB, and L cells were repressed coordinately by arginine in the culture medium, whereas activities of these enzymes increased when arginine was replaced by citrulline in the medium (118). Similarly, AS activity increased in cultured human lymphoblasts and in human RPMI 2650 cells (an epithelial tumor cell line) when arginine was replaced by citrulline in the culture medium, but AL activity was unchanged (57, 128). However, arginine deficiency causes no changes in activities of any of the urea cycle enzymes in cultured rat hepatocytes or in a rat hepatoma cell line (123). Thus, the mechanism for responding to variations in arginine concentration appears to be cell specific.

Changes in AS activity in cultured RPMI 2650 cells reflect changes in AS mRNA abundance (58). Measurements of AS mRNA precursor abundance, nuclear transcription rates, and expression of transfected AS minigenes indicate that the metabolite regulation is largely at the level of transcription (58). Because leucine starvation has effects similar to arginine starvation, the suggestion has been made that mammalian cells may have control mechanisms similar to those involved in general control of amino acid biosynthesis in *Saccharomyces cerevisiae* (58).

When cultured in the presence of canavanine, a toxic arginine analog, canavanine-resistant variants of human lymphoblast lines and RPMI 2650 cells, which express approximately 200-fold higher levels of AS, can be obtained (2, 128). AS activity in the canavanine-resistant cells is similar to that in normal liver. No change in AS gene copy number occurred in the canavanine-resistant cells, thus demonstrating that elevated AS expression was not caused by AS gene amplification (2, 129). Interestingly, AS expression in the canavanine-resistant cells was no longer subject to regulation by arginine (60, 128). Although the mechanism responsible for elevated AS expression in these cells remains unknown, the available data are consistent with a model that involves a positively acting factor (mechanism) that acts in *trans* (10, 11).

TRANSCRIPTIONAL CONTROL MECHANISMS

Although the extent to which changes in enzyme abundance reflect regulatory events at transcriptional, posttranscriptional, translational, or posttranslational steps has not been determined fully for the urea cycle enzymes in most instances, the evidence to date nonetheless indicates that altered enzyme abundance is primarily a consequence of altered mRNA abundance, which, in turn, is likely due to corresponding changes in transcription rates. This does not rule out the possibility of regulation at posttranscriptional steps under certain conditions, and translational control of CPS-I expression has been reported (138, 139). Although transcriptional control of urea cycle enzyme gene expression has been demonstrated in several laboratories, relatively little information is available regarding the types or organization of DNA regulatory elements present in the urea cycle enzyme genes. However, the presence of certain hormonal response or tissue-specific expression elements can be inferred from the expression patterns of these genes. Although a variety of potential regulatory elements in the 5' flanking regions of some of the urea cycle enzyme genes have been identified by computer analysis, this review emphasizes DNA elements that have activity in cell transfection or in vitro transcription assays or that exhibit specific binding to nuclear proteins.

The promoter region of the rat CPS-I gene contains TATA and CAAT consensus elements at positions -21 and -82, respectively (69). CPS-I promoter activity in liver nuclear extracts is dependent on a C/EBP element at position -109 and an unidentified element(s) between -161 and -1200 (52). No consensus cAMP response elements (CREs) or glucocorticoid response elements (GREs) were readily apparent within 1 kb of the transcription start site (69). Functional analyses of CPS-I promoter elements by transfection assays have not been reported.

The promoters of the rat, mouse, and human OTC genes are atypical in that they appear to lack consensus TATA and CAAT elements at the usual positions (44, 133, 140). Approximately 1.3 kb of the 5' flanking region of the OTC gene is sufficient to direct expression of a transgene specifically within liver and intestine of mice (93). Liver-specific expression of the rat OTC gene, as judged by cell transfection assays, is conferred by 222 bp of the 5' flanking region, which apparently contains at least one negative and two positive regulatory elements (92), one of which binds a protein possibly related to the COUP transcription factor (136). In addition, a liver-specific enhancer element located at -11 kb contains binding sites for a factor related to C/EBP (92).

Isolation and characterization of the human AS promoter was complicated by the fact that the human genome contains one expressed AS gene and 14 pseudogenes (34). The human AS promoter contains a TATA element approximately at position -30, but no apparent CAAT element has been observed (33, 61). Studies of AS promoter function have been concerned with the mechanisms of repression by arginine and with over-expression in canavanine-resistant cells. AS minigene constructs containing both 5' flanking sequences and intragenic sequences are subject to arginine repression (9, 60). As this repression also occurs in a construct containing as little as 149 bp of 5' flanking sequence (9), intragenic DNA sequences may possibly be involved in the regulation by arginine. Intragenic sequences also appear to be involved in regulated AS expression in canavanine-resistant cells (11). AS DNA constructs that exhibit overexpression in canavanine-resistant cells have not been identified. Overexpression of the endogenous AS gene in the resistant cells may involve epigenetic changes in AS DNA or chromatin structure that are not easily duplicated with transfected DNA constructs.

The promoter region of the human AL gene contains no TATA element, but four potential Sp1 binding sites are located within 200 bp of the transcription start site (1). No functional analysis of the AL promoter region has been reported.

Promoters of both rat and human liver arginase contain TATA and CAAT elements approximately at positions −25 and −60 to −70, respectively (100, 131). Potential GREs are found in the rat and human arginase promoter regions (100, 131), and a potential CRE has been noted for the human arginase 5' flanking region (131). A rat liver arginase construct containing nucleotides −90 to +286 is as efficient in in vitro transcription assays using rat liver nuclear extracts as is a construct containing nucleotides from −2.7 kb to +286 (132). A C/EBP-related factor footprints a region from −95 to −82 in the arginase promoter, thus suggesting a possible functional role for this element. However, because it was not reported whether the −90 to +286 construct retained this footprint, it is not clear whether this apparent C/EBP element is indeed functional. Deletion analysis indicates at least one positive DNA regulatory element is located in the region −90 to −51. This segment contains a region that can be footprinted by CTF/NF-1- and Sp1-related factors in a mutually exclusive fashion (132).

As the urea cycle enzyme genes generally exhibit coordinate expression in vivo, it is reasonable to believe that the 5' flanking regions of these genes share some DNA regulatory elements in common. Thus, at least three of these genes—CPS-I, OTC, and arginase—have DNA elements that are recognized by C/EBP or C/EBP-related factors (52, 92, 132). From their patterns of hormonal response, one can infer that, except for OTC, these genes likely contain functional CREs and GREs. However, the delayed response to dexamethasone and the requirement for ongoing protein synthesis in cultured hepatocytes suggest that these GREs require protein factors in addition to the glucocorticoid receptor in order to function (97). This hypothesis was proposed also for the phosphoenolpyruvate carboxykinase gene (95, 96), and evidence supporting this hypothesis has been reported (55). Based on their tissue-specific and developmental profiles, these genes are likely to have DNA elements that are recognized in common by one or more liver-specific transcription factors. Finally, at least four of the urea cycle enzyme genes—OTC, AS, AL, and arginase—share a common homologous sequence of unknown function (28, 100). As this element, termed the "urea cycle element" (UCE), is found also in the ornithine aminotransferase promoter (28), possibly this element may be involved in the coordinate expression of a larger group of genes, all involved in nitrogen metabolism. However, this hypothesis has yet to be tested, and the functional significance of the UCE in the urea cycle enzyme genes, if any, is unknown.

The presence of additional regulatory elements in at least two of the urea cycle enzyme genes—CPS-I and AS—is suggested by loss of expression of these, as well as of other liver-specific genes, in somatic cell hybrids of hepatoma and fibroblast cells (18, 110). In the case of AS, this loss of expression, termed extinction, involves at least two genetic loci that act in *trans:* tissue-specific extinguisher-1 (TSE1), which maps to mouse chromosome 11 and human chromosome 17, and at least one other locus that has not been identified (134). A recent report states that TSE1 encodes a regulatory subunit of cAMP-dependent protein kinase (8, 62). The regulatory DNA elements of the AS gene that are involved in extinction have not yet been identified but presumably include elements recognized by either a CRE-binding protein (CREB) or some other factor that is phosphorylated by protein kinase A.

Regulatory mutants affecting urea cycle enzyme gene expression have been sought in hope that they may shed light on various control mechanisms. For example, researchers have identified mouse strains that carry mutations affecting the normal postnatal expression of the urea cycle enzymes as well as of other genes in liver (56, 88, 110). One group of mice carries radiation-induced deletions of a portion of chromosome 7 (88, 110). Whereas heterozygotes are phenotypically indistinguishable from wild-type normal mice, certain of these deletion mutants permit near-normal fetal development in the homozygous condition, but homozygotes die within a few hours of birth. Levels of the urea cycle enzymes, their mRNAs, and transcription rates are significantly below normal in liver of homozygous neonates, although the reductions in transcription rate are insufficient to fully account for the reduction in mRNA abundance (88, 110). The reductions in expression of the urea cycle enzymes in homozygotes are essentially coordinate in degree. Whether gene expression in liver of homozygotes is also abnormal prior to birth is not known. Expression of arginine biosynthetic enzymes in kidney and intestine of neonatal homozygotes is unaffected, suggesting that the mutants may have a defect in some regulatory pathway that is primarily affected in liver (88). The nature of the defect underlying the aberrant expression of the urea cycle enzymes and other liver-specific genes in the mutant mice has not been elucidated. The putative regulatory gene encompassed by the deletions must act in *trans,* as the deletions are unlinked to the structural genes whose expression is affected and the homozygotes appear to have defects at both transcriptional and posttranscriptional levels of control (88). The lack of a gene dosage effect suggests that the defect more likely represents the absence of a positive regulatory factor rather than increased expression of a negative regulatory factor.

Activities of the urea cycle enzymes are reduced also in liver of C3H-H-2°-jsv mice (56). In contrast to the previously described mutants, reductions in activities of the urea cycle enzymes in the jsv mice are not coordinate in degree nor are significant reductions observed until 25 days of age, subsequent to development of a fatty liver. Most likely the reductions in urea cycle enzyme activities are secondary to other metabolic defects, as the ureagenic capacity of hepatocytes was markedly reduced in several animal models with fatty liver (75).

Finally, epigenetic events that affect expression of OTC, the only urea cycle enzyme gene located on the X chromosome, have been noted. Rat hepatoma cell lines that do not express OTC spontaneously give rise to OTC-expressing variants at a low frequency (24, 37). Although this frequency can be greatly increased by treatment with 5-azacytidine (24, 37), thus implying that the methylation state of the OTC gene influences its expression, a direct analysis of the methylation status of the OTC gene in expressing and nonexpressing cells has not been reported. Furthermore, in mice an increasing proportion of liver cells with a silent OTC gene on an inactive X chromosome region become OTC expressors with age (144), perhaps signifying important changes in gene expression patterns as females age. We do not know whether the changes in OTC expression in vivo involve DNA methylation.

CONCLUDING REMARKS

Genomic and cDNA clones for all the urea cycle enzymes have been isolated only within the past few years. Thus, the signals and mechanisms regulating the nutritional, hormonal, and developmental expression of these genes in liver and other organs are just beginning to be elucidated. For example, responses of these genes to agents such as cAMP, insulin, an glucocorticoid—or even to the complete absence of hormones—are being characterized and found to exhibit gene-specific patterns of expression (97). The finding of unique response patterns is intriguing because levels of these enzymes and their mRNAs in liver coordinately adapt to many physiologic changes in vivo (88, 89, 115, 116). What are the signals and mechanisms whereby coordinate expression of these genes is achieved in vivo?

The identification of *cis*-acting DNA elements and *trans*-acting regulatory factors undoubtedly will receive increasing attention, as transcription is very likely a major regulated step in expression of the genes encoding the urea and arginine biosynthetic enzymes. Integrating the results of such studies in a meaningful fashion will be challenging because at present the physiologic expression of a gene cannot be predicted merely by cataloging its individual regulatory elements. For example, in the case of genes that respond to both cAMP and glucocorticoids, it is unclear why some genes exhibit a synergistic response to a combination of these agents while others do not. In addition, a thorough understanding of the physiologic expression of these genes will require knowledge of how the regulatory factors themselves are regulated.

The cloned DNAs are powerful tools that can be used to transfect cultured cells or generate transgenic animals in order to address important questions about the physiologic roles of these enzymes: What are the rate-limiting factors for ureagenesis under various physiologic and pathophysiologic conditions? Does the urea cycle play a role in regulating pH homeostasis (4)? Are the arginine biosynthetic enzymes rate-limiting for the production of nitric oxide in brain, macrophages, or other cell types (81)? How important is the spatial expression of the urea cycle enzymes in liver in regulating ammonia, urea, and glutamine metabolism (45)? Clinically, this information may be useful in the management of certain pathophysiologic conditions, and the cloned genes can be used also to treat inherited defects in these enzymes by gene therapy.

Recent years have seen greater appreciation of the complex inter- and intra-organ interplay in the metabolism of ammonia, urea, and arginine. Much is yet to be learned about the regulated metabolism of these essential compounds, as well as the interrelation between urea and arginine biosynthesis and other metabolic pathways. The means to explore challenging problems in these areas are at hand, and the future holds great promise for exciting advances in our understanding both of urea and arginine biosynthesis in particular and of metabolic regulation in general.

ACKNOWLEDGMENTS

This work was supported in part by NIH grant DK33144.

Literature Cited

1. Abramson, R. D., Barbosa, P., Kalumuck, K., O'Brien, W. E. 1991. Characterization of the human argininosuccinate lyase gene and analysis of exon skipping. *Genomics* 10:126–32
2. Amos, J. A., Fleming, B. C., Gusella, J. F., Jacoby, L. B. 1984. Relative argininosuccinate synthetase mRNA levels and gene copy number in canavanine-resistant lymphoblasts. *Biochim. Biophys. Acta* 782:247–53
3. Anderson, P. M. 1991. Glutamine-dependent urea synthesis in elasmobranch fishes. *Biochem. Cell Biol.* 69:317–19
4. Atkinson, D. E., Bourke, E. 1987. Metabolic aspects of the regulation of systemic pH. *Am. J. Physiol.* 252:F947–56
5. Beaudet, A. L., O'Brien, W. E., Bock, H. G. O., Freytag, S. O., Su, T.-S. 1986. The human argininosuccinate synthetase locus and citrullinemia. *Adv. Hum. Genet.* 15:161–96
6. Bond, J. S., Failla, M. L., Unger, D. F. 1983. Elevated manganese concentration and arginase activity in livers of streptozotocin-induced diabetic rats. *J. Biol. Chem.* 258:8004–9
7. Borsook, H., Dubnoff, J. W. 1941. The conversion of citrulline to arginine in kidney. *J. Biol. Chem.* 141:717–38
8. Boshart, M., Weih, F., Nichols, M., Schutz, G. 1991. The tissue-specific extinguisher locus TSE1 encodes a regulatory subunit of cAMP-dependent protein kinase. *Cell* 66:849–59
9. Boyce, F. M., Anderson, G. M., Rusk, C. D., Freytag, S. O. 1986. Human argininosuccinate synthetase minigenes are subject to arginine-mediated repression but not to trans induction. *Mol. Cell. Biol.* 6:1244–52
10. Boyce, F. M., Freytag, S. O. 1989. Regulation of human argininosuccinate synthetase gene: Induction by positive-acting nuclear mechanism in canavanine-resistant cell variants. *Somat. Cell Mol. Genet.* 15:113–21
11. Boyce, F. M., Pogulis, R. J., Freytag, S. O. 1989. Paradoxical regulation of human argininosuccinate synthetase cDNA minigene in opposition to endogenous gene: Evidence for intragen-

ic control sequences. *Somat. Cell Mol. Genet.* 15:123–29

12. Brebnor, L. D., Grimm, J., Balinsky, J. B. 1981. Regulation of urea cycle enzymes in transplantable hepatomas and in the livers of tumor-bearing rats and humans. *Cancer Res.* 41:2692–99

13. Brebnor, L., Phillips, E., Balinsky, J. B. 1981. Control of urea cycle enzymes in rat liver by glucagon. *Enzyme* 26:265–70

14. Brown, C. L., Houghton, B. J., Souhami, R. L., Richards, P. 1972. The effects of low-protein diet and uraemia upon urea-cycle enzymes and transaminases in rats. *Clin. Sci.* 43:371–76

15. Brusilow, S. W., Horwich, A. L. 1989. Urea cycle enzymes. In *The Metabolic Basis of Inherited Disease*, ed. C. R. Scriver, A. L. Beaudet, W. S. Sly, D. Valle, pp. 629–63. New York: McGraw-Hill. 3006 pp. 6th ed.

16. Carey, G. P., Kime, Z., Rogers, Q. R., Morris, J. G., Hargrove, D., et al. 1987. An arginine-deficient diet in humans does not evoke hyperammonia or orotic aciduria. *J. Nutr.* 117:1734–39

17. Chan, W., Wang, M., Kopple, J. D., Swendsaid, M. E. 1974. Citrulline levels and urea cycle enzymes in uremic rats. *J. Nutr.* 104:678–83

18. Chin, A. C., Fournier, R. E. K. 1987. A genetic analysis of extinction: Transregulation of 16 liver-specific genes in hepatoma-fibroblast hybrid cells. *Proc. Natl. Acad. Sci. USA* 84:1614–18

19. Christowitz, D., Mattheyse, F. J., Balinsky, J. B. 1981. Dietary and hormonal regulation of urea cycle enzymes in rat liver. *Enzyme* 26:113–21

20. Cohen, P. P., Brucker, R. F., Morris, S. M. 1978. Cellular and molecular aspects of thyroid hormone action during amphibian metamorphosis. In *Hormonal Proteins and Peptides. Thyroid Hormones*, ed. C. H. Li, 6:273–381. New York: Academic. 446 pp.

21. Cohen, P. P., Hayano, M. 1946. The conversion of citrulline to arginine (transimination) by tissue slices and homogenates. *J. Biol. Chem.* 166:239–50

22. Das, T. K., Waterlow, J. C. 1974. The rate of adaptation of urea cycle enzymes, aminotransferases and glutamic dehydrogenase to changes in dietary protein intake. *Br. J. Nutr.* 32:353–73

23. de Groot, C. J., van Zonneveld, A. J., Mooren, P. G., Zonneveld, D., van den Dool, A., et al. 1984. Regulation of mRNA levels of rat liver carbamoylphosphate synthetase by glucocorticosteroids and cyclic AMP as estimated

with a specific cDNA. *Biochem. Biophys. Res. Commun.* 124:882–88

24. Delers, A., Szpirer, J., Szpirer, C., Saggioro, D. 1984. Spontaneous and 5-azacytidine-induced reexpression of ornithine carbamoyl transferase in hepatoma cells. *Mol. Cell. Biol.* 4:809–12

25. Dhanakoti, S. N., Brosnan, J. T., Herzberg, G. R., Brosnan, M. E. 1990. Renal arginine synthesis: studies in vitro and in vivo. *Am. J. Physiol.* 259:E437–42

26. Dizikes, G. J., Spector, E. B., Cederbaum, S. D. 1986. Cloning of rat liver arginase cDNA and elucidation of regulation of arginase gene expression in H4 rat hepatoma cells. *Somat. Cell Mol. Genet.* 12:374–84

27. Dubois, N., Cavard, C., Chasse, J.-F., Kamoun, P., Briand, P. 1988. Compared expression levels of ornithine transcarbamylase and carbamylphosphate synthetase in liver and small intestine of normal and mutant mice. *Biochim. Biophys. Acta* 950:321–28

28. Engelhardt, J. F., Steel, G., Valle, D. 1990. Transcriptional analysis of the human ornithine aminotransferase promoter. *J. Biol. Chem.* 266:752–58

29. Featherston, W. R., Rogers, Q. R., Freedland, R. A. 1973. Relative importance of kidney and liver in synthesis of arginine by the rat. *Am. J. Physiol.* 224:127–29

30. Felipo, V., Minana, M.-D., Grisolia, S. 1991. Control of urea synthesis and ammonia utilization in protein deprivation and refeeding. *Arch. Biochem. Biophys.* 285:351–56

31. Folin, O. 1905. Laws governing the chemical composition of urine. *Am. J. Physiol.* 13:66–113

32. Freedland, R. A., Avery, E. H., Taylor, A. R. 1968. Effect of thyroid hormones on metabolism. II. The effect of adrenalectomy or hypophysectomy on responses of rat liver enzyme activity to L-thyroxine injection. *Can. J. Biochem.* 46:141–50

33. Freytag, S. O., Beaudet, A. L., Bock, H. G. O., O'Brien, W. E. 1984. Molecular structure of the human argininosuccinate synthetase gene: Occurrence of alternative mRNA splicing. *Mol. Cell. Biol.* 4:1978–84

34. Freytag, S. O., Bock, H. G. O., Beaudet, A. L., O'Brien, W. E. 1984. Molecular structures of human argininosuccinate synthetase pseudogenes. Evolutionary and mechanistic implications. *J. Biol. Chem.* 259:3160–66

35. Gebhardt, R., Mecke, D. 1979. Per-

missive effect of dexamethasone on glucagon induction of urea-cycle enzymes in perifused primary monolayer cultures of rat hepatocytes. *Eur. J. Biochem.* 97:29–35

36. Goss, S. J. 1984. Arginine synthesis by hepatomas in vitro. II. Isolation and characterization of Morris hepatoma variants unable to convert ornithine to arginine, and modulation of urea-cycle enzymes by dexamethasone and cyclic-AMP. *J. Cell Sci.* 68:305–19

37. Goss, S. J. 1984. The associated reactivation of two X-linked genes. The spontaneous and azacytidine-induced reexpression of ornithine transcarbamoylase and glucose-6-phosphate dehydrogenase in a rat hepatoma. *J. Cell Sci.* 72:241–57

38. Grillo, M. A. 1964. Urea cycle in the liver of the hyperthyroid rat. *Clin. Chim. Acta* 10:259–61

39. Grillo, M. A., Fossa, T. 1966. Urea synthesis in the liver of the hypothyroid rat. *Clin. Chim. Acta* 13:383–86

40. Haggerty, D. F., Spector, E. B., Lynch, M., Kern, R., Frank, L. B., et al. 1982. Regulation by glucocorticoids of arginase and argininosuccinate synthetase in cultured rat hepatoma cells. *J. Biol. Chem.* 257:2246–53

41. Haggerty, D. F., Spector, E. B., Lynch, M., Kern, R., Frank, L. B., et al. 1983. Regulation of expression of genes for enzymes of the mammalian urea cycle in permanent cell-culture lines of hepatic and non-hepatic origin. *Mol. Cell. Biochem.* 53/54: 57–76

42. Halperin, M. L., Jungas, R. L., Chemma-Dhadli, S., Brosnan, J. T. 1987. Disposal of the daily acid load: an integrated function of the liver, lungs, and kidneys. *Trends Biochem. Sci.* 12:197–99

43. Hamano, Y., Kodama, H., Yanagisawa, M., Haraguchi, Y., Mori, M., et al. 1988. Immunocytochemical localization of ornithine transcarbamylase in rat intestinal mucosa. Light and electron microscopic study. *J. Histochem. Cytochem.* 36:29–35

44. Hata, A., Tsuzuki, T., Shimada, K., Takiguchi, M., Mori, M., et al. 1986. Isolation and characterization of the human ornithine transcarbamylase gene: Structure of the 5'-end region. *J. Biochem.* 100:717–25

45. Haussinger, D. 1990. Nitrogen metabolism in liver: structural and functional organization and physiological relevance. *Biochem. J.* 267:281–90

46. Hayase, K., Yokogoshi, H., Yoshida, A. 1980. Effect of dietary proteins and amino acid deficiencies on urinary excretion of nitrogen and the urea synthesizing system in rats. *J. Nutr.* 110:1327–37

47. Hecker, M., Sessa, W. C., Harris, H. J., Anggard, E. E., Vane, J. R. 1990. The metabolism of L-arginine and its significance for the biosynthesis of endothelium-derived relaxing factor: Cultured endothelial cells recycle L-citrulline to L-arginine. *Proc. Natl. Acad. Sci. USA* 87:8612–16

48. Heitmann, R. N., Bergman, E. N. 1980. Integration of amino acid metabolism in sheep: effects of fasting and acidosis. *Am. J. Physiol.* 238:E248–54

49. Henning, S. J., Leeper, L. L. 1982. Coordinate loss of glucocorticoid responsiveness by intestinal enzymes during postnatal development. *Am. J. Physiol.* 242:G89–94

50. Herzfeld, A., Raper, S. M. 1976. Enzymes of ornithine metabolism in adult and developing rat intestine. *Biochim. Biophys. Acta* 428:600–10

51. Hoogenraad, N., Totino, N., Elmer, H., Wraight, C., Alewood, P., et al. 1985. Inhibition of intestinal citrulline synthesis causes severe growth retardation in rats. *Am. J. Physiol.* 249: G792–99

52. Howell, B. W., Lagace, M., Shore, G. C. 1989. Activity of the carbamyl phosphate synthetase I promoter in liver nuclear extracts is dependent on a cis-acting C/EBP recognition element. *Mol. Cell. Biol.* 9:2928–33

53. Hurwitz, R., Kretchmer, N. 1986. Development of arginine-synthesizing enzymes in mouse intestine. *Am. J. Physiol.* 251:G103–10

54. Hutchinson, J. H., Jolley, R. L., Labby, D. H. 1964. Studies of rat liver and kidney enzymes. I. Response to massive intragastric doses of chronically administered nitrogenous substances. *Am. J. Clin. Nutr.* 14:291–301

55. Imai, E., Stromstedt, P.-E., Quinn, P. G., Carlstedt-Duke, J., Gustafsson, J.-A., et al. 1990. Characterization of a complex glucocorticoid response unit in the phosphoenolpyruvate carboxykinase gene. *Mol. Cell. Biol.* 10:4712–19

56. Imamura, Y., Saheki, T., Arakawa, H., Noda, T., Koizuma, T., et al. 1990. Urea cycle disorder in C3H-H-2° mice with juvenile steatosis of viscera. *FEBS Lett.* 260:119–21

57. Irr, J., Jacoby, L. 1978. Control of argininosuccinate synthetase by arginine in human lymphoblasts. *Somat. Cell Genet.* 4:111–24

58. Jackson, M. J., Allen, S. J., Beaudet, A. L., O'Brien, W. E. 1988. Metabolite regulation of argininosuccinate synthetase in cultured human cells. *J. Biol. Chem.* 263:16388–94

59. Jackson, M. J., Beaudet, A. L., O'Brien, W. E. 1986. Mammalian urea cycle enzymes. *Annu. Rev. Genet.* 20:431–64

60. Jackson, M. J., O'Brien, W. E., Beaudet, A. L. 1986. Arginine-mediated regulation of an argininosuccinate synthetase minigene in normal and canavanine-resistant human cells. *Mol. Cell. Biol.* 6:2257–61

61. Jinno, Y., Matuo, S., Nomiyama, H., Shimada, K., Matsuda, I. 1985. Novel structure of the 5' end region of the human argininosuccinate synthetase gene. *J. Biochem.* 98:1395–1403

62. Jones, K. W., Shapero, M. H., Chevrette, M., Fournier, R. E. K. 1991. Subtractive hybridization cloning of a tissue-specific extinguisher: TSE1 encodes a regulatory subunit of protein kinase A. *Cell* 66:861–72

63. Jorda, A., Cabo, J., Grisolia, S. 1981. Changes in the levels of urea cycle enzymes and in metabolites thereof in diabetes. *Enzyme* 26:240–44

64. Kitagawa, Y. 1987. Hormonal regulation of carbamoyl-phosphate synthetase I synthesis in primary cultured hepatocytes and Reuber hepatoma H-35. Defective regulation in hepatoma cells. *Eur. J. Biochem.* 167:19–25

65. Kitagawa, Y., Ryall, J., Nguyen, M., Shore, G. C. 1985. Expression of carbamoyl-phosphate synthetase I mRNA in Reuber hepatoma H-35 cells. Regulation by glucocorticoid and insulin. *Biochim. Biophys. Acta* 825:148–53

66. Kitagawa, Y., Sugimoto, E. 1985. Interaction between glucocorticoids, 8-bromoadenosine 3',5'-monophosphate, and insulin in regulation of synthesis of carbamoyl-phosphate synthetase I in Reuber hepatoma H-35. *Eur. J. Biochem.* 150:249–54

67. Krebs, H. A., Henseleit, K. 1932. Untersuchungen uber die Harnstoffbildung im Tierkorper. *Hoppe-Seyler's Z. Physiol. Chem.* 210:33–66

68. Kuhn, N. J., Talbot, J., Ward, S. 1991. pH-sensitive control of arginase by Mn(II) ions at submicromolar concentrations. *Arch. Biochem. Biophys.* 186:217–21

69. Lagace, M., Howell, B. W., Burak, R., Lusty, C. J., Shore, G. C. 1987. Rat carbamyl-phosphate synthetase I gene. Promoter sequence and tissue-specific

transcriptional regulation in vitro. *J. Biol. Chem.* 262:10415–18

70. Levillain, O., Hus-Citharel, A., Morel, F., Bankir, L. 1990. Localization of arginine synthesis along the rat nephron. *Am. J. Physiol.* 259:F916–23

71. Lightbody, H. D., Kleinman, A. 1939. Variations produced by food differences in the concentration of arginase in the livers of white rats. *J. Biol. Chem.* 129:71–78

72. Lin, R. C., Snodgrass, P. J., Rabier, D. 1982. Induction of urea cycle enzymes by glucagon and dexamethasone in monolayer cultures of adult rat hepatocytes. *J. Biol. Chem.* 257:5061–67

73. Malo, C., Qureshi, I. A., Letarte, J. 1986. Postnatal maturation of enterocytes in sparse-fur mutant mice. *Am. J. Physiol.* 250:G177–84

74. Marti, J., Portoles, M., Jimenez-Nacher, I., Cabo, J., Jorda, A. 1988. Effect of thyroid hormones on urea biosynthesis and related processes in rat liver. *Endocrinology* 123:2167–74

75. Maswoswe, S. M., Tremblay, G. C. 1989. Biosynthesis of hippurate, urea and pyrimidines in the fatty liver: Studies with rats fed orotic acid or a diet deficient in choline and inositol, and with genetically obese (Zucker) rats. *J. Nutr.* 119:273–79

76. McIntyre, P., DeMartinis, M. L., Hoogenraad, N. 1983. Changes in carbamyl phosphate synthetase and ornithine transcarbamylase levels during development and in response to changes in diet. Application of the electrophoretic transfer technique. *Biochem. Int.* 6:365–73

77. McLean, P., Gurney, M. W. 1963. Effect of adrenalectomy and of growth hormone on enzymes concerned with urea synthesis in rat liver. *Biochem. J.* 87:96–104

78. McLean, P., Novello, F. 1965. Influence of pancreatic hormones on enzymes concerned with urea synthesis in rat liver. *Biochem. J.* 94:410–22

79. Meijer, A. J., Lamers, W. H., Chamuleau, R. A. F. M. 1990. Nitrogen metabolism and ornithine cycle function. *Physiol. Rev.* 70:701–48

80. Milland, J., Tsykin, A., Thomas, T., Aldred, A. R., Cole, T., et al. 1990. Gene expression in regenerating and acute-phase liver. *Am. J. Physiol.* 259:G340–47

81. Moncada, S., Palmer, R. M. J., Higgs, E. A. 1991. Nitric oxide: Physiology, pathophysiology, and pharmacology. *Pharmacol. Rev.* 43:109–42

82. Moorman, A. F. M., de Boer, P. A. J., Charles, R., Lamers, W. H. 1990. Diet- and hormone-induced reversal of the carbamoylphosphate synthetase mRNA gradient in the rat liver lobulus. *FEBS Lett.* 276:9–13

83. Mori, M., Miura, S., Tatibana, M., Cohen, P. P. 1981. Cell-free translation of carbamyl phosphate synthetase I and ornithine transcarbamylase messenger RNAs of rat liver. Effect of dietary protein and fasting on translatable mRNA levels. *J. Biol. Chem.* 256:4127–32

84. Morris, J. G., Rogers, Q. R. 1978. Ammonia intoxication in the near-adult cat as a result of a dietary deficiency of arginine. *Science* 199:431–32

85. Morris, S. M. Jr. 1987. Thyroxine elicits divergent changes in mRNA levels for two urea cycle enzymes and one gluconeogenic enzyme in tadpole liver. *Arch. Biochem. Biophys.* 259:144–48

86. Morris, S. M. Jr., Kepka, D. M., Sweeney, W. E. Jr., Avner, E. D. 1989. Abundance of mRNAs encoding urea cycle enzymes in fetal and neonatal mouse liver. *Arch. Biochem. Biophys.* 269:175–80

87. Morris, S. M. Jr., Moncman, C. L., Holub, J. S., Hod, Y. 1989. Nutritional and hormonal regulation of mRNA abundance for arginine biosynthetic enzymes in kidney. *Arch. Biochem. Biophys.* 273:230–37

88. Morris, S. M. Jr., Moncman, C. L., Kepka, D. M., Nebes, V. L., Diven, W. R., et al. 1988. Effects of deletions in mouse chromosome 7 on expression of genes encoding the urea-cycle enzymes and phosphoenolpyruvate carboxykinase (GTP) in liver, kidney, and intestine. *Biochem. Genet.* 26:769–81

89. Morris, S. M. Jr., Moncman, C. L., Rand, K. D., Dizikes, G. J., Ceder- baum, S. D., et al. 1987. Regulation of mRNA levels for five urea cycle enzymes in rat liver by diet, cyclic AMP, and glucocorticoids. *Arch. Biochem. Biophys.* 256:343–53

90. Morris, S. M. Jr., Sweeney, W. E. Jr., Kepka, D. M., O'Brien, W. E., Avner, E. D. 1991. Localization of arginine biosynthetic enzymes in renal proximal tubules and abundance of mRNA during development. *Pediatr. Res.* 19:151–54

91. Murakami, A., Kitagawa, Y., Sugimo- to, E. 1983. Induction of carbamoyl- phosphate synthetase I in Reuber hepato- ma H-35 by dexamethasone. *Biochim. Biophys. Acta* 740:38–45

92. Murakami, T., Nishiyori, A., Taki- guchi, M., Mori, M. 1990. Promoter

and 11-kilobase upstream enhancer ele- ments responsible for hepatoma cell- specific expression of the rat ornithine transcarbamylase gene. *Mol. Cell. Biol.* 10:1180–91

93. Murakami, T., Takiguchi, M., Inomoto, T., Yamamura, K.-I., Mori, M. 1989. Tissue- and developmental stage-specif- ic expression of the rat ornithine carba- moyltransferase gene in transgenic mice. *Dev. Genet.* 10:393–401

94. Nebes, V. L. 1988. *Regulation of urea cycle enzyme mRNA levels and relative transcription rates by a cAMP analog, dexamethasone, and insulin.* PhD thesis. Univ. Pittsburgh, Penn. 113 pp.

95. Nebes, V. L., DeFranco, D., Morris, S. M. Jr. 1990. Differential induction of transcription for glucocorticoid-respon- sive genes in cultured rat hepatocytes. *Biochem. Biophys. Res. Commun.* 166:133–38

96. Nebes, V. L., Morris, S. M. Jr. 1987. Induction of mRNA for phosphoenol- pyruvate carboxykinase (GTP) by dexa- methasone in cultured rat hepatocytes re- quires on-going protein synthesis. *Biochem. J.* 246:237–40

97. Nebes, V. L., Morris, S. M. Jr. 1988. Regulation of messenger ribonucleic acid levels for five urea cycle enzymes in cultured rat hepatocytes. Require- ments for cyclic adenosine monophos- phate, glucocorticoids, and ongoing pro- tein synthesis. *Mol. Endocrinol.* 2:444– 51

98. Nicoletti, M., Guerri, C., Grisolia, S. 1977. Turnover of carbamyl-phosphate synthase, of other mitochondrial en- zymes and of rat tissues. Effect of diet and of thyroidectomy. *Eur. J. Biochem.* 75:583–92

99. Nuzum, C. T., Snodgrass, P. J. 1971. Urea cycle enzyme adaptation to dietary protein in primates. *Science* 172:1042– 43

100. Ohtake, A., Takiguchi, M., Shigeto, Y., Amaya, Y., Kawamoto, S., et al. 1988. Structural organization of the gene for rat liver-type arginase. *J. Biol. Chem.* 263:2245–49

101. Ohtake, Y., Clemens, M. G. 1991. In- terrelationship between hepatic ureagen- esis and gluconeogenesis in early sepsis. *Am. J. Physiol.* 260:E453–58

102. Palekar, A. G., Collipp, P. J., Mad- daiah, V. T. 1981. Growth hormone and rat liver mitochondria effects on urea cycle enzymes. *Biochem. Biophys. Res. Commun.* 100:1604–10

103. Powers, S. G., Meister, A. 1988. Urea synthesis and ammonia metabolism. In

The Liver: Biology and Pathobiology, ed. I. Arias, W. B. Jakoby, H. Popper, D. Schachter, D. A. Shafritz, pp. 317–29. New York: Raven. 1377 pp. 2nd ed.

104. Ratner, S., Morell, H., Carvalho, E. 1960. Enzymes of arginine metabolism in brain. *Arch. Biochem. Biophys.* 91:280–89

105. Ratner, S., Petrack, B. 1953. The mechanism of arginine synthesis from citrulline in kidney. *J. Biol. Chem.* 200:175–85

106. Rogers, Q. R., Freedland, R. A., Symmons, R. A. 1972. In vivo synthesis and utilization of arginine in the rat. *Am. J. Physiol.* 223:236–40

107. Rogers, Q. R., Morris, J. G., Freedland, R. A. 1977. Lack of hepatic enzymatic adaptation to low and high levels of dietary protein in the adult cat. *Enzyme* 22:348–56

108. Rosebrough, R. W., Mitchell, A. D., Richards, M. P., Steele, N. C., McMurtry, J. P. 1987. Effect of dietary protein status on urea metabolism and hepatic arginase activity of the pig. *Nutr. Res.* 7:547–56

109. Rozen, R., Noel, C., Shore, G. C. 1983. Effects of glucagon on biosynthesis of the mitochondrial enzyme, carbamoyl-phosphate synthase I, in primary hepatocytes and Morris hepatoma 5123D. *Biochim. Biophys. Acta* 741:47–54

110. Ruppert, S., Boshart, M., Bosch, F. X., Schmid, W., Fournier, R. E. K., et al. 1990. Two genetically defined trans-acting loci coordinating regulate overlapping sets of liver-specific genes. *Cell* 61:895–504

111. Ryall, J., Nguyen, M., Bendayan, M., Shore, G. C. 1985. Expression of nuclear genes encoding the urea cycle enzymes, carbamoyl-phosphate synthetase I and ornithine carbamoyl transferase, in rat liver and intestinal mucosa. *Eur. J. Biochem.* 152:287–92

112. Ryall, J., Rachubinski, R. A., Nguyen, M., Rozen, R., Broglie, K. E., et al. 1984. Regulation and expression of carbamyl phosphate synthetase I mRNA in developing rat liver and Morris hepatoma 5123D. *J. Biol. Chem.* 259:9172–76

113. Ryall, J. C., Quantz, M. A., Shore, G. C. 1986. Rat liver and intestinal mucosa differ in the developmental pattern and hormonal regulation of carbamoyl-phosphate synthetase I and ornithine carbamoyl transferase gene expression. *Eur. J. Biochem.* 156:453–58

114. Saheki, T., Katsunuma, T., Sase, M. 1977. Regulation of urea synthesis in rat liver. Changes of ornithine and acetylglutamate concentrations in the livers of rats subjected to dietary transitions. *J. Biochem.* 82:551–58

115. Schimke, R. T. 1962. Adaptive characteristics of urea cycle enzymes in the rat. *J. Biol. Chem.* 237:459–68

116. Schimke, R. T. 1962. Differential effects of fasting and protein-free diets on levels of urea cycle enzymes in rat liver. *J. Biol. Chem.* 237:1921–24

117. Schimke, R. T. 1963. Studies on factors affecting the levels of urea cycle enzymes in rat liver. *J. Biol. Chem.* 238:1012–18

118. Schimke, R. T. 1964. Enzymes of arginine metabolism in mammalian cell culture. I. Repression of argininosuccinate synthetase and argininosuccinase. *J. Biol. Chem.* 238:136–45

119. Schimke, R. T. 1964. The importance of both synthesis and degradation in the control of arginase levels in rat liver. *J. Biol Chem.* 239:3808–17

120. Scull, C. W., Rose, W. D. 1930. Arginine metabolism. I. The relation of the arginine content of the diet to the increments in tissue arginine during growth. *J. Biol. Chem.* 89:109–23

121. Semon, B. A., Leung, P. M. B., Rogers, Q. R., Gietzen, D. W. 1989. Plasma ammonia, plasma brain and liver amino acids and urea cycle enzyme activities in rats fed ammonium acetate. *J. Nutr.* 119:166–74

122. Snodgrass, P. J., Lin, R. C. 1981. Induction of urea cycle enzymes of rat liver by amino acids. *J. Nutr.* 111:586–601

123. Snodgrass, P. J., Lin, R. C. 1987. Differing effects of arginine deficiency on the urea cycle enzymes of rat liver, cultured hepatocytes, and hepatoma cells. *J. Nutr.* 117:1827–37

124. Snodgrass, P. J., Lin, R. C., Muller, W. A., Aoki, T. T. 1978. Induction of urea cycle enzymes of rat liver by glucagon. *J. Biol. Chem.* 253:1748–53

125. Sochor, M., McLean, P., Brown, J., Greenbaum, A. L. 1981. Regulation of pathways of ornithine metabolism. Effects of thyroid hormone and diabetes on the activity of enzymes at the 'ornithine crossroads' in rat liver. *Enzyme* 26:15–23

126. Spector, E. B., Rice, S. C. H., Cederbaum, S. D. 1983. Immunologic studies of arginase in tissues of normal human adult and arginase-deficient patients. *Pediatr. Res.* 17:941–44

127. Spolarics, Z., Bond, J. S. 1989. Comparison of biochemical properties of liv-

er arginase from streptozotocin-induced diabetic and control mice. *Arch. Biochem. Biophys.* 274:426–33

128. Su, T.-S., Beaudet, A. L., O'Brien, W. E. 1981. Increased translatable messenger ribonucleic acid for argininosuccinate synthetase in canavanine-resistant human cells. *Biochemistry* 20:2956–60

129. Su, T.-S., Bock, H. G. O., O'Brien, W. E., Beaudet, A. L. 1981. Cloning of cDNA for argininosuccinate synthetase mRNA and study of enzyme overproduction in a human cell line. *J. Biol. Chem.* 256:11826–31

130. Szepesi, B., Avery, E. H., Freedland, R. A. 1970. Role of kidney in gluconeogenesis and amino acid catabolism. *Am. J. Physiol.* 219:1627–31

131. Takiguchi, M., Haraguchi, Y., Mori, M. 1988. Human liver-type arginase gene: structure of the gene and analysis of the promoter region. *Nucleic Acids Res.* 16:8789–8802

132. Takiguchi, M., Mori, M. 1991. In vitro analysis of the rat liver-type arginase promoter. *J. Biol. Chem.* 266:9186–93

133. Takiguchi, M., Murakami, T., Miura, S., Mori, M. 1987. Structure of the rat ornithine carbamoyltransferase gene, a large, X chromosome-linked gene with an atypical promoter. *Proc. Natl. Acad. Sci. USA* 84:6136–40

134. Thayer, M. J., Fournier, R. E. K. 1989. Hormonal regulation of TSE1-repressed genes: Evidence for multiple genetic controls in extinction. *Mol. Cell. Biol.* 9:2837–46

135. Tizianello, A., De Ferrari, G., Garibotto, G., Guerri, G., Robaudo, C. 1980. Renal metabolism of amino acids and ammonia in subjects with normal renal function and in patients with chronic renal insufficiency. *J. Clin. Invest.* 65:1162–73

136. Tsai, S. Y., Sagami, I., Wang, H., Tsai, M., O'Malley, B. S. 1987. Interactions between a DNA–binding transcription factor (COUP) and a non-DNA binding factor (S300-II). *Cell* 50:701–9

137. Tsuda, M., Shikata, Y., Katsunuma, T. 1979. Effect of dietary proteins on the

turnover of rat liver argininosuccinate synthetase. *J. Biochem.* 85:699–704

138. van den Bogaert, A. J. W., Lamers, W. H., Moorman, A. F. M. 1992. Translational control of glutamine synthetase and carbamoylphosphate synthetase in the rat perinatal period. *J. Biol. Chem.* In press

139. van Roon, M. A., Zonneveld, D., Charles, R., Lamers, W. H. 1988. Accumulation of carbamoylphosphate-synthetase and phosphoenolpyruvate-carboxykinase mRNA in embryonic rat hepatocytes. Evidence for translational control during the initial phases of hepatocyte-specific gene expression in vitro. *Eur. J. Biochem.* 178:191–96

140. Veres, G., Craigen, W. J., Caskey, C. T. 1986. The 5' flanking region of the ornithine transcarbamylase gene contains DNA sequences regulating tissue-specific expression. *J. Biol. Chem.* 261:7588–91

141. Visek, W. 1986. Arginine needs, physiological state and usual diets. A reevaluation. *J. Nutr.* 116:36–46

142. Wallace, R., Knecht, E., Grisolia, S. 1986. Turnover of rat liver ornithine transcarbamylase. *FEBS Lett.* 208:427–30

143. Walser, M. 1986. Roles of urea production, ammonium excretion, and amino acid oxidation in acid-base balance. *Am. J. Physiol.* 250:F181–88

144. Wareham, K. A., Lyon, M. F., Glenister, P. H., Williams, E. D. 1987. Age-related reactivation of an X-linked gene. *Nature* 327:725–27

145. Watford, M. 1989. Channeling in the urea cycle: a metabolon spanning two compartments. *Trends Biochem. Sci.* 14:313–14

146. Windmueller, H. G., Spaeth, A. E. 1981. Source and fate of circulating citrulline. *Am. J. Physiol.* 241:E473–80

147. Wraight, C., Lingelbach, K., Hoogenraad, N. 1985. Comparison of ornithine transcarbamylase from rat liver and intestine. Evidence for differential regulation of enzyme levels. *Eur. J. Biochem.* 153:239–42

Annu. Rev. Nutr. 1992. 12:103–17

THE INFLUENCE OF MATERNAL NUTRITION ON LACTATION

Kathleen Maher Rasmussen

Division of Nutritional Sciences, Cornell University, Ithaca, New York 14853

KEY WORDS: obesity, overnutrition, undernutrition, milk composition, milk production, infant growth, rats

CONTENTS

INTRODUCTION

The first objective of this review is to examine the effects of general over- or undernutrition on lactational performance. There are many ways to evaluate lactational performance, and for this review the focus is on infant milk and nutrient intake and infant growth. In each section, data from studies in experimental species are used to develop hypotheses that are then explored in the data from studies of lactating women. The second objective of this review is to go beyond the recent book, *Nutrition During Lactation* (18), to examine the effect of maternal nutritional status on lactational performance within a

103

framework for evaluating causality. A biologically based conceptual model is developed to guide this analysis.

CONCEPTUAL FRAMEWORK

Important factors in the relationship between maternal dietary intake and various measures of lactational performance are identified in Figure 1. To simplify the diagram, I have omitted any possible effects that a change in maternal dietary intake may have on either maternal physical activity or the thermic effect of food or maternal and infant illness. Also omitted are the clear effects of maternal dietary intake before and during pregnancy on maternal adipose tissue and nutrient stores at parturition, lactational capacity, and infant size at birth. This diagram assumes that infants are exclusively breast-fed because the effect of direct supplementation of the infant, which would increase infant growth and decrease milk intake, is omitted. The diagram is both more specific and inclusive than the causal sequence (maternal nutritional status → lactational performance → infant growth) that has traditionally guided research in this field.

It is important to distinguish between maternal dietary intake (what the mother actually consumes) and nutritional status (manifestations of that consumption, such as blood nutrient concentrations or measures of body composition). In general, interventions have selected lactating women based on

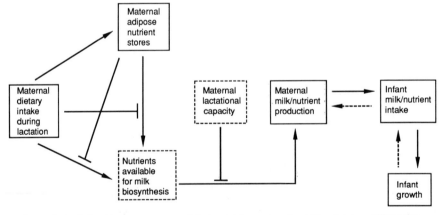

Figure 1   Relationships among maternal dietary intake, maternal nutritional status, milk production, and infant growth at any one time during lactation. Shown in *dotted boxes* are two important, but usually unmeasured, variables: nutrients available for milk biosynthesis and maternal lactational capacity (see text). *Solid lines with arrowheads* denote direct transfer of nutrients or energy-producing compounds. *Dotted line with arrowheads* denote the direct influence of one factor on another that does not occur via nutrient flux. *Solid lines ending with a bar* denote possible modification of effect on the *arrow* on which they abut.

some measure of nutritional status and then have sought to change dietary intake. The proportion of ingested nutrients partitioned for milk biosynthesis may depend on maternal nutrient stores. Nutrient stores also may be mobilized to contribute to the nutrients available for milk biosynthesis, and it is likely that the extent of nutrient mobilization is conditioned upon dietary intake. In experimental species, nutrients available for milk biosynthesis can be calculated from values for blood nutrient concentrations and from blood flow to the mammary glands.

The amount of milk a woman actually produces is limited by her lactational capacity, defined here as a woman's ability to produce milk at any given time. Lactational capacity is probably a function of genetic heritage, age (17), and parity (16) as well as breast enlargement during pregnancy (16) and nutritional history (42). It can respond to changing circumstances (improved dietary intake, increased infant demand) and, thus, might have a different value at some later time during lactation. Lactational capacity has usually been approximated by measuring milk production (infant milk intake plus residual milk). For well-nourished women, lactational capacity is likely to be greater than milk production, which, in turn, is usually greater than short-term infant milk intake [by 100 ml per day in one recent study (8)]. For poorly nourished women, lactational capacity, milk production, and infant milk intake are expected to (25) have more similar values. In the single study (25) that has examined this proposition, it was, in fact, not the case.

The amount of milk a woman actually produces also responds to infant milk intake, which itself responds to infant size and growth rate (infant vigor, suckling stimulus). In recent years, studies have emphasized that the infant's demand determines milk production via the prolactin response to suckling (8, 10, 23, 50). However, this concept is only valid when maternal lactational capacity is not limiting for milk production.

This model leads to the prediction that, given adequate infant demand, lactational capacity would not be limiting in obese women, who are likely to have sufficient dietary intake or nutrient stores to maintain lactation. This model also leads to the prediction that, among undernourished women, milk production might be limited by either lactational capacity or the availability of substrates for milk biosynthesis or both, even when infant demand is adequate.

EFFECT OF MATERNAL OVERNUTRITION ON LACTATIONAL PERFORMANCE

Several different experimental approaches have been used in animals and provide useful information about the effects of overnutrition on lactational performance (milk volume and composition, growth of the suckling young). Only very limited information is available from human subjects.

Studies in Experimental Animals

EFFECTS OF OVERFEEDING DURING LACTATION AMONG OBESE AN-
IMALS   In rats, maternal obesity can have negative effects on the pups.
Forty-four per cent of the obese rats studied by Rolls et al (40) were unable to
maintain their litters after 6 days of life. These rat dams were made obese by
consumption of the "cafeteria diet" (stock diet plus access to salami, crackers,
and cookies) before conception and during pregnancy and lactation. The pups
who survived were significantly smaller than those of the controls. Similar
findings in rats have been reported by others (51) who used a protein-
supplemented, high-fat diet to produce maternal obesity before conception.

These obese animals weighed at least 30% more at conception than ad
libitum–fed controls (41). In these investigations, overweight is assumed to
be overfat: in general, data on body composition were not obtained. Obese
dams did not develop the hyperphagia characteristic of lactation and, as a
result, consumed less energy than did control animals (39). Among those
obese dams whose pups survived until weaning, the heavier the mothers were
at delivery, the more weight they lost during lactation. Although the weight
lost probably consisted mostly of fat (48), the amount of maternal weight lost
during lactation was unrelated to the weight of the pups at weaning (39).

The immediate predictors of pup growth in this model have been investigated. Studies have shown that obese dams produce 25% less milk (35),
but this milk had 26% more energy (37) because it contained nearly double
the usual proportion of lipid. The milk of the obese dams also contained 21%
lower protein and 10% lower lactose concentrations as well as a reduced
proportion of medium-chain fatty acids and an increased proportion of long-
chain fatty acids (36). These changes in milk composition resulted primarily
from cafeteria feeding during lactation (27). The higher fat concentration also
may have resulted from the increased rate of mobilization of adipose tissue
stores during lactation that is characteristic of obese animals (38). That the
pups grew less well with an apparently adequate energy intake suggests that
the combination of the other changes in milk composition with the reduction
in volume resulted in a total nutrient intake by the pups that was limiting for
their growth.

The effect of obesity on lactational performance as measured by pup weight
also has been studied in golden hamsters. They were overfed with a high-fat
diet (stock diet plus access to sunflower seeds and a 1:3 mixture of peanut
butter:vegetable shortening) before mating, during pregnancy, and during
lactation (11). The overfed hamsters weighed about 30% more at delivery and
lost weight at the same rate during lactation as controls and, thus, ended
lactation with the same difference in body weight that had existed at delivery.
In contrast to observations in rats, no effect of maternal body weight on litter
growth or survival was observed.

The cafeteria diet has a low protein-energy ratio and, from the available data, diet composition cannot be excluded as the cause of poor lactational performance of the rats or the difference in results between rats and hamsters. Investigation of the effects on lactational performance of other experimental paradigms for producing obesity in laboratory species is warranted. Further exploration of this relationship is of public health interest because of the high proportion of American women of reproductive age who are overweight or obese (52) and national health goals to increase breast-feeding rates (7).

EFFECTS OF OVERFEEDING DURING LACTATION AMONG LEAN AN-IMALS   Conflicting results about lactational performance have been obtained from studies of the effects of supplementing lean (i.e. control or nonobese) rats during lactation. Investigators have employed cafeteria feeding (37, 41) or a purified diet supplemented with a homogenous mixture of eggs and oil (34). In the studies that employed cafeteria feeding, the supplemented rats increased their energy intake compared to controls fed a stock diet (40, 41), but their pups did not grow as well as those of the controls (41). These supplemented rats produced milk with a lower protein concentration and a higher fat concentration (resulting in a higher energy density) than the milk of control rats (37). Their milk contained a lower proportion of medium-chain and a higher proportion of long-chain fatty acids than did the milk of control rats. The milk production of lean rats fed the cafeteria diet only during lactation has not been reported.

In contrast, when lean rats were fed a purified diet supplemented with eggs and oil, they gained weight, reduced their fat mobilization during lactation, produced more milk, and had larger pups than control rats (34). Milk composition was not examined in this experiment. The authors attributed these positive findings to the higher protein-energy ratio of their supplement compared to the supplemental foods used in cafeteria feeding. These results (34) indicate that ad libitum–fed lactating animals are capable of increasing their food intake with beneficial effects on lactational performance. These results also suggest that, in the amounts usually consumed, the stock diet does not maximize either milk production or litter growth. Thus, a reexamination of the nutrient requirements of the rat and the possible formulation of a more appropriate stock diet for use during lactation merits attention.

EFFECTS OF UNDERFEEDING OBESE ANIMALS DURING LACTATION   When obese rats were switched at parturition from cafeteria feeding to a stock diet, they consumed less energy and lost about twice as much weight as obese rats maintained on the cafeteria diet (41). Their milk production has not been reported. Milk protein concentrations were higher than those of obese rats continued on cafeteria feeding; fat concentration and energy density did not differ between these two groups (37). Their pups grew much less well than those of the obese rats maintained on the cafeteria diet who, as described above, grew less well than lean controls fed the stock diet (41). Thus, although these animals had free access to a theoretically adequate diet during lactation as well as adequate adipose tissue reserves at parturition (which they mobilized during lactation), pup growth remained poor. This finding was not attributable to low milk protein concentration or energy density, but the role of milk production in causing the poor pup growth remains unknown.

This experimental approach represents an animal model for dieting during lactation, a subject of concern to women who were overweight before conception or who gained excessive amounts of weight during pregnancy. As such, these data tell a cautionary tale. However, the poor growth of the pups of the dieting rats may result from intrauterine factors attributable to the low protein-energy ratio of the cafeteria diet fed during pregnancy or may result from inadequate dietary intake during lactation. Again, other means of producing obesity might be informative and deserve exploration.

Studies in Women

In contrast to the interest in and concern about obesity and the outcome of pregnancy among women, little attention has been given to any possible relation between preexisting or pregnancy-related obesity and any aspect of lactation.

Among well-nourished women studied longitudinally between 1 and 4 months postpartum, dietary intake accounted for 13% of the variability in infant milk intake (4). Anthropometric indicators of nutritional status were not associated with lactational performance in these women. Among well-nourished women studied longitudinally from 3 to 12 months postpartum (26), some associations were found between maternal nutritional status and milk composition. These relationships were stronger at later stages of lactation.

Butte & Garza (3) commented that their failure to detect an association between maternal nutritional status and lactational performance might have resulted from (a) the relative homogeneity of their study population in body size, (b) the relatively greater importance of diet compared to tissue reserves for maintaining energy balance during lactation, (c) variability in energy needs for maintenance and activity, or (d) imprecision in the anthropometric measurements used to estimate body composition. These limitations also apply to the other longitudinal studies of well-nourished women that have been conducted.

The effect of short-term caloric restriction has been investigated in one group of well-nourished women 6–24 weeks after delivery (47). Women who chose to restrict their dietary intake for one week were somewhat heavier and had gained more weight during pregnancy than those who chose not to do so.

During the period of restricted intake, milk volume and composition were maintained, but in the week afterward, infant milk intake and weight gain were less than in the pre-dieting period. The negative effects of dieting were greater among the mothers who restricted their intake to < 1500 kcal/day.

In summary, the available data from human subjects have not included enough overweight or obese subjects to examine overnutrition as a separate predictor of lactational performance. Within the range of maternal body weight that has been examined, infant milk intake is minimally responsive to variations in dietary intake or anthropometric indicators of maternal nutritional status.

EFFECT OF MATERNAL UNDERNUTRITION ON LACTATIONAL PERFORMANCE

Much more extensive information exists about the effects of various forms of undernutrition on lactational performance from both experimental species and human subjects. In animal models, the experimental approaches used parallel those described above for studying the effects of overnutrition. In human subjects, data are available from both observational and experimental studies, and the latter have included both nonrandomized and randomized designs. Emphasis is placed on the last of these research designs.

Studies in Experimental Animals

It is well-known that various dietary regimens (including chronically or acutely restricted food intake, low protein or carbohydrate intake, and poor protein quality) compromise lactational performance (milk volume and composition, pup growth) in rats (15, 19, 22, 44–46, 50). Of particular interest are the conditions under which these negative effects occur and to what extent they can be reversed by improving dietary intake during lactation. In this section, data are selected to focus on these two issues because they are the key to understanding the data from human subjects. Emphasis is placed on animal models of chronic food restriction because they are more applicable to the nutritional stress encountered by lactating women living under poor circumstances than are models of particular nutrient deficits. In this context, chronic food restriction refers to animals fed limited amounts of a nutritious diet beginning before conception and continuing until peak lactation (day 14). Acute food restriction refers to animals fed limited amounts of a nutritious diet for shorter periods (usually 7–14 days during lactation).

The extent of compromise in lactational performance depends on the degree and duration of food restriction. Litters of rats subjected to acute (15, 21) or chronic (21, 43, 50, 53) food restriction ingest less milk. This occurs at chronic maternal dietary intakes of 70% or less of ad libitum intake or at acute dietary intakes of 60% or less of ad libitum intake. Chronic dietary restriction to 70% (43) or 50% (21) of ad libitum intake affected milk composition: protein and fat concentrations rose and lactose concentration fell; caloric density increased. No reduction in pup nutrient intake (nitrogen or energy) was observed when dams were chronically restricted to 70% of ad libitum intake (43), but at 50% of ad libitum intake, pup caloric intake was only 32% of control values (21). Pup growth was reduced at chronic maternal dietary intakes of 75% or less of ad libitum intake (15, 43, 53).

The effects of acute food restriction have also been modelled in baboons (33). Infant milk intake was not significantly reduced in animals fed 80% of ad libitum intake between 2 and 10 weeks postpartum but was 37% lower in those fed 60% of ad libitum intake. Milk composition was not affected by either level of dietary restriction. Infants of baboons in both food-restricted groups grew less well than those of baboons in the control group. Baboons in the 60% group lost four times as much weight during the experimental period as the other groups. To protect milk production, animals compensated for their reduced intake by increasing the efficiency of energy utilization by 17–25%, principally by reducing energy expenditure.

The effect of giving food supplements to undernourished lactating women has been modelled in rats by permitting chronically food-restricted dams to eat ad libitum from delivery until peak lactation (21, 32, 50). Milk intake by the litters of these dams increased dramatically compared to that of litters of rats whose food intake remained restricted and equaled that of litters of ad libitum-fed animals (21). Milk composition values for the refed animals were intermediate between those of chronically food-restricted rats and ad libitum-fed controls. As a result, the total caloric intake of the pups of the refed dams was the highest of these dietary treatment groups. Although the growth rate of the pups of the refed dams initially lagged that of control pups, by peak lactation their growth rate was the same or higher. Thus, provision of adequate dietary intake during lactation reversed the negative effects of prior chronic undernutrition on lactational performance.

The effect of refeeding on maternal nutritional status was also dramatic, but less complete. Refed dams gained weight during lactation and increased their proportion of body fat compared to that of chronically food-restricted rats (20); however, the refed rats remained significantly smaller, lighter, and leaner than control animals. Food consumption by the refed dams was similar to that of ad libitum-fed controls (49). A dietary intake similar to that of the controls represents a large food supplement because the refed rats weighed only 56% as much as the ad libitum–fed rats at parturition.

Important to an overall interpretation of the studies from rats is an understanding of the condition of the pups at birth. Chronically food-restricted dams deliver fewer and smaller pups than ad libitum–fed controls (2, 20, 31).

In addition, food restriction during pregnancy reduces the weight and compromises the development of the mammary gland (42). Thus, at parturition both maternal lactational capacity and the suckling stimulus of the pups is lower in chronically undernourished rats than in controls.

These studies in animal models show that maternal nutritional status at delivery as well as dietary intake during lactation influence lactational performance separately and jointly. Furthermore, these data show that the effects of prior maternal food restriction on milk and nutrient intake by the nursing litter and on litter growth are reversible with ad libitum feeding. These results from animal models are likely to be more dramatic than those that could be obtained from human subjects because: (a) some of the animals have been subjected to more severe undernutrition than is common among lactating women and, thus, the animals have more potential to respond to improvements in their dietary intake, and (b) their response is not constrained by the confounding factors present in studies of lactating women (e.g. direct supplementation of the nursing infant, sharing of the supplement with other family members, or changes in labor patterns).

Studies in Women

Studies of the effects of general undernutrition on women's lactational performance were reviewed recently (18). Evidence from observational studies is mixed. Although women living under poor circumstances are reported to eat less than women with better living conditions, they don't necessarily produce less milk (29). However, among such women, seasonal food shortages are associated with decreases in infant milk intake (27).

Data are available from two types of food supplementation trials: those with or without a randomized design. For reasons related to the study design and methods employed, the data from the three nonrandomized interventions that have provided food supplements to free-living women in rural areas of Mexico (5), the Gambia (28, 30), and India (13) are difficult to interpret (18).

Lessons learned from these early experimental studies include the importance of: (a) studying women who, because of their own poor nutritional status or inadequate current dietary intake, are likely to benefit from the supplement by improving their own health or nutritional status or by improving the nutrient intake or growth of their infants; (b) making this assessment at the time of peak lactational stress; (c) excluding the effect of concurrent supplementation of the infant (which is likely to decrease the infant's demand for milk and, thereby, milk production); and (d) assessing both milk volume and composition because maternal supplementation may change either or both factors and is likely to act through their product, the infant's total nutrient intake, to improve infant growth.

More recently, two randomized studies of the effect of food supplementation on lactational performance have been completed, one in Burma (24) and the other in Guatemala (14). Both experiments have designs that are suitable for causal inference and report that milk intake increased among infants of the women who received the food supplements. However, some aspects of both studies remain unexplained.

The study in Burma (24) included 21 women, 1–4 months postpartum, who were selected because their weight-for-height was <80% of an international standard. The supplement, a curry of animal protein cooked in oil, was home-delivered to the experimental group twice daily for 14 days. Supplemented women increased their dietary intakes, which were not low initially (2425 kcal per day), by a net of 900 kcal and 39 g of protein.

Infant milk intake was unchanged among control women, who received no treatment at all, but increased by 102 ml per day among supplemented women. No effect of the supplement on the protein concentration of breast milk was noted. Despite their increase in milk intake, infants of the supplemented mothers did not gain significantly more weight during the experimental period than infants of control mothers, which the authors attributed to the short duration of the experiment. The supplement also increased maternal body fat content as assessed from the sum of skinfold thicknesses at four sites.

This study is important because it shows that both maternal nutritional status and infant milk intake can be improved simultaneously. However, various aspects of this brief report are puzzling. One is how such a large net increase in dietary intake was achieved by women whose home dietary intakes were as high as these. Also puzzling is why women with these reported home dietary intakes were nonetheless so underweight. The usual activity pattern of these women was not discussed by the authors.

The study in Guatemala has only been reported as an abstract (14). It included 111 women who were selected because they had a low calf-circumference value during the last trimester of pregnancy. The subjects received either a low-energy (140 kcal per day) or a high-energy (500 kcal per day) supplement in the form of cookies delivered to their homes on weekdays for 20 weeks (from week 5 to week 25 of lactation).

In this study, infant milk intake was 47 g per day higher in the high-energy than the low-energy group at week 20 of lactation. A higher proportion of infants in the high-energy group was exclusively breast-fed at weeks 10 and 20 of lactation. This study is the first to document a causal relationship between maternal food supplementation and breast-feeding pattern. However, a complete understanding of this experiment must await full publication of its results.

In interpreting this study, one must remember that women in the low-energy group received a supplement. Therefore, they could have improved their lactational performance as well as their own nutritional status during the intervention period compared to unsupplemented women in the community. Because such women were not studied by González-Cossío and her colleagues (14), the absolute degree to which the low-energy group may have benefited will remain unknown. However, it is likely that the difference between the two treated groups underestimates the full impact of food supplementation in this population.

A crude comparison across studies of the effect of food supplementation on infant milk intake can be constructed from the available data. In Burma (24), daily milk intake increased by about 12 g per 100 additional kcal. In Guatemala (14), this increase was very similar: 13 g per 100 kcal from supplement in the high-energy groups compared to the low-energy group.

The data from the study in Burma (24) reveal benefits of supplementation for the mother: the increase in maternal weight did not reach statistical significance, but sum of skinfold thickness did. Results from the nonrandomized study conducted in The Gambia are congruent: supplemented women gained weight, were less likely to report being ill (28), and were said to perform more agricultural labor (6).

The positive effects of food supplementation on infant milk intake found in the two randomized intervention trials support a causal link of maternal dietary intake → infant milk intake (Figure 1). These positive effects contrast sharply with mixed effects found in the nonrandomized studies. This suggests that confounding factors may have compromised the ability of the nonrandomized studies to demonstrate an impact of food supplementation on infant milk intake. Finding effects is likely to be conditional on conducting impact evaluations in populations with the potential to benefit from supplementation. Finding effects also is likely to depend on selecting an amount and type of supplement that both addresses the underlying nutritional deficits in the lactating women studied and will be accepted by them.

To evaluate the link of infant milk/nutrient intake → infant growth in this causal sequence requires demonstrable improvements in the infant's total nutrient intake (Figure 1). There is every reason to expect that infants whose nutrient intake increases will grow more rapidly (12). The data now available show that any improvements in infant growth that result from maternal food supplementation during lactation are too small or too variable to be detected with the number of subjects studied. This may be because improvements in infant nutrient intake have been too small to produce discernibly better growth during the relatively short intervals in which growth has been evaluated.

It is necessary to establish association, time order, and direction for a full evaluation of a possible causal sequence. Direction is especially problematical for the link infant milk/nutrient intake → infant growth. It is likely that the effect of improving maternal dietary intake operates *first* in this direction.

However, because milk production responds to the suckling stimulus (10), once milk production begins to improve and the baby grows more rapidly, the infant's demand will likely cause *further* increases in milk production as long as maternal lactational capacity is not limiting for milk production. This direction of action cannot be excluded from the available data and deserves further study.

As illustrated in Figure 1, the availability of adequate nutrients for milk biosynthesis is a necessary but not sufficient condition for increasing milk production: milk production will increase only with adequate infant demand. Similarly, adequate infant demand is a necessary but not sufficient condition for increasing milk production: milk production will increase only with adequate lactational capacity. Thus, for a valid test of the effect of improving maternal dietary intake on infant milk intake and growth, factors constraining lactational capacity must be removed *and* the infant must be capable of growing better in response to increased milk intake. The maximum benefit to the infant of maternal food supplementation will be achieved when the infant's response to increased milk intake elicits further improvements in milk intake. Factors that constrain this improvement in milk intake (e.g. a premature infant with low suckling vigor or fewer daily breast-feedings because better-fed mothers are away from home for longer hours to perform more agricultural work) reduce the possibility that improving maternal dietary intake will improve infant growth directly via increased milk intake.

CONCLUSIONS

Although cafeteria feeding may represent an accurate animal model for producing obesity in human subjects, the negative effects of long- or short-term use of this diet on lactational performance may relate to its low protein-energy ratio. Thus, other animal models of obesity should be investigated. The proportion of women of childbearing age in the United States and other developed countries who are overweight or obese indicates that the effect of maternal overnutrition on lactational performance deserves more systematic study.

In experimental animals, chronic undernutrition has deleterious effects on milk and nutrient intake of the young as well as on their growth. Improving previously poor maternal dietary intake during lactation increases milk intake by the young and corrects their growth deficit; it also improves maternal nutritional status, but only incompletely.

The newer data from randomized intervention trials among undernourished women show that improving maternal dietary intake increases infant milk intake. These data are less persuasive in demonstrating a direct causal effect of maternal food supplementation on infant growth. The finding from one study that food supplementation increased the proportion of exclusively breast-fed infants has important, positive implications for infant health and justifies further experimental studies on this subject.

ACKNOWLEDGMENTS

The thoughtful comments on drafts of this paper by Marie Drake, Cutberto Garza, Jean-Pierre Habicht, Rebecca Kliewer, Michelle McGuire, Grace Marquis, Kerry Schulze, Rebecca Stoltzfus, and Anna Winkvist are gratefully acknowledged.

Literature Cited

1. Allen J. C., Keller, R. P., Archer, P., Neville, M.C. 1991. Studies in human lactation: milk composition and daily secretion rats of macronutrients in the first year of lactation. *Am. J. Clin. Nutr.* 54:69–80
2. Brigham, H. E. 1987. *The effect of food restriction during the reproductive cycle on organ weight and organ blood flow in the rat.* MNS thesis. Cornell Univ., Ithaca, New York. 107 pp.
3. Butte, N. F., Garza, C. 1986. Anthropometry in the appraisal of lactation performance among well-nourished women. In *Human Lactation 2: Maternal and Environmental Factors,* ed. M. Hamosh, A. S. Goldman, pp. 61–67. New York: Plenum. 657 pp.
4. Butte, N. F., Garza, C., Stuff, J. E., Smith, E. O'B., Nichols, B. L. 1984. Effect of maternal diet and body composition on lactational performance. *Am. J. Clin. Nutr.* 39:296–306
5. Chávez, A., Martínez, C. 1980. Effects of maternal undernutrition and dietary supplementation on milk production. In *Maternal Nutrition During Pregnancy and Lactation,* ed. H. Aebi, R. G. Whitehead, pp. 274–84. Bern: Hans Huber. 354 pp.
6. Coward, W. A., Paul, A. A., Prentice, A. M. 1984. The impact of malnutrition on human lactation: observations from community studies. *Fed. Proc.* 43:2432–37
7. Department of Health and Human Services. 1991. *Healthy People 2000, DHHS Publ. No. (PHS) 91-50212.* Washington, DC: US Gov. Print. Off. 692 pp.
8. Dewey, K. G., Heinig, J., Nommsen, L. A., Lönnerdal, B. 1991. Maternal versus infant factors related to breast milk intake and residual milk volume. The DARLING Study. *Pediatrics* 87:829–37

9. Dewey, K. G., Lönnerdal, B. 1983. Milk and nutrient intake of breast-fed infants from 1 to 6 months: relation to growth and fatness. *J. Pediatr. Gastroenterol. Nutr.* 3:497–506
10. Dewey, K. G., Lönnerdal, B. 1986. Infant self-regulation of breast milk intake. *Acta Pædiatr. Scand.* 75:893–98
11. Fleming, A. S., Miceli, M. 1983. Effects of diet on feeding and body weight regulation during pregnancy and lactation in the golden hamster *(Mesocricetus auratus). Behav. Neurosci.* 97:246–54
12. Forsum, E., Sadurskis, A. 1986. Growth, body composition and breast milk intake of Swedish infants during early life. *Early Hum. Dev.* 14:121–29
13. Girija, A., Geervani, P., Rao, G. N. 1984. Influence of dietary supplementation during lactation on lactational performance. *J. Trop. Pediatr.* 30:140–44
14. González-Cossío, T., Habicht, J-P., Delgado, H., Rasmussen, K. M. 1991. Food supplementation during lactation increases infant milk intake and the proportion of exclusive breastfeeding. *FASEB J.* 5:A917
15. Grigor, M. R., Allan, J. E., Carrington, J. M., Carne, A., Geursen, A., et al. 1987. Effect of dietary protein and food restriction on milk production and composition, maternal tissues and enzymes in lactating rats. *J. Nutr.* 117:1247–58
16. Hytten, F. E. 1954. Clinical and chemical studies in human lactation. VI. The functional capacity of the breast. *Br. Med. J.* 1:912–15
17. Hytten, F. E. 1954. Clinical and chemical studies in human lactation. VIII. Relationship of the age, physique, and nutritional status of the mother to the yield and composition of her milk. *Br. Med. J.* 2:844–45
18. Institute of Medicine. 1991. *Nutrition During Lactation.* Rep. Subcommittee

Nutr. During Lactation, Comm. Nutr. Status During Pregnancy and Lactation, Food and Nutr. Board. Washington, DC: Natl. Acad. Press. 309 pp.

19. Jansen, G. R., Grayson, C., Hunsaker, H. 1987. Wheat gluten during pregnancy and lactation: effects on mammary gland development and pup viability. *Am. J. Clin. Nutr.* 46:250–57

20. Kliewer, R. L. 1986. *The effects of malnutrition during the reproductive cycle on galactopoietic hormone values and lactational performance in the rat.* MS thesis. Cornell Univ., Ithaca, New York. 119 pp.

21. Kliewer, R. L., Rasmussen, K. M. 1987. Malnutrition during the reproductive cycle: effects on galactopoietic hormones and lactational performance in the rat. *Am. J. Clin. Nutr.* 46:926–35

22. Koski, K. G., Hill, F. W., Lönnerdal, B. 1990. Altered lactational performance in rats fed low carbohydrate diets and its effect on growth of neonatal rat pups. *J. Nutr.* 120:1028–36

23. Lunn, P. G. 1985. Maternal nutrition and lactational infertility: the baby in the driving seat. In *Maternal Nutrition and Lactational Infertility*, ed. J. Dobbing, pp. 41–53. New York: Raven. 149 pp.

24. Naing, K-M., Oo, T-T. 1987. Effect of dietary supplementation on lactation performance of undernourished Burmese mothers. *Food Nutr. Bull.* 9:59–61

25. Naing, K-M., Oo, T-T., Thein, K., Hlaing, N-N. 1980. Study on lactation performance of Burmese mothers. *Am. J. Clin. Nutr.* 33:2665–68

26. Nommsen, L. A., Lovelady, C. A., Heinig, M. J., Lönnerdal, B., Dewey, K. G. 1991. Determinants of energy, protein, lipid, and lactose concentrations in human milk during the first 12 mo of lactation: The DARLING Study. *Am. J. Clin. Nutr.* 53:457–65

27. Prentice, A. M. 1980. Variations in maternal dietary intake, birthweight and breast-milk output in the Gambia. See Ref. 5, pp. 167–83

28. Prentice, A. M., Lunn, P. G., Watkinson, M., Whitehead, R. G. 1983. Dietary supplementation of lactating Gambian women. II. Effect on maternal health, nutrition status and biochemistry. *Hum. Nutr: Clin. Nutr.* 37C:65–74

29. Prentice, A. M., Paul, A., Prentice, A., Black, A., Cole, T., et al. 1986. Cross-cultural differences in lactational performance. See Ref. 3, pp. 13–44

30. Prentice, A. M., Roberts, S. B., Prentice, A., Paul, A. A., Watkinson, M., et

al. 1983. Dietary supplementation of lactating Gambian women. I. Effect on breast-milk volume and quality. *Hum. Nutr: Clin. Nutr.* 37C:53–64

31. Rasmussen, K. M., Fischbeck, K. L. 1987. Effect of repeated reproductive cycles on pregnancy outcome in ad libitum-fed and chronically food-restricted rats. *J. Nutr.* 117:1959–66

32. Rasmussen, K. M., Warman, N. L. 1983. Effect of maternal malnutrition during the reproductive cycle on growth and nutritional status of suckling rat pups. *Am. J. Clin. Nutr.* 38:77–83

33. Roberts, S. B., Cole, T. J., Coward, W. A. 1985. Lactational performance in relation to energy intake in the baboon. *Am. J. Clin. Nutr.* 41:1270–76

34. Roberts, S. B., Coward, W. A. 1985. Dietary supplementation increases milk output in the rat. *Br. J. Nutr.* 53:1–9

35. Rolls, B. A., Barley, J. B., Gurr, M. I. 1983. The influence of dietary obesity on milk production in the rat. *Proc. Nutr. Soc.* 42:83A

36. Rolls, B. A., Edwards-Webb, J. D., Gurr, M. I. 1981. The influence of dietary obesity on milk composition in the rat. *Proc. Nutr. Soc.* 40:60A

37. Rolls, B. A., Gurr, M. I., van Duijvenvoorde, P. M., Rolls, B. J., Rowe, E. A. 1986. Lactation in lean and obese rats: Effect of cafeteria feeding and of dietary obesity on milk composition. *Physiol. Behav.* 38:185–90

38. Rolls, B. J., Rowe, E. A. 1982. Pregnancy and lactation in the obese rat: Effects on maternal and pup weights. *Physiol. Behav.* 28:393–400

39. Rolls, B. J., Rowe, E. A. 1981. Effects of obesity on maintenance of body weight in the lactating rat. *Proc. Nutr. Soc.* 40:61A

40. Rolls, B. J., Rowe, E. A., Fahrbach, S. E., Agius, L., Williamson, D. H. 1980. Obesity and high energy diets reduce survival and growth rates of rat pups. *Proc. Nutr. Soc.* 39:51A

41. Rolls, B. J., van Duijvenvoorde, P. M., Rowe, E. A. 1984. Effects of diet and obesity on body weight regulation during pregnancy and lactation in the rat. *Physiol. Behav.* 32:161–68

42. Rosso, P., Keyou, G. E., Bassi, J. A., Slusser, W. M. 1981. Effect of malnutrition during pregnancy on the development of the mammary glands of rats. *J. Nutr.* 111:1937–41

43. Sadurskis, A., Sohlström, A., Kabir, N., Forsum, E. 1991. Energy restriction and the partitioning of energy among the costs of reproduction in rats in relation to

growth of the progeny. *J. Nutr.* 121:1798–1810

44. Sampson, D. A., Hunsaker, H. A., Jansen, G. R. 1986. Dietary protein quality and food intake: Effects on lactation and on protein synthesis and tissue composition in mammary tissue and liver in rats. *J. Nutr.* 116:365–75

45. Sampson, D. A., Jansen, G. R. 1983. The effect of dietary protein quality and feeding level on milk secretion and mammary protein synthesis in the rat. *J. Pediatr. Gastroenterol. Nutr.* 4:274–83

46. Sampson, D. A., Jansen, G. R. 1984. Protein and energy nutriton during lactation. *Annu. Rev. Nutr.* 4:43–67

47. Strode, M. A., Dewey, K. G., Lönnerdal, B. 1986. Effects of short-term caloric restriction on lactational performance of well-nourished women. *Acta Pædiatr. Scand.* 75:222–29

48. van Duijvenvoorde, P. M., Rolls, B. J. 1985. Body fat regulation during pregnancy and lactation: the roles of diet and insulin. *Biochem. Soc. Trans.* 13:824–25

49. Warman, N. L. 1983. *The effects of malnutrition during the reproductive cycle on the nutritional status of rat dams and their pups.* MNS thesis. Cornell Univ., Ithaca, New York. 98 pp.

50. Warman, N. L., Rasmussen, K. M. 1983. Effects of malnutrition during the reproductive cycle on nutritional status and lactational performance of rat dams. *Nutr. Res.* 3:527–45

51. Wehmer, F., Bertino, M., Jen, K-L. C. 1979. The effects of high fat diet on reproduction in female rats. *Behav. Neural Biol.* 27:120–24

52. Williamson, D. F., Kahn, H.S., Remington, P. L., Anda, R. F. 1990. The 10-year incidence of overweight and major weight gain in US adults. *Arch. Intern. Med.* 150:665–72

53. Young, C. M., Rasmussen, K. M. 1985. Effects of varying degrees of chronic dietary restriction in rat dams on reproductive and lactational performance and body composition in dams and their pups. *Am. J. Clin. Nutr.* 41:979–87

Annu. Rev. Nutr. 1992. 12:119 37

DIETARY IMPACT OF FOOD PROCESSING[1]

Mendel Friedman

Food Safety Research Unit, USDA-ARS Western Regional Research Center, 800 Buchanan Street, Albany, California 94710

KEY WORDS: food browning, food processing, browning prevention, food safety, nutrition

---

CONTENTS

INTRODUCTION

A variety of processing methods are used to make foods edible, to permit storage, to alter texture and flavor, to sterilize and pasteurize food, and to destroy toxic microorganisms. These methods include baking, cooking, freezing, frying, and roasting. Many such efforts have both beneficial and harmful

---

[1]The US government has the right to retain a nonexclusive, royalty-free license in and to any copyright covering this paper.

effects. It is a paradox of nature that the processing of foods and feeds can improve nutrition, quality, safety, and taste, and yet occasionally can lead to the formation of antinutritional and toxic compounds. These multifaceted consequences of food processing arise from molecular interactions among nutrients and with other food ingredients. Billions of new compounds can, in principle, be formed from such interactions among the approximately 60 known nutrients.

Beneficial and adverse effects of food processing are of increasing importance to food science, nutrition, and human health. A better understanding of the molecular changes during food processing and the resulting nutritional and safety consequences is needed to optimize beneficial effects such as bioavailability, food quality, and food safety, and to minimize the formation and facilitate the inactivation of deleterious compounds. Such an understanding will encompass multidisciplinary studies of the chemistry, biochemistry, nutrition, and toxicology of food ingredients. This limited review uses examples largely based on our own studies to illustrate general concepts. It describes the nutritional impact of two major food processing conditions: pH and heat. The references cited offer the reader an entry into the comprehensive, but widely scattered, relevant literature (1–107).

EFFECT OF pH

Racemization of Amino Acids

Since the early part of this century, alkali and heat treatments have been known to racemize amino acids (9). As a result of food processing using these treatments, D-amino acids are continuously consumed by animals and man. Because all of the amino acid residues in a protein undergo racemization simultaneously, but at differing rates, assessment of the extent of racemization in a food protein requires quantitative measurement of at least 36 optical isomers, 18 L and 18 D. Analytically, this is a difficult problem not yet solved (48, 54, 55, 74, 75, 77).

Racemization of an amino acid proceeds by removal of a proton from the α-carbon atom to form a carbanion intermediate. The trigonal carbon atom of the carbanion, having lost the original asymmetry of the α-carbon, recombines with a proton from the environment to regenerate a tetrahedral structure. The reaction is written as

$$\text{L-amino acid} \underset{k'_{rac}}{\overset{k_{rac}}{\rightleftharpoons}} \text{D-amino acid} \qquad 1.$$

where k_{rac} and k'_{rac} are the first-order rate constants for the forward and reverse racemization of the stereoisomers.

The product is racemic if recombination can take place equally well on either side of the carbanion, giving an equimolar mixture of L- and D-isomers. Recombination may be biased if the molecule has more than one asymmetric center, resulting in an equilibrium mixture slightly different from a 1:1 enantiomeric ratio.

Because the structural and electronic factors that facilitate the formation and stabilization of the carbanion intermediate are unique for each amino acid, it follows that the reaction rate for the isomerization of each amino acid is also unique. Thus, the inductive strengths of the R-substituents have been invoked to explain differing racemization rates in the various amino acids. Plotting racemization for individual amino acids in casein and soybean proteins against the inductive parameters clearly demonstrates strong correlations (74, 75, 77).

Two pathways are available for the biological utilization of D-amino acids: (a) racemases or epimerases may convert D-amino acids directly to L-isomers or to (DL) mixtures; or (b) D-amino-acid oxidases may catalyze oxidative deamination of the α-amino group to form α-keto acids, which can then be specifically reaminated to the L-form. Although both pathways may operate in microorganisms, only the second has been demonstrated in mammals.

The amounts and specificities of D-amino acid oxidase are known to vary in different animal species. In some, the oxidase system may be rate limiting in the utilization of a D-amino acid as a source of the L-isomer. In this case, the kinetics of transamination of D-enantiomers would be too slow to support optimal growth. In addition, growth depression could result from nutritionally antagonistic or toxic manifestations of D-enantiomers exerting a metabolic burden on the organism.

The nutritional utilization of different D-amino acids varies widely, both in animals and humans (3, 42–44, 47, 48, 50, 54, 55, 77). In addition, some D-amino acids may be deleterious. For example, although D-phenylalanine is nutritionally available as a source of L-phenylalanine, our studies have shown that high concentrations of D-tyrosine inhibit the growth of mice (43). The antimetabolic effect of D-tyrosine can be minimized by increasing the L-phenylalanine content (protein bound, or free) of the diet. Similarly, L-cysteine has a sparing effect on L-methionine when fed to mice (44); however, D-cysteine does not. The wide variation in the utilization of D-amino acids is exemplified by the fact that D-lysine is not utilized as a source of the L-isomer for growth. The utilization of methionine is dose dependent, reaching 76% of the value obtained with L-methionine. Both D-serine and the mixture of L-L and L-D isomers of lysinoalanine induce histological changes in the rat kidneys. D-tyrosine, D-serine, and lysinoalanine are produced in significant amounts under the influence of even short periods of alkaline treatment.

Unresolved is whether the biological effects of D-amino acids vary depending on whether they are consumed in the free state or as part of a food protein.

Indications are that L-D, D-L, and D-D peptide bonds in food proteins may not hydrolyze as readily as naturally occurring L-L peptide bonds (50, 58, 82). Possible metabolic interaction, antagonism, or synergism among D-amino acids in vivo also merits further study. The described results with mice complement related studies with other species and contribute to the understanding of nutritional and toxicological consequences of ingesting D-amino acids. Such an understanding will make it possible to devise food processing conditions to minimize or prevent the formation of undesirable D-amino acids in food proteins and to prepare better and safer foods.

Lysinoalanine Cross-links

Lysinoalanine   [HOOCCH(NH$_2$)CH$_2$CH$_2$CH$_2$CH$_2$NHCH$_2$CH(NH$_2$)COOH] (LAL), is an unnatural amino acid that has been identified in hydrolyzates of processed food proteins, in particular those subjected to alkali (16, 25, 53, 58, 75).

Detailed studies revealed that base-catalyzed synthesis of lysinoalanine proceeds by the addition of the ϵ-NH$_2$ group of lysine to the double bond of a dehydroalanine residue. This residue is derived from cysteine and/or serine. From a nutritional standpoint, lysinoalanine formation results in a decrease of the essential amino lysine, and the semiessential amino acid cystine, as well as in a decrease in digestibility of the modified protein (51, 53, 58).

In rats, studies have found histological changes in the kidneys related to dietary exposure to this substance, either isolated or as part of intact proteins (104). The lesions are located in the epithelial cells of the straight portion of the proximal renal tubules and are characterized by enlargement of the nucleus and cytoplasm, increased nucleoprotein content, and disturbances in DNA synthesis and mitosis.

Because of these observations, concern has arisen about the safety of foods that contain LAL and related dehydroalanine-derived amino acids known to produce similar lesions. However, since the mechanism by which these compounds damage the rat kidney is unknown, it is difficult to assess the risk to human health caused by their presence in the diet.

LAL has two asymmetric carbon atoms, making possible four separate diastereoisomeric forms: LL, LD, DL, and DD. Its structure suggests that it should have excellent chelating potential for metal ions, a property that may be relevant to its toxic action. Accordingly, we have examined LAL for its affinity towards a series of metal ions, of which copper (II) was chelated the most strongly (40b, 57, 87). On this basis, we have suggested a possible mechanism for kidney damage in the rat involving LAL's interaction with copper within the epithelial cells of the proximal tubules.

Of the four isomers of LAL, LL and LD are derived from L-lysine and the other two from D-lysine. Since L-lysine is the natural amino acid present in proteins, most of the LAL formed during food processing can be expected to be a mixture of LL and LD. However, since exposure of food proteins to heat and alkali may racemize a small fraction of L-lysine to the D-isomer, treated food proteins may also contain small amounts of DL- and DD-LAL.

As described in detail elsewhere (57, 87), it is possible to predict the equilibria in vivo between histidine, the major low-molecular weight copper carrier in plasma, and competing chelating agents such as LAL (Equation 2).

$$CuHis_2 + H_2LAL \rightleftharpoons CuLAL + 2HHis \qquad\qquad 2.$$

A mathematical analysis of the equilibrium shown in Equation 2 was used to calculate LAL plasma levels needed to displace histidine as the major copper carrier in vivo. The calculated values are 27 μM for LD-LAL, 100 μM for LL-LAL, and 49 μM for the mixture of the two.

The above considerations suggest that LD-LAL would be a better competitor for copper (II) in vivo than the LL-isomer, i.e. it should take about one fourth as much LD-LAL as LL-LAL to displace the same amount of histidine from copper-histidine. This difference could explain the greater observed toxicity of the LD-LAL. The apparent direct relationship between the observed affinities of the two LAL isomers for copper (II) ions in vitro and their relative toxic manifestation in the rat kidney is consistent with our hypothesis that LAL exerts its biological effect through chelation of copper in body fluids and tissues. Limited studies on the binding of LL- and LD-lysinoalanines to cobalt (II), zinc (II), and other metal ions imply that lysinoalanine could also influence cobalt utilization in vivo. Animal studies are needed to confirm the predicted role of lysinoalanine in metal ion transport, utilization, and histopathology (13a, 19, 21, 81, 89a).

EFFECT OF HEAT

Maillard Browning

Maillard-type reactions of primary amino groups with reducing sugars, and other nonenzymatic browning reactions with nonreducing carbohydrates, cause deterioration of food during storage and commercial or domestic food processing. The loss in nutritional quality, and potentially in safety, is attributed to some or all of the following factors: (a) destruction of essential amino acids, (b) decrease in digestibility, and (c) production of antinutritional and toxic compounds (1a, 1b, 8, 13a, 14, 17, 18, 22–29, 34, 35, 38–40, 52a, 60, 61, 70–73, 76, 78–82, 88, 90, 98, 102, 106, 107).

Although extensive efforts have been made to elucidate the chemistry of both desirable and undesirable compositional changes during browning, parallel studies on the nutritional and toxicological consequences of browning are limited. This is understandable since, in principle, each combination of a specific amino acid or protein with a particular carbohydrate needs to be investigated to understand the scope of the problem. Reported studies in this area include (a) influence of damage to essential amino acids, especially lysine, on nutritional quality; (b) effects of fortifying browning products with essential amino acids on recovery of nutritional quality; (c) nutritional damage as a function of processing conditions; (d) biological utilization of character-ized browning compounds, such as ϵ-N-deoxy-fructosyl-L-lysine; and (e) formation of mutagenic and clastogenic products.

A number of investigators have examined the effects of the Maillard browning reaction on digestibility and nutritional quality (14, 17, 18, 61–64, 73, 82, 83, 85). Our studies (61) show that loss of nutritional quality of heat-treated casein is related to decreased nitrogen digestibility rather than to simple destruction of essential amino acids. The influence of glucose and starch was minimal compared to observed effects of heat on casein alone under the conditions used. Glucose and perhaps starch augment protein degradation and loss of nutritional quality under moderate, dry-heat con-ditions. Further studies are needed to explain the molecular basis for the extent and nature of the heat-induced destruction of essential amino acids and the formation of undigestible browned and cross-linked products. These changes impair intestinal absorption and nutritional quality in general. Toxic compounds formed under these conditions might also modulate nutritional quality. Thus, such studies should differentiate antinutritional and toxicologi-cal interrelationships and develop means for preventing or minimizing the formation of deleterious compounds in foods.

Ascorbate Browning

When a nutritionally complete, low-protein basal diet containing 10% casein was supplemented with 20% protein from unheated casein, wheat gluten, or soybean, test mice exhibited a significantly increased weight gain (52, 52a, 107). In contrast, weight gain was markedly reduced when the supplement was soy protein or gluten heated at 200° or 215°C for 72 min in the dry state (simulated crust baking). Baked casein was nonnutritive. Adding carbohy-drates to gluten during heating prevented subsequent growth inhibition. After heating with sodium ascorbate (but not L-ascorbic acid), soy protein (at 200°C) and gluten (at 215°C) completely prevented growth when added to the basal diet. Growth inhibition was also aggravated by a heated casein-ascorbate mixture, but less than with the other proteins. The extent of nutritive damage increased sharply with heating temperature in the range 180 to 215°C, and with sodium ascorbate concentration in the range 1 to 20%.

In a related study, Oste & Friedman (85) showed that sodium ascorbate heated with amino acids, especially tryptophan, results in the formation of antinutritional compounds.

The reduced weight gain of mice fed a nutritionally adequate diet supplemented with these materials suggests that heating induces the formation of nutritionally antagonistic or toxic compounds that interfere with essential metabolic pathways such as digestion, transport, absorption, and utilization of nutrients. Further studies of the chemical basis of these effects may be more conveniently performed with tryptophan or other amino acid/ascorbate mixtures than with the more complex protein/ascorbate blends, since the heat-induced products may be easier to isolate and characterize.

Our heating experiments used proportionately much more sodium ascorbate than is used in thermal food processing to improve bread-dough characteristics such as loaf volume and bread texture, and to inhibit nitrosamine formation in bacon. However, since our findings do not rule out possible cumulative biological effects, additional studies are needed to determine whether consumption of low levels of the heat-derived compounds can be a human health hazard.

Our results suggest that deleterious material formed during heating of gluten or soy protein, and to a lesser extent casein, may represent the degradation of protein to nitrogenous materials without nutritional value. At the nominal protein level fed, such materials would represent a severe metabolic burden (toxic effect) to the animal, which must then eliminate them. The protective effect of carbohydrates in diminishing the formation of toxic gluten is interpreted as a thermochemical volatilization of deleterious products, while sodium ascorbate appears to reduce vaporization.

These considerations suggest the need (a) to characterize the compound(s) in heated protein- and amino acid-sodium ascorbate mixtures that may be responsible for the observed growth inhibition; (b) to determine the safety of the pure compounds in laboratory animals and measure their prevalence in commercial foods in order to define possible human risk; (c) to carry out studies with related food ingredients such as sodium citrate, sodium gluconate, and sodium glutamate in order to define the mechanism of this type of growth inhibition; and (d) to devise processing conditions to prevent the formation of the growth inhibitors in food (10, 13, 69).

Food Allergenicity

As noted earlier, carbohydrates interact with proteins to form Maillard browning products. Oste and colleagues (83, 84) studied the effects of these transformations on the antigenicity of the Kunitz soybean trypsin inhibitor (KTI) with two monoclonal antibodies. They report that solid mixtures of KTI and carbohydrates were heated in an oven at 120°C, dialyzed, freeze-dried, and analyzed by enzyme-linked immunosorbent assay (ELISA). Glucose, lactose, and maltose decreased the antigenicity of KTI to levels 60–80% lower than those observed in a control sample heated without carbohydrate. Starch was less effective than the three reducing sugars. The decrease was rapid, occurring within 10 min when glucose was heated with KTI, with retention of 60% of the chemically available lysine. Longer heating times increased browning and reduced the level of available lysine in KTI, without further reducing antigenicity. The results suggest that relatively mild conditions of heating food proteins with carbohydrates can reduce the antigenicity of the protein and possibly modify sites known to elicit allergenic responses.

The nature and extent of browning reactions, as well as the magnitude of antigenicity changes, are probably highly dependent on the experimental conditions. In addition, the relative importance of the Maillard reaction and the reactions of nonreducing carbohydrates merit further study. Nevertheless, the results of our studies lead us to hypothesize that the early stages of the Maillard reaction can significantly affect protein antigenicity. Note that these reactions can also introduce new antigenic determinants into a food protein (6, 11, 12).

The Schiff's base formed in the first step of the Maillard reaction is biologically available (14, 17). Therefore, it does not adversely affect protein nutrition. Many foods present a favorable medium for the Maillard reaction, including allergenic foods such as milk (67b). These results suggest that the antigenic sites of food proteins responsible for adverse allergenic responses, by eliciting production of IgE and possibly other isotypes that trigger allergic reactions, could be selectively altered by modification with reducing carbohydrates under mild conditions. Chemical and structural modification during food processing could at least partly account for the observation that cow's milk is less antigenic in vivo after heat treatment and for apparent differences in the allergenicities between liquid and powdered soybean infant formulas. Conditions might be developed to exploit such beneficial effects of nonenzymatic browning and related food-processing-induced changes.

Inactivation of Inhibitors of Digestive Enzymes

Adverse effects of ingestion of raw soybean meal have been attributed to the presence of soybean inhibitors of chymotrypsin and trypsin and to other factors. To improve nutritional quality and safety of soy products, inhibitors are partly inactivated by heat treatment during food processing. The approach used is exemplified by the following summary of a study by Friedman et al (37).

The content and heat stability of protease inhibitors of a standard cultivar (Williams 82) and an isoline (L81-4590) lacking the Kunitz trypsin inhibitor (KTI) were measured by using enzyme inhibition and enzyme-linked immunosorbent assays (ELISA). Steam heating of the isoline flour (121°C, 20 min) resulted in a near-zero level of trypsin inhibitory activity, while 20% remained in the Williams 82 sample. The raw soy flour prepared from the isoline was nutritionally superior to the raw flour prepared from the standard variety, as measured by protein efficiency ratio (PER) and pancreatic weights. The increased PER was likely due to the lower level of trypsin inhibitor activity in the isoline. Steam heating the flours for up to 30 min at 121°C progressively increased the PER for both strains. Less heat was needed to inactivate the inhibitors in the isoline than in the standard cultivar.

Related studies showed that treating raw soy flour with cysteine, N-acetyl-L-cysteine, or reduced glutathione introduces new half-cystine residues into native proteins, with a corresponding improvement of nutritional quality and safety (46). The proteins are modified through formation of mixed disulfide bonds among added thiols, protease inhibitors, and structural protein molecules. This leads to decreased inhibitory activity and increased protein digestibility and nutritive value (40a). The SH-containing amino acids also facilitate heat inactivation of hemagglutinins (lectins) in lima bean flour. Exposure of raw soy flour to sodium sulfite was also nutritionally beneficial (40a, 40b, 45, 46).

Naturally occurring enzyme inhibitors, such as Bowman-Birk inhibitor (BBI) in which every sixth amino acid residues is cystine, also have beneficial effects such as prevention of development of colon cancer in mice (97). Although the molecular basis for such beneficial effects needs to be ascertained, one possibility is that the inhibitors or inhibitor-protease complexes act as free radical traps, whereby the free electrons on damaging oxygen radicals are transferred or dissipated to the sulfur atoms of the sulfur-rich inhibitors or complexes. [For reviews of free radical chemistry of proteins, see Friedman (20) and Swallow (96).] These considerations suggest the need for further studies to enrich our knowledge about possible beneficial effects of plant protease inhibitors in relation to sulfur amino acids (60).

BIOAVAILABILITY OF AMINO ACIDS

Lysine and Derivatives

Wheat gluten, the major protein in many baking formulations, is considered a poor-quality protein, primarily because it has insufficient amounts of two essential amino acids: lysine, the first-limiting amino acid, and threonine, the second-limiting one. To compensate for the poor quality of most cereal proteins such as gluten, the minimum recommended daily allowance (RDA) for these proteins has been set at 65 g, compared to 45 g for good-quality proteins such as casein (63).

As noted by Ziderman & Friedman (106), during baking, the mixture of protein, carbohydrate, and water plus additives in dough is exposed to two distinct transformations. Desiccation of the surface on exposure to temperatures reaching 215°C produces the crust. The crust encloses part of the dough in steam phase at approximately 100°C, resulting in the formation of the crumb.

Because lysine's ϵ-amino group interacts with food constituents to make it nutritionally less available (94), the baking process further reduces the dietary availability and utilization of lysine, especially in the crust, which makes up about 40% of the bread by weight. Many such interactions have been described including (a) the reaction of the amino group with carbonyl groups of sugars and fatty acids to form Maillard browning products; (b) the formation of cross-linked amino acids such as lanthionine, lysinoalanine, and glutamyllysine; (c) the interaction with tannins and quinones; and (d) steric blocking of the action of digestive enzymes by newly introduced cross-links, as well as native ones such as disulfide bonds. Because these reactions of lysine with other dietary components may lead to protein damage and to the formation of physiologically active compounds, an important objective of food science and nutrition is to overcome these effects (1b, 21–36, 38, 39, 99, 102, 103).

In principle, it is possible to enhance the nutritional quality of bread by amino acid fortification (1b). A major problem encountered when free lysine is used to fortify foods is that the added amino acid can itself participate in browning and other side reactions. Because ϵ-acyl-lysine derivatives are less susceptible to Maillard reactions than is free lysine (14, 17) our objective was to compare lysine and glutamyllysine as nutritional sources of lysine for mice fed bread crust, crumb, and whole bread co-baked with these amino acids. Since the ϵ-NH$_2$ group of N$^\epsilon$(γ-L-glutamyl)-L-lysine (glutamyllysine) is blocked in the form of an isopeptide bond with the γ-COOH group of glutamic acid, expectations were that this peptide should also undergo less damage than lysine during baking.

To assess whether the glutamyllysine can serve as a nutritional source of lysine, we compared the growth of mice fed (a) an amino acid diet in which lysine was replaced by four dietary levels of glutamyllysine; (b) wheat gluten diets fortified with lysine; (c) a wheat bread–based diet (10% protein) supplemented before feeding with lysine or glutamyllysine, not co-baked; and (d) bread diets baked with these levels of lysine or glutamyllysine (40). For the amino acid diet, the relative growth response to glutamyllysine was about half that of lysine. The effect of added lysine on the nutritional improvement of wheat gluten depended on both lysine and gluten concentrations in the diet. With 10 and 15% gluten, 0.37% lysine hydrochloride produced markedly increased weight gain. Further increase in lysine hydrochloride to 0.75% proved somewhat detrimental to weight gain. Lysine hydrochloride addition improved growth at 20 to 25% gluten in the diet and did not prove detrimental at 0.75%. For whole bread, glutamyllysine served nearly as well as lysine to improve weight gain. The nutritive value of bread crust, fortified or not, was markedly less than that of crumb or whole bread. Other data showed that lysine or glutamyllysine at the highest level of fortification, 0.3%, improved the protein quality (PER) of crumb over that of either crust or whole bread, indicating a possible greater availability of the second-limiting amino acid, threonine, in crumb. These data and additional metabolic studies with [U-^{14}C]glutamyllysine suggest that glutamyllysine, cobaked or not, is metabolized in the kidneys and utilized in vivo as a source of lysine; it and related peptides merit further study as sources of lysine in low-lysine foods (105).

Amino acids are used both metabolically, as building blocks for protein biosynthesis, and catabolically, as energy sources. Catabolism for most amino acids proceeds through transamination pathways; the exceptions are lysine and threonine. These nutritionally limiting amino acids are catabolized by nonaminotransferase-specific enzymes: threonine dehydratase acts on threonine and lysine ketoglutarate reductase on lysine. The concentrations of these enzymes in the liver of rats are subject to adaptive responses that control the utilization of these two amino acids (63). Although both enzymes are induced by feeding diets high in protein, rats differ in the mechanism of the adaptive response to high-protein diets and to diets whose threonine or lysine content is less than needed for growth. Thus, reductase falls to very low levels in the liver of rats fed wheat gluten. This appears to be an adaptive response conserving body lysine. At the same time, catabolism of body proteins increases, producing endogenous lysine needed for survival. These considerations imply that as the level of wheat gluten in the diet decreases, lysine is no longer the limiting amino acid. Total protein or some other amino acid then becomes limiting.

In contrast to the apparent mechanism of lysine catabolism, threonine dehydratase does not appear to be substrate induced (63). Therefore, when lysine is the limiting amino acid, the catabolic enzyme falls to low levels and lysine is apparently conserved at the expense of body proteins. Loss of tissue proteins is much less when a diet low in threonine is fed, since the level of threonine dehydratase does not seem to be significantly affected by the protein or threonine content of the diet. Additional studies are needed to establish whether the catabolic enzyme patterns in mice parallel those of rats.

Our results also show that mice provide a good animal model to study protein quality of native, fortified, and processed wheat proteins. Mouse bioassays have a major advantage in applications to label foods for protein nutritional quality. They require about one fifth of the test material needed for rats and can be completed in 14 instead of 28 days (68, 92). They are especially useful to evaluate nutritional and safety impacts of new food ingredients formed during processing (2, 13, 14, 41–44, 47, 50, 51, 59, 61, 92, 92a, 92b) and of new plants and plant parts (35a, 36), when amount of material available for bioassays is limited.

Methionine and Derivatives

The low content of the essential amino acid L-methionine limits the nutritive value of many food proteins of plant origin. These include soybeans and other legumes. The problem is further compounded for two reasons. First, during food processing and storage L-methionine and other amino acids are chemically modified, further reducing nutritional quality. In the case of methionine, such modifications include oxidation to methionine sulfoxide and methionine sulfone, racemization to D-methionine, and degradation to compounds with undesirable flavors. Second, protein-bound methionine in some plant foods is poorly utilized, presumably because of poor digestibility (1a, 7, 45, 47, 95, 99).

A related aspect is the widespread use of L-methionine to fortify low-methionine foods in order to improve nutritional quality. Because of the reported antinutritional or toxic manifestations of high levels of free methionine in the diet, a need exists to find out whether methionine analogs and derivatives lack the apparent toxicity of L-methionine and whether they can be used as methionine substitutes in the diet.

As part of a program to evaluate the nutritional and toxicological potential of novel amino acids formed during food processing, we compared weight gain in mice fed amino acid diets containing graded levels of L-methionine and 16 methionine derivatives, isomeric dipeptides, and analogs (47). Because the mice received no other source of sulfur amino acids, the results reflect the ability of each of the compounds to meet the animals' entire metabolic demand for dietary sulfur amino acids, relative to that for L-methionine.

Linear response was closely approximated for concentrations below those yielding maximum growth. Derivatization of L-methionine generally lowered potency, calculated as the ratio of the slopes of the two dose-response curves. However, the three isomeric dipeptides (L-L-, L-D-, and D-L-methionyl-methionine), N-acetyl- and N-formyl-L-methionine, L-methionine sulfoxide, and D-methionine were well utilized. The double derivative, N-acetyl-L-methionine sulfoxide, reduced potency below 60%. D-methionine sulfoxide, N-acetyl-D-methionine, and D-methionyl-D-methionine possessed potencies between 4 and 40%. The calcium salts of L- and D-α-hydroxy analogs of methionine had potencies of 55.4 and 85.7%, respectively. Several of the analogs were less growth-inhibiting or toxic at high concentrations in the diet than was L-methionine. These results imply that some methionine dipeptides or analogs may be better candidates for foritfying foods than L-methionine.

The data for mice demonstrate that (a) the assay is highly reproducible, exhibits excellent dose-response characteristics, and yields useful estimates of relative potency for the 16 methionine analogs; and (b) somewhat rigorous control of concentration may be required for dietary supplementation with L-methionine in order to achieve maximum nutritional benefit while preventing toxicity problems. This constraint may be alleviated or avoided by using one or more analogs as alternatives. Whether these compounds will also alleviate the reported adverse flavor aspects of sulfur amino acid supplementation associated with methionine when it is added to foods awaits further study.

Tryptophan and Derivatives

The essential amino acid tryptophan contributes to normal growth and protein synthesis and participates in numerous biochemical processes. Since tryptophan is a nutritionally second-limiting amino acid in maize, and since cereals and processed foods are increasingly used to meet human dietary needs, it is of paramount importance to develop an understanding of thermally induced changes in tryptophan in order to improve the quality and safety of our food supply (38).

The stability of free or protein-bound tryptophan during processing and storage depends on temperature and the presence of oxygen or other oxidizing agents, especially lipid peroxides, and radiation. In the absence of oxidizing agents, tryptophan is a stable amino acid, even in strongly basic or acidic conditions. Free or bound tryptophan is relatively stable during heat treatments such as industrial or home cooking in the presence of air or steam sterilization. Only severe treatments cause a significant degradation of this amino acid. In the presence of carbonyl compounds and/or at high temperatures, however, carboline formation occurs. Both carbolines and tryptophan-derived nitroso compounds are potential carcinogens. Tryptophan losses during food processing cannot always be monitored because of the lack of reliable analytical methods.

The losses in tryptophan bioavailability during heat treatment such as home cooking or industrial sterilization appear less important than other detrimental effects, particularly on lysine or methionine. Some of the reported variabilities in the utilization of D-tryptophan could be due to the fact that the value (potency), of D-tryptophan as a nutritional source of L-tryptophan is strongly dose dependent (38).

Possible consequences for nutrition, food safety, and human health of halogenated tryptophans (38), light-induced tryptophan adducts (93a); tryptophan-derived carbolines (38), and tryptophan-induced eosinophilia myalgia (96a) are active areas of current research.

PREVENTION OF ADVERSE EFFECTS

Sulfur-containing amino acids such as cysteine, N-acetylcysteine, and the tripeptide glutathione play key roles in the biotransformation of xenobiotics by actively participating in their detoxification. These antioxidant and antitoxic effects are due to a multiplicity of mechanisms including their ability to act as (a) reducing agents, (b) scavengers of reactive oxygen (free-radical species), (c) strong nucleophiles that can trap electrophilic compounds and intermediates, (d) precursors for intracellular reduced glutathione, and (e) inducers of cellular detoxification.

For these reasons, fruitful results were expected from evaluation of the effectiveness of sulfur amino acids and sulfur-rich proteins to (a) prevent the formation of toxic browning products by trapping intermediates, and (b) reduce the toxicity of browning products in animals by preventing activation of such compounds to biologically active forms (10, 35, 46, 56, 79, 80).

These expectations were realized, as evidenced by the following observations on the prevention of both enzymatic and nonenzymatic browning by sulfur amino acids (20, 35, 46, 56, 79, 80, 103). To demonstrate whether SH-containing sulfur amino acids minimize nonenzymatic browning, β-alanine, N-α-acetyl-L-lysine, glycyl-glycine, and a mixture of amino acids were each heated with glucose in the absence and presence of the following potential inhibitors: N-acetyl-L-cysteine, L-cysteine, reduced glutathione, sodium bisulfite, and urea. Inhibition was measured as a function of temperature, time of heating, and concentration of reactants. The results suggest that it should be possible to devise conditions to inhibit browning in amino acid–carbohydrate solutions used for parenteral nutrition (59, 72, 73, 89, 91, 93).

Reflectance measurements were used to compare the relative effectiveness of a series of compounds in inhibition browning in freshly prepared and commercial fruit juices including apple, grape, grapefruit, orange, and pineapple juices (79). For comparison, related studies were also carried out with several protein-containing foods such as casein, barley flour, soy flour, and nonfat dry milk, and the commercial infant formula "Isomil." The results revealed that under certain conditions, SH-containing N-acetyl-L-cysteine and the tripeptide reduced glutathione may be as effective as sodium sulfite in preventing both enzymatic and nonenzymatic browning.

In a related study designed to develop sulfite alternatives (15), Russet Burbank potatoes, Washington golden delicious apples, and Washington red delicious apples were subjected to enzymatic browning in air and evacuated plastic pouches in the absence and presence of browning inhibitors (80). Studies on the effects of concentration of inhibitors, storage conditions, and pH revealed that N-acetyl-L-cysteine and reduced glutathione were nearly as effective as sodium sulfite in preventing browning of both apples and potatoes (66, 86, 101).

CONCLUSIONS

This overview shows that pH, heat, and oxygen have both beneficial and adverse effects on many nutrients. To maximize beneficial effects, future studies should emphasize the prevention of browning and the consequent antinutritional and toxicological manifestations of browning products in whole foods. Many of the safety concerns cited, especially those of genotoxic potential, are based on in vitro data that may not always be relevant to in vivo effects following the consumption of whole food products containing the browning-derived constituents. The presence of other dietary constituents in the food and the process of digestion and metabolism can be expected to decrease or increase the adverse manifestations of browning products. Most urgent is the need to develop food processing conditions to prevent the formation of carcinogenic heterocyclic amines (38, 65, 101). For nutrition and food safety, possible consequences of chelation of nutritionally essential trace materials to processing-induced food ingredients (17, 21, 40b, 64, 67, 81, 87, 89a), beneficial effects of processing on food allergy and the immune system (4, 5, 11, 12, 49, 62, 67b, 83, 84), and differentiating adverse and beneficial effects of heat and oxygen on lipids (67c, 69a, 86a) and vitamins (1, 6a, 52, 52a, 59, 64, 67a, 69, 75a, 78, 85, 88, 90, 93) also merit study.

ACKNOWLEDGMENTS

I thank my colleagues whose names appear on the cited references for excellent scientific collaboration.

Literature Cited

1. Adrian, J., Frangne, R. 1991. Synthesis and availability of niacin in roasted coffee. See Ref. 34, pp. 49–50
1a. Begbie, R., Pusztai, A. 1989. The resistance to proteolytic breakdown of some plant (seed) proteins and their effects on nutrient utilization and gut metabolism. See Ref. 33, pp. 233–63
1b. Betschart, A. A. 1978. Improving protein quality of bread—nutritional benefits and realities. See Ref. 26, pp. 701–34
2. Birch, G. G., Parker, K. J., eds. 1982. *Food and Health: Science and Technology.* London: Applied Publ. 532 pp.
3. Borg, B. S., Wahlstrom, R. C. 1989. Species and isomeric variation in the utilization of amino acids. See Ref. 31, pp. 155–72
4. Bounos, G., Kongshavn, P. A. L. 1989. Influence of protein type in nutritionally adequate diets on the development of immunity. See Ref. 32, pp. 235–45
5. Brandon, D. L., Bates, A. H., Friedman, M. 1991. ELISA analysis of soybean trypsin inhibitors in processed foods. See Ref. 34, pp. 321–37
6. Burks, A. W., Williams, L. W., Helm, R. M., Thresher, W., Brooks, J. R., Sampson, H. A. 1991. Identification of

soy protein allergens in patients with atopic dermatitis and positive soy challenges: determination of change in allergenicity after heating or enzyme digestion. See Ref. 34, pp. 295–320
6a. Carter, E. G. A., Carpenter, K. J., Friedman, M. 1982. The nutritional value of some niacin analogs for rats. *Nutr. Rep. Int.* 25:389–97
7. Chang, K. C., Kendrick, J. G., Marshall, H. F., Satterlee, L. D. 1985. Effect of partial methionine oxidation on the nutritional quality of soy isolate and casein. *J. Food Sci.* 50:849–59
8. Chuyen, N. V., Utsunomiya, N., Kato, H. 1991. Nutritional and physiological effects of casein modified by glucose under various conditions on growing adult rats. *Agric. Biol. Chem.* 55:659–64
9. Dakin, H. D., Dudley, H. W. 1913. The action of enzymes on racemized proteins and their fate in the animal body. *J. Biol. Chem.* 15:271–76
10. De Flora, S., Benicelli, C., Serra, D., Izzotti, A., Cesarone, C. F. 1989. Role of glutathione and N-acetylcysteine as inhibitors of mutagenesis and carcinogenesis. See Ref. 31, pp. 19–54
11. Djurtoft, R., Pedersen, H.S., Aabin, B., Barkholt, V. 1991. Studies of food

allergens: soybean and egg protein. See Ref. 34, pp. 281–93

12. Doke, S., Nakamura, R., Watanabe, K. 1990. Allergenicity of malondialdehyde-protein complex in soybean sensitive individuals. See Ref. 18, pp. 315–20

13. Dunn, J. A., Ahmed, M. U., Murtiashaw, M. H., Richardson, J. M., Walla, M. D., et al. 1990. Reaction of ascorbate with lysine and protein under autooxidizing conditions: formation of N-ε-(carboxymethyl)lysine by reaction between lysine and products of autooxidation of ascorbate. *Biochemistry* 29:10964–70

13a. Ebersdobler, H. F. 1989. Protein reactions during food processing and storage—their relevance to human nutrition. In *Nutritional Impact of Food Processing*, ed. J. C. Somogyi, H. R. Muller, pp. 140–55. Basel: Karger

14. Ebersdobler, H. F., Ohmann, M., Buhl, K. 1991. Utilization of early Maillard reaction products by humans. See Ref. 34, pp. 363–70

15. Fan, A. M., Book, S. A. 1989. Sulfite hypersensitivity: a review of current issues. *J. Appl. Nutr.* 39:71–78

16. Finley, J. W., Snow, J. T., Johnston, P., Friedman, M. 1978. Inhibition of lysinoalanine formation in food proteins. *J. Food Sci.* 43:619–21

17. Finot, P. A. 1990. Metabolism and physiological effects of Maillard reaction products. See Ref. 18, pp. 259–72

18. Finot, P. A., Aeschbacher, H. U., Hurrell, R. F., Liardon, R., eds. 1990. *The Maillard Reaction in Food Processing, Human Nutrition, and Physiology.* Basel, Switzerland: Birkhauser. 516 pp.

19. Finot, P. A., Furniss, D. E. 1986. Nephrocytomegaly in rats induced by Maillard reaction products: the involvement of metal ions. In *Amino-Carbonyl Reactions in Food and Biological Systems,* ed. M. Fumimaki, pp. 493–502. Amsterdam: Elsevier

20. Friedman, M. 1973. *The Chemistry and Biochemistry of the Sulfhydryl Group in Amino Acids, Peptides, and Proteins.* Oxford, England: Pergammon. 485 pp.

21. Friedman, M., ed. 1974. *Protein-Metal Interactions.* New York: Plenum. 692 pp.

22. Friedman, M., ed. 1975. *Protein Nutritional Quality of Foods and Feeds, Part 1. Assay Methods—Biological, Biochemical, Chemical.* New York: Marcel Dekker. 626 pp.

23. Friedman, M., ed. 1975. *Protein Nutritional Quality of Foods and Feeds, Part 2. Quality Factors—Plant Breeding, Composition, Processing and Antinutri-*

ents. New York: Marcel Dekker. 674 pp.

24. Friedman, M., ed. 1977. *Protein Crosslinking: Biochemical and Molecular Aspects.* New York: Plenum. 760 pp.

25. Friedman, M., ed. 1977. *Protein Crosslinking: Nutritional and Medical Consequences.* New York: Plenum. 740 pp.

26. Friedman, M., ed. 1978. *Nutritional Improvement of Food and Feed Proteins.* New York: Plenum. 882 pp.

27. Friedman, M. 1982. Chemically reactive and unreactive lysine as an index of browning. *Diabetes* 31(Suppl. 3):5–14

28. Friedman, M. 1982. Lysinoalanine formation in soybean proteins: kinetics and mechanism. In *Mechanisms of Food Protein Deterioration,* ed. J. P. Cherry, pp. 231–73. Washington, DC: ACS Symp. Ser.

29. Friedman, M., ed. 1984. *Nutritional and Toxicological Aspects of Food Safety.* New York: Plenum. 596 pp.

30. Friedman, M., ed. 1986. *Nutritional and Toxicological Significance of Enzyme Inhibitors in Foods.* New York: Plenum. 570 pp.

31. Friedman, M., ed. 1989. *Absorption and Utilization of Amino Acids, Vol. 1.* Boca Raton, Fla: CRC Press. 257 pp.

32. Friedman, M., ed. 1989. *Absorption and Utilization of Amino Acids, Vol. 2.* Boca Raton, Fla: CRC Press. 301 pp.

33. Friedman, M., ed. 1989. *Absorption and Utilization of Amino Acids, Vol. 3.* Boca Raton, Fla: CRC Press. 321 pp.

34. Friedman, M., ed. 1991. *Nutritional and Toxicological Consequences of Food Processing.* New York: Plenum. 542 pp.

35. Friedman, M. 1991. Prevention of adverse effects of food browning. See Ref. 34, pp. 171–216

35a. Friedman, M. 1992. Composition and safety evaluation of potato berries, potato and tomato seeds, potatoes, and potato alkaloids. In *Evaluation of Food Safety,* ed. J. W. Finley, A. Armstrong, pp. 429–62. Washington, DC: Am. Chem. Soc.

36. Friedman, M., Atsmon, D. 1988. Comparison of grain composition and nutritional quality in wild barley (*Hordeum spontaneum*) and in a standard cultivar. *J. Agric. Food Chem.* 36:1167–72

37. Friedman, M., Brandon, D. L., Bates, A. H., Hymowitz, T. 1991. Comparison of a commercial soybean cultivar and an isoline lacking the Kunitz trypsin inhibitor: composition, nutritional value, and effects of heating. *J. Agric. Food Chem.* 39:337–45

38. Friedman, M., Cuq, J. L. 1988. Chemistry, analysis, nutritional value, and toxicology of tryptophan. A review. *J. Agric. Food Chem.* 36:1079–93

39. Friedman, M., Dao, L. 1990. Effect of autoclaving and conventional and microwave baking on the ergot alkaloid and cholorogenic acid content of morning glory seeds. *J. Agric. Food Chem.* 38:805–8

40. Friedman, M., Finot, P. A. 1990. Nutritional improvement of bread with lysine and glutamyl-lysine. *J. Agric. Food Chem.* 38:2011–20

40a. Friedman, M., Grosjean, O. K., Gumbmann, M. R. 1984. Nutritional improvement of soybean flour. *J. Nutr.* 114:2241–46

40b. Friedman, M., Grosjean, O. K., Zahnley, J. C. 1986. Inactivation of metalloenzymes by food constituents. *Food and Chem. Toxicol.* 24:897–902

41. Friedman, M., Gumbmann, M. R. 1979. Biological availability of epsilon-N-methyl-L-lysine, 1-N-methyl-L-histidine, and 3-N-methyl-L-histidine in mice. *Nutr. Rep. Int.* 19:437–43

42. Friedman, M., Gumbmann, M. R. 1981. Bioavailability of some lysine derivatives in mice. *J. Nutr.* 111:1362–69

43. Friedman, M., Gumbmann, M. R. 1984. The nutritive value and safety of D-phenylalanine and D-tyrosine in mice. *J. Nutr.* 114:2087–89

44. Friedman, M., Gumbmann, M. R. 1984. The utilization and safety of isomeric sulfur amino acids in mice. *J. Nutr.* 114:2301–10

45. Friedman, M., Gumbmann, M. R. 1986. Nutritional improvement of soy flour through inactivation of trypsin inhibitors by sodium sulfite. *J. Food Sci.* 51:1239–41

46. Friedman, M., Gumbmann, M. R. 1986. Nutritional improvement of legume proteins through disulfide interchange. See Ref. 30, pp. 357–89

47. Friedman, M., Gumbmann, M. R. 1988. Nutritional value and safety of methionine derivatives, isomeric dipeptides, and hydroxy analogs in mice. *J. Nutr.* 118:388–97

48. Friedman, M., Gumbmann, M. R. 1989. Dietary significance of D-amino acids. See Ref. 31, pp. 173–90

49. Friedman, M., Gumbmann, M. R., Brandon, D. L., Bates, A. H. 1989. Inactivation and analysis of soybean inhibitors of digestive enzymes. In *Food Proteins*, ed. W. G. Soucie, J. Kinsella, pp. 296–328. Champaign, Ill: Am. Oil Chem. Soc.

50. Friedman, M., Gumbmann, M. R., Masters, P. M. 1984. Protein-alkali reactions: chemistry, toxicology, and nutritional consequences. See Ref. 29, pp. 367–412

51. Friedman, M., Gumbmann, M. R., Savoie, L. 1982. The nutritional value of lysinoalanine as a source of lysine. *Nutr. Rep. Int.* 26:937–43

52. Friedman, M., Gumbmann, M. R., Ziderman, I. I. 1987. Nutritional value and safety in mice of proteins and their admixtures with carbohydrates and vitamin C after heating. *J. Nutr.* 117:508–18

52a. Friedman, M., Gumbmann, M. R., Ziderman, I. I. 1988. Nutritional and toxicological consequences of browning during simulated crust baking. In *Protein Quality and Effects of Processing*, ed. R. D. Phillips, J. W. Finley, pp. 189–218. New York: Marcel Dekker.

53. Friedman, M., Levin, C. E., Noma, A. T. 1984. Factors governing lysinoalanine formation in soy proteins. *J. Food Sci.* 49:1282–88

54. Friedman, M., Liardon, R. 1985. Racemization kinetics of amino acid residues in alkali-treated soybean proteins. *J. Agric. Food Chem.* 33:666–72

55. Friedman, M., Masters, P. M. 1982. Kinetics of racemization of amino acid residues in casein. *J. Food Sci.* 47:760–64

56. Friedman, M., Molnar-Perl, I. 1990. Inhibition of food browning by sulfur amino acids. Part 1. Heated amino acid-glucose systems. *J. Agric. Food Chem.* 38:1642–47

57. Friedman, M., Pearce, K. N. 1989. Copper (II) and cobalt (II) affinities of LL- and LD-lysinoalanine diasteromers: implications for food safety and nutrition. *J. Agric. Food Chem.* 37:123–27

58. Friedman, M., Zahnley, J. C., Masters, P. M. 1981. Relationship between in vitro digestibility of casein and its content of lysinoalanine and D-amino acids. *J. Food Sci.* 46:127–31, 134

59. Grun, I. U., Barbeau, W. E., Chrisley, B. Mc., Driskell, J. A. 1991. Determination of vitamin B_6, available lysine, and ε-pyridoxyllysine in a new instant baby food product. *J. Agric. Food Chem.* 39:102–8

60. Gumbmann, M. R., Friedman, M. 1987. Effect of sulfur amino acid supplementation of raw soy flour on the growth and pancreatic weights of rats. *J. Nutr.* 117:1018–23

61. Gumbmann, M. R., Friedman, M., Smith, G. A. 1983. The nutritional values and digestibilities of heat-damaged

casein and casein-carbohydrate mixtures. *Nutr. Rep. Int.* 28:355–61
62. Hathcock, J. N. 1991. Residue trypsin inhibitor: data needs for risk assessment. See Ref. 34, pp. 273–79
63. Hegsted, D. M. 1977. Protein quality and its determination. See Ref. 102, pp. 347–62
64. Hurrell, R. F. 1982. Interaction of food components during processing. See Ref. 2, pp. 369–88
65. Jagerstadt, M., Skog, K. 1991. Formation of meat mutagens. See Ref. 34, pp. 83–105
66. Jiang, Z., Ooraikul, B. V. 1989. Reduction of nonenzymatic browning in potato chips and french fries with glucose oxidase. *J. Food Process Preserv.* 13:175–86
67. Johnson, P. E. 1991. Effect of processing and preparation on mineral utilization. See Ref. 34, pp. 483–98
67a. Jonsson, L. 1991. Thermal degradation of carotenes and influence on their physiological functions. See Ref. 34, pp. 75–82
67b. Jost, R., Monti, J. C., Pahud, J. J. 1991. Reduction of whey protein allergenicity by processing. See Ref. 34, pp. 309–20
67c. Kanazawa, K. 1991. Hepatotoxicity caused by dietary secondary products originating from lipid peroxidation. See Ref. 34, pp. 237–53
68. Keith, M. O., Bell, M. J. 1988. Digestibility of nitrogen and amino acids in selected protein sources fed to mice. *J. Nutr.* 118:561–68
69. Kincal, N. S., Giray, C. 1987. Kinetics of ascorbic acid degradation during potato blanching. *Int. J. Food Sci. Technol.* 22:249–54
69a. Kinsella, J. E. 1991. Dietary N-3 polyunsaturated fatty acids of fish oils, autooxidation ex vivo and peroxidation in vivo: implications. See Ref. 34, pp. 255–68
70. Kratzer, F. H., Bersch, S., Vohra, P. 1990. Evaluation of heat-damage to protein by Coomassie blue G dye-binding. *J. Food Sci.* 55:805–7
71. Kratzer, F. H., Bersch, S., Vohra, P., Ernst, R. A. 1990. Chemical and biological evaluation of soya-bean flakes autoclaved for different durations. *Anim. Feed Sci. Technol.* 31:247–59
72. Labuza, T. P., Massaro, S. A. 1990. Browning and amino acid loss in model total parenteral nutrition solutions. *J. Food Sci.* 55:821–26
73. Lee, T. C., Pintauro, S. J., Chichester, C. O. 1982. Nutritional and toxicological effects of nonenzymatic Maillard browning. *Diabetes* 31(Suppl. 3):37–46
74. Liardon, R., Friedman, M. 1987. Effect of peptide bond cleavage on the racemization of amino acid residues in proteins. *J. Agric. Food Chem.* 35:661–67
75. Liardon, R., Friedman, M., Philippossian, G. 1991. Isomeric composition of protein bound lysinoalanine. *J. Agric. Food Chem.* 39:531–37
75a. Loscher, J., Kroh, L., Westphal, G., Vogel, J. 1991. L-ascorbic acid—a carbonyl component of non-enzymatic browning reactions. *Z. Lebensm. Unters. Forsch.* 192:323–27
76. MacGregor, J. T., Tucker, J. D., Ziderman, I. I., Wehr, C. M., Wilson, R. E., Friedman, M. 1989. Nonclastogenicity in mouse bone marrow of fructose/lysine and other sugar/amino acid browning products with *in vitro* genotoxicity. *Food Chem. Toxicol.* 27:715–21
77. Masters, P. M., Friedman, M. 1980. Amino acid racemization in alkali-treated food proteins—chemistry, toxicology, and nutritional consequences. See Ref. 102, pp. 165–94
78. Mauron, J. 1982. Methodology to detect nutritional damage during thermal processing. See Ref. 2, pp. 389–412
79. Molnar-Perl, I., Friedman, M. 1990. Inhibition of food browning by sulfur amino acids. Part 2. Fruit juices and protein-containing foods. *J. Agric. Food Chem.* 38:1648–51
80. Molnar-Perl, I., Friedman, M. 1990. Inhibition of food browning by sulfur amino acids. Part 3. Apples and potatoes. *J. Agric. Food Chem.* 38:1652–56
81. O'Brien, J., Morissey, P. A., Flynn, A. 1988. Nephrocalcinosis and disturbances of mineral balance in rats fed Maillard reaction products. See Ref. 98, pp. 177–85
82. Oste, R. E. 1991. Digestibility of processed food proteins. See Ref. 34, pp. 371–88
83. Oste, R. E., Brandon, D. L., Bates, A. H., Friedman, M. 1990. Effect of the Maillard reaction of the Kunitz soybean trypsin inhibitor on its interaction with monoclonal antibodies. *J. Agric. Food Chem.* 38:258–61
84. Oste, R. E., Brandon, D. L., Bates, A. H., Friedman, M. 1990. Antibody binding to Maillard reacted protein. See Ref. 18, pp. 303–8
85. Oste, R. E., Friedman, M. 1990. Safety of heated amino acid/sodium ascorbate blends. *J. Agric. Food Chem.* 38:1687–90
86. Oszmianski, J., Lee, C. Y. 1990. In-

hibition of polyphenol oxidase activity and browning by honey. *J. Agric. Food Chem.* 38:1892–95

86a. Pariza, M. W., Ha, Y. L., Benjamin, H., Sword, J. T., Gruter, A., et al. 1991. Formation and action of anticarcinogenic fatty acids. See Ref. 34, pp. 269–72

87. Pearce, K. N., Friedman, M. 1988. The binding of copper (II) and other metal ions by lysinoalanine and related compounds and its significance for food safety. *J. Agric. Food Chem.* 36:707–817

88. Poiffait, A., Adrian, J. 1991. Interaction between casein and vitamin A during food processing. See Ref. 34, pp. 61–73

89. Rassin, D. K. 1989. Amino acid metabolism in total parenteral nutrition during development. See Ref. 32, pp. 71–86

89a. Rehner, G., Walter, Th. 1991. Effect of Maillard products and lysinoalanine on the bioavailability of iron, copper, and zinc. *Z. Ernahrungswiss.* 30:50–55

90. Quattrucci, E. 1988. Heat treatments and nutritional significance of Maillard reaction products. See Ref. 98, pp. 113–23

91. Sarwar, G. 1991. Amino acid ratings of different forms of infant formulas based on varying degrees of processing. See Ref. 34, pp. 389–402

92. Sarwar, G., Blair, R., Friedman, M., Gumbmann, M. R., Hackler, L. R., et al. 1985. Comparison of interlaboratory variation in amino acid analysis and rat growth assays for evaluation of protein quality. *J. Assoc. Off. Anal. Chem.* 68:52–56

92a. Sarwar, G., Blair, R., Friedman, M., Gumbmann, M. R., Hackler, L. R., et al. 1984. Inter- and intra-laboratory variability in rat growth assays for estimating protein quality in foods. *J. Assoc. Anal. Chem.* 67:976–81

92b. Sarwar, G., Christensen, D. A., Finlayson, A. J., Friedman, M., Hackler, L. R., et al. 1983. Intra- and inter-laboratory variation in amino acid analysis. *J. Food Sci.* 48:526–31

93. Schmidl, M. K., Massaro, S. S., Labuza, T. P. 1988. Parenteral and enteral food systems. *Food Technol.* 42:77–82, 85–87

93a. Silva, E., Salim-Hanna, M., Edwards, A. N., Becker, M. I., De Ioannes, E. 1991. A light-induced tryptophan-riboflavin binding: biological implications. See Ref. 34, pp. 33–48

94. Smith, G. A., Friedman, M. 1984.

Effect of carbohydrates and heat on the amino acid composition and chemically available lysine content of casein. *J. Food Sci.* 49:817–21

95. Smolin, L. A., Benevenga, N. J. 1989. Methionine, homocyst(e)ine, cyst(e)ine—metabolic interrelationships. See Ref. 31, pp. 158–87

96. Swallow, J. A. 1991. Wholesomeness and safety of irradiated food. See Ref. 34, 11–31

96a. Sygert, L. A., Maes, E. F., Sewell, L. E., Miller, L., Falk, H., Kilbourne, E. M. 1990. Eosinophilia myalgia syndrome—results of a national survey. *J. Am. Med. Assoc.* 264:1698–1703

97. Troll, W. 1986. Protease inhibitors: their role as modifiers of the carcinogenic process. See Ref. 30, pp. 153–56

98. Walker, R., Quattrucci, E., eds. *Nutritional and Toxicological Aspects of Food Safety.* London: Taylor & Francis. 375 pp.

99. Wang, C. H., Damodaran, S. 1990. Thermal destruction of cysteine and cystine residues of soy protein under conditions of gelation. *J. Food Sci.* 55: 1077–80

100. Watanabe, N., Ohtsuka, M., Takahashi, S., I., Sakano, Y., Fujimoto, D. 1987. Enzymatic deglycation of fructosyllysine. *Agric. Biol. Chem.* 51:1063–64

101. Weisburger, J. H. 1991. Carcinogens in food and cancer prevention. See Ref. 34, pp. 137–51

102. Whitaker, J. R., Fujimaki, M., eds. 1980. *Chemical Deterioration of Proteins.* Washington, DC: ACS Symp. Ser. 268 pp.

103. Wong, D. W. S. 1989. *Mechanism and Theory in Food Chemistry.* New York: Van Nostrand Reinhold. 428 pp.

104. Woodard, J. C., Short, D. D., Alvarez, M. R., Reyniers, J. 1975. Biologic effects of lysinoalanine. See Ref. 23, pp. 595–615

105. Yasumoto, K., Suzuki, F. 1990. Aspartyl- and glutamyl-lysine crosslinks formation and their nutritional availability. *J. Nutr. Vitaminol.* 36:S71–S77

106. Ziderman, I. I., Friedman, M. 1985. Thermal and compositional changes of dry wheat gluten-carbohydrate mixtures during simulated crust baking. *J. Agric. Food Chem.* 33:1096–1102

107. Ziderman, I. I., Grogorski, K. S., Lopez, S. V., Friedman, M. 1989. Thermal interaction of vitamin C with proteins in relation to nonenzymatic browning of foods and Maillard reactions. *J. Agric. Food Chem.* 37:1480–86

Annu. Rev. Nutr. 1992. 12:139–59

DIETARY CAROTENES, VITAMIN C, AND VITAMIN E AS PROTECTIVE ANTIOXIDANTS IN HUMAN CANCERS[1]

Tim Byers and Geraldine Perry

Centers for Disease Control, Epidemiology Branch, Division of Nutrition, National Center for Chronic Disease Prevention and Health Promotion, Atlanta, Georgia 30333

KEY WORDS: nutrition, diet, prevention, β-carotene

CONTENTS

INTRODUCTION

A massive research effort over the last decade has improved our understanding of the role of antioxidants in protecting the human body against cancer.

---

[1]The US government has the right to retain a nonexclusive, royalty-free license in and to any copyright covering this paper.

Consequently, cancer chemoprevention through supplementation and fortification of the diet with micronutrient antioxidants could become an effective strategy for cancer control before the close of this century (7, 63). The scientific literature on antioxidants and cancer is extensive and spans many disciplines, from basic chemistry to experimental research and epidemiology. Therefore, an exhaustive examination of the literature on the role of antioxidants in preventing cancer is beyond the scope of this short review. In 1989, an expert committee of the National Research Council (NRC) published an extensive review of the relationship between diet and health, including the relationship between antioxidants in the diet and cancer risk (65). Our review focuses principally on the research reported since 1987, the year before the NRC panel concluded its work. Selected major reviews published since 1987 on the topic of antioxidants and cancer prevention are listed in Table 1.

DIETARY ANTIOXIDANTS: THEORETICAL ROLES IN CANCER PREVENTION

Some dietary factor, or set of factors, apparently plays an important part in the etiology of cancer (65). Cancer rates vary considerably between countries, and migrants tend to acquire the cancer risk of persons in their new country within a generation or two. One of the most important changes in the experience of migrants is a change in dietary practices. Factors in the changing diet of migrants hypothesized to be related to cancer risk include carcinogens found in foods, cancer-promoting factors such as high levels of dietary fat, and factors in the diet that may be anticarcinogenic as well (4). Such protective factors might include dietary fiber, trace minerals, and other micronutrients, including many of the vitamins, provitamins, and other compounds that have chemical properties of antioxidants (95).

The human body is under constant assault by reactive oxygen molecules. Reactive oxygen molecules (free radicals and singlet oxygen) are formed as a natural consequence of normal biochemical activity. Reactive oxygen can damage the body in many ways: by denaturing proteins, by damaging nucleic acids, and by saturating the double bonds of fatty acids in lipid membranes, thereby altering membrane structure and function (95). Because oxidative damage can be life threatening, the body has many overlapping defense mechanisms to protect against oxidation (28). These defenses include both mineral-dependent enzymes and small molecules that act as scavengers of reactive oxygen species. One such mineral-dependent enzyme is glutathione peroxidase, the selenium-dependent free radical scavenger. Small molecules that act as antioxidants include water-soluble compounds such as vitamin C, glutathione, and uric acid, as well as lipid-soluble molecules such as the carotenoids and vitamin E. This review focuses on the antioxidant properties

Table 1 Selected major reviews of dietary antioxidants and cancer risk

Author (Ref.)	Dietary antioxidants reviewed[a]				Description
	Carot.	Vit. C	Vit. E	Others	
NRC (65)	X	X	X	X	Comprehensive review by expert panel of both animal and human studies through 1987
Bertram et al (7)	X	X	X	X	Thorough discussion of the scientific basis for chemo prevention trials
Ziegler (116)	X				Review of the epidemiologic evidence regarding carotenoids and cancer
Moon (64)	X			X	Review of the animal experimental evidence regarding carotenoids and cancer
Schneider & Shaw (83)	X	X		X	Cervical cancer epidemiology review, including dietary factors
Vogel & McPherson (102)	X	X		X	Colon cancer epidemiology review, including dietary factors
London & Willet (55)	X	X	X		Review of dietary factors in breast cancer epidemiology
Henson et al (38)		X			Comprehensive review of cancer-relevant vitamin C physiology
Ziegler (115)	X			X	Review of epidemiologic evidence regarding fruits and vegetables, emphasizing carotenoids
Knekt et al (49)			X		Review of animal and epidemiologic studies on vitamin E and cancer
Block et al (10)		X			Comprehensive review of epidemiology regarding vitamin C and cancer
Krinsky (51)	X				Comprehensive review of carotenoids and cancer in laboratory experiments
Weisburger (105)	X	X	X	X	Review of mechanisms that might explain epidemiologic findings on antioxidants
Dorgan & Schatekin (29)	X	X	X	X	Review of epidemiologic studies regarding cancer and several micronutrients
Byers (18)	X	X	X	X	Review of dietary antioxidants in lung cancer epidemiology

[a] Carot. = carotenoids; vit. C = vitamin C; vit E = vitamin E; others = various other antioxidants.

of the three most intensively studied antioxidant compounds found in the diet: carotenoids, vitamin C, and vitamin E.

The carotenoids are a set of several hundred pigmented, fat-soluble antioxidants found in fruits and vegetables (64). Approximately 10% of the carotenoids are convertable to vitamin A (retinol) by enzymatic cleavage in the body. Because vitamin A is essential for the maintenance of normal epithelial cellular differentiation, the carotenoids are important in cancer prevention, in part because of their provitamin A activity. Perhaps more importantly, the carotenoids may reduce cancer risk because of their role as antioxidants in the body (73). Antioxidation may be the most important effect of carotenoids in cancer prevention.

Vitamin C is an essential nutrient that likely protects against cancer by several mechanisms (38), including its role in promoting the formation of collagen in the body and in inhibiting the formation of N-nitroso compounds in the stomach. Perhaps more importantly, vitamin C is the most abundant water-soluble antioxidant in the body (32). Recently it has become apparent that both vitamin C and the carotenoids may also have beneficial effects on immune function, thereby reducing cancer risk by enhancing tumor surveillance by the immune system (6, 38, 78).

Vitamin E is a fat-soluble compound found in a wide variety of foods and serves many physiologic functions (9). Vitamin E may protect against cancer by several mechanisms (49). Like Vitamin C it may inhibit the formation of N-nitroso compounds in the stomach. Vitamin E also protects selenium against reduction (40) and protects polyunsaturated fatty acids in lipid membranes from oxidative damage (39). Vitamin E is thought to be the most important antioxidant found within lipid membranes in the body (74).

LABORATORY ANIMAL RESEARCH

Animal experimentation has played an important role in the development of our understanding of the role of antioxidants in preventing cancer. Animal experiments have enabled researchers to demonstrate the biologic plausibility of a preventive role for several antioxidants compounds. Animal experiments have also been designed to examine the efficacy of antioxidants at various stages of the development of cancer. Because oxidative damage to nucleic acids can lead to genetic damage, and hence to the initiation of cancer, it is reasonable to hypothesize that antioxidants are most likely to be protective at the initiation stage. The results of animal experiments have clearly shown, however, that antioxidants are most protective during the later promotional phases of cancer development (51, 64). Animal experiments have demonstrated beneficial effects of antioxidants on cellular membrane integrity and on immune system function. These observations have helped to direct epidemiologic research on the effects of antioxidants in the later promotional phases of cancer.

Animal experimental models have many limitations, however. Many animal tumors in experimental systems are either induced by high doses of strong carcinogens or they occur sporadically only in particular genetic strains of animals. In addition, findings from one animal tumor model are often not generalizable to other species, and the behavior of animal tumors often differs in significant ways from the behavior of human cancer. In most cases, for instance, animal tumors used to model human cancer do not metastasize. Nonetheless, many laboratory animal experiments have been conducted. Following is a brief summary of findings from experimentation with antioxidant supplementation in animal tumor models.

Carotenoids

Many investigators have demonstrated the ability of β-carotene, independent of its role in the formation of vitamin A, to protect against cancer in animals (51). The effectiveness of β-carotene has been confirmed in different animal species, at different cancer sites, and in several different cancer model systems using different inducing agents (8, 51). Carotenoids have been shown to be protective against the development of skin tumors in rats after the administration of 7,12-dimethylbenz (α)anthracene (DMBA) (85), and in mice after the administration of benzo(α)pyrene (81). High levels of carotene in the diet of mice prior to their irradiation with ultraviolet light prolonged the time to appearance of tumors (58), while β-carotene supplementation after the stage of initiation reduced the incidence of tumor development (59).

Nevertheless, the effects of β-carotene in experimental tumor models can be sensitive to experimental conditions. For example, salivary gland tumors initiated by DMBA appeared less frequently and were of lower weight in rats fed high levels of β-carotene in one study (2), but the same investigators were later unable to replicate those findings (3). They speculated that subtle effects owing to different amounts of DMBA used in initiation may have made a difference in the β-carotene effect. β-Carotene has been shown to have a protective effect against colonic tumors induced by 1,2-dimethylhydrazine (DMH) in mice (97). Gastric dysplasia was altered by β-carotene supplementation in rats whose tumors were initiated with N-methyl-N'-nitro-N-nitrosaguanidine, but supplementation did not affect the subsequent rate of cancer development (82).

The effects of β-carotene in animal tumor model systems are therefore varied. Animal experiments have shown the plausibility of an anticancer role for β-carotene in the later promotional and progression phases. However, effects appear to be sensitive to experimental conditions and are not universally seen at all cancer sites.

Vitamin C

The ability of ascorbic acid to inhibit the formation of carcinogenic nitrosamines is the best documented cancer-protecting effect of vitamin C (10, 21). Vitamin C may have additional cancer-preventing effects that are independent of this mechanism, however. Like β-carotene, vitamin C has been shown to reduce the rate of tumor development in mouse skin cancer models employing either ultraviolet light (30) or DMBA-croton oil models (86).

As is true for β-carotene, however, findings in other organ systems are mixed for vitamin C. Investigators reported that ascorbate supplementation did not affect tumor development in chemical-induced bladder (90) or mammary cancers (1). Colon tumors induced by DMH were shown to be inhibited in rats in one study (76), but were enhanced in another (87). DMH-induced colon cancers were unaffected by ascorbate (46, 87). Some experiments have demonstrated the possibility of adverse effects of ascorbate supplementation. In rats treated with N,N-dibutylnitrosamines, ascorbate increased bladder tumor yield (34). When both nutrients were supplemented together, vitamin C was found to nullify the beneficial effects of selenium in mammary tumorigenesis (44).

Vitamin E

Vitamin E shares with vitamin C the ability to inhibit nitrosamine formation in the stomach (21). Vitamin E has also been shown to have cancer-preventing effects in several other animal tumor model systems. Skin cancer induced by DMBA is inhibited by vitamin E supplementation in mice (72). Likewise, vitamin E has been shown to reduce the incidence of DMBA-induced cancers of the oral cavity (88) and mammary gland (104). This effect may be modified by the level of fat in the diet (47, 60). Experimental diets extremely deficient in vitamin E have been shown to increase tumor initiation of DMH-induced colon tumors (96) and DMBA-induced mammary tumors (43), but the relevance to the human condition of states of very extreme vitamin E deficiency is unclear.

EPIDEMIOLOGIC STUDIES

The literature on the epidemiology of antioxidants in cancer is complex, largely because of the complexity of the human diet (17, 37). Ecologic studies are inquiries in which the average diet of a set of populations is correlated with cancer rates. Typically, ecologic studies are conducted by comparing diet and cancer across different countries. Although these studies are useful for generating hypotheses, they are susceptible to the ecologic fallacy: the appearance of spurious associations between diet and cancer that are due to truly causal correlated factors. Because of this limitation, ecologic studies are not reviewed in detail here. Instead, we review those studies in which researchers have investigated cancer risk as related to dietary or supplemental antioxidant intake in individual study subjects. Epidemiologic studies of individuals include case-control studies, cohort studies, and randomized controlled trials.

Dietary Assessment

The greatest limitation of case-control and cohort studies is the difficulty of accurately ascertaining dietary intake. The only practical method for estimating a person's usual intake of foods and nutrients over a period of time is the food frequency method (107). By asking subjects to report their usual frequency of intake of selected foods that are key indicators of nutrients of interest, one can create "nutrient indices," quantitative estimates of intake that are useful for establishing the rank-order of intake in the study group. Correct ranking of study subjects is essential in order to accurately estimate relative risks for the extremes of intake (for example, the lowest compared with the highest quartiles of intake) in case-control and cohort studies. Several validation studies have shown that food frequency techniques can be used to accurately classify people into extreme quantiles of intake of antioxidant micronutrients. The validity of food frequency techniques has been established by comparing food frequency estimates of dietary intake with intake estimates from multiple diet records (109) and with nutrient levels circulating in the blood (111).

The major sources of carotenes, vitamin C, and vitamin E in the American diet are listed in Table 2. The twenty top contributors for these three nutrients are listed along with the cumulative percent of the total nutrient intake in the diet contained in the listed foods. Approximately two thirds of the nutrient content of the diet is contained in only five foods for carotene and in ten foods for vitamin C. However, more than twenty foods are needed to account for two thirds of the vitamin E in the diet. In many studies, investigators have not specifically computed nutrient indices, but they have ascertained and reported many of these key indicator foods. Therefore, many studies are useful for elucidating the role of antioxidant micronutrients in cancer, even though their findings are not described in those terms. This is particularly true for vitamin C and carotenes, which are highly concentrated in a relatively short list of fruits and vegetables. Because the types of foods that contain vitamin E are less distinct and less discernible in the diet (for example, vegetable oils and margarines), inferring vitamin E intake in the diet from a limited set of foods in a food frequency interview may be more difficult.

Case-Control and Cohort Diet Studies

Case-control studies compare diets of individuals having cancer with those of individuals without cancer, when both are sampled from the same population. The major problem with case-control studies is the possibility of bias, either

Table 2 Major sources of carotenes, vitamin C, and vitamin E in the American diet

Carotenes[a]		Vitamin C[b]		Vitamin E[a]	
Food	Cumulative %[c]	Food	Cumulative %[c]	Food	Cumulative %[c]
1. Carrots	37.8	1. Orange juice	26.5	1. Mayonnaise	14.6
2. Tomatoes	51.0	2. Grapefruit (and juice)	33.7	2. Potato chips	18.8
3. Sweet potatoes	56.7	3. Tomatoes (and juice)	39.9	3. Apples	22.9
4. Yellow squash	62.3	4. Fortified fruit drinks	45.7	4. Nuts	27.0
5. Spinach (cooked)	67.9	5. Oranges	50.6	5. Peanut butter	30.9
6. Cantaloupe	71.7	6. Potatoes (not fried)	54.8	6. Oil and vinegar	34.2
7. Mixed vegetables	75.4	7. Potatoes (fried)	58.9	7. Tomatoes	37.4
8. Romaine lettuce	78.3	8. Green salad	62.4	8. Margarine	40.5
9. Broccoli	80.6	9. Other fruit juices	65.2	9. Sweet roll	43.2
10. Spinach (raw)	83.0	10. Broccoli	67.2	10. Tomato sauce	45.9
11. Tomato sauce	84.4	11. Coleslaw, cabbage	69.1	11. Sweet potatoes	48.3
12. Margarine	85.7	12. Spaghetti and sauce	71.0	12. Eggs	50.4
13. Orange juice	87.0	13. Orange juice substitute	72.8	13. Cold cereal	52.5
14. Iceberg lettuce	88.1	14. Cold cereal	74.6	14. Shrimp	54.6
15. Pizza	89.1	15. Hot dogs, lunch meat	76.3	15. Cake	56.6
16. Cheese	90.1	16. Cantaloupe	77.9	16. Cabbage	58.5
17. String beans	91.0	17. Whole milk	79.4	17. Iceberg lettuce	60.3
18. Peas	91.9	18. Greens	80.8	18. Tuna	62.2
19. Oranges	92.6	19. Strawberries	82.1	19. Cheese	63.9
20. Whole milk	93.4	20. Fortified cold cereal	83.0	20. Whole milk	65.1

[a] From Romieu et al (79).
[b] From Block et al (11).
[c] The cumulative percent of total nutrient intake in the diet provided by the listed foods.

in the selection of participants or in the estimation of diet. Although several validation studies have shown that people can accurately report their diets, the possibility of biased error in dietary reports related to the recent experience of cancer diagnosis and treatment has not been thoroughly studied. An additional problem with case-control studies is that either cancer or its treatment can affect levels of biologic markers of nutrients, thus rendering them virtually useless for drawing causal inference. Prospective studies, in which diet or biological markers of nutrition are ascertained years before diagnosis, have many advantages over case-control studies. However, prospective studies are often plagued by small numbers of incident cases and/or by very cursory measures of diet.

Findings from selected case-control and cohort studies that have examined the relationship between dietary antioxidants and cancer risk are listed in Table 3 (carotenes), Table 4 (vitamin C), and Table 5 (vitamin E). Only those studies published since 1987 that specifically presented findings according to indices of nutrient intake and that included at least 100 cases (or 30 cases if a cohort study) are included in these tables.

The risk of lung cancer is generally elevated for those with lower levels of carotene intake (Table 3). This is consistent with findings from many earlier studies (18). All of these studies have been adjusted for tobacco exposures. Findings are more mixed for dietary vitamin C (Table 4), though there may be more consistency in the observation of higher risk at lower levels of intake in men than in women. Because cigarette smoking lowers levels of both circulating carotenes and vitamin C, low dietary intake may be a risk factor for low circulating levels of antioxidants, compounds in high demand after tobacco use. The observation of higher risk of lung cancer with infrequent intake of fruits and vegetables has been made in several studies (18, 25, 50, 66). Of note is the finding from a cohort study of over one million Americans that those reporting infrequent intakes of fruits were at higher risk of subsequent lung cancer mortality (56).

Breast cancer is less consistently related to carotene intake than is lung cancer, though there may be a weak relationship. Low intakes of vitamin C were associated with elevated breast cancer risk in only two studies. Those two studies were conducted in the Soviet Union (113) and China (53). This suggests that vitamin C deprivation might increase breast cancer risk, but only at a lower threshold of intake than is commonly seen in the United States or Western Europe. Other studies have shown evidence of a protective effect of fruit and vegetable consumption (77), but this finding is not universal (66).

Several studies have examined dietary antioxidants and cancers of the gastrointestinal tract. Stomach cancer is associated with low intakes of vitamin C, as is cancer of the pancreas. The mechanism for the effect of dietary antioxidants may well be the inhibition of nitrosamine formation by ascorbate

Table 3 Selected[a] case-control and cohort studies of the relationship between dietary carotenes and cancer risk

Author (Ref.)	Cancer site	Study design (n:N)[b]	Extremes compared by relative risk[c]	Relative risk[c]
Byers et al (19)	Lung (M)	C:C (296:587)	Quartiles	1.8
	Lung (F)	C:C (154:315)	Quartiles	1.3
Kromhout (52)	Lung (M)	Cohort (63:878)	Quartiles	1.5
Bond et al (12)	Lung (M)	C:C (308:308)	Tertiles	1.2
Wu et al (112)	Lung (F)	C:C (220:220)	Quartiles	2.5
Fontham et al (31)	Lung (M)	C:C (866:982)	Tertiles	1.2
	Lung (F)	C:C (287:292)	Tertiles	1.0
Le Marchand et al (54)	Lung (M)	C:C (230:597)	Quartiles	2.0
	Lung (F)	C:C (102:268)	Quartiles	2.9
Mettlin (62)	Lung (M,F)	C:C (569:569)	Quintiles	2.0
Jain et al (45)	Lung (M,F)	C:C (839:772)	Quartiles	1.1
Hsing et al (42)	Prostate (M)	Cohort (149:17,633)	Quartiles	1.1
Paganini-Hill et al (70)	Prostate (M)	Cohort (92:10,433)	Tertiles	1.0
	Breast (F)	Cohort (123:10,473)	Tertiles	1.2
Howe et al (41)	Breast (F)	C:C (12 studies)	Quartiles	1.2
Zaridze et al (113)	Breast (F)	C:C (139:139)	Quartiles	4.8
Lee et al (53)	Breast (F)	C:C (200:420)	Tertiles	3.4
Van'T Veer et al (100)	Breast (F)	C:C (133:289)	Below vs above 2 mg	1.1
Richardson et al (77)	Breast (F)	C:C (409:515)	Tertiles	1.0
Maclure & Willett (57)	Kidney (M,F)	C:C (203:207)	Quintiles	1.2
Brock et al (14)	Cervix (F)	C:C (117:196)	Quartiles	1.0
Verreault et al (101)	Cervix (F)	C:C (189:227)	Quartiles	1.7
vanEenwyk et al (99)	Cervix (F)	C:C (102:102)	Quartiles	2.8
Shu et al (89)	Ovary (F)	C:C (172:172)	Quartiles	1.0

Table 3 *(Continued)*

Author (Ref.)	Cancer site	Study design (n:N)[b]	Extremes compared by relative risk[c]	Relative risk[c]
Ghadirian et al (35)	Pancreas (M,F)	C:C (179:239)	Quartiles	1.4
Bueno de Mesquita et al (15)	Pancreas (M,F)	C:C (164:480)	Quintiles	1.7
Stryker et al (93)	Skin (M,F)	C:C (204:248)	Quintiles	1.4
Nomura et al (68)	Bladder (M)	C:C (195:390)	Quartiles	1.4
Paganini-Hill et al (70)	Colon (M)	Cohort (52:10,473)	Tertiles	1.1
	Colon (F)	Cohort (58:10,473)	Tertiles	0.9
Freudenheim et al (33)	Colon (M)	C:C (205:205)	Quartiles	1.2
	Colon (F)	C:C (223:224)	Quartiles	1.5
	Rectum (M)	C:C (293:277)	Quartiles	2.9
	Rectum (F)	C:C (151:146)	Tertiles	2.1
Buiatti et al (16)	Stomach (M,F)	C:C (923:1159)	Quintiles	1.4
Sturgeon et al (94)	Vulva (F)	C:C (201:343)	Quintiles	0.8

[a] Only includes studies published since 1987 that present findings by nutrient indices and have at least 100 cases (or 30 if it is a cohort study).
[b] C:C = case-control, (n:N) = the number of cases:controls or the number of incident cases:cohort size.
[c] Relative risk is the ratio of risk for persons in the lowest category of intake compared to those in the highest category of intake.

in the low-pH gastric environment. The dramatic decline in gastric cancer incidence in the past eighty years in developed countries may be due in large part to better vitamin C nutrition year-round (65, 67). Little association has been found between vitamin C and colon cancer, but rectal cancer risk was increased in persons with low intakes of carotenes and/or vitamin C in one study (33). Several investigators have examined the risk of stomach and colon cancers as related to fruit and vegetable intake. In general, these studies tended to indicate elevated risk for those who ate fruits and vegetables infrequently (22, 66, 106), though such effects are not always seen (110). Whether this is an effect of deficiencies in fiber (98), micronutrients, and/or due to the higher fat content of diets of those who infrequently eat fruits and vegetables, is not clear.

Studies of cervical cancer have been generally consistent in demonstrating higher risk at lower levels of intake of carotenes, vitamin C, and vitamin E

Table 4 Selected[a] case control and cohort studies of the relationship between dietary vitamin C and cancer risk

Author (Ref.)	Cancer site	Study design (n:N)[b]	Extremes compared by relative risk[c]	Relative risk[c]
Byers et al (19)	Lung (M)	C:C (296:587)	Quartiles	1.2
	Lung (F)	C:C (154:315)	Quartiles	1.1
Kromhout (52)	Lung (M)	Cohort (63:878)	Quartiles	2.8
Jain et al (45)	Lung (M,F)	C:C (839:772)	Quartiles	0.9
Le Marchand et al (54)	Lung (M)	C:C (230:597)	Quartiles	2.3
	Lung (F)	C:C (102:268)	Tertiles	0.7
Fontham et al (31)	Lung (M)	C:C (866:982)	Tertiles	1.6
	Lung (F)	C:C (287:292)	Tertiles	0.9
Zatonski et al (114)	Pancreas (M,F)	C:C (110:195)	Quartiles	2.7
Ghadirian et al (35)	Pancreas (M,F)	C:C (179:239)	Quartiles	1.8
Bueno de Mesquita et al (15)	Pancreas (M,F)	C:C (164:480)	Quintiles	1.2
Howe et al (41)	Breast (F)	C:C (12 studies)	Quartiles	1.5
Zaridze et al (113)	Breast (F)	C:C (139:139)	Quartiles	3.1
Nomura et al (68)	Bladder (M)	C:C (195:390)	Quartiles	0.8
Freudenheim et al (33)	Colon (M)	C:C (205:205)	Quartiles	1.2
	Colon (F)	C:C (223:224)	Quartiles	0.9
	Rectum (M)	C:C (293:277)	Quartiles	2.3
	Rectum (F)	C:C (151:146)	Tertiles	5.0
Buiatti et al (16)	Stomach (M,F)	C:C (923:1159)	Quintiles	2.0
Brock et al (14)	Cervix (F)	C:C (117:196)	Quartiles	1.7
Verreault et al (101)	Cervix (F)	C:C (189:227)	Quartiles	2.0
Sturgeon et al (94)	Vulva (F)	C:C (201:324)	Quartiles	0.9
Shu et al (89)	Ovary (F)	C:C (172:172)	Quartiles	1.1

[a] Only includes studies published since 1987 that present findings by nutrient indices and have at least 100 cases (or 30 if it is a cohort study).

[b] C:C = case-control, (n:N) = the number of cases:controls or the number of incident cases:cohort size.

[c] Relative risk is the ratio of risk for persons in the lowest category of intake compared with those in the highest category of intake.

Table 5 Selected[a] case control and cohort studies of the relationship between dietary vitamin E and cancer risk

Author (Ref.)	Cancer site	Study design (n:N)[b]	Extremes compared by relative risk[c]	Relative risk[c]
Byers et al (19)	Lung (M)	C:C (296:587)	Quartiles	1.3
	Lung (F)	C:C (154:315)	Quartiles	1.1
Lee et al (53)	Breast (F)	C:C (200:420)	Tertiles	1.7
Richardson et al (77)	Breast (F)	C:C (409:515)	Tertiles	0.8
Verreault et al (101)	Cervix (F)	C:C (189:227)	Quartiles	2.5
Ghadirian et al (35)	Pancreas (M,F)	C:C (179:239)	Quartiles	0.8
Stryker et al (93)	Skin (M,F)	C:C (204:248)	Quintiles	1.4
Buiatti et al (16)	Stomach (M,F)	C:C (923:1159)	Quintiles	2.0

[a] Only includes studies published since 1987 that present findings by nutrient indices and have at least 100 cases (or 30 if it is a cohort study).
[b] "C:C" = case-control, (n:N) = the number of cases:controls or the number of incident cases:cohort size.
[c] Relative risk is the ratio of risk for persons in the lowest category of intake compared with those in the highest category of intake.

(Table 5). However, cervical dysplasia, the premalignant phase of cervical cancer, was not affected by β-carotene supplementation in a therapeutic trial (27). Too few studies of diet and cancers of other sites have been conducted to support any general conclusions.

Studies of Blood Nutrients

In many cohort studies, blood was collected and assayed for nutrients either at the time of the baseline examination or later, using frozen serum sampled in nested case-control designs. Findings from those cohort studies that have examined antioxidant levels in the blood as related to subsequent cancer risk are presented in Table 6. Only those studies with 30 or more incident cancers are included. A general pattern of increased risk with low levels of circulating antioxidants is observed. This pattern is particularly apparent for β-carotene in lung cancer, whereas findings for vitamin C and vitamin E in lung cancer are less convincing. Blood levels are only weakly and inconsistently associated with cancers of the breast, colon, and rectum.

Though studies of blood nutrient markers are useful, they also have important limitations. Dietary intake is just one of many determinants of circulating blood nutrient levels. A single determination of the level of any antioxidant micronutrient at a point in time may therefore be a poor indication of

Table 6 Selected[a] cohort studies of the relationship between antioxidant levels in the blood and subsequent cancer risk

Author (Ref.)	Cancer site	Years of follow-up (n:N)[b]	Dietary anti-oxidants	Categories compared by relative risk[c]	Relative risk[c]
Willett et al (108)	All	5 (111:210)	Total carotene	Quintiles	1.5
	All	5 (111:210)	Vitamin E	Quintiles	1.2
Salonen et al (80)	All	4 (51:51)	Vitamin E	Below vs above 33rd percentile	1.6
Wald et al (103)	All	10 (271:533)	β-Carotene	Quinitiles	1.7
Nomura et al (69)	Lung	10 (74:302)	β-Carotene	Quintiles	2.2
Connett et al (24)	Lung	10 (66:131)	Total carotene	Quintiles	1.8
	Lung	10 (66:131)	β-Carotene	Quintiles	2.3
Stahelin et al (91)	Lung	12 (68:2421)	Carotene	Below vs above 25th percentile	1.8
	Lung	12 (68:2421)	Vitamin E	Below vs above 25th percentile	1.5
	Lung	12 (68:2421)	Vitamin C	Below vs above 22.7 mmole/liter	0.8
Menkes et al (61)	Lung	8 (99:196)	β-Carotene	Quintiles	2.2
	Lung	8 (99:196)	Vitamin E	Quintiles	2.5
Comstock et al (23)	Breast	8 (30:59)	β-Carotene	Quintiles	0.9
	Breast	8 (30:59)	Vitamin E	Quintiles	0.6
	Rectum	8 (34:68)	β-Carotene	Quintiles	0.8
	Rectum	8 (34:68)	Vitamin E	Quintiles	0.6
Schober et al (84)	Colon	8 (72:143)	β-Carotene	Quintiles	1.2
	Colon	8 (72:143)	Vitamin E	Quintiles	1.5
Knekt et al (50)	Stomach	10 (48:841)	Vitamin E	Quintiles	1.6
	Lung	10 (144:841)	Vitamin E	Quintiles	1.4
	Prostate	10 (37:841)	Vitamin E	Quintiles	2.3
	Skin	10 (49:841)	Vitamin E	Quintiles	1.9

[a] Only studies with 30 or more incident cases are included.
[b] (n:N) = the number of incident cases per number of matched cohort members assayed for comparison.
[c] Relative risk is the ratio of risk for persons in the lowest category of intake compared with those in the highest category of intake.

long-term dietary exposures. Nonetheless, cohort study findings based on blood levels of antioxidant micronutrients are generally consistent with the findings from dietary epidemiology studies, which suggest that fruit and vegetable intake can prevent many cancers in humans.

Randomized Controlled Trials

Randomized controlled trials, in which antioxidants are administered blindly to individuals at risk for cancer, are the most definitive means of testing hypotheses related to specific antioxidant compounds. Trials have their own limitations, however. If the protective effects of antioxidants found in foods are dependent on the complex chemical mixtures of whole foods, then supplementation with only a single antioxidant may be an insufficient test of the hypothesis that food-borne antioxidants are cancer-preventive. In addition, randomized controlled trials are often conducted over a relatively short period of time in individuals at high risk for cancer. Trials now underway will therefore be very useful if they yield positive findings, but if they yield negative findings, interpreting the role of food-borne antioxidants over a lifetime for individuals at average risk may be difficult.

Table 7 lists the randomized controlled trials that have been designed to investigate the potential of reducing cancer risk by daily antioxidant supplementation (13). Only two trials have been completed. β-Carotene supplementation had no effect on the rate of new basal cell skin cancers in patients who had previously had a basal cell excised (36), and the rate of recurrence of rectal polyps was unaffected by combined supplementation with vitamins C and E in a small trial on patients who had previously undergone total colectomies for familial polyposis (26). Results from the other trials listed in Table 7 may not be reported for several years. Of special note is the National Cancer Institute trial of increasing fruit and vegetable intake to prevent adenomatous polyps in individuals who have had a polyp resection. This is the first randomized controlled trial in which researchers are using a fundamental modification of the diet as the intervention (seven servings of fruits and vegetables per day) rather than a nutritional supplement.

CONCLUSIONS AND RECOMMENDATIONS

Antioxidant micronutrients, especially carotenes, vitamin C, and vitamin E, appear to play many important roles in protecting the body against cancer. They block the formation of chemical carcinogens in the stomach, protect DNA and lipid membranes from oxidative damage, and enhance immune function. Nevertheless, many important questions need to be answered before either micronutrient supplementation or food fortification can be recommended as a cancer prevention strategy to the general population. Why do micronutrient effects differ by organ site, what are the optimal doses at which

Table 7 Randomized controlled trials of antioxidants in cancer prevention

Cancer site	Antioxidants	Study populations	Investigator/results
Lung	β-Carotene	Cigarette smokers	Lewis Kuller, University of Pittsburg
Lung	β-Carotene	Asbestos workers	Jerry McLarty, University of Texas
Lung	β-Carotene (and retinol)	Asbestos workers	Gilbert Omenn, Fred Hutchinson Center, Washington
Lung	β-Carotene (and retinol)	Cigarette smokers	Gary Goodman, Fred Hutchinson Center, Washington
Lung	β-Carotene and vitamin E	Cigarette smokers	Demetrius Albanes, National Cancer Institute
Colon	Fruits and vegetables	Patients with adenomatous polyps	Arthur Schatzkin, National Cancer Institute
Colon	β-Carotene and vitamins C & E	Patients with adenomatous polyps	Robert E. Greenberg, Dartmouth, New Hampshire
Rectum	Vitamins C & E (and fiber)	Patients with familial polyposis	Jerome DeCosse, Memorial Hospital, New York (26): There was no benefit of combined Vitamin C and E on rectal polyp recurrence after colectomy
Skin	β-Carotene	Albinos in Africa	Jeff Luande, Muhimbili Medical Center, Tanzania
Skin	β-Carotene	Patients with basal cell carcinoma	Robert E. Greenberg, Dartmouth, New Hampshire (36): There was no benefit of β-carotene on basal cell cancer recurrence
Skin	β-Carotene and vitamins C & E	Patients with basal cell carcinoma	Bijan Safai, Memorial Hospital, New York
All sites	β-Carotene	Physicians	Charles Hennekens, Harvard School of Public Health
Esophagus	β-Carotene and vitamin E (and minerals and multivitamins)	Residents in high risk area in China	Philip Taylor, National Cancer Institute

risk is reduced, and are there potential adverse effects are questions that need to be answered. Animal experiments can help us to better understand possible dose-response relationships and to study potential toxicities, but firm answers to important questions about the effects of micronutrients in humans will emerge only from human studies. Randomized controlled trials will be helpful, but they cannot answer questions about the effects of diet on individuals at average risk for cancer. Prospective studies that include better measures of diet need to be designed. Repeated questioning of cohort members using a combination of food frequency questionnaires, interviews, and diet recalls could be employed to construct a set of dietary indicators that would more accurately classify cohort members according to their usual dietary habits.

Although many important questions remain before dietary supplementation and/or food fortification can be recommended to the population, there is a strong scientific basis for current US recommendations that emphasize frequent fruit and vegetable consumption (5, 92). Fruits and vegetables seem to protect the human body against cancer, perhaps because they protect the body against oxidative damage. Achieving the current US Year 2000 goal of doubling the frequency of consumption of fruits and vegetables to five servings per day (75) is therefore likely to have an important effect on the cancer risk of Americans. To achieve this goal will require the cooperation of many in our society to make significant positive changes in the diet of the population (71).

Literature Cited

1. Abul-Hajj, Y. J., Kelliher, M. 1982. Failure of ascorbic acid to inhibit growth of transplantable and dimethylbenzanthracene induced rat mammary tumors. *Cancer Lett.* 17:67

2. Alam, B. S., Alam, S. Q. 1987. The effect of different levels of dietary β-carotene on DMBA-induced salivary gland tumors. *Nutr. Cancer* 9:93–101

3. Alam, B. S., Alam, S. Q., Weir, J. C. 1988. Effects of excess vitamin A and canthaxanthin on salivary gland tumors. *Nutr. Cancer* 11:233–41

4. Ames, B. N. 1983. Dietary carcinogens and anticarcinogens. *Science* 221:1256–64

5. Bal, D. G., Foerster, S. B. 1991. Changing the American diet. Impact on cancer prevention policy recommendations and program implications for the American Cancer Society. *Cancer* 67:2671–80

6. Bendich, A. 1989. Carotenoids and the immune response. *J. Nutr.* 119:112–15

7. Bertram, J. S., Kolonel, L. N., Meyskens, F. L. 1987. Rationale and strategies for chemoprevention of cancer in humans. *Cancer Res.* 47:3012–31

8. Birt, D. F. 1986. Update on the effects of vitamins A, C, and E and selenium on carcinogenesis. *Proc. Soc. Exp. Biol. Med.* 183:311–20

9. Bjorneboe, A., Bjorneboe, G-E., Drevon, C. A. 1990. Absorption, transport and distribution of vitamin E. *J. Nutr.* 120:233–42

10. Block, G. 1991. Vitamin C and cancer prevention: the epidemiologic evaluation. *Am. J. Clin. Nutr.* 53:270S–82S

11. Block, G., Dresser, C. M., Hartman, A. M., Carroll, M. D. 1985. Nutrient sources in the American diet: quantitative data from the NHANES II survey: Vitamins and minerals. *Am. J. Epidemiol.* 122:13–26

12. Bond, G. G., Thompson, F. E., Cook, R. R. 1987. Dietary vitamin A and lung cancer: results of a case-control study among chemical workers. *Nutr. Cancer* 9:109–21

13. Boone, C. W., Kelloff, G. J., Malone, W. E. 1990. Identification of candidate

cancer chemoprevention agents and their evaluation in animal models and human cancer trials, a review. *Cancer Res.* 50:2–9

14. Brock, K. E., Berry, G., Mock, P. A., MacLennan, R., Truswell, A. S., Brinton, L. A. 1988. Nutrients in diet and plasma and risk of in situ cervical cancer. *J. Natl. Cancer Inst.* 30:580–85

15. Bueno de Mesquita, H. B., Maisonneuve, P., Moerman, C. J. 1991. Intake of foods and nutrients and cancer of the exocrine pancreas: a population-based case-control study in The Netherlands. *Int. J. Cancer* 48:540–49

16. Buiatti, E., Palli, D., Bianchi, S., Decarli, A., Amadori, D., et al. 1991. A case-control study of gastric cancer and diet in Italy. III. Risk patterns by histologic type. *Int. J. Cancer* 48:369–74

17. Byers, T. 1988. Diet and cancer. Any progress in the interim? *Cancer* 62:1713–24

18. Byers, T. 1991. Diet as a factor in the etiology and prevention of lung cancer. In *Lung Biology in Health and Disease,* ed. J. Samet. New York: Marcel Dekker. In press

19. Byers, T. E., Graham, S., Haughey, B. P., Marshall, J. R., Swanson, M. K. 1987. Diet and lung cancer risk: findings from the Western New York Diet Study. *Am. J. Epidemiol.* 125:351–63

20. Deleted in proof

21. Chen, L. H., Boissonneault, G. A., Glauert, H. P. 1988. Vitamin C, vitamin E and cancer. *Anticancer Res.* 8:739–48

22. Chyou, P-H., Nomura, A. M. Y., Hankin, J. H., Stemmermann, G. N. 1990. A case-cohort study of diet and stomach cancer. *Cancer Res.* 50:7501–4

23. Comstock, G. W., Helzlsouer, K. J., Bush, T. L. 1991. Prediagnostic serum levels of carotenoids and vitamin E as related to subsequent cancer in Washington County, Maryland. *Am. J. Clin. Nutr.* 53:260S–64S

24. Connett, J. E., Kuller, L. H., Kjelsberg, M. O., Polk, B. F., Collins, G., et al. 1989. Relationship between carotenoids and cancer. The multiple risk factor intervention trial (MRFIT) study. *Cancer* 64:126–34

25. Dartigues, J. F., Davis, F., Gros, N., Moise, A., Bois, G., et al. 1990. Dietary vitamin A, beta carotene and risk of epidermoid lung cancer in South-Western France. *Eur. J. Epidemiol.* 9:261–65

26. DeCosse, J. J., Miller, H. H., Lesser, M. L. 1989. Effects of wheat fiber and vitamins C and E on rectal polyps in patients with familial adenomatous polyposis. *J. Natl. Cancer Inst.* 81:1290–97

27. De Vet, H. C. W., Knipschild, P. G., Willebrand, D., Schouten, H. J. A., Sturmans, F. 1991. The effect of beta-carotene on the regression and progression of cervical dysplasia: A clinical experiment. *J. Clin. Epidemiol.* 44(3): 273–83

28. Di Mascio, P., Murphy, M. E., Sies, H. 1991. Antioxidant defense systems: the role of carotenoids, tocopherols, and thiols. *Am. J. Clin. Nutr.* 53:194S–200S

29. Dorgan, J. F., Schatzkin, A. 1991. Antioxidant micronutrients in cancer prevention. *Hematol. Oncol. Clin. North Am.* 5(1):43–68

30. Dunham, W. B., Zuckerkandle, E., Reynolds, R., Willoughby, R., Marcuson, R., et al. 1982. Effects of intake of L-ascorbic acid on the incidence of dermal neoplasms induced in mice by ultraviolet light. *Proc. Natl. Acad. Sci. USA* 79:7532–36

31. Fontham, E. T., Pickle, L. W., Haenszel, W., Correa, P., Lin, Y., Falk, R. T. 1988. Dietary vitamins A and C and lung cancer risk in Louisiana. *Cancer* 62:2267–73

32. Frei, B., England, L., Ames, B. N. 1989. Ascorbate is an outstanding antioxidant in human blood plasma. *Proc. Natl. Acad. Sci. USA* 86:6377–81

33. Freudenheim, J. L., Graham, S., Marshall, J. R., Haughey, B. P., Cholewinski, S., et al. 1991. Folate intake and carcinogenesis of the colon and rectum. *Int. J. Epidemiol.* 20(2):368–74

34. Fukushima, S., Imaida, K., Sakata, T., Okamura, T., Shibata, M., Ito, M. 1983. Promoting effects of sodium L-ascorbate on two-stage urinary bladder carcinogenesis in rats. *Cancer Res.* 43:4454–57

35. Ghadirian, P., Simard, A., Baillargeon, J., Maisonneuve, P., Boyle, P. 1991. Nutritional factors and pancreatic cancer in the francophone community in Montreal, Canada. *Int. J. Cancer* 47:1–6

36. Greenberg, E. R., Baron, J. A., Stukel, T. A., Stevens, M. M., Mandel, J. S., et al. 1990. A clinical trial of beta carotene to prevent basal-cell and squamous-cell cancers of the skin. *New Engl. J. Med.* 323:789–95

37. Hebert, J. R., Miller, D. R. 1988. Methodologic considerations for investigating the diet-cancer link. *Am. J. Clin. Nutr.* 47:1068–77

38. Henson, D. E., Block, G., Levine, M. 1991. Ascorbic acid: Biologic functions and relation to cancer. *J. Natl. Cancer Inst.* 83(8):547–50

39. Horrobin, D. F. 1991. Is the main problem in free radical damage caused by radiation, oxygen and other toxins the loss of membrane essential fatty acids rather than the accumulation of toxic materials? *Med. Hypotheses* 35:23–26

40. Horvath, P. M., Ip, C. 1983. Synergistic effect of vitamin E and selenium in the chemoprevention of mammary carcinogenesis in rats. *Cancer Res.* 43:5335–41

41. Howe, G. R., Hirohata, T., Hislop, T. G., Iscovich, J. M., Yuan, J-M., et al. 1990. Dietary factors and risk of breast cancer: Combined analysis of 12 case-control studies. *J. Natl. Cancer Inst.* 82:561–69

42. Hsing, A. W., McLaughlin, J. K., Schuman, L. M., Bjelke, E., Gridley, G., et al. 1990. Diet, tobacco use, and fatal prostate cancer: results from the Lutheran brotherhood cohort study. *Cancer Res.* 50:6836–40

43. Ip, C. 1982. Dietary vitamin E intake and mammary carcinogenesis in rats. *Carcinogenesis* 3:1453–56

44. Ip, C. 1986. Interaction of vitamin C and selenium supplementation in the modification of mammary carcinogenesis in rats. *J. Natl. Cancer Inst.* 77:299–303

45. Jain, M., Burch, J. D., Howe, G. R., Risch, H. A., Miller, A. B. 1990. Dietary factors and risk of lung cancer: results from a case-control study, Toronto, 1981–85. *Int. J. Cancer* 45:287–93

46. Jones, F. E., Komorowski, R. A., Condon, R. E. 1981. Chemoprevention of 1,2-dimethylhydrazine-induced large bowel neoplasma. *Surg. Forum* 32:435

47. King, M. M., McCay, P. B. 1983. Modulation of tumor incidence and possible mechanism of inhibition of mammary carcinogenesis by dietary antioxidants. *Cancer Res.* 43:2485–90

48. Knekt, P., Aromaa, A., Maatela, J., Aaran, R. K., Nikkari, T., et al. 1988. Serum vitamin E and risk of cancer among Finnish men during a 10-year follow-up. *Am. J. Epidemiol.* 127:28–41

49. Knekt, P., Aromaa, A., Maatela, J., Aaran, R. K., Nikkari, T., et al. 1991. Vitamin E and cancer prevention. *Am. J. Clin. Nutr.* 53:283S–86S

50. Knekt, P., Jarvinen, R., Seppanen, R., Rissanen, A., Aromaa, A., et al. 1991. Dietary antioxidants and the risk of lung cancer. *Am. J. Epidemiol.* 134:471–79

51. Krinsky, N. I. 1991. Effects of carotenoids in cellular and animal systems. *Am. J. Clin. Nutr.* 53:238S–46S

52. Kromhout, D. 1987. Essential micronutrients in relation to carcinogenesis. *Am. J. Clin. Nutr.* 45:1361–67

53. Lee, H. P., Gourley, L., Duffy, S. W., Esteve, J., Lee, J., et al. 1991. Dietary effects on breast-cancer risk in Singapore. *Lancet* 337:1197–1200

54. Le Marchand, L., Yoshizawa, C. N., Kolonel, L. N., Hankin, J. H., Goodman, M. T. 1989. Vegetable consumption and lung cancer risk: a population-based case-control study in Hawaii. *J. Natl. Cancer Inst.* 81:1158–64

55. London, S., Willett, W. 1989. Diet and the risk of breast cancer. *Hematol. Oncol. Clin. North Am.* 3(4):559–76

56. Longde, W., Hammond, E. C. 1985. Lung cancer, fruit, green salad, and vitamin pills. *Chin. Med. J.* 98:206–10

57. Maclure, M., Willett, W. 1990. A case-control study of diet and risk of renal adenocarcinoma. *Epidemiology* 1:430–40

58. Mathews-Roth, M. M., Krinsky, N. I. 1985. Carotenoid dose level and protection against UV-B induced skin tumors. *Photochem. Photobiol.* 42:35–38

59. Mathews-Roth, M. M., Krinsky, N. I. 1987. Carotenoids affect development of UV-B induced skin cancer. *Photochem. Photobiol.* 46:507–9

60. McCay, P. B., King, M. M., Pitha, J. V. 1981. Evidence that the effectiveness of antioxidants as inhibitors of 7,12-dimethylbenz(α)anthracene-induced mammary tumors is a function of dietary fat composition. *Cancer Res.* 41:3745–48

61. Menkes, M. S., Comstock, G. W., Vuilleumier, J. P., Helsing, K. J., Rider, A. A., et al. 1986. Serum beta-carotene, vitamins A and E, selenium, and the risk of lung cancer. *New Engl. J. Med.* 315:1250–54

62. Mettlin, C. 1989. Milk drinking, other beverage habits, and lung cancer risks. *Int. J. Cancer* 43:608–12

63. Meyskens, F. L. 1990. Coming of age—the chemoprevention of cancer. *New Engl. J. Med.* 323(12):825–27

64. Moon, R. C. 1989. Comparative aspects of carotenoids and retinoids as chemopreventive agents for cancer. *J. Nutr.* 119:127–34

65. National Research Council Committee on Diet and Health, Food and Nutrition Board, Commission on Life Sciences. 1989. *Diet and Health.* Washington, DC: Natl. Acad. Press

66. Negri, E., La Vecchia, C., Franceschi, S., D'Avanzo, B., Parazzini, F. 1991. Vegetable and fruit consumption and cancer risk. *Int. J. Cancer* 48:350–54

67. Nomura, A., Grove, J. S., Stemmer-

mann, G. N., Severson, R. K. 1990. A prospective study of stomach cancer and its relation to diet, cigarettes, and alcohol consumption. *Cancer Res.* 50:627–31

68. Nomura, A. M. Y., Kolonel, L. N., Hankin, J. H., Yoshizawa, C. N. 1991. Dietary factors in cancer of the lower urinary tract. *Int. J. Cancer* 48:199–205

69. Nomura, A. M. Y., Stemmermann, G. N., Heilbrun, L. K., Saldeld, R. M., Vuilleumier, J. P. 1985. Serum vitamin levels and the risk of cancer of specific sites in men of Japanese ancestry in Hawaii. *Cancer Res.* 45:2369–72

70. Paganini-Hill, A., Chao, A., Ross, R. K., Henderson, B. E. 1987. Vitamin A, β-carotene and the risk of cancer. *J. Natl. Cancer Inst.* 79:433–48

71. Patterson, B. H., Block, G. 1988. Food choices and the cancer guidelines. *Am. J. Public Health* 78:282–86

72. Perchellet, J. P., Owen, M. D., Posey, T. D., Orten, D. K., Schneider, B. A. 1985. Inhibitory effects of glutathione level raising agents and D-α-tocopherol on ornithine decarboxylase induction and mouse skin tumor promotion by 12-O-tetradecanoylphorbol-13-acetate. *Carcinogenesis* 6:567–73

73. Peto, R., Doll, R., Buckley, J. D., Sporn, M. B. 1981. Can dietary beta-carotene materially reduce human cancer rates? *Nature* 290:201–8

74. Pryor, W. A. 1991. Can vitamin E protect humans against the pathological effects of ozone in smog? *Am. J. Clin. Nutr.* 53:702–22

75. Public Health Service. 1991. *Healthy People 2000. National Health Promotion and Disease Prevention Objectives—Full report, with commentary. DHHS Publ. (PHS) 91–50212.* Washington, DC: US Dept. Health Hum. Serv., PHS

76. Reddy, B. S., Hirota, N., Katayama, S. 1982. Effect of dietary sodium ascorbate on 1,2-dimethylhydrazine- or methylnitrosurea-induced colon carcinogenesis in rats. *Carcinogenesis* 3:1097

77. Richardson, S., Gerber, M., Cenee, S. 1991. The role of fat, animal protein and some vitamin consumption in breast cancer: a case-control study in southern France. *Int. J. Cancer* 48:1–9

78. Ringer, T. V., DeLoof, M. J., Winterrowd, G. E., Francom, S. F., Gaylor, S. K., et al. 1991. Beta-carotene's effects on serum lipoproteins and immunologic indices in humans. *Am. J. Clin. Nutr.* 53:688–94

79. Romieu, I., Stampfer, M. J., Stryker, W. S., Hernandez, M., Kaplan, L.

1990. Food predictors of plasma beta-carotene and alpha-tocopherol: validation of a food frequency questionnaire. *Am. J. Epidemiol.* 131(5):864–76

80. Salonen, J. T., Salonen, R., Lappetelainen, R., Maenpaa, P. H., Alfthan, G., et al. 1985. Risk of cancer in relation to serum concentrations of selenium and vitamins A and E: matched case-control analysis of prospective data. *Br. Med. J.* 290:417–20

81. Santamaria, L., Bianchi, A. 1989. Cancer chemoprevention by supplemental carotenoids in animals and humans. *Prevent. Med.* 18:603–23

82. Santamaria, L., Bianchi, A., Ravetto, C., Arnaboldi, A., Santagati, G., Andreoni, L. 1987. Prevention of gastric cancer induced by N-methyl-N'-nitro-N-nitrosoguanidine in rats fed supplemental carotenoids. *J. Nutr. Growth Cancer* 4:175–81

83. Schneider, A., Shah, K. 1989. The role of vitamins in the etiology of cervical neoplasia: an epidemiological review. *Arch. Gynecol. Obstet.* 246:1–13

84. Schober, S. E., Comstock, G. W., Helsing, K. J., Salkeld, R. M., Morris, J. S., et al. 1987. Serologic precursors of cancer. 1. Prediagnostic serum nutrients and colon cancer risk. *Am. J. Epidemiol.* 126(6):1033–41

85. Seifter, E., Rettura, G., Levenson, S. M. 1984. Supplemental β-carotene (BC); prophylactic action against 7,12-dimethylbenz(α)anthracene (DMBA) carcinogenesis. *Fed. Proc.* 43:662 (Abstr.)

86. Shamberger, R. 1972. Increase of peroxidation in carcinogenesis. *J. Natl. Cancer Inst.* 48:1491

87. Shirai, T., Ikawa, E., Hirose, M., Thamavit, W., Ito, N. 1985. Modification by five antioxidants of 1,1-dimethylhydrazine-initiated colon carcinogenesis in F 344 rats. *Carcinogenesis* 6:637–39

88. Shklar, G. 1982. Oral mucosal carcinogenesis in hamster: inhibition by vitamin E. *J. Natl. Cancer Inst.* 68:791–97

89. Shu, X. O., Gao, Y. T., Yuan, J. M., Ziegler, R. G., Brinton, L. A. 1989. Dietary factors and epithelial ovarian cancer. *Br. J. Cancer* 59:92–96

90. Soloway, M. S., Cohen, S. M., Dekernion, J. B., Persky, L. 1975. Failure of ascorbic acid to inhibit FANFT-induced bladder cancer. *J. Urol.* 113:483–86

91. Stahelin, H. B., Gey, K. F., Eichholzer, M., Ludin, E., Bernasconi, F., et al. 1991. Plasma antioxidant vitamins and subsequent cancer mortality in the 12-

year follow-up of the prospective basel study. *Am. J. Epidemiol.* 133(8):766–75
92. Steinmetz, K. A., Potter, J. D. 1991. Vegetables, fruit, and cancer. I. Epidemiology. *Cancer Causes and Control* 2:325–57
93. Stryker, W. S., Stampfer, M. J., Stein, E. A., Kaplan, L., Louis, T. A., et al. 1990. Diet, plasma levels of beta-carotene and alpha-tocopherol, and risk of malignant melanoma. *Am. J. Epidemiol.* 131(4):597–611
94. Sturgeon, S. R., Ziegler, R. G., Brinton, L. A., Nasca, P. C., Mallin, K., et al. 1991. Diet and the risk of vulvar cancer. *Ann. Epidemiol.* 1:427–37
95. Sun, Y. 1990. Free radicals, antioxidant enzymes, and carcinogenesis. *Free Radic. Biol. & Med.* 8:583–99
96. Suniyoshi, H. 1985. Effects of vitamin E deficiency on 1,2-dimethylhydrazine-induced intestinal carcinogenesis in rats. *Hiroshima J. Med. Sci.* 34:363–69
97. Temple, N. J., Basu, T. K. 1987. Protective effect of β-carotene against colon tumors in mice. *J. Natl. Cancer Inst.* 78:1211–14
98. Trock, B., Lanza, E., Greenwald, P. 1990. Dietary fiber, vegetables, and colon cancer: critical review and meta-analyses of the epidemiologic evidence. *J. Natl. Cancer Inst.* 82:650–61
99. VanEenwyk, J., Davis, F. G., Bowen, P. E. 1991. Dietary and serum carotenoids and cervical intraepithelial neoplasia. *Int. J. Cancer* 48:34–38
100. Van'T Veer, P., van Leer, E. M., Rietdijk, A., Kok, F. J., Schouten, E. G., et al. 1991. Combination of dietary factors in relation to breast-cancer occurrence. *Int. J. Cancer* 47:649–53
101. Verreault, R., Chu, J., Mandelson, M., Shy, K. 1989. A case-control study of diet and invasive cervical cancer. *Int. J. Cancer* 43:1050–54
102. Vogel, V. G., McPherson, R. S. 1989. Dietary epidemiology of colon cancer. *Hematol. Oncol. Clin. North. Am.* 3(1):35–62
103. Wald, N. J., Thompson, S. G., Densem, J. W., Boreham, H., Bailey, A. 1988. Serum beta-carotene and subsequent risk of cancer: results from the BUPA study. *Br. J. Cancer* 57:428–33
104. Wang, Y. M., Howell, S. K., Kimball, J. C., Tsai, C. C., Stao, J., et al. 1982. Alpha-tocopherol as a potential modifier of daunomycin carcinogenicity in Sprague-Dawley rats. In *Molecular Interrelations of Nutrition and Cancer,* ed. M. S. Arnott, Y. M. Wang, pp. 369–79. New York: Academic

105. Weisburger, J. H. 1991. Nutritional approach to cancer prevention with emphasis on vitamins, antioxidants, and carotenoids. *Am. J. Clin. Nutr.* 53: 226S–37S
106. Whittemore, A. S., Wu-Williams, A. H., Lee, M., Zheng, S., Gallagher, R. P., et al. 1990. Diet, physical activity, and colorectal cancer among Chinese in North America and China. *J. Natl. Cancer Inst.* 82:915–26
107. Willett, W. C. 1990. Reproducibility and validity of food frequency questionnaire. In *Nutritional Epidemiology,* pp. 92–126. New York: Oxford Univ. Press
108. Willett, W. C., Polk, F., Underwood, B. A., Stampfer, M. J., Pressel, S., et al. 1984. Relation of serum vitamins A and E and carotenoids to the risk of cancer. *New Engl. J. Med.* 310:430–34
109. Willett, W. C., Reynolds, R. D., Cottrell-Hoehner, S., Sampson, L., Browne, M. L. 1987. Validity of semiquantitative food frequency questionnaire: comparison with a one-year diet record. *J. Am. Diet. Assoc.* 87:43–47
110. Willett, W. C., Stampfer, M. J., Colditz, G. A., Rosner, B. A., Speizer, F. E. 1990. Relation of meat, fat, and fiber intake to the risk of colon cancer in a prospective study among women. *New Engl. J. Med.* 323:1664–72
111. Willett, W. C., Stampfer, M. J., Underwood, B. A., Speizer, F. E., Rosner, B., et al. 1983. Validation of a dietary questionnaire with plasma carotenoid and α-tocopherol levels. *Am. J. Clin. Nutr.* 38:631–39
112. Wu, A. H., Henderson, B. E., Pike, M. C., Yu, M. C. 1985. Smoking and other risk factors for lung cancer in women. *J. Natl. Cancer Inst.* 74:747–51
113. Zaridze, D., Lifanova, Y., Maximovitch, D., Day, N. E., Duffy, S. W. 1991. Diet, alcohol consumption and reproductive factors in a case-control study of breast cancer in Moscow. *Int. J. Cancer* 48:493–501
114. Zatonski, W., Przewozniak, K., Howe, G. R., Maisonneuve, P., Walker, A. M., et al. 1991. Nutritional factors and pancreatic cancer: a case-control study from South-West Poland. *Int. J. Cancer* 48:390–94
115. Ziegler, R. G. 1991. Vegetables, fruit, and carotenoids and the risk of cancer. *Am. J. Clin. Nutr.* 53:251S–59S
116. Ziegler, R. G. 1989. A review of epidemiologic evidence that carotenoids reduce the risk of cancer. *J. Nutr.* 119:116–22

Annu. Rev. Nutr. 1992. 12·161–81

RETINOIDS AND CANCER PREVENTION

Donald L. Hill

Biochemistry Research, Southern Research Institute, Birmingham, Alabama 35255

Clinton J. Grubbs

Department of Nutrition Sciences, University of Alabama at Birmingham, Alabama 35294

KEY WORDS: carcinogen, skin, mammary, leukemia, bladder

CONTENTS

Perspectives

Prevention of cancer may soon be a possibility for people who, because of genetic disposition or environmental exposure to carcinogens, have a high risk for developing the disease. Compounds representing several diverse chemical classes are being tested in animals and in humans for their effectiveness in combatting cancer (chemoprevention). One such class is the ret-

161

0199-9885/92/0715-0161$02.00

inoids, which consists of the natural vitamin A compounds [retinol, retinal, and all-*trans*-retinoic acid (RA)] and their derivatives and analogs (Figure 1).

When administered to humans, retinoids have demonstrated activity in preventing cancers of the skin (67, 71, 101), head and neck (59), lungs (98), and bladder (129). Further, they are effective in the treatment of leukoplakia (58), a preneoplastic disease; acute promyelocytic leukemia (60); and myelodysplastic syndromes (12). At present, in six different clinical trials, retinoids are being evaluated in patients with the preneoplastic diseases of cervical dysplasia, asbestosis, and actinic keratoses and in high-risk groups including cigarette smokers, women previously treated for breast cancer, and patients with a predisposition to develop basal cell carcinomas (18). The promise of retinoids in such trials is based, to some extent, on animal experiments, in which these compounds demonstrate activity in preventing cancer of the skin, forestomach, liver, mammary gland, and bladder (see 88).

To detect preventive activity of retinoids, chemical carcinogens are ordinarily employed to induce cancer in experimental animals. These carcinogens vary in structure. There are polycyclic hydrocarbons, such as 7,12-

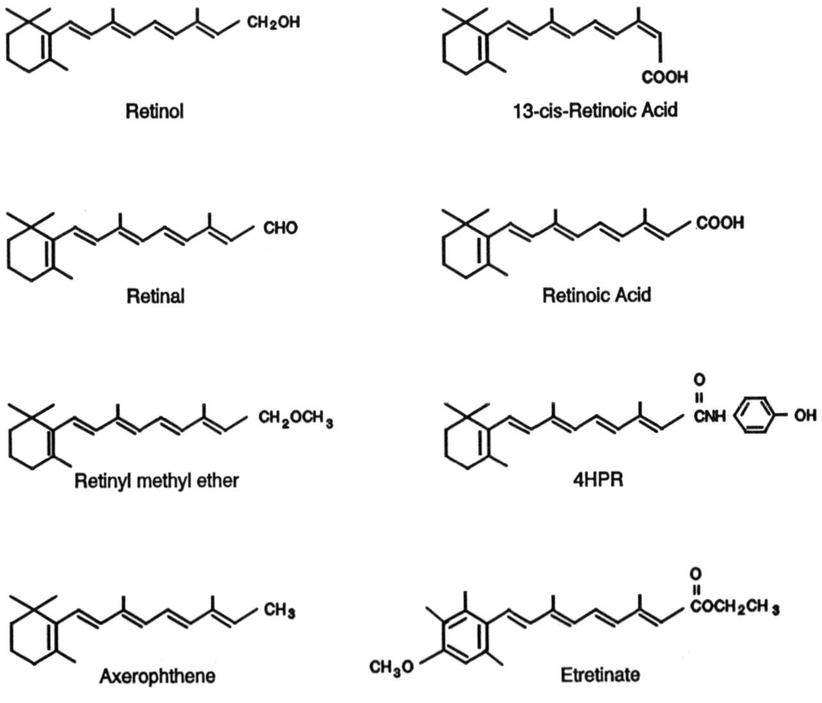

Figure 1   Structures of selected retinoids.

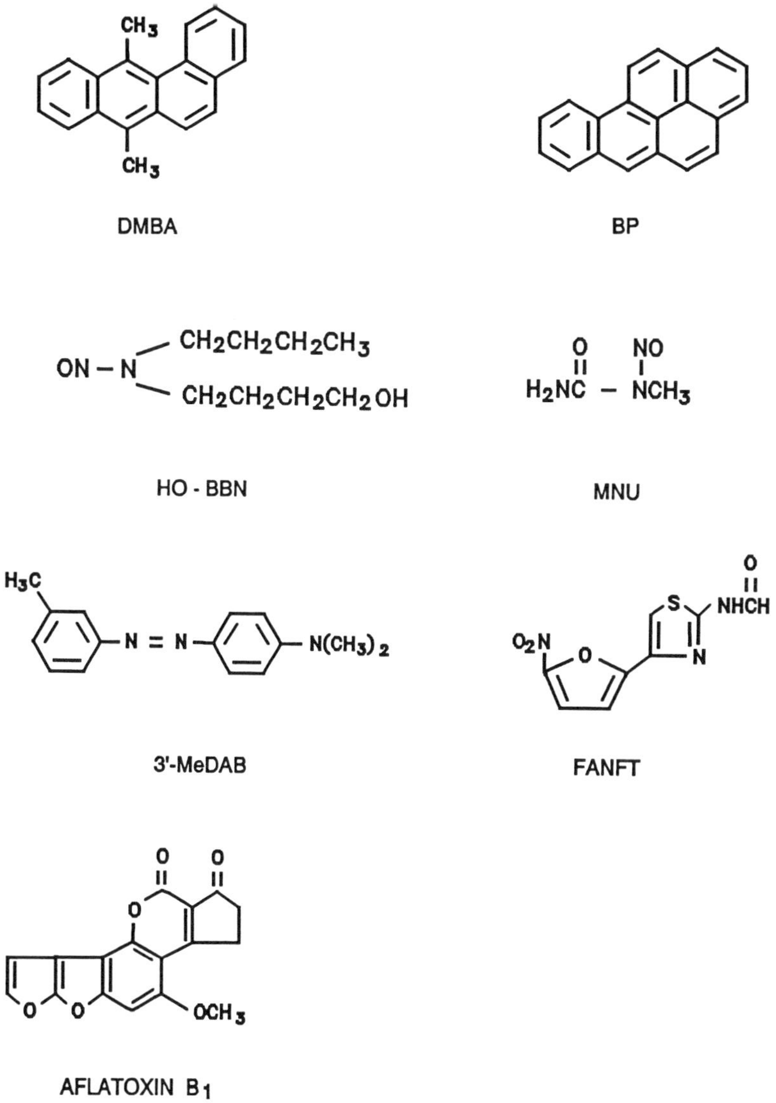

DMBA

BP

HO - BBN

MNU

3'-MeDAB

FANFT

AFLATOXIN B₁

Figure 2 Structures of selected carcinogens.

dimethylbenz(α)anthracene (DMBA) and benzo(α)pyrene (BP); nitrosamines, such as N-butyl-N-(4-hydroxybutyl)nitrosamine (HO-BBN); nitro-
soureas, such as methylnitrosourea (MNU); azobenzenes, such as 3'-
dimethylaminoazobenzene (3'-MeDAB); and chemicals with complex structures, such as N-[4-(5-nitro-2-furyl)-2-thiazolyl]formamide (FANFT) and
aflatoxin B_1 (Figure 2).

Carcinogenesis occurs in cells through a process involving initiation, promotion, and progression. During initiation, an irreversible change takes place, involving production of a mutation in the genome usually as a result of interaction of the carcinogen with cellular DNA. Promotion occurs when initiated cells are converted to the tumor phenotype by a chemical that is not an initiator and, by itself, not a carcinogen. In the final stage, the tumor becomes highly malignant and grows rapidly.

The action of retinoids in preventing cancers of the skin and mammary gland is generally thought to be accomplished via an antipromoting effect (46, 138), but their mechanism of action in preventing other types of cancer is not yet known. Retinoids are also immunostimulants, a fact that may account for their observed activity against established cancers (see 55). Both cell-mediated cytotoxicity and that of natural killer cells are enhanced by RA. In some systems, retinoids cause leukemic cells to differentiate, irreversibly converting them to morphologically mature granulocytes with the same functional markers as the mature neutrophil (19).

At the molecular level, retinoids seem to modify gene expression through the mediation of intracellular-binding proteins and nuclear receptors (111). The mode of action of retinol and RA in the control of differentiation and tumorigenesis apparently involves cellular retinol-binding protein (CRBP) and cellular RA-binding protein (CRABP) (6, 109). These proteins may be involved in the transport of retinoids to nuclear sites, where they interact with their receptors (RARs) (45, 111). Binding of retinoids to CRABP and to RARs requires a free terminal carboxyl group (109); a hydroxyl or lipophilic terminal group is required for binding to CRBP (56, 97).

In general, retinoids that are active in the prevention of murine leukemias and lymphomas and murine papillomas and carcinomas of the skin have a sidechain that possesses, or is readily convertible to, a free carboxylic acid group. The presence of such a group appears to be necessary, but not sufficient, for chemopreventive activity in the skin. In contrast, retinol and its derivatives and other retinoids lacking a terminal functional group are most active in preventing cancer of the mammary gland. However, to exert their activity, these retinoids are not likely to be converted to compounds with free carboxyl groups, because RA and 13-*cis*-RA have demonstrated little chemopreventive activity against mammary cancer (90).

Because retinamides show activity in preventing breast and bladder cancer, do not bind to either of these proteins, and are not readily converted to RA (118), other metabolic conversions may be involved in the formation of active metabolites of these compounds. Alternatively, there may be a separate class of binding proteins or receptors for retinamides and other retinoids that do not bind to the known RARs.

We conclude that at least three distinct classes of retinoids, each with its own biochemical properties, are active in chemoprevention: one characterized by a terminal carboxylic acid group that is active in preventing skin cancer, one by a hydroxyl or nonpolar terminal group that is active in preventing mammary cancer, and one by a terminal amide group that is active in preventing bladder and mammary cancer.

Pharmacokinetics also appears to be a factor in the chemopreventive potential of retinoids (see 55, 56, 62). RA, 13-*cis*-RA, *N*-(4-hydroxyphenyl)retinamide (4-HPR), and *N*-(2-hydroxyethyl)retinamide are effective in the prevention of bladder cancer, and, in this tissue, relative to other tissues, they have prolonged half-lives. For chemoprevention of lung cancer, however, neither 13-*cis*-RA, *N*-(2-hydroxyethyl)retinamide, nor 4-HPR has appreciable activity; 13-*cis*-RA is likewise inactive in the prevention of colon cancer. Accordingly, loss of these compounds from the respective tissues is relatively rapid. Further, retinyl methyl ether, 4-HPR, and axerophthene, none of which is readily converted to a structure with a free carboxyl group, show activity in preventing breast cancer and accumulate in breast tissue. *N*-(4-Methoxyphenyl)retinamide, the major metabolite of 4-HPR, also accumulates in the mammary gland of rats dosed with this metabolite. Accumulation of retinoids in the mammary gland is not related entirely to their lipophilicity, for retinyl butyl ether, which is less polar than retinyl methyl ether but has less chemopreventive activity than retinyl methyl ether, does not accumulate to the same extent. In the following sections we discuss the cancer preventive activity of retinoids in various tissues and organs.

Skin

A model system involving Swiss or CD1 mice is used extensively in evaluation of retinoids for the prevention of papillomas and basal cell carcinomas. In this system, DMBA is applied to the skin, and later the test retinoid is applied along with croton oil or the active ingredient in this oil, 12-*O*-tetradecanoylphorbol-13-acetate (TPA). Retinoids active in this system are etretinate, RA, retinal, retinol, retinyl acetate, retinyl palmitate, 13-*cis*-RA, 5,6-epoxy-RA, 5,6-dihydro-RA, various arotinoids (see 30, 55, 136, 139, 140), and various 3-substituted 4-oxoretinoic acids (117); other retinoids tested have little or no effect. Administered in the feed of mice, retinyl palmitate, but not 13-*cis*-RA, is active in preventing papillomas (42). With DMBA as an initiator and either anthralin (31) or 7-bromomethylbenz(α)anthracene (138) as a promoter, RA applied topically reduces the frequency of papilloma formation. When BP or *N*-methyl-*N*'-nitro-*N*-nitrosoguanidine is the initiator and no promoter is given, however, topically applied 13-*cis*-RA does not prevent papilloma formation (40). Similarly, when BP is administered without a promoter, orally administered retinyl palmitate and RA are not effective in preventing skin papillomas and carcinomas (114). After BP has been applied to the skin of mice as an initiator and TPA and 13-*cis*-RA have been applied for several weeks, inhibition of tumor promotion is stable in the absence of further promotion by TPA (41).

When DMBA is applied without a promoter to the skin of CD1 mice, RA does not reduce the frequency of papilloma formation (136). In fact, the frequency of papillomas can increase (51). In contrast, when 3-methylcholanthrene (3MC) is administered without a promoter, retinyl palmitate and 13-*cis*-RA are effective in reducing the frequency of skin papillomas and carcinomas (1). Several experiments have been performed with SENCAR mice, which develop cancer more readily than other strains. In the model system involving DMBA and TPA, RA is active in reducing the frequency of papillomas (30). RA is also active in this system in which mezerein is substituted for TPA (137). When no promoter is administered along with DMBA, however, RA and 13-*cis*-RA, given in the feed or applied topically, increase the frequency of papilloma formation (78). Administered in the feed, 4-HPR decreases the frequency of papillomas; however, applied topically, it has no effect on the frequency (77). When SENCAR mice are dosed with BP and anthralin as initiator and promoter, respectively, 13-*cis*-RA has no effect on the frequency of papilloma formation (40).

A related system involves induction of papillomas with DMBA and croton oil or TPA on the skin of Swiss mice and treatment of the established tumors with retinoids administered orally or intraperitoneally. Retinoids with demonstrated activity in this system are RA, retinyl palmitate, etretinate, motretinide, ethyl retinoate, acitretin, 7,8-dihydro-RA, 8,9-dihydroacitretin, 4,5-dihydroacitretin, various arotinoids, and various fluorinated aromatic retinoids (see 22, 55, 74, 99, 100). Other retinoids tested have little or no effect. Regression of established basal cell carcinomas has been noted in mice treated with etretinate (16).

When administered orally to mice previously exposed to ultraviolet light, neither retinyl palmitate, etretinate, RA, nor 13-*cis*-RA has any effect on the frequency of formation of papillomas and squamous cell carcinomas (63, 65, 144). Further, when RA is applied to the skin of these mice, an increased frequency of squamous cell carcinomas is noted (34, 36). Although a recent report (39) indicates that retinyl palmitate administered continually before, during, and after exposure to ultraviolet light, reduced the total volume of papillomas that are formed, retinoids show little promise in the prevention of skin cancers caused by ultraviolet light.

For humans, topical application of RA can cause the regression of dysplastic nevi (33) and cutaneous lesions of malignant melanoma (69). Some basal cell carcinomas (101), squamous cell carcinomas (see 71), and cutaneous lesions of malignant melanoma (87) respond to oral administration of 13-*cis*-RA. Patients with xeroderma pigmentosum develop skin cancers at a slower rate when they are dosed orally with 13-*cis*-RA (67); and humans with actinic keratosis, a premalignant lesion, show therapeutic responses to etretinate (91).

Oral, Head, and Neck Cancer

Reports are conflicting about the efficacy of retinyl esters applied directly to the cheek pouch of hamsters in which buccal tumors are induced by administration of DMBA. Some reports indicate that these compounds increase the incidence of oral tumors (see 84), but one indicates that retinyl palmitate delays the induction of such tumors (64). Overall, the results are not promising for further experiments involving topical application of retinyl esters. Conversely, three reports note that oral administration of 13-*cis*-RA can produce favorable results. Not only can the incidence of buccal tumors induced by DMBA be reduced (120), but the same result can be obtained for those tumors induced by application of this carcinogen to the tongue (43, 119). These results demonstrate that selection of the appropriate retinoid and the appropriate route of administration are important factors in experiments involving chemoprevention by retinoids.

For humans, 13-*cis*-RA is used to treat oral leukoplakia, a disease that has a low, but definite, incidence of malignant transformation. The observed responses, 9 of 11 (116) and 16 of 24 (58), demonstrate the compound is moderately effective. In a recent and promising development, patients who had been successfully treated for head and neck cancer were placed in a clinical trial to evaluate the preventive effect of 13-*cis*-RA versus that of a placebo (59). Although no difference was noted between the two groups in recurrence of the tumor at the primary site, there were significantly fewer occurrences of second primary tumors (2 of 49 in the treated group versus 12 of 51 in the control group).

Leukemias and Lymphomas

The demonstration that retinoids can have an effect on cancers of the blood in experimental animals and humans is a relatively recent development. C57B1/10W mice exposed to X rays and placed on a diet containing 13-*cis*-RA developed fewer thymic lymphomas than did controls (103); on a similar diet, AKR mice, in which thymic lymphomas spontaneously appear, developed fewer lymphomas than did controls (104). In both cases, however, the administered dose prevented the weight gain seen in controls, which led to the conclusion that the observed effect on lymphomas could be related to retinoid toxicity. In mice, intraperitoneal administration of etretinate reduced the incidence of spontaneously occurring lymphoma (20); and administration of retinyl palmitate to newborn Swiss mice, the mothers of which had been exposed to ethylnitrosourea, prevented the development of leukemias in about half of the offspring (150). In the latter report, there was no mention of retinoid toxicity. Retinyl palmitate, applied to the skin of mice after application of an extract of pepper, reduced the number of lymphomas developing in the spleens of these animals; oral administratin of retinyl palmitate was ineffective following skin application or feeding of the pepper extract (121). No decrease in the incidence of leukemias was noted in Long-Evans rats dosed with etretinate, during and/or after dosing with DMBA (11).

More promising results have been demonstrated in treatment of a type of human leukemia. All 24 patients with acute promyelocytic leukemia, some of whom were unresponsive to chemotherapy, experienced complete responses after receiving doses of RA (60). In subsequent experiments, 14 complete responses in 22 patients (21) and 9 complete responses in 11 patients were noted (142). Similarly, dosing with 13-*cis*-RA produced responses in about one third of patients with myelodysplastic syndromes (see 12). Nevertheless, dosing with 4-HPR produced no responses in 14 patients with myelodysplastic syndromes and may, in fact, have enhanced the leukemic progression in these individuals (38). 13-*cis*-Retinoic acid is apparently ineffective in the treatment of acute myeloid leukemia in elderly patients (68) and acute nonlymphocytic leukemia in pediatric patients (9). These results demonstrate that retinoids can be active in blood diseases but that different retinoids can have differing activities, even in humans. The retinamides especially should be administered with caution.

Mammary Gland

Reports from various laboratories demonstrate prevention by retinoids of mammary tumors induced in rats by either of several different carcinogens. With DMBA as the carcinogen, retinyl acetate and retinyl methyl ether are reasonably effective (see 55). Other retinoids that cause a reduction in frequency of mammary cancers are 4-HPR (17, 46, 79); temarotene, an arotinoid without a functional end group on the sidechain (131); and either of two other arotinoids (49). 13-*cis*-RA, at nontoxic doses, has little or no chemopreventive effect for mammary cancer (2). One report indicates that 4-HPR is not effective in reducing the frequency of mammary tumors induced by either DMBA or MNU (122), but these negative results may be due to the fact that the investigators used a rat chow different from that used by others.

With MNU as the carcinogen, retinyl acetate and retinyl methyl ether are effective in preventing mammary cancers (see 55, 117, 146). Other active retinoids are 4-HPR (see 46, 55); retinyl propynyl ether (117); and axerophthene, the hydrocarbon analog of retinol (133). Retinoids reported to be without activity are the methyl ether analog of etretinate and 13-*cis*-RA (117, 134). Another report demonstrates that retinyl acetate, fed prior to but not after MNU, causes an increase in the number of adenocarcinomas (46).

Nevertheless, when this retinoid is fed continuously, the number of adenocarcinomas that develop is greatly reduced.

Retinyl acetate is also effective, in rats, in reducing the frequency of adenocarcinomas that are induced by exposure to X rays (128), by BP (80), or by estrogens (57) as well as those that occur spontaneously (128).

In contrast to the encouraging results with rats, mice that have been dosed with DMBA develop mammary tumors that are not prevented by either retinyl acetate or 4-HPR (145). Further, retinyl acetate is not effective in preventing the spontaneous mammary cancers that develop in C3H-A mice (76). If mice are on a diet containing retinyl acetate during the time that they are being dosed with estrogen plus progesterone, the frequency of mammary tumors is increased (147).

In the mammary gland, the retinol-binding protein, CRBP, may be involved in mediating the chemopreventive activity of retinoids, for most compounds that are active in this system also bind to this protein (56). Although 4-HPR is a notable exception, this discrepancy may not invalidate the relationship, since an unidentified metabolite, which may bind to CRBP, is thought to be the active form of 4-HPR (85).

Lung

An early report indicated that retinyl palmitate, when administered to hamsters, prevented tumors induced in the respiratory tract by instillation of BP + ferric oxide (108). Although this experiment has not been repeated exactly, attempts to demonstrate a response with retinyl acetate in similar situations were unsuccessful (8, 124, 125). Another observation not confirmed is that 13-cis-RA prevented the formation of lung cancer caused by BP + ferric oxide (102). In this experiment, the number of cancers was small, and no detailed description of the results has appeared.

Attempts to use retinoids to prevent tracheal cancer caused by instillation of MNU into hamsters were not successful (see 55). In contrast, hamsters dosed with diethylnitrosamine (DEN) developed fewer lung carcinomas when they were placed on a diet containing 4-HPR (86), and hamsters dosed with 2,6-dimethylnitrosomorpholine (DMNM) developed fewer lung adenomas and carcinomas when they were placed on a diet containing 13-cis-RA (127).

An attempt to prevent lung cancer caused by administration of dibutylnitrosamine to rats by weekly dosing with retinyl palmitate was unsuccessful (114), and development of lung adenomas in the offspring of mice dosed with ethylnitrosourea (ENU) on day 14 of pregnancy was not affected by placing retinyl palmitate in the drinking water (150). Mice dosed with ethyl carbamate, however, developed fewer lung tumors when they are also administered N-homocysteine thiolactonyl retinamide (82) or N-homocysteine thiolactonyl retinamido cobalamin (83).

Although the experiments with animals have not demonstrated great promise for the use of retinoids in the prevention of lung cancer, two clinical trials have been encouraging. Etretinate, administered over a six-month period, reduced the degree of bronchial metaplasia in humans with extensive exposure to cigarette smoke (44); and patients with resected stage 1a lung cancer had fewer recurrences if they were dosed with retinyl palmitate for 14 months (98).

Liver

Retinoids may be active in preventing hepatic cancer. Retinyl acetate reduced the incidence of spontaneous liver tumors developing in C3H-A^{vy} mice (76), and RA reduced the incidence of liver tumors induced in Sprague-Dawley rats by 3'-MeDAB (29). Acitretin prevented spontaneous hepatomas in C3H/HeNCrj mice and 3'-MeDAB-induced tumors in rats (94), and retinyl acetate delayed the elevation of hepatic γ-GTPase and the appearance of preneoplastic nodules (75). 4-HPR did not reduce the incidence of spontaneously developing liver tumors in C3H/He mice, but did reduce the incidence of tumors induced by diethylnitrosamine in BALB/c mice (61). RA did not prevent liver tumors in BDF mice dosed with diethylnitrosamine (81). In Lewis rats, N-(2-hydroxyethyl)retinamide reduced the incidence of liver cell carcinomas (27).

Some disturbing reports regarding retinamides and liver cancer have appeared. Female rats dosed with the carcinogen azaserine and either of three retinamides had an increased incidence of liver cancer (72). Further, after 72 weeks of administration of 13-cis-N-ethylretinamide, hepatocellular carcinomas and adenomas developed in mice dosed with this compound only or with HO-BBN and the retinoid (54). No such results, however, were seen when N-ethylretinamide was administered for 1–2 years (53). In a separate experiment, both N-ethylretinamide and 13-cis-N-ethylretinamide increased the incidence of liver tumors in mice (81). In hamsters, N-(2-hydroxyethyl)retinamide may have prevented (14) or, along with other retinamides and 13-cis-RA, enhanced (13) the development of liver adenomas.

Whether or not only retinamides cause such toxicity and whether or not such an effect is limited to rodents remains to be determined. Nevertheless, for prevention of cancer, retinamides should be used with caution. Considerable experimental work must be performed before a biochemical understanding of these results can be achieved.

Pancreas

Various retinamides reduce the incidence of pancreatic carcinomas in rats dosed with azaserine (see 27). Retinamides reduce the frequency of acidophilic foci, which are thought to be preneoplastic lesions, in the pancreata of these rats (106). The results are not as convincing, however, for hamsters dosed with N-nitrosobis(2-oxopropyl)amine (BOP). In these animals, retinamides may reduce the incidence (73), have no effect (14), or may enhance (13) the frequency of pancreatic carcinomas. Nevertheless, 13-cis-RA may reduce the incidence of pancreatic tumors in hamsters dosed with DMNM (127). Again, retinamides should be used with caution in experiments designed to prevent cancer.

Bladder

The activity of retinoids in preventing carcinogen-induced bladder cancer was established by studies in several different laboratories, which involved two species of animals and two different chemical carcinogens. Various retinoids were used to reduce the occurrence of papillomas, transitional-cell carcinomas, and squamous-cell carcinomas induced in rats and mice by HO-BBN (see 52, 53, 55, 89). The most effective retinoids in rats dosed with MNU were 13-cis-RA, RA, retinyl acetate, 4-HPR, and some alkyl retinamides; other alkyl retinamides had no measurable activity (126, 130). A nine-week delay in starting the feeding of 13-cis-RA to rats dosed with HO-BBN did not diminish its inhibition of bladder carcinogenesis (7). One group of investigators found that N-(2-hydroxyethyl)retinamide had chemopreventive activity in the bladders of rats (132); another group could not confirm such activity (105). Likewise, etretinate was reported to be both active (93) and inactive (113) in rats.

In contrast, transitional cell neoplasms induced by FANFT were not responsive to either 13-cis-RA, retinyl palmitate, N-ethylretinamide, or N-(2-hydroxyethyl)retinamide (see 26), and bladder cancers induced in rats dosed with dibutylnitrosamine were not prevented by retinyl palmitate (114).

In humans, oral administration of RA resulted in 4 complete and 7 partial remissions in 15 patients with recurrent papillomas of the bladder (35); oral administration of etretinate produced 6 complete and 5 partial remissions in 15 similar patients (5). In a double-blind, randomized clinical trial that had proceeded for 24 months, etretinate reduced the incidence of recurrent bladder tumors (129). Recurrences were observed in 9 of 16 patients dosed with a placebo but in only 4 of 14 dosed with etretinate. Retinamides have not yet been tested in humans for such activity. In view of the hepatocarcinogenicity of retinamides on long-term administration to rodents, these compounds should not be tested further until it is determined if such an effect is limited to rodents.

Colon

The most encouraging reports regarding the prevention of colon cancer by retinoids have shown a reduction in the number of tumors per rat in rats dosed simultaneously with 1,2-dimethylhydrazine (DMH) and retinyl palmitate (107); a modest delay in the time to tumor development in rats dosed simultaneously with DMH and 13-*cis*-RA (96); and a reduction in the frequency of colon tumors in rats dosed simultaneously with DMH and 13-*cis*-RA (95). In another test (123), oral administration of retinyl acetate or 4-HPR to rats dosed with DMH caused a decrease in the number of tumor-bearing rats compared to that of controls, but retinoid-dosed animals showed decreased food consumption and decreased body weight gain owing to the toxicity of the retinoids. In the same study, several other retinoids, administered in the feed to rats dosed with either MNU or DMH, did not reduce the frequency of colon adenomas or carcinomas, and intrarectal administration of retinyl palmitate before dosing with MNU increased the percentage of tumor-bearing rats compared to that of controls. Further negative results were reported in several attempts to prevent colon cancer in rats by administration of retinoids (see 32). Thus, although several different retinoids have been tested for the prevention of colon cancer, the promise for success is modest.

Other

Retinyl palmitate added to the feed reduced the incidence of sarcomas induced in mice by Moloney murine sarcoma virus (115), and intraperitoneal administration of etretinate reduced the incidence of sarcomas induced in hamsters by Rous sarcoma virus (37). Intramuscular etretinate also reduced the incidence of papillomas induced on rabbit skin by application of Shope papilloma virus (37). Retinyl palmitate, administered orally and simultaneously with BP, had no effect on the production of spindle-cell sarcomas in rats (114); but intraperitoneal administration of RA to mice reduced the incidence of fibrosarcomas produced by 3MC (23). The effects of retinoids in preventing sarcomas may be related to their immunoenhancing effects. Topical application of retinoids in such experiments can not be recommended.

Orally administered retinyl palmitate reduced the incidence of forestomach papillomas induced in hamsters by intratracheal application of BP + ferric oxide (108, 124, 125). Further, hamsters dosed with DMBA or BP and retinyl palmitate developed fewer forestomach carcinomas than did those dosed with the carcinogen alone (24). Nevertheless, 13-*cis*-RA had no effect on the formation of forestomach papillomas in hamsters dosed with DMNM (127); etretinate did not reduce the incidence of forestomach papillomas that occur spontaneously in mice (151); and retinyl acetate, administered in the drinking water, increased the incidence of forestomach papillomas produced by butylated hydroxyanisole (50). Further experiments are required before a general conclusion can be drawn about the effectiveness of retinoids in these model systems.

Neither 13-*cis*-RA nor motretinide reduced the incidence of kidney tumors in rats dosed with DMN (48); and 13-*cis*-RA was inactive in reducing the incidence of kidney tumors in hamsters dosed with DMNM (127). Neither RA nor retinyl palmitate demonstrated activity in preventing tumors induced by DMBA in the salivary glands of rats (3, 4). Retinyl palmitate (114) and 13-*cis*-RA (47) were also ineffective in preventing tumors induced by MNU in the central and peripheral nervous systems of rats.

In regard to prevention of esophageal cancer by retinoids, a promising report stated that administration of retinyl palmitate to hamsters also dosed with DMBA reduced the incidence of esophageal lesions (24), but the experiments described have yet to be confirmed. Attempts by others to demonstrate preventive effects of retinyl acetate, 13-*cis*-RA, and etretinate for esophageal tumors produced by N-nitrosomethyl benzylamine (NMBA) have been uniformly unsuccessful (see 28, 141). A randomized intervention trial in China showed, for a combination of retinol, riboflavin, and zinc, no effect on premalignant esophageal lesions in humans (92).

Retinyl palmitate reportedly reduced the incidence of carcinomas of the cervix and vagina of hamsters dosed with DMBA (24), but apparently no further experiments of this type have been performed. In a clinical trial, however, RA, topically applied to the cervix, produced responses in 12 of 36 patients with cervical dysplasia (143).

Summary and Conclusions

As indicated above, in some cases the effects of retinoids appear to be species-specific. Although retinyl acetate and 4-HPR are ineffective in preventing mammary cancer induced by DMBA or occurring spontaneously in mice (76, 145), these retinoids prevent carcinogen-induced mammary cancer in rats. In contrast, retinoids have modest chemopreventive activity for bladder cancer in various strains of both mice and rats and may have some therapeutic and preventive effects in human bladder (see 129). Retinyl palmitate is reported to reduce the incidence of esophageal lesions in hamsters (24); however, retinyl acetate may increase the incidence of esophageal tumors in rats (141). Although 13-*cis*-RA reduces the incidence of spontaneous thymic lymphomas in AKR mice and C57B1/10W mice exposed to X rays (103, 104) and has some therapeutic effect on myelodysplastic syndromes in humans (see 12), 4-HPR may enhance leukemic progression in patients with this syndrome (38). For treatment of this syndrome, selection of the proper retinoid appears to be important. Topically applied retinyl palmitate reduces the incidence of cervical cancer in hamsters (24), and topically applied RA has a therapeutic effect on cervical dysplasia in humans (143). Retinamides have a modest chemopreventive effect against pancreatic cancer in rats dosed with azaserine (see 27); these compounds are reported both to increase and to decrease the incidence of pancreatic cancer in hamsters (13, 73).

Retinoids may, or may not, be carcinogen-specific in different species. Some are effective in preventing mammary cancer in rats, regardless of which carcinogen is used (see 55). Applied to mouse skin, retinoids are active with either DMBA or BP as the carcinogen and 12-tetradecanoyl phorbol-13-acetate (TPA) as the promoter (41, 135). Nevertheless, retinoids are not effective in preventing skin papillomas and carcinomas caused by UV light (see 65). There is no comparable system for humans, although retinoids demonstrate activity against basal cell carcinomas (101), squamous cell carcinomas (67, 71, 87), and actinic keratoses (91) on the skin of humans. Fewer bladder tumors develop in rats dosed with HO-BBN when they are put on diets containing certain retinoids (see 55), but those dosed with FANFT are not affected (26). Similarly, retinyl acetate is reported to be active against liver tumors induced by 3'-MeDAB but not against those induced by aflatoxin B_1 (29, 95). In contrast, forestomach carcinomas induced in hamsters by either DMBA or BP are prevented by retinyl palmitate (24, 108, 124, 125).

The route of administration of retinoids may also be important. In the prevention of skin tumors, retinoids are effective when administered topically, intraperitoneally, or orally (15, 100, 135). 4-HPR, however, must be administered orally, apparently so that it can be converted to an active metabolite (77). If retinyl palmitate or retinyl acetate is administered topically, an increased frequency of cancer in the buccal pouch of hamsters dosed with DMBA is observed (see 84); however, the frequency of cancer is decreased with oral administration of 13-*cis*-RA (43, 119, 120). Administration of retinyl palmitate or 4-HPR intrarectally increased the incidence of rectal tumors; administered orally, these retinoids had no effect (123).

The time and extent of dosing with retinoids is often important. To be most effective in preventing skin cancer, retinoids should be applied shortly before the promoter (135). Administration of retinoids prior to, but not after, the carcinogen, can result in more mammary tumors than if no retinoid is administered (46). Continued administration of retinoid, relative to administration of no retinoid, however, greatly reduces the number of tumors that appear.

Retinoids are often tissue specific in their effects. As described above, retinoids appear to be effective in preventing cancer of the skin, oral cavity, blood, mammary gland, pancreas, and bladder. Convincing data is lacking, however, to indicate that retinoids have substantial effect in preventing cancer of the lung, esophagus, or colon. Data for the liver and forestomach are equivocal. A major limitation to the use of retinoids in humans is their toxicity, which includes embryopathy (10). Administered to mice and rats, retinoids with a free carboxyl group and the all-*trans* configuration are generally more toxic than others without these structural features (70, 110). Also, in general, the retinoids with teratogenic activity are those with the all-*trans* configuration and with either a free carboxyl group or a structure that can readily be transformed to such a group (67a, 112, 148). Thus, all-*trans*- and 13-*cis*-N-ethylretinamide and 13-*cis*-N-(2-hydroxyethyl)retinamide have little teratogenic activity, and 13-*cis*-4-HPR has only 1/20 the embryotoxicity of RA (148, 149). 4-HPR also has limited teratogenic activity (66).

We now know more about the types of retinoids that are needed to achieve better chemoprevention. The results obtained to date are encouraging. In humans, retinoids appear to be effective in preventing cancer in some individuals. More success will likely be realized by continued development of short-term tests that are predictive of preventive activity in human organs with the highest cancer incidence; better procedures for classification of retinoids according to their mode of action; and more effective and less toxic analogs of retinoids with established activity. Of nearly 400 reported tests of retinoid activity in preventing cancer, about three fourths have involved one of six common retinoids: retinyl palmitate, retinyl acetate, retinoic acid, 13-*cis*-retinoic acid, etretinate, or 4-HPR, each of which has some undesirable toxic effects. New, effective, and less toxic retinoids are needed.

Literature Cited

1. Abdel-Galil, A. M., Wrba, H., El-Mofty, M. M. 1984. Prevention of 3-methylcholanthrene-induced skin tumors in mice by simultaneous application of 13-*cis*-retinoic acid. *Exp. Pathol.* 25: 97–102

2. Abou-Issa, H., Koolemans-Beynen, A., Minton, J. P., Webb, T. E. 1989. Synergistic interaction between 13-*cis*-retinoic acid and glucarate: activity against rat mammary tumor induction and MCF-7 cells. *Biochem. Biophys. Res. Commun.* 163:1364–69

3. Alam, B. S., Alam, S. Q., Weir, J. C. Jr. 1988. Effects of excess vitamin A and canthraxanthin on salivary gland tumors. *Nutr. Cancer* 11:233–41

4. Alam, S. Q., Alam, B. S. 1983. Chemopreventive effects of β-carotene and 13-*cis*-retinoic acid on salivary gland tumors. *Fed. Proc. Fed. Am. Soc. Exp. Biol.* 42:1313

5. Alfthan, O., Tarkkanen, J., Gröhn, P., Heinonen, E., Pyrhönen, S., et al. 1983. Tigason (etretinate) in prevention of recurrence of superficial bladder tumors. A double-blind clinical trial. *Eur. Urol.* 9:6–9

6. Bashor, M. M., Toft, D. O., Chytil, F. 1973. In vitro binding of retinol to rat-tissue components. *Proc. Natl. Acad. Sci. USA* 70:3483–87

7. Becci, P. J., Thompson, H. J., Grubbs, C. J., Brown, C. C., Moon, R. C. 1979. Effect of delay in administration of 13-*cis*-retinoic acid on the inhibition of urinary bladder carcinogenesis in the rat. *Cancer Res.* 39:3141–44

8. Beems, R. B. 1984. Modifying effect of vitamin A on benzo[a]pyrene-induced respiratory tract tumours in hamsters. *Carcinogenesis* 5:1057–60

9. Bell, B. A., Findley, H. W., Krischer, J., Whitehead, V. M., Holbrook, T., et al. 1991. Phase II study of 13-*cis*-retinoic acid in pediatric patients with acute nonlymphocytic leukemia—a pediatric oncology group study. *J. Immunother.* 10:77–83

10. Benke, P. J. 1984. The isotretinoin teratogen syndrome. *J. Am. Med. Assoc.* 251:3267–69

11. Berger, M. R., Schmähl, D. 1986. Protection by the alkyllysophospholipid, 1-octadecyl-2-methoxy-*rac*-glycero-3-phosphocholine, but not by the retinoid etretinate against leukemia development in DMBA-treated Long-Evans rats. *Cancer Lett.* 30:73–78

12. Besa, E. C., Abrahm, J. L., Bartholomew, M. J., Hyzinski, M., Nowell, P. C. 1990. Treatment with 13-*cis*-retinoic acid in transfusion-dependent patients with myelodysplastic syndrome and

decreased toxicity with addition of α-tocopherol. *Am. J. Med.* 89:739–47

13. Birt, D. F., Davies, M. H., Pour, P. M., Salmasi, S. 1983. Lack of inhibition by retinoids of bis(2-oxopropyl)nitrosamine-induced carcinogenesis in Syrian hamsters. *Carcinogenesis* 4:1215–20

14. Birt, D. F., Sayed, S., Davies, M. H., Pour, P. 1981. Sex differences in the effects of retinoids on carcinogenesis by *N*-nitrosobis(2-oxopropyl)amine in Syrian hamsters. *Cancer Lett.* 14:13–21

15. Bollag, W. 1971. Therapy of chemically induced skin tumors of mice with vitamin A palmitate and vitamin A acid. *Experientia* 27:90–92

16. Bollag, W. 1974. Therapeutic effects of an aromatic retinoic acid analog on chemically induced skin papillomas and carcinomas of mice. *Eur. J. Cancer* 10:731–37

17. Bollag, W., Hartmann, H-R. 1987. Inhibition of rat mammary carcinogenesis by an arotinoid without a polar end group (Ro 15-0778). *Eur. J. Cancer Clin. Oncol.* 23:131–35

18. Boone, C. W., Kelloff, G. J., Malone, W. E. 1990. Identification of candidate cancer chemopreventive agents and their evaluation in animal models and human clinical trials: a review. *Cancer Res.* 50:2–9

19. Breitman, T. R., Selinock, S. E., Collins, S. J. 1980. Induction of differentiation of the human promyelocytic leukemia cell line (HL-60). *Proc. Natl. Acad. Sci. USA* 77:2936–40

20. Bruley-Rosset, M., Hercend, T., Martinez, J., Rappoport, H., Mathé, G. 1981. Prevention of spontaneous tumors of aged mice by immunopharmacologic manipulation: study of immune antitumor mechanisms. *J. Natl. Cancer Inst.* 66:1113–19

21. Castaigne, S., Chomienne, C., Daniel,, M. T., Ballerini, P., Berger, R., et al. 1990. All-*trans*-retinoic acid as a differentiation therapy for acute promyelocytic leukemia. I. Clinical results. *Blood* 76:1704–9

22. Chan, K. K., Specian, A. C. Jr., Pawson, B. A. 1981. Fluorinated retinoic acids and their analogues. 2. Synthesis and biological activity of aromatic 4-fluoro analogues. *J. Med. Chem.* 24: 101–4

23. Chauvenet, P. H., Paque, R. E. 1982. Effect of all-*trans*-retinoic acid on induction, lethality and immunogenicity of murine methylcholanthrene-induced fibrosarcomas. *Int. J. Cancer* 30:187–92

24. Chu, E. W., Malmgren, R. A. 1965. An inhibitory effect of vitamin A on the induction of tumors of forestomach and cervix in the Syrian hamster by carcinogenic polycyclic hydrocarbons. *Cancer Res.* 25:884–95

25. Deleted in proof

26. Croft, W. A., Croft, M. A., Paulus, K. P., Williams, J. H., Wang, C. Y., et al. 1981. Synthetic retinamides: effect on urinary bladder carcinogenesis by FANFT in Fischer rats. *Carcinogenesis* 2:515–17

27. Curphey, T. J., Kuhlmann, E. T., Roebuck, B. D., Longnecker, D. S. 1988. Inhibition of pancreatic and liver carcinogenesis in rats by retinoid- and selenium-supplemented diets. *Pancreas* 3:36–40

28. Daniel, E., Stoner, G. 1990. The effects of ellagic acid and 13-*cis*-retinoic acid on *N* - nitrosobenzylmethylamine - induced esophageal tumorigenesis in rats. *Cancer Lett.* 56:117–24

29. Daoud, A., Griffin, A. C. 1980. Effect of retinoic acid, butylated hydroxytoluene, selenium and sorbic acid on azo-dye hepatocarcinogenesis. *Cancer Lett.* 9:299–304

30. Dawson, M. I., Chao, W.-R. 1988. Comparison of the inhibitory effects of retinoids on 12-*O*-tetradecanoylphorbol-13-acetate-promoted tumor formation in CD-1 and SENCAR mice. *Cancer Lett.* 40:7–12

31. Dawson, M. I., Chao, W.-R., Helmes, C. T. 1987. Inhibition by retinoids of anthralin-induced mouse epidermal ornithine decarboxylase activity and anthralin-promoted skin tumor formation. *Cancer Res.* 47:6210–15

32. Decaëns, C., Rosa, B., Bara, J., Daher, N., Burtin, P. 1983. Effect of 13-*cis*-retinoic acid on early precancerous antigenic goblet-cell modifications and induction of cancer during 1,2-dimethylhydrazine carcinogenesis in rats. *Carcinogenesis* 4:1175–78

33. Edwards, L., Jaffe, P. 1990. The effect of topical tretinoin on dysplastic nevi. *Arch. Dermatol.* 126:494–99

34. Epstein, J. H. 1977. Chemicals and photocarcinogenesis. *Aust. J. Dermatol.* 18:57–61

35. Evard, J. P., Bollag, W. 1972. Konservative Behandlung der rezidivierenden Harnblasenpapillomatose mit Vitamin-A-Säure. *Schweiz. Med. Wochenschr.* 102:1880–83

36. Forbes, P. D., Urbach, F., Davies, R. E. 1979. Enhancement of experimental photocarcinogenesis by topical retinoic acid. *Cancer Lett.* 7:85–90

37. Frankel, J. W., Horton, E. J., Winters,

A. L., Samis, H. V., Ito, Y. 1980. Inhibition of viral tumorigenesis by a retinoic acid analog. *Curr. Chemother. Infect. Dis., Proc. Int. Conf. Chemother., 11th, 1979*, p. 1505–6

38. Garewal, H. S., List, A., Meyskens, F., Buzaid, A., Greenberg, B., et al. 1989. Phase II trial of fenretinide [*N*-(4-hydroxyphenyl)retinamide] in myelodysplasia: possible retinoid-induced disease acceleration. *Leuk. Res.* 13:339–43

39. Gensler, H. L., Aickin, M., Peng, Y. M. 1990. Cumulative reduction of primary skin tumor growth in UV-irradiated mice by the combination of retinyl plamitate and canthaxathin. *Cancer Lett.* 53:27–31

40. Gensler, H. L., Bowden, G. T. 1984. Influence of 13-*cis*-retinoic acid on mouse skin tumor initiation and promotion. *Cancer Lett.* 22:71–75

41. Gensler, H. L., Sim, D. A., Bowden, G. T. 1986. Influence of the duration of topical 13-*cis*-retinoic acid treatment on inhibition of mouse skin tumor promotion. *Cancer Res.* 46:2767–70

42. Gensler, H. L., Watson, R. R., Moriguchi, S., Bowden, G. T. 1987. Effects of dietary retinyl palmitate or 13-*cis*-retinoic acid on the promotion of tumors in mouse skin. *Cancer Res.* 47:967–70

43. Goodwin, W. J., Bordash, G. D., Huijing, F., Altman, N. 1986. Inhibition of hamster tongue carcinogenesis by selenium and retinoic acid. *Ann. Otol. Rhinol. Laryngol.* 95:162–66

44. Gouveia, J., Hercend, T., Lemaigre, G., Mathé, G., Gros, F., et al. 1982. Degree of bronchial metaplasia in heavy smokers and its regression after treatment with a retinoid. *Lancet* 1:710–12

45. Green, S., Chambon, P. 1988. Nuclear receptors enhance our understanding of transcription regulation. *Trends Genet.* 4:309–14

46. Grubbs, C. J., Eto, I., Juliana, M. M., Hardin, J. M., Whitaker, L. M. 1990. Effect of retinyl acetate and 4-hydroxyphenylretinamide on initiation of chemically induced mammary tumors. *Anticancer Res.* 10:661–66

47. Grubbs, C. J., Hill, D. L., Farnell, D. R., Kalin, J. R., McDonough, K. C. 1985. Effect of long-term administration of retinoids on rats exposed transplacentally to ethylnitrosourea. *Anticancer Res.* 5:205–10

48. Hard, G. C., Ogiu, T. 1984. Null effects of vitamin A analogs on the dimethylnitrosamine kidney tumor model. *Carcinogenesis* 5:665–69

49. Hartmann, H. R., Bollag, W. 1985. The effects of arotinoids on rat mammary carcinogenesis. *Cancer Chemother. Pharmacol.* 15:141–43

50. Hasegawa, R., Takahashi, M., Furukawa, F., Toyoda, K., Sato, H., et al. 1988. Co-carcinogenic effect of retinyl acetate on forestomach carcinogenesis of male F344 rats induced with butylated hydroxyanisole. *Jpn. J. Cancer Res.* 79:320–28

51. Hennings, H., Wenk, M. L., Donohoe, R. 1982. Retinoic acid promotion of papilloma formation in mouse skin. *Cancer Lett.* 16:1–5

52. Hicks, R. M. 1983. The scientific basis for regarding vitamin A and its analogues as anticarcinogenic agents. *Proc. Nutr. Soc.* 42:83–93

53. Hicks, R. M., Chowaniec, J., Turton, J. A., Massey, E. D., Harvey, A. 1982. The effect of dietary retinoids on experimentally induced carcinogenesis in the rat bladder. In *Molecular Interrelations of Nutrition and Cancer,* ed. M. S. Arnott, J. van Eys, Y.-M. Wang, pp. 419–47. New York: Raven

54. Hicks, R. M., Turton, J. 1986. Retinoids and cancer. *Biochem. Soc. Trans.* 14:939–42

55. Hill, D. L., Grubbs, C. J. 1982. Retinoids as chemopreventive and anticancer agents in intact animals. *Anticancer Res.* 2:111–24

56. Hill, D. L., Sani, B. P. 1991. Metabolic disposition and development of new chemopreventive retinoids. *Drug Metab. Rev.* 23:313–38

57. Holtzman, S. 1988. Retinyl acetate inhibits estrogen-induced mammary carcinogenesis. *Carcinogenesis* 9:305–7

58. Hong, W. K., Endicott, J., Itri, L. M., Doos, W., Batsakis, J. G., et al. 1986. 13-*cis*-Retinoic acid in the treatment of oral leukoplakia. *New Engl. J. Med.* 315:1501–5

59. Hong, W. K., Lippman, S. M., Itri, L. M., Karp, D. D., Lee, J. S., et al. 1990. Prevention of second primary tumors with isotretinoin in squamous-cell carcinoma of the head and neck. *New Engl. J. Med.* 323:795–801

60. Huang, M., Ye, Y., Chen, S., Chai, J., Lu, J-X., et al. 1988. Use of all-*trans*-retinoic acid in the treatment of acute promyelocytic leukemia. *Blood* 72:567–72

61. Hultin, T. A., Filla, M. S., Detrisac, C. J., Moon, R. C. 1988. Pharmacogenetic differences in retinoid metabolism and chemoprevention of liver cancer in inbred rats. *Proc. Am. Assoc. Cancer Res.* 29:135

62. Hultin, T. A., May, C. M., Moon, R. C. 1989. *N* - (4 - Hydroxyphenyl) - all-

trans-retinamide pharmacokinetics in female rats and mice. *Drug. Metab. Dispos.* 14:714–17

63. Israili, Z. H., Razdan, R., Willis, I. 1982. Effect of vitamin A and analogs on the induction of squamous cell carcinoma in hairless mice caused by solar simulated UV-light. *Proc. Am. Assoc. Cancer Res.* 23:204

64. Kandarkar, S. V., Sirsat, S. M. 1983. Influence of excess of retinoid on DMBA carcinogenesis. *Neoplasma* 30:43–50

65. Kelly, G. E., Meikle, W. D., Sheil, A. G. R. 1989. Effects of oral retinoid (vitamin A and etretinate) therapy on photocarcinogenesis in hairless mice. *Photochem. Photobiol.* 50:213–15

66. Kenel, M. F., Krayer, J. H., Merz, E. A., Pritchard, J. F. 1988. Teratogenicity of *N*-(4-hydroxyphenyl)-all-*trans*-retinamide in rats and rabbits. *Teratog. Carcinog. Mutagen.* 8:1–11

67. Kraemer, K. H., DiGiovanna, J. J., Moshell, A. N., Tarone, R. E., Peck, G. L. 1988. Prevention of skin cancer in xeroderma pigmentosum with the use of oral isotretinoin. *New Engl. J. Med.* 318:1633–37

67a. Kraft, J. C., Kochhar, D. M., Scott, W. J., Nau, H. 1987. Low teratogenicity of 13-*cis*-retinoic acid (isotretinoin) in the mouse corresponds to low embryo concentrations during organogenesis: comparison to the all-*trans* isomer. *Toxicol. Appl. Pharmacol.* 87:474–82

68. Kramer, Z. B., Boros, L., Wiernik, P. H., Andersen, J., Bennett, J. M., et al. 1991. 13-*cis*-Retinoic acid in the treatment of elderly patients with acute myeloid leukemia. *Cancer* 67:1484–86

69. Levine, N., Meyskens, F. L. 1980. Topical vitamin-A-acid therapy for cutaneous metastatic melanoma. *Lancet* 2:224–26

70. Lindamood, C. III, Dillehay, D. L., Lamon, E. W., Giles, H. D., Shealy, Y. F., et al. 1988. Toxicologic and immunologic evaluations of *N*-(all-*trans*-retinoyl-DL-leucine and *N*-(all-*trans*-retinoyl)glycine. *Toxicol. Appl. Pharmacol.* 96:279–95

71. Lippman, S. M., Kessler, J. F., Al-Sarraf, M., Alberts, D. S., Itri, L. M., et al. 1988. Treatment of advanced squamous cell carcinoma of the head and neck with isotretinoin: a phase II randomized trial. *Invest. New Drugs* 6:51–56

72. Longnecker, D., Kuhlmann, E. T., Curphey, T. J. 1983. Divergent effects of retinoids on pancreatic and liver carcinogenesis in azaserine-treated rats. *Cancer Res.* 43:3219–25

73. Longnecker, D., Kuhlmann, E. T., Curphey, T. J. 1983. Effects of four retinoids in *N*-nitrosobis(2-oxopropyl) amine-treated hamsters. *Cancer Res.* 43:3226–30

74. Lovey, A. J., Pawson, B. A. 1982. Fluorinated retinoic acids and their analogues. 3. Synthesis and biological activity of aromatic 6-fluoro analogues. *J. Med. Chem.* 25:71–75

75. Mack, D. O., Reed, V. L., Smith, L. D. 1990. Retinyl acetate inhibition of 3'-methyl-4-dimethyl-aminoazobenzene induced hepatic neoplasia. *Int. J. Biochem.* 22:359–65

76. Maiorana, A., Gullino, P. M. 1980. Effect of retinyl acetate on the incidence of mammary carcinomas and hepatomas in mice. *J. Natl. Cancer Inst.* 64:655–63

77. McCormick, D. L., Bagg, B. J., Hultin, T. A. 1987. Importance of systemic metabolism in the modulation of skin tumor induction in mice by the retinoid *N* - (4 - hydroxyphenyl)retinamide (4-HPR). *Proc. Am. Assoc. Cancer Res.* 28:145

78. McCormick, D. L., Bagg, B. J., Hultin, T. A. 1987. Comparative activity of dietary or topical exposure to three retinoids in the promotion of skin tumor induction in mice. *Cancer Res.* 47:5989–93

79. McCormick, D. L., Becci, P. J., Moon, R. C. 1982. Inhibition of mammary and urinary bladder carcinogenesis by a retinoid and a maleic anhydride-divinyl ether copolymer (MVE-2). *Carcinogenesis* 3:1473–77

80. McCormick, D. L., Burns, F. J., Albert, R. E. 1981. Inhibition of benzo[*a*]pyrene-induced mammary carcinogenesis by retinyl acetate. *J. Natl. Cancer Inst.* 66:559–64

81. McCormick, D. L., Long, R. E. 1990. Promotion of diethylnitrosamine-induced hepatic carcinogenesis in mice by all-*trans*-retinoic acid and two synthetic retinamides. *Toxicologist* 10:232

82. McCully, K. S., Vezeridis, M. P. 1987. Chemopreventive and antineoplastic activity of *N*-homocysteine thiolactonyl retinamide. *Carcinogenesis* 8:1559–62

83. McCully, K. S., Vezeridis, M. P. 1989. Antineoplastic activity of *N*-maleamide homocysteine thiolactone amide encapsulated within liposomes. *Proc. Soc. Exp. Biol. Med.* 191:346–51

84. McGaughey, C., Jensen, J. L. 1980. Effects of the differentiating agents (inducers) dimethylacetamide, di- and tetramethylurea on epidermal tumor promotion by retinyl (vitamin A) acetate and croton oil in hamster cheek pouch. *Oncology* 37:65–70

85. Mehta, R. G., Hultin, T. A., Moon, R. C. 1988. Metabolism of the chemopreventive retinoid N-(4-hydroxyphenyl)retinamide by mammary gland in organ culture. *Biochem. J.* 256:579–84

86. Mehta, R. G., Rao, K. V. N., Detrisac, C. J., Kelloff, G. J., Moon, R. C. 1988. Inhibition of diethylnitrosamine-induced lung carcinogenesis by retinoids. *Proc. Am. Assoc. Cancer Res.* 29:129

87. Meyskens, F. L. Jr., Gilmartin, E., Alberts, D. S., Levine, N. S., Brooks, R., et al. 1982. Activity of isotretinoin against squamous cell cancers and preneoplastic lesions. *Cancer Treat. Rep.* 66:1315–19

88. Moon, R. C., Itri, L. M. 1984. Retinoids and cancer. In *The Retinoids,* ed. M. B. Sporn, A. B. Roberts, D. W. Goodman, 2:327–71. New York: Academic

89. Moon, R. C., McCormick, D. L., Becci, P. J., Shealy, Y. F., Frickel, F., et al. 1982. Influence of 15 retinoic acid amides on urinary bladder carcinogenesis in the mouse. *Carcinogenesis* 3:1469–72

90. Moon, R. C., McCormick, D. L., Mehta, R. G. 1983. Inhibition of carcinogenesis by retinoids. *Cancer Res.* 43:2469S–75S

91. Moriarity, M., Dunn, J., Darragh, A., Lambe, R., Brick, I. 1982. Etretinate in treatment of actinic keratosis. *Lancet* 1:364–65

92. Munoz, N., Wahrendorf, J., Bang, L. J., Crespi, M., Thurnham, D. I., et al. 1985. No effect of riboflavine, retinol, and zinc on prevalence of precancerous lesions of oesophagus. *Lancet* 2:111–14

93. Murasaki, G., Miyata, Y., Babaya, K., Armi, M., Fukushima, S., et al. 1980. Inhibitory effect of an aromatic retinoic acid analog on urinary bladder carcinogenesis in rats treated with N-butyl-N-(4-hydroxybutyl)nitrosamine. *Gann* 71:333–40

94. Muto, Y., Moriwaki, H. 1984. Antitumor activity of vitamin A and its derivatives. *J. Natl. Cancer Inst.* 73:1389–93

95. Newberne, P. M., Suphakarn, V. 1977. Preventive role of vitamin A in colon carcinogenesis in rats. *Cancer* 40:2553–56

96. O'Dwyer, P. J., Ravikumar, T. S., McCabe, D. P., Steele, G. Jr. 1987. Effect of 13-*cis*-retinoic acid on tumor prevention, tumor growth, and metastasis in experimental colon cancer. *J. Surg. Res.* 43:550–57

97. Ong, D. E., Chytil, F. 1975. Specificity of cellular retinol-binding protein for compounds with vitamin A activity. *Nature* 255:74–75

98. Pastorino, U., Soresi, E., Clerici, M., Chiesa, G., Belloni, P. A., et al. 1988. Lung cancer chemoprevention with retinol palmitate. Preliminary data from a randomized trial on stage Ia non small-cell lung cancer. *Acta Oncol.* 27:773–82

99. Pawson, B. A., Chan, K-K., DeNoble, J., Han, R-J. L., Piermattie, V., et al. 1979. Fluorinated retinoic acids and their analogues. 1. Synthesis and biological activity of (4-methoxy-2,3,6-trimethylphenyl)nonatetraenoic acid analogues. *J. Med. Chem.* 22:1059–67

100. Pawson, B. A., Cheung, H.-C., Han, R-J. L., Trown, P. W., Buck, M., et al. 1977. Dihydroretinoic acids and their derivatives. Synthesis and biological activity. *J. Med. Chem.* 20:918–25

101. Peck, G. L., DiGiovanna, J. J., Sarnoff, D. S., Gross, E. G., Butkus, D., et al. 1988. Treatment and prevention of basal cell carcinoma with oral isotretinoin. *J. Am. Acad. Dermatol.* 19:176–85

102. Port, C. D., Sporn, M. B., Kaufman, D. G. 1975. Prevention of lung cancer in hamsters by 13-*cis*-retinoic acid. *Proc. Am. Assoc. Cancer Res.* 16:21

103. Przybyszewska, M. 1985. A protective role of 13-*cis*-retinoic acid in thymic lymphoma induction. *Arch. Immunol. Ther. Exp.* 33:811–15

104. Przybyszewska, M., Szaniawska, B., Janik, P. 1986. Effect of 13-*cis*-retinoic acid on the spontaneous thymic lymphoma development in AKR mice. *Neoplasma* 33:341–44

105. Quander, R. V., Leary, S. L., Strandberg, J. D., Yarbrough, B. A., Squire, R. A. 1985. Long-term effect of 2-hydroxyethylretinamide on urinary bladder carcinogenesis and tumor transplantation in Fischer 344 rats. *Cancer Res.* 45:5235–29

106. Roebuck, B. D., Baumgartner, K. J., Thron, C. D., Longnecker, D. S. 1984. Inhibition by retinoids of the growth of azaserine-induced foci in the rat pancreas. *J. Natl. Cancer Inst.* 73:233–36

107. Rogers, A. E., Herndon, B. J., Newberne, P. M. 1973. Induction by dimethylhydrazine of intestinal carcinoma in normal rats and rats fed high or low levels of vitamin A. *Cancer Res.* 33:1003–9

108. Saffiotti, U., Montesano, R., Sellakumar, A. R., Borg, S. A. 1967. Experimental cancer of the lung. Inhibition by vitamin A of the induction of tracheobronchial squamous metaplasia and squamous cell tumors. *Cancer* 20:857–64

109. Sani, B. P., Hill, D. L. 1976. A retinoic acid-binding protein from chick embryo skin. *Cancer Res.* 36:409–13
110. Sani, B. P., Meeks, R. G. 1983. Subacute toxicity of all-*trans*- and 13-*cis*-isomers of *N*-ethyl retinamide, *N*-2-hydroxyethyl retinamide, and *N*-4-hydroxyphenyl retinamide. *Toxicol. Appl. Pharmacol.* 70:228–35
111. Sani, B. P., Singh, R. K., Reddy, L. G., Gaub, M-P. 1990. Isolation, partial purification and characterization of nuclear retinoic acid receptors from chick skin. *Arch. Biochem. Biophys.* 283:107–13
112. Satre, M. A., Penner, J. D., Kocchar, D. M. 1989. Pharmacokinetic assessment of teratologically effective concentrations of an endogenous retinoic acid metabolite. *Teratology* 39:341–48
113. Schmähl, D., Habs, M. 1978. Experiments on the influence of an aromatic retinoid on the chemical carcinogenesis in rats by butyl-butanol-nitrosamine and 1,2-dimethylhydrazine. *Arzneim. Forsch.* 28:49–51
114. Schmähl, D., Krüger, C., Preissler, P. 1972. Versuche zur Krebsprophylaxe mit Vitamin A. *Arzneim. Forsch.* 22:946–49
115. Seifter, E., Rettura, G., Padawer, J., Demetriou, A. A., Levenson, S. 1976. Antipyretic and antiviral action of vitamin A in Moloney sarcoma virus- and poxvirus-inoculated mice. *J. Natl. Cancer Inst.* 57:355–59
116. Shah, J. P., Strong, E. W., DeCosse, J. J., Itri, L., Sellers, P. 1983. Effect of retinoids on oral leukoplakia. *Am. J. Surg.* 146:466–70
117. Shealy, Y. F. 1989. Synthesis and evaluation of some new retinoids for cancer chemoprevention. *Prev. Med.* 18:624–45
118. Shih, T. W., Shealy, Y. F., Hill, D. L. 1988. Enzymatic hydrolysis of retinamides. *Drug Metab. Dispos.* 16:337–40
119. Shklar, G., Marefat, P., Kornhouser, A., Trickler, D. P., Wallace, K. D. 1980. Retinoid inhibition of lingual carcinogenesis. *Oral Surg. Oral Med. Oral Pathol.* 49:325–32
120. Shklar, G., Schwartz, J., Grau, D., Trickler, D. P., Wallace, K. D. 1980. Inhibition of hamster buccal pouch carcinogenesis by 13-*cis*-retinoic acid. *Oral Surg. Oral Med. Oral Pathol.* 50:45–52
121. Shwaireb, M. H., Wrba, H., El-Mofty, M. M., Dutter, A. 1990. Carcinogenesis induced by black pepper (*Piper nigrum*) and modulated by vitamin A. *Exp. Pathol.* 40:233–38
122. Silverman, J., Katayama, S., Radok, P., Levenstein, M. J., Weisburger, J. H. 1983. Effect of short-term administration of *N*-(4-hydroxyphenyl)-all-*trans*-retinamide on chemically induced mammary tumors. *Nutr. Cancer* 4:186–91
123. Silverman, J., Katayama, S., Zelenakas, K., Lauber, J., Musser, T. K., et al. 1981. Effect of retinoids on the induction of colon cancer in F344 rats by *N*-methyl-*N*-nitrosourea or by 1,2-dimethylhydrazine. *Carcinogenesis* 2:1167–72
124. Smith, D. M., Rogers, A. E., Herndon, B. J., Newberne, P. M. 1975. Vitamin A (retinyl acetate) and benzo(*a*)pyrene-induced respiratory tract carcinogenesis in hamsters fed a commercial diet. *Cancer Res.* 35:11–16
125. Smith, D. M., Rogers, A. E., Newberne, P. M. 1975. Vitamin A and benzo(*a*)pyrene carcinogenesis in the respiratory tract of hamsters fed a semisynthetic diet. *Cancer Res.* 35:1485–88
126. Sporn, M. B., Squire, R. A., Brown, C. C., Smith, J. M., Wenk, M. L., et al. 1977. 13-*cis*-Retinoic acid: inhibition of bladder carcinogenesis in the rat. *Science* 195:487–89
127. Stinson, S. F., Reznik, G., Levitt, M. H., Wenk, M., Saffiotti, U. 1987. Contrasting effects of 13-*cis*-retinoic acid (CRA) on the induction of neoplasms at different sites by 2,6-dimethylnitrosomorpholine (DMNM) in hamsters. *Proc. Am. Assoc. Cancer Res.* 28:135
128. Stone, J. P., Shellabarger, C. J., Holtzman, S. 1987. Life-span retinyl acetate inhibition of spontaneous and induced rat mammary carcinogenesis. *Proc. Am. Assoc. Cancer Res.* 28:141
129. Studer, U. E., Biedermann, C., Chollet, D., Karrer, P., Kraft, R., et al. 1984. Prevention of recurrent superficial bladder tumors by oral etretinate: preliminary results of a randomized, double blind multicenter trial in Switzerland. *J. Urol.* 131:47–49
130. Tannenbaum, M., Tannenbaum, S., Richelo, B. N., Trown, P. W. 1979. Effects of 13-*cis*- and all-*trans*-retinoic acid on the development of bladder cancer in rats. *Fed. Proc. Fed. Am. Soc. Exp. Biol.* 38:1073
131. Teelmann, K., Bollag, W. 1988. Therapeutic effect of arotinoid Ro 15-0778 on chemically induced rat mammary carcinoma. *Eur. J. Cancer Clin. Oncol.* 24:1205–9
132. Thompson, H. J., Becci, P. J., Grubbs, C. J., Shealy, Y. F., Stanek, E. J., et al. 1981. Inhibition of urinary bladder cancer by *N*-(ethyl)-all-*trans*-retinamide and

N-(2-hydroxyethyl)-all-*trans*-retinamide in rats and mice. *Cancer Res.* 41:933–36

133. Thompson, H. J., Becci, P. J., Moon, R. C., Sporn, M. B., Newton, D. L., et al. 1980. Inhibition of 1-methyl-1-nitrosourea-induced mammary carcinogenesis in the rat by the retinoid axerophthene. *Arzneim. Forsch.* 30:1127–29
134. Thompson, H. J., Grubbs, C. J., Becci, P. J., Sporn, M. B., Moon, R. C. 1978. Effect of retinoids on *N*-methyl-*N*-nitrosourea (MNU)-induced mammary cancer. *Fed. Proc. Fed. Am. Soc. Exp. Biol.* 36:261
135. Verma, A. K., Boutwell, R. K. 1977. Vitamin A acid (retinoic acid), a potent inhibitor of 12-*O*-tetradecanoyl-phorbol-13-acetate-induced ornithine decarboxylase activity in mouse epidermis. *Cancer Res.* 37:2196–2201
136. Verma, A. K., Conrad, E. A., Boutwell, R. K. 1982. Differential effects of retinoic acid and 7,8-benzoflavone on the induction of mouse skin tumors by the complete carcinogenesis process and by the initiation-promotion regimen. *Cancer Res.* 42:3519–25
137. Verma, A. K., Erickson, D. 1986. Retinoic acid (RA) and multi-stage carcinogenesis in mouse skin: inhibition of stage I and stage II tumor promotion and ornithine decarboxylase (ODC)-gene-transcription. *Proc. Am. Assoc. Cancer Res.* 27:147
138. Verma, A. K., Garcia, C. T., Ashendel, C. L., Boutwell, R. K. 1983. Inhibition of 7-bromomethylbenz[*a*]anthracene-promoted mouse skin tumor formation by retinoic acid and dexamethasone. *Cancer Res.* 43:3045–59
139. Verma, A. K., Shapas, B. G., Rice, H. M., Boutwell, R. K. 1979. Correlation of the inhibition by retinoids of tumor promoter-induced mouse epidermal ornithine decarboxylase activity and of skin tumor promotion. *Cancer Res.* 39:419–25
140. Verma, A. K., Slaga, T. J., Wertz, P. W., Meuller, G. C., Boutwell, R. K. 1980. Inhibition of skin tumor promotion by retinoic acid and its metabolite 5,6-epoxyretinoic acid. *Cancer Res.* 40:2367–71
141. Wargovich, M. J., Hong, W. K. 1990. Promotion of nitrosomethylbenzylamine-induced squamous cell carcinomas of the esophagus by retinyl acetate. *Proc. Am. Assoc. Cancer Res.* 31:163
142. Warrell, R. P. Jr., Frankel, S. R., Miller, W. H. Jr., Scheinberg, D. A., Itri, L. M., et al. 1991. Differentiation ther-

apy of acute promyelocytic leukemia with tretinoin (all-*trans*-retinoic acid). *New Engl. J. Med.* 324:1385–93
143. Weiner, S. A., Surwit, E. A., Graham, V. E., Meyskens, F. L. Jr. 1986. A phase I trial of topically applied *trans*-retinoic acid in cervical dysplasia-clinical efficacy. *Invest. New Drugs* 4:241–44
144. Weiss, V. C., Cambazard, F., Ronan, S., Ghosh, L., Buys, C. M., et al. 1984. Effect of systemic retinoids (isotretinoin and etretinate) on PUVA-induced carcinogenesis in albino hairless mice. *Dermatologica* 169:236
145. Welsch, C. W., DeHoog, J. V., Moon, R. C. 1984. Lack of an effect of dietary retinoids in chemical carcinogenesis on the mouse mammary gland: inverse relationship between mammary tumor cell anaplasia and retinoid efficacy. *Carcinogenesis* 5:1301–4
146. Welsch, C. W., DeHoog, J. V., Scieszka, K. M., Aylsworth, C. F. 1984. Retinoid feeding, hormone inhibition, and/or immune stimulation and the progression of *N*-methyl-*N*-nitrosourea-induced rat mammary carcinoma: suppression by retinoids of peptide hormone-induced tumor cell proliferation in vivo and in vitro. *Cancer Res.* 44:166–71
147. Welsch, C. W., Goodrich-Smith, M., Brown, C. K., Crowe, N. 1981. Enhancement by retinyl acetate of hormone-induced mammary tumorigenesis in female GR/A mice. *J. Natl. Cancer Inst.* 67:935–38
148. Willhite, C. C., Dawson, M. I., Williams, K. J. 1984. Structure-activity relationships of retinoids in developmental toxicology. I. Studies on the nature of the polar terminus of the vitamin A molecule. *Toxicol. Appl. Pharmacol.* 74:397–410
149. Willhite, C. C., Shealy, Y. R. 1984. Amelioration of embryotoxicity by structural modification of the terminal group of cancer chemopreventive retinoids. *J. Natl. Cancer Inst.* 72:689–95
150. Wrba, H., Dutter, A., Hacker-Rieder, A. 1983. Influence of vitamin A on the formation of ethylnitrosourea (ENU)-induced leukemias. *Arch. Geschwulstforsh.* 53:89–92
151. Yokoyama, M., Kitamura, Y., Kohrogi, T., Miyoshi, I. 1982. Necessity of bile for and lack of inhibitory effect of retinoid on development of forestomach papillomas in nontreated mutant mice of the W/Wv genotype. *Cancer Res.* 42:3806–9

Annu. Rev. Nutr. 1992. 12:183 206

NUTRIENT TRANSPORT PATHWAYS ACROSS THE EPITHELIUM OF THE PLACENTA

C. H. Smith and A. J. Moe

Department of Pediatrics, Washington University School of Medicine, St. Louis, Missouri 63110

V. Ganapathy

Department of Biochemistry and Molecular Biology, Medical College of Georgia, Augusta, Georgia 30912-2100

KEY WORDS: trophoblast, fetal nutrition, membrane transport, amino acids, vitamin and mineral transport, ion transport, sugar transport

CONTENTS

0199-9885/92/0715-0183$02.00

INTRODUCTION

Fetal growth and development depends on a continuous nutrient supply from maternal blood (9, 103). In the human placenta the supply process includes uptake from maternal blood and transfer across the syncytiotrophoblast and cytotrophoblast layers, the underlying basal lamina, the fetal connective tissue space, and the fetal capillary endothelium. Although paracellular mechanisms exist (85, 132), the quantitative and qualitative importance of these pathways to overall nutrient flux is yet to be determined. Flux across the syncytiotrophoblast and its plasma membranes represents the rate-limiting steps in maternal-fetal transfer of most important nutrients. Most sites of known mechanisms of cellular transport are localized to the maternal- and fetal-facing plasma membrane surfaces of the syncytiotrophoblast (124).

The placental syncytiotrophoblast is a continuous epithelial layer. It covers the maternal surface of the human placenta (124, 154) and forms by fusion and terminal differentiation of the underlying cytotrophoblast. The syncytiotrophoblast is a polarized epithelium that resembles the epithelia of kidney and intestine, except that lateral cell borders are absent and the epithelium forms a multinucleated syncytium. The microvillous membrane of the syncytiotrophoblast is in contact with the maternal blood while the basal membrane faces the fetal circulation. These two membranes can be isolated from the same placenta by well-established procedures of mild shearing and centrifugation (12, 82) or by a newer method using homogenization (65). Alkaline phosphatase and 5' nucleotidase are localized to the microvillous membrane, which obtains its "finger-like" structure from highly ordered actin-containing microfilaments (13, 84). The basal membrane contains the sodium potassium ATPase and β-adrenergic receptor coupled to adenyl cyclase (82).

Regulation of syncytiotrophoblast transport mechanisms is potentially of great importance in the control of fetal nutrient supply. Regulation of transport may be affected either by intrinsic mechanisms, in which substrate or cellular proteins interact with transporters, or by extracellular or circulating hormones or effectors (83, 104, 125, 153, 155, 157) and their receptors in the trophoblast (11, 21, 106). In addition to physiologic regulators, considerable evidence suggests that environmental substances such as ethanol and cannabinoids can alter transport of various nutrients (47, 113). The two trophoblast membranes can communicate across the syncytial cytoplasm by cyclic AMP generation at the basal membrane in response to stimuli originating in the fetus, which in turn stimulates a protein kinase in the microvillous membrane (1). This and other communication pathways involving other receptors and protein kinases at the two membranes provide potential mechanisms by which stimuli originating in the fetus or the mother can regulate transport of substrates across the maternal- or fetal-facing membranes.

The mechanisms of nutrient transfer from mother to fetus have been investigated by a variety of methods. These include in vivo, chronic maternal and fetal blood sampling in sheep and other animals (9, 103), perfusion of the delivered human placenta or its cotyledons (116, 136), in vitro incubation of placental tissue fragments (128), isolation and incubation of maternal- and fetal-facing plasma membranes from human placenta (12, 65, 82), and, most recently, investigations in our and other laboratories using cultured human trophoblast. Each of these methods can potentially contribute important and ultimately complementary data on the mechanisms of placental solute transfer. This review summarizes investigations of the cellular and membrane mechanisms underlying the placental transport of a variety of important nutrients.

ORGANIC NUTRIENTS

Monosaccharides

Glucose is a major fetal and placental fuel whose metabolism accounts for a substantial fraction of fetal oxygen consumption (9, 103). The stereospecificity and saturability of transplacental glucose transfer indicates that it is a mediated process. Transfer mechanisms mediating the facilitated diffusion of glucose have been characterized in microvillous and basal plasma membranes in our own and other laboratories (67, 71, 73). The transporters interact strongly with D-aldohexoses and other sugars that can assume the C-1 chair configuration. The placental microvillous and basal membrane carriers differ from those of the apical surface of intestine and kidney, which possess sodium-dependent concentrative glucose transporters. Thus, while kidney and intestine may concentrate glucose against a concentration gradient, the placental syncytiotrophoblast lacks such ability. Some (but not all) studies have reported regulation of microvillous membrane glucose uptake by insulin under certain conditions and at concentrations (10^{-9}M) somewhat above the high physiologic range (24, 71).

Cytochalasin B-labelling in two laboratories has identified the microvillous membrane transporter as a protein of approximately 50 kDa (68, 72). More recent molccular studies have suggested that the human placenta contains either the isoform GLUT 1 or a mixture of GLUT 1 and GLUT 3 (10). K_m values for placental glucose transporters have been reported as 25 mM (71, 73) and 3–5 mM (67), a range more in keeping with the K_m of GLUT 1 and GLUT 3. The higher K_m values were found in studies using equilibrium exchange, a procedure that may more closely parallel in vivo conditions and that in general results in higher K_m and V_{max} values than a zero trans procedure. Indeed the lack of saturation of the overall maternal-fetal transfer process at glucose concentrations severalfold higher than in maternal blood (66) is in agreement with the higher range of K_m values.

The glucose transporters of the two plasma membranes supply glucose for fetal and placental metabolism under conditions in which the glucose concentration in the fetal circulation is approximately 70% of that in the maternal circulation (9, 103). Using kinetic parameters and surface area determinations, an investigation has estimated that the glucose transfer capacity of the microvillous membrane is manyfold higher than would be required to supply fetal and placental needs for glucose (71). Thus, in vivo the trophoblast cytosol is likely to be nearly equilibrated with maternal plasma glucose. The maternal-fetal glucose gradient may arise from limitation of overall transsyncytial transfer by the lower transport capacity of the basal membrane with its smaller surface area (73).

Amino Acids

The transfer of amino acids across the syncytiotrophoblast involves mediated transport mechanisms at the microvillous and basal membrane and possibly diffusion. Analysis of the process is complex and incomplete because of the large number of amino acids, the overlapping specificity of amino acid transport mechanisms, and the utilization of amino acids for nutrition of the trophoblast itself. Plasma membrane transport systems have been the subject of a number of recent reviews that describe their specificity, sodium-dependence, and other important properties (31, 147).

NEUTRAL AMINO ACIDS   The microvillous membrane of placenta utilizes transport systems common to many cell types. This is in contrast to intestinal or renal brush borders, which utilize specialized neutral amino acid transport systems (131) that may be similar to system $B^{0,+}$ described in developing mouse blastocysts (147). Reported placental systems include (a) system A—a sodium-dependent transporter that interacts strongly with alanine, serine, methylaminoisobutyric acid (MeAIB), and proline and is a ubiquitous concentrative transporter of amino acids; (b) system N—a sodium-dependent system that interacts strongly with histidine and glutamine; and (c) various sodium-independent systems (17, 74, 79). One of the sodium-independent transporters interacts strongly with leucine, tryptophan, tyrosine, and phenylalanine and resembles system L (52, 88). A second sodium-independent system apparently reacts strongly with alanine and serine and weakly with the branched-chain and aromatic amino acids (74). System T and system ASC have been reported to be absent in microvillous membrane (52, 74). Recent evidence, however, suggests that storage conditions of the membrane vesicles may influence the transport systems subsequently identified (78).

The transport systems of placental basal membrane resemble those of the basolateral membranes of intestine that possess sodium-dependent systems similar to the classical A and ASC systems as well as a sodium-independent L system (99). At least three mediated processes are present: (a) a sodium-dependent A-like system shared by MeAIB that is similar to the system employed by microvillous membrane; (b) a second sodium-dependent ASC-like system that interacts with alanine, serine, and apparently cysteine but is resistant to MeAIB; (c) a sodium-independent system that interacts with leucine, phenylalanine, and BCH and resembles system L (61) and potentially another sodium-independent system that transports tyrosine (88). To our knowledge, basal membrane has not been investigated for system N. Some rather ambitious investigations attempted to elucidate a broad scope of transporters on either membrane using a small number of inhibitors employed at high concentrations (87, 89). The potential for noncompetitive inhibition and system overlap under these conditions makes these studies difficult to interpret.

Transporters in the two membranes must act in concert to bring about the concentrative transfer of amino acids from mother to fetus. The physiological sodium gradient drives the sodium-dependent systems of the two plasma membranes toward uptake of amino acids from maternal and fetal circulations into the syncytium. With the larger surface area of the microvillous membrane (140) and its higher system A transport activity, the A system of microvillous membrane is likely to be responsible for generating the high syncytial concentration of alanine (74) (estimated as 7-fold greater than the concentration in maternal blood and 4-fold greater than the concentration in fetal blood) (108).

Transport from the trophoblast to the fetus across the basal membrane may occur via both system L and nonmediated pathways (61). The roles of the sodium-dependent systems in basal membrane are not established. They may, as suggested for similar systems in the intestine (99), provide mechanisms for trophoblast nutrition or for amino acids to leave the fetus under some metabolic conditions.

Cellular regulation of amino acid transport must involve regulation of transport systems of either of the two surface membranes. System A is known to be regulated by two mechanisms in placental tissue. One requires protein synthesis and provides a threefold or greater increase in V_{max} (127), and the other, transinhibition, is a feedback process by which intracellular substrate can inhibit uptake (129). The high trophoblast concentration of A system substrates established by microvillous membrane uptake could reduce uptake by system A in basal membrane. Both of these regulatory mechanisms are specific to system A and do not affect system L or system ASC. Transinhibition may serve to maintain a constancy of trophoblast amino acid concentrations with variation in maternal metabolic environment. Insulin, another regulator, is known to increase system A activity in muscle. Insulin is apparently without effect in placental villous tissue (128), although recent data from trophoblast primary cultures suggests that insulin at a concentration $(10^{-9}M)$ slightly above the high normal range increases the V_{max} of aminoiso-butyrate uptake (Peter Karl and Stanley Fisher, personal communication).

ANIONIC AMINO ACIDS   Anionic amino acids are an exception to the general pattern because they are not concentrated in the fetal circulation and are not transferred between maternal and fetal circulations (95, 115, 130). Anionic amino acids are taken up from either or both circulations, and in vivo millimolar concentrations of aspartate and glutamate are present within the placenta, whereas concentrations in maternal and fetal blood are in the micromolar range (108). Investigations of microvillous membrane in another laboratory (62) and of both microvillous and basal membrane in our own laboratory have demonstrated a common high affinity transporter (60, 102). This transporter is sodium-dependent, potassium-stimulated, pH sensitive, and electrogenic. Its characteristics resemble those of the X^-_{AG} system found in the apical surface of other epithelia and in nonepithelial tissues.

The observed absence of transport between mother and fetus apparently results not from the lack of uptake pathways from either circulation by the trophoblast but from the absence of an appropriate egress mechanism for release. Thus, evidence from the plasma membrane transport systems as well as physiologic studies strongly suggest that glutamate and aspartate are metabolized in the placenta rather than transported to the fetus. In vivo this metabolism may provide some benefit to the fetus, perhaps as part of an interorgan substrate cycle for fetal nitrogen metabolism or as a mechanism for protection of the developing fetus from glutamate toxicity (references cited in 60 and 102).

CATIONIC AMINO ACIDS   The cationic amino acids lysine and arginine are concentrated in the fetal circulation and within the placenta. Their transport by apical microvillous membrane has not been investigated in detail, but we do know from perfusion studies that they are taken up from both surfaces into the trophoblast (152). In our laboratory two sodium-independent systems were found in basal membrane (50). One of these is the ubiquitous y^+ system, which is a relatively high-capacity, low-affinity transporter and probably serves as the major transporter of cationic amino acids across the basal membrane. The other transporter resembles the $b^{0,+}$ system, which has been described in developing mouse blastocysts (148). Both systems interact with lysine and arginine but have different specificities and widely different con-centration dependence properties. This investigation provided the first known evidence of the existence of a system resembling $b^{0,+}$ outside of the develop-ing mouse blastocyst. The very low K_m and V_{max} of this transporter suggests that in the placenta it may serve a particular function other than bulk transport of cationic amino acids. Alternatively, in the basal membrane it may represent a residual of an earlier developmental stage of the trophoblast.

β-AMINO ACIDS   The amino acid in highest concentration in placenta is the β-amino acid taurine (108). The fetus requires an exogenous supply of taurine. The active transport of taurine by the placenta yields fetal concentrations that are greater than maternal concentrations (90). Taurine is taken up by microvillous membrane by a β-amino acid carrier that interacts with other β-amino acids, but not with the neutral α-amino acids (100). This carrier requires Na^+ and Cl^- and is stimulated by gradients of these ions. The transport process is electrogenic, and the sodium-taurine stoichiometry is $2:1$ or $3:1$. The transporter is highly specific for β-amino acids with a K_m in the low micromolar range. A carrier with similar properties is present in various cell membranes and in the JAR placental choriocarcinoma cell line in which it is under the control of protein kinase C (90). This carrier produces a steep gradient between the placenta and maternal and fetal blood. Taurine is transferred down this concentration gradient across the basal membrane by an uncharacterized mechanism. Taurine transport with generally similar properties has been demonstrated in perfused human placentas, although there has been disagreement between two laboratories concerning selectivity (59, 76).

Monocarboxylates and Dicarboxylates

Placental metabolism of glucose generates lactate at a high rate even under well-oxygenated conditions (58, 101), and the lactate is delivered into fetal and maternal circulations. This indicates that the brush border as well as the basal membranes of the syncytiotrophoblast possess mechanisms necessary for the transfer of lactate. The lactate carrier of the brush border membrane is Na^+-independent, but it is stimulated by a transmembrane H^+ gradient (6). The operational mechanism appears to be lactate-H^+ cotransport. The system is specific for monocarboxylates such as lactate, pyruvate, and β-OH-butyrate. Dicarboxylates in general do not interact with the transporter. The transporter probably functions symmetrically, with the directionality of lactate transfer governed by the magnitude and direction of lactate and H^+ gradients across the brush border membrane. Since there is evidence that placenta generates lactic acid in vivo and delivers it into the maternal circulation, the transporter appears to function normally to eliminate lactate and H^+ from the syncytiotrophoblast. This process may play a very important role in the maintenance of intracellular pH in the syncytiotrophoblast because lactic acid, unless promptly removed, will acidify the cell. The lactate transport mechanism of the basal membrane has not yet been studied.

    The brush border membrane of the placental syncytiotrophoblast also possesses a high-affinity transport system for dicarboxylates that is distinct from the transport system serving monocarboxylates (55). The dicarboxylate transporter is energized by a transmembrane electrochemical Na^+ gradient. The substrates of the system include many intermediates of the tricarboxylic acid cycle such as succinate, malate, fumarate, α-oxoglutarate, and citrate. Pyruvate and lactate are not substrates. Because of a favorable electrochemical Na^+ gradient across the brush border membrane, the transporter may play a role in providing these metabolic intermediates that are utilized by the placenta and/or the fetus. The normal plasma concentrations of succinate, α-oxoglutarate, and citrate, for example, are in the range of 40–130 μM. Therefore, the placenta can extract these compounds from the maternal circulation via the brush border membrane dicarboxylate transporter.

Lipids and Related Compounds

Placental transport of fatty acids and triglycerides along with related metabolites including choline, inositol, ethanolamine, carnitine, cholesterol, steroid hormones, and fat-soluble vitamins has been expertly reviewed by Coleman (34). Essentially, lipoprotein lipase on the maternal surface of the syncytiotrophoblast hydrolyzes triacylglycerol carried by maternal very low density lipoprotein (VLDL). The free fatty acids are taken up by the trophoblast by unknown mechanisms that are presumably similar to those of other cells and that utilize specific membrane proteins. The trophoblast may use the fatty acids or transfer them to the fetus. The rate of fatty acid synthesis in the placenta is also quite high. LDL cholesterol is taken up by receptor-mediated endocytosis and released in lysosomes (34). Apolipoprotein E, which promotes receptor-mediated lipoprotein uptake, is synthesized and secreted by the trophoblast (111). Recently, an anion exchanger that mediates taurocholate uptake has been described in the basal membrane from human trophoblast. The transporter is electroneutral and presumably driven by bicarbonate exchange (45, 97). This system may provide an important route for fetal excretion of bile acids.

Water-Soluble Vitamins

ASCORBATE   Vitamin C under physiologic conditions exists in an oxidized (dehydroascorbate) or reduced form (ascorbate). Dehydroascorbate is taken up more rapidly by syncytiotrophoblast microvillous membrane vesicles (69) and placental fragments (30) than is ascorbate. Uptake is sodium independent, and evidence has been reported both for and against mediation by the placental glucose transporter (30, 69, 70). The studies showing an effect of glucose on ascorbate transport (69, 70) used substrate concentrations one or more orders of magnitude higher than ascorbate concentrations found in maternal plasma, whereas physiologic concentrations were used in the study that failed to find a glucose interaction (30). The dehydroascorbate entering the placental syncytiotrophoblast is metabolized to the more useful ascorbate form and released into the fetal circulation (30). The mechanism for crossing the basal membrane of the syncytiotrophoblast is unknown.

FOLATE   Folate transport has been studied extensively in perfused guinea pig placenta (137, 138). Uptake occurs on both sides of the placenta by sodium-independent high affinity transporters, which can be inhibited by 5-methyltetrahydrofolate (major form in plasma) but not by methotrexate (137, 138). A putative transport protein has been identified in homogenates of human placental villi (2) and may be related to two homologous membrane-associated folate-binding proteins that have recently been cloned (110). These placental folate-binding proteins may derive from different cell or membrane locations or possible contamination by maternal tissue (110).

RIBOFLAVIN   A low-capacity riboflavin transport system investigated in perfused human placenta (38–40) transfers riboflavin faster in the maternal-fetal than in the fetal-maternal direction (38, 40). This transfer was saturable and the radioactivity in the fetal perfusate was riboflavin rather than either of its two most common metabolites, flavin adenine dinucleotide (FAD) or flavin mononucleotide (FMN) (38). Riboflavin was concentrated in the placenta and partially metabolized to FAD and FMN (39).

These studies demonstrate that the transfer of riboflavin across the human placenta is mediated (39) and suggest the possibility of a transporter in the microvillous membrane of the syncytiotrophoblast capable of concentrating riboflavin against a concentration gradient. Riboflavin exit into the fetal circulation may be driven by a concentration gradient between the syncytial cytoplasm and the fetal circulation.

THIAMIN   Thiamin transport possesses many of the same features as placental riboflavin transport. In perfused human placentas the transfer index for thiamin was greater for transfer toward the fetus than for transfer in the reverse direction, which approximated diffusion (41). Thiamin transport was saturable at low concentrations, a transplacental gradient was established, and the tissue thiamin concentration exceeded both perfusates (41). Schenker et al (114) using the same model showed inhibition of thiamin transport by thiamin analogs. Results of these studies are consistent with the existence of a thiamin transporter in the microvillous membrane of the syncytiotrophoblast capable of uptake against a concentration gradient.

OTHER VITAMINS   Concentrations of cobalamin (vitamin B_{12}), biotin, vitamin B_6, pantothenic acid, and nicotinic acid are higher in umbilical cord blood at delivery than they are in maternal blood. This finding is in agreement with the studies of vitamins discussed previously and with the idea that water-soluble vitamins in general are transported across the placenta by mediated processes (3). In addition, human placental tissue concentrates vitamin B_{12} (105), presumably by interaction with saturable high-affinity binding sites for the transcobalamin-vitamin B_{12} complex, on the plasma membrane (49). Recently a high affinity (K_m in low μM range) Na^+-stimulated biotin transporter has been found in microvillous membrane vesicles and placental trophoblast (77).

Lipid-Soluble Vitamins

VITAMIN A   Vitamin A is normally transported in the blood as a complex of retinol and retinol-binding protein (139). The mechanisms of transplacental flux of vitamin A are still not well-understood. Studies in rat (139) and monkey (144) have indicated that maternal retinol-binding protein crosses the placenta. The human placenta expresses surface receptors that bind retinol-binding protein (141), and the transplacental flux of the retinol-binding protein complex may be the main pathway for vitamin A transport across the placenta early in pregnancy. Fetal retinol-binding protein production occurs later in gestation (139). A second proposed mechanism for placental vitamin A transport is that the maternal retinol-binding protein binds to its receptor and releases retinol into the syncytiotrophoblast without the protein being internalized (144). The placenta then forms a retinyl ester that is secreted into the fetal circulation as lipoprotein. This hypothesis is supported by high concentrations of retinol found in lipoproteins in fetal circulations of monkeys (144). Similarly, a considerable fraction of retinol was esterified by human placenta and subsequently released unesterified in vitro (141).

VITAMIN D   Conflicting information is available concerning placental vitamin D transport. Maternal and fetal blood concentrations of $1,25(OH)_2D_3$ and vitamin D-binding protein have been positively correlated in the rat (4). Concentrations of both $25(OH)D_3$ and $1,25(OH)_2D_3$ were found to be correlated between maternal and cord blood in humans (16). These data suggest the diffusion of vitamin D across the placenta. However, some investigators have not found a correlation between maternal and fetal blood concentrations and suggest that vitamin D is poorly transferred across the placenta (42) and instead may be produced by the placenta to meet the needs of the fetus. Unidirectional transfer of $25(OH)D_3$ and $1,25(OH)_2D_3$ has been demonstrated in a perfused placenta system (112). The clearance index of $1,25(OH)_2D_3$ was tenfold higher than that of $25(OH)D_3$ when vitamin D-binding protein was present in the perfusate.

VITAMINS E AND K   Concentrations of lipid-soluble vitamins E and K are higher in maternal plasma than in umbilical cord plasma (15, 120). A study in rabbits indicates that α-tocopherol (vitamin E) does not cross the placenta in any significant amount (15). Studies in the rat (57) and human (80) have found the placenta to be a substantial barrier to vitamin K transport.

Nucleosides

Nucleosides are precursors of nucleotides, which form the building blocks of nucleic acids but additionally may be involved in other biological functions. In particular, adenosine has been shown to be a vasoconstrictor in placenta (126). Study of adenosine transport is complicated by cellular (151) as well as brush border membrane vesicle (8) metabolism of adenosine. Recently, Yudilevich (8) and co-workers were able to inhibit brush border adenosine metabolism with 2'-deoxycoformycin and thereby investigate transport. Adenosine uptake in brush border and basal membrane vesicles from human placenta was by a high-affinity mediated system that is nucleoside sensitive (8).

INORGANIC NUTRIENTS

Macro Minerals and Ions

CALCIUM   Increasingly large quantities of calcium are required to support the mineralization of the growing fetal skeleton (7 mmol per day in the third trimester). Concentrations of total and ionic calcium in cord blood exceed those in maternal blood, indicating that placental transfer is an active process (109). While large quantities of calcium pass through the placenta, the syncytiotrophoblast maintains a cytosolic free calcium in the nano- to micro-molar range, several orders of magnitude lower than the circulating concentrations (86). By analogy with other calcium-transporting epithelia, three types of mechanisms are likely to be used by the syncytiotrophoblast to accomplish this task: (*a*) an entry mechanism at the microvillous membrane, (*b*) a cytoplasmic mechanism for accomplishing transplacental flux while maintaining low free intracellular calcium concentrations, and (*c*) an exit mechanism to deliver calcium to the fetal circulation (20).

Recent work in our laboratory has demonstrated that mediated transport in human placental microvillous membrane possesses properties similar to those found in intestine and kidney (75). The K_m value of the higher capacity system is in the low millimolar range and suggests partial saturation at physiologic concentrations of maternal blood. The millimolar K_m, the lack of inhibition by known calcium channel blockers, and other characteristics suggest that the identified mechanism(s) are facilitated diffusion transporters rather than channels or pores. Calcium also binds to the microvillous membrane by a saturable process that may protect this portion of the trophoblast from the high concentrations found in maternal blood. The relationship of calcium-binding activity to the microfilaments of the microvilli and to cytoplasmic calcium-binding proteins needs further investigation.

The processes mediating transcellular or cytosolic transfer have not been clearly defined in the placental syncytiotrophoblast. It has been suggested that vesicular packet mechanisms for transcellular or cytosolic transfer exist in other epithelia, but they have not been characterized in placental tissue (20, 146). Calcium-binding proteins have been identified in at least two laboratories. The first of these is a low molecular weight (10 kDa) protein in the rodent, similar to that in intestine. This protein is likely to be located within the endoderm of the intraplacental yolk sac, which may play a role in transplacental calcium transport in this species (26). Higher molecular weight binding proteins for calcium have been described in human placenta, and it has been suggested that they are associated with a calcium-pumping mechanism in endoplasmic reticulum or plasma membrane (143).

Several investigators have attempted to characterize ATP-dependent transport pumps or calcium-dependent ATPase activities in membranes from human placenta. Many of these earlier studies described either nonspecific divalent cation-dependent ATPases or calcium-dependent ATPases with a K_m very much greater than estimates of intracellular calcium concentrations (references reviewed in 81). We have characterized an ATP-dependent transport process in isolated basal membrane whose characteristics are similar to known ATP-dependent calcium pumps in other plasma membranes (46). We have also demonstrated ATP hydrolysis by this transporter as well as by nonspecific divalent cation-dependent ATPases activated by magnesium as well as by calcium (81). The transporter subunits react with polyclonal and monoclonal antibodies prepared against the erythrocyte calcium transporter (140 kDa) and are phosphorylated under the same conditions as the red cell transporter (81). Transporter antigenic material is located at the fetal-facing surface of the trophoblast in human placenta, the innermost (fetal-facing) layer of trophoblast in rat labyrinthine placenta, and in the endoderm of the intraplacental yolk sac in the rat placenta (14). These findings are consistent with the presence of an ATP-dependent pump situated to deliver calcium to the fetal circulation.

The studies described here provide evidence for a reasonable cellular pathway for the transfer of calcium from mother to fetus by the placental syncytiotrophoblast. At least three important questions remain: (a) To what extent are calcium entry and exit processes polarized between the two plasma membranes of the syncytiotrophoblast as they are in other transporting epithelia? (b) What is the rate-limiting step in transplacental calcium transfer? In intestinal and renal epithelia the calcium-binding protein apparently functions to increase the rate of transfer of calcium through an area of low free calcium concentration, and this cytosolic transfer is believed to be rate limiting (20). Studies using inhibitors (135) or examining the development of mRNAs for calcium transport ATPases suggest that the ATP-dependent transport step is not rate limiting but that the binding protein-facilitated transfer may be (K. Thornburg and R. Boyd, personal communication). (c) What is the relative importance of transcellular and potential paracellular transfer of calcium across the placental syncytiotrophoblast layer? The concentration of calcium within the fetal circulation strongly indicates that transcellular transfer is a major process, but the relative transcellular and paracellular transfer rates are difficult to evaluate in perfusion studies with human tissue (132).

MAGNESIUM   The paucity of data concerning magnesium transport—not only in the placenta but in general—is primarily due to the lack of a long half-lived radioisotope. Two recent studies using the perfused rat placenta indicate that maternal-fetal Mg^{2+} transfer is transcellular, Na^+-dependent, and independent of Ca^{2+} transport (118, 119). Perfusion with the carbonic anhydrase inhibitor acetazolamide did not affect maternal-fetal Mg^{2+} clearance, which suggests that intracellular HCO_3^- or H^+ concentrations do not alter Mg^{2+} transport (119).

PHOSPHATE   Fetal serum concentrations of phosphate are higher than maternal concentrations (134), which means that the large supply of phosphate needed by the fetus in the last three months of pregnancy must be transported against a concentration gradient. Studies in whole perfused guinea pig placentas revealed a sodium-dependent transport process (133, 134) that has been localized to the microvillous membrane of the syncytiotrophoblast in human placenta (22, 94). In vesicles, sodium-stimulated phosphate uptake was concentrative and electrogenic with a pH optima of 7.0 (22) and a sodium coupling stoichiometry of 2:1 (94). Increasing temperature lowered the K_m and increased the V_{max} (94). The presence of insulin (25) or parathyroid hormone (23) reportedly decreased phosphate uptake. The parathyroid hormone effect was presumably mediated via increased cAMP (23). However, interpretation of these results is uncertain, since in both studies (23, 25) hormone concentrations were well above normal physiologic concentrations. Phosphate transport across the basal membrane could be driven by the concentration gradient between the syncytiotrophoblast and the fetal circulation.

SODIUM   Transport of sodium across the brush border membrane of the placental syncytiotrophoblast occurs by at least three mechanisms: Na^+-H^+ exchange, cotransport with inorganic anions and organic solutes, and Na^+ conductance. The Na^+-H^+ exchanger catalyzes an electroneutral coupling of

Na^+ influx from the maternal circulation into the syncytiotrophoblast with H^+ efflux in the opposite direction (5, 29). The recent demonstration of the presence of a Cl^--HCO_3^- exchanger in the placental brush border membrane (56, 63) suggests the possibility of a functional coupling between it and the Na^+-H^+ exchanger. Therefore, in addition to participating in the transport of Na^+, the activity of the Na^+-H^+ exchanger is likely to contribute to Cl^- entry and HCO_3^- exit from the syncytiotrophoblast. The external substrate-binding site of the Na^+-H^+ exchanger can interact with Na^+ as well as with other cations such as Li^+, NH_4^+, and H^+. The K_m for Na^+ is approximately 10 mM, which implies that under physiologic conditions the exchanger normally operates at or near saturation. The activity of the Na^+-H^+ exchanger can be inhibited by amiloride and its analogs and also by harmaline, cimetidine, and clonidine (5, 51, 54). Chemical modification studies have indicated that histidyl, carboxyl, and thiol groups of the exchanger protein are essential for the catalytic activity (53, 93).

Two distinct types of Na^+-H^+ exchangers are known to be present in the animal cell plasma membrane and are referred to as the "housekeeping" (NHE-1) and "epithelial" (NHE-2) types. The Na^+-H^+ exchanger of the placental brush border membrane belongs to the "housekeeping" type (91). In other cells, this type has been shown to be involved in the regulation of intracellular pH and in the mediation of at least some of the biological actions of several mitogens and growth factors. These findings indicate that the Na^+-H^+ exchanger of the placental brush border membrane may participate in functions that are essential to the normal growth and development of the placenta. The recent finding that the brush border exchanger possesses an allosteric activator site for H^+ on the cytosolic side (64) suggests that the removal of H^+ from the cell by the exchanger can be finely controlled in response to changes in the intracellular concentration of H^+. The exchanger is thus an ideal system to participate in the regulation of intracellular pH.

Cotransport with other molecules also contributes significantly to the entry of Na^+ into the syncytiotrophoblast, because a number of transport systems in the placental brush border membrane utilize a transmembrane Na^+ gradient as the driving force. Various amino acid transporters (156), dicarboxylate transporter (55), serotonin transporter (7), and phosphate transporter (22) are a few examples. The third mechanism for the entry of Na^+ into the syncytiotrophoblast is a conductive pathway (29) in which Na^+ moves down its electrochemical gradient.

The exit of Na^+ from the syncytiotrophoblast into the fetal circulation occurs across the basal membrane. The primary mechanism responsible for this process is the Na^+-K^+ pump, which is localized predominantly in the basal membrane (82). Active extrusion of Na^+ via the Na^+-K^+ pump generates an inwardly directed Na^+ gradient across the brush border membrane, which in turn facilitates the Na^+ entry mechanisms. Thus, the vectorial movement of Na^+ in the direction of maternal to fetal circulation is made possible by the differential localization of the entry and the exit mechanisms for Na^+ at the two poles of the syncytiotrophoblast.

Our most recent studies have shown that the basal membrane of the syncytiotrophoblast also contains Na^+-H^+ exchanger activity (unpublished data). Interestingly, this exchanger is pharmacologically distinguishable from the exchanger identified in the brush border membrane of this cell. The brush border membrane Na^+-H^+ exchanger belongs to the "housekeeping" type (NHE-1), whereas the basal membrane Na^+-H^+ exchanger belongs to the "epithelial" or "apical" type (NHE-2). However, the activity of the exchanger in the basal membrane is about five times less than the activity of the exchanger in the brush border membrane. The role of the basal membrane Na^+-H^+ exchanger in the physiology of the syncytiotrophoblast is not readily apparent at this time.

POTASSIUM   In contrast to the transport of Na^+, very little information is available on the processes responsible for the transport of K^+ across the placenta. Apparently, the K^+-transporting mechanisms such as the K^+-Cl^- cotransport and the Na^+-K^+-Cl^- cotransport, which are known to catalyze K^+ entry across the plasma membrane of certain cell types, are not present in the placental brush border membrane (63). Potassium extrusion by the Na^+/K^+-dependent anionic amino acid transporter may occur (60, 102).

HYDROGEN ION   The movement of H^+ across the placenta is important to the maintenance of acid-base balance in the fetus and the placenta. Four potential mechanisms are known. (a) The Na^+-H^+ exchanger of the placental brush border membrane may transport H^+ by the Na^+ gradient across the brush border membrane. (b) A proton pump whose presence has recently been described in placental brush border membrane (92) may also participate in the removal of H^+ from the cell. This H^+ pump belongs to the vacuolar type, and active extrusion of H^+ by the pump is energized by the hydrolysis of ATP. (c) A third mechanism is the coupled transport of H^+ and an organic ion such as occurs via the lactate-H^+ symport (see above). (d) A fourth potential mechanism for H^+ transfer is likely protein-mediated (28). This conductive pathway is voltage dependent and is allosterically activated by external monovalent anions such as I^-, Br^-, and Cl^-. Due to the presence of an inside-negative membrane potential across the syncytiotrophoblast brush border membrane in vivo, this pathway likely functions normally in the uptake of H^+.

IRON   The fetus obtains its supply of iron from the maternal circulation (48). The probable first step in placental iron transfer is receptor-mediated endocytosis of ferric transferrin. Transferrin receptors have been found on the surface membrane of the human placental syncytiotrophoblast (96, 117, 150). The 94-kDa molecular weight glycoprotein binds diferric transferrin with higher affinity than it binds apotransferrin (142). Studies have indicated that very little transferrin crosses the placenta (35). Most likely, iron is released in low pH endosomal compartments and the apotransferrin is recycled to the plasma membrane (43).

How iron traverses the cell and is released into the fetal circulation remains largely unknown, although the mechanism probably involves ferritin and/or other heme proteins. A transferrin receptor has been identified on the basal membrane of the syncytiotrophoblast, apparently identical to the receptor on the microvillous membrane (149). The function of this receptor is unknown, but its location suggests that it may be involved in maternal-fetal transfer of iron or perhaps that it has alternative functions unrelated to iron transport (149).

Aspects of transferrin cycling have been studied in various cell lines (145), but only recently has an appropriate model for normal trophoblast been available. Douglas & King have used this model to determine that transferrin receptors on the trophoblast surface function to bind, internalize, and recycle ^{125}I-labelled transferrin with a calculated cycle time of 11 min (43). Iron was rapidly lost from transferrin upon internalization, and most of the iron became associated with ferritin (43). Iron release from normal cultured trophoblast was gradual ($t_{1/2}>40$ h) and was associated with a low molecular weight fraction, transferrin, and ferritin (43).

CHLORIDE AND IODIDE   At least three distinct mechanisms may function in the transfer of Cl^- across the placental brush border membrane. (a) Many investigators have reported the existence of an anion exchanger that accepts Cl^- as a substrate (56, 63, 123). The preferred exchange anion for this system appears to be HCO_3^- rather than OH^- (56), and influx of Cl^- via the exchanger is coupled to the efflux of HCO_3^-. A functional coupling between this transport system and the Na^+-H^+ exchanger is presumed to occur in vivo. Several monovalent anions compete with Cl^- for the exchange process including Br^-, I^-, NO_3^-, and SCN^- (123). The interaction of I^- with the exchanger is interesting because it may provide the necessary mechanism for active transport of this nutritionally essential element into the syncytiotrophoblast. (b) A conductive voltage-dependent pathway inhibited by the Cl^--channel blocker diphenylamine-2-carboxylate is known to exist (63, 123). (c) Chloride-coupled transport systems mediating the transport of certain organic solutes may also function in chloride entry. Examples are the taurine (77, 100) and serotonin transporters (7, 36). Both transporters exhibit an absolute requirement for NaCl, and the catalytic mechanism appears to be the cotransport of Na^+ and Cl^- with the substrate.

Micro Minerals and Ions

SULFATE/SELENATE/CHROMATE/MOLYBDATE   Sulfate is an essential nutrient for optimal growth and development of the fetus, and evidence indicates that this anion is actively transported across the placenta (33). The entry into the syncytiotrophoblast across the brush border membrane as well as the exit from the cell across the basal membrane occur by anion exchange mechanisms. The brush border membrane SO_4^{2-} transporter is Na^+-independent and can catalyze uphill transport of SO_4^{2-} in response to a transmembrane HCO_3^- gradient, thus indicating the presence of a SO_4^{2-}/HCO_3^- exchange (32). Monovalent anions such as Cl^-, I^-, and SCN^- do not interfere with SO_4^{2-} transfer by this system (18). In contrast, divalent anions such as SeO_4^{2-}, CrO_4^{2-}, and MoO_4^{2-} interact significantly (19), suggesting that the essential trace elements selenium, chromium, and molybdenum may be transported to the fetus via this system. Recently, direct evidence has been obtained for the transport of SeO_4^{2-} catalyzed by the brush border membrane SO_4^{2-} transporter (121). The basal membrane SO_4^{2-} transporter also functions as an anion exchanger, but this process may not be identical with its counterpart in the brush border membrane (27).

TRACE ELEMENTS   Zinc, an essential trace element, is important for growth and development. The presence of sulfhydryl groups may be important to zinc uptake by trophoblast cells (98). Transfer of zinc was increased by the presence of ligands in the perfusate of perfused guinea pig placenta (107).

   Pregnant mice given aluminum orally or intraperitoneally accumulated aluminum in fetuses and placentas in concentrations 2 to 10 times those of controls (37). The doses in this study were high (both animals in the high intraperitoneal group died). Oral doses did not significantly raise maternal liver concentrations, so it is unclear whether aluminum intake would significantly raise exposure to the placenta. The heavy metals cadmium, mercury, and lead accumulate in placenta and fetal tissue, but nothing is known about their mechanisms of placental transport (122). Similarly, radioactive vanadium accumulated in fetuses of rats given intravenous injections (44). In general, the placenta does not appear to present a barrier to most trace elements, but the mechanisms of transport are unknown.

CONCLUSIONS

Substantial progress has been made in elucidating the cellular transport mechanisms of placental trophoblast. Three principal model systems have recently served as tools: placental perfusion, isolated maternal- and fetal-facing surface membranes, and, most recently, isolated trophoblast and trophoblast-derived cells. Most of the major transport systems are analogous to those of other epithelial and nonepithelial cells, and yet their distinctive arrangement and operation in the syncytiotrophoblast is likely to be important in the functions of the placenta. These transport processes have the potential to be regulated both by intrinsic interaction with substrates and cellular nutrients and by extrinsic mechanisms involving membrane receptors and local and systemic effector substances.

The application of these and other approaches to cellular investigation will continue to develop more complete knowledge of trophoblast transport mechanisms. The integration of this knowledge with that from in vivo investigations will allow us to understand the cellular basis of trophoblast and placental function in utero and the pathways it provides for the nutrition of the fetus.

ACKNOWLEDGMENTS

We thank the many investigators who shared their work with us prior to publication. We are grateful to Barbara Hartman for her skillful assistance with the manuscript. This review was supported by research grants from the National Institutes of Health: HD 07562 (to C.S.), HD 27258 (to A.M.), and HD 21103 and HD 24451 (to V.G.).

Literature Cited

1. Albe, K. R., Witkin, H. J., Kelley, L. K., Smith, C. H. 1983. Protein kinases of the human placental microvillous membrane. Their potential role in intrasyncytial communication. *Exp. Cell Res.* 147:167–76

2. Antony, A. C., Utley, C., Van Horne, K. C., Kolhouse, J. F. 1981. Isolation and characterization of a folate receptor from human placenta. *J. Biol. Chem.* 256:9684–92

3. Baker, H., Frank, O., Deangelis, B., Feingold, S., Kaminetzky, H. A. 1981. Role of placenta in maternal-fetal vitamin transfer in humans. *Am. J. Obstet. Gynecol.* 141:792–96

4. Baker, H., Frank, O., Thomson, A. D., Langer, A., Munves, E. D., et al. 1975. Vitamin profile of 174 mothers and newborns at parturition. *Am. J. Clin. Nutr.* 28:56–67

5. Balkovetz, D. F., Leibach, F. H., Mahesh, V. B., Devoe, L. D., Cragoe, E. J. Jr., Ganapathy, V. 1986. The Na^+-H^+ exchanger of human placental brush border membrane. Identification and characterization. *Am. J. Physiol.* 251:C852–60

6. Balkovetz, D. F., Leibach, F. H., Mahesh, V. B., Ganapathy, V. 1988. A proton gradient is the driving force for uphill transport of lactate in human placental brush border membrane vesicles. *J. Biol. Chem.* 263:13823–30

7. Balkovetz, D. F., Tiruppathi, C., Leibach, F. H., Mahesh, V. B., Ganapathy, V. 1989. Evidence for an imipramine-sensitive serotonin transporter in human placental brush border membranes. *J. Biol. Chem.* 264:2195–98

8. Barros, L. F., Bustamante, J. C., Yudilevich, D. L., Jarvis, S. M. 1991. Adenosine transport and nitrobenzylthioinosine binding in human placental membrane vesicles from brush-border and basal sides of the trophoblast. *J. Membr. Biol.* 119:151–61

9. Battaglia, F. C., Meschia, G. 1986. In *An Introduction to Fetal Physiology*, ed. F. C. Battaglia, G. Meschia, pp. 1–27. Orlando, Fla: Academic.

10. Bell, G. I., Kayano, T., Buse, J. B., Burant, C. F., Takeda, J., et al. 1990. Molecular biology of mammalian glucose transporters. *Diabetes Care* 13:198–208

11. Blay, J., Hollenberg, M. D. 1989. The nature and function of polypeptide growth factor receptors in the human placenta. *J. Dev. Physiol.* 12:237–48

12. Booth, A. G., Olaniyan, R. O., Vanderpuye, O. A. 1980. An improved method for the preparation of human placental

syncytiotrophoblast microvilli. *Placenta* 1:327–36
13. Booth, A. G., Vanderpuye, O. A. 1983. Structure of human placental microvilli. In *Brush Border Membranes, Ciba Found. Symp. 95*, pp. 180–94. London: Pitman Books
14. Borke, J. L., Caride, A., Verma, A. K., Kelley, L. K., Smith, C. H., et al. 1989. Calcium pump epitopes in placental trophoblast basal plasma membranes. *Am. J. Physiol.* 257:C341–46
15. Bortolotti, A., Traina, G. L., Barzago, M. M., Celardo, A., Bonati, M. 1990. Placental transfer and tissue distribution of vitamin E in pregnant rabbits. *Biopharm. Drug Dispos.* 11:679–88
16. Bouillon, R., Van Assche, F. A., Van Baelen, H., Heyns, W., DeMoor, P. 1981. Influence of the vitamin D-binding protein on the serum concentration of 1,25-dihydroxyvitamin D_3. Significance of the free 1,25-dihydroxyvitamin D_3 concentration. *J. Clin. Invest.* 67:589–96
17. Boyd, C. A. R., Lund, E. K. 1981. L-Proline transport by brush border membrane vesicles prepared from human placenta. *J. Physiol.* 315:9–19
18. Boyd, C. A. R., Shennan, D. B. 1986. Human placental sulfate transport: studies on chorionic trophoblast brush border membrane vesicles. *J. Physiol.* 377:15–24
19. Boyd, C. A. R., Shennan, D. B. 1986. Sulphate transport into vesicles prepared from human placental brush border membranes: inhibition by trace element oxides. *J. Physiol.* 379:367–76
20. Bronner, F. 1990. Transcellular calcium transport. In *Intracellular Calcium Regulation,* ed. F. Bronner, pp. 415–37. New York: Liss
21. Brunette, M. G. 1988. Calcium transport through the placenta. *Can. J. Physiol. Pharmacol.* 66:1261–69
22. Brunette, M. G., Allard, S. 1985. Phosphate uptake by syncytial brush border membranes of human placenta. *Pediatr. Res.* 19:1179–82
23. Brunette, M. G., Auger, D., Lafond, J. 1989. Effect of parathyroid hormone on PO_4 transport through the human placenta microvilli. *Pediatr. Res.* 25:15–18
24. Brunette, M. G., Lajeunesse, D., Leclerc, M., Lafond, J. 1990. Effect of insulin on D-glucose transport by human placental brush border membranes. *Mol. Cell. Endocrinol.* 69:59–68
25. Brunette, M. G., Leclerc, M., Ramachandran, C., Lafond, J., Lajeunesse, D. 1989. Influence of insulin

on phosphate uptake by brush border membranes from human placenta. *Mol. Cell. Endocrinol.* 63:57–65
26. Bruns, M. E., Kleeman, E., Bruns, D. E. 1986. Vitamin D-dependent calcium-binding protein of mouse yolk sac. Biochemical and immunochemical properties and responses to 1,25-dihydroxycholecalciferol. *J. Biol. Chem.* 261:7485–90
27. Bustamante, J. C., Yudilevich, D. L., Boyd, C. A. R. 1988. A new form of asymmetry in epithelia: kintics of apical and basal sulphate transport in human placenta. *Q. J. Exp. Physiol.* 73:1013–16
28. Cabrini, G., Illsley, N. P., Verkman, A. S. 1986. External anions regulate stilbene-sensitive proton transport in placental brush border vesicles. *Biochemistry* 25:6300–5
29. Chipperfield, A. R., Langridge-Smith, J. E., Steele, L. W. 1988. Sodium entry into human placental microvillous (maternal) plasma membrane vesicles. *Q. J. Exp. Physiol.* 73:399–411
30. Choi, J.-L., Rose, R. C. 1989. Transport and metabolism of ascorbic acid in human placenta. *Am. J. Physiol.* 257:C110–13
31. Christensen, H. N. 1984. Organic ion transport during seven decades—the amino acids. *Biochim. Biophys. Acta* 779:255–69
32. Cole, D. E. C. 1984. Sulfate transport in brush border membrane vesicles prepared from human placental syncytiotrophoblast. *Biochem. Biophys. Res. Commun.* 123:223–29
33. Cole, D. E. C., Baldwin, L. S., Stirk, L. J. 1984. Increased inorganic sulfate in mother and fetus at parturition: evidence for a fetal-to-maternal gradient. *Am. J. Obstet. Gynecol.* 148:596–99
34. Coleman, R. A. 1989. The role of the placenta in lipid metabolism and transport. *Semin. Perinatol.* 13:180–91
35. Contractor, S. F., Eaton, B. M. 1986. Role of transferrin in iron transport between maternal and fetal circulations of a perfused lobule of human placenta. *Cell Biochem. Funct.* 4:69–74
36. Cool, D. R., Leibach, F. H., Ganapathy, V. 1990. Modulation of serotonin uptake kinetics by ions and ion gradients in human placental brush border membrane vesicles. *Biochemistry* 29:1818–22
37. Cranmer, J. M., Wilkins, J. D., Cannon, D. J., Smith, L. 1986. Fetal-placental-maternal uptake of aluminum in mice following gestational exposure:

Effect of dose and route of administration. *Neurotoxicology* 7:601–8

38. Dancis, J., Lehanka, J., Levitz, M. 1985. Transfer of riboflavin by the perfused human placenta. *Pediatr. Res.* 19:1143–46

39. Dancis, J., Lehanka, J., Levitz, M. 1988. Placental transport of riboflavin: Differential rates of uptake at the maternal and fetal surfaces of the perfused human placenta. *Am. J. Obstet. Gynecol.* 158:204–10

40. Dancis, J., Lehanka, J., Levitz, M., Schneider, H. 1986. Establishment of gradients of riboflavin, L-lysine and α-aminoisobutyric acid across the perfused human placenta. *J. Reprod. Med.* 31:293–96

41. Dancis, J., Wilson, D., Hoskins, I. A., Levitz, M. 1988. Placental transfer of thiamine in the human subject: In vitro perfusion studies and maternal-cord plasma concentrations. *Am. J. Obstet. Gynecol.* 159:1435–39

42. Delvin, E. E., Glorieux, F. H., Salle, B. L., David, L., Varenne, J. P. 1982. Control of vitamin D metabolism in preterm infants: Feto-maternal relationships. *Arch. Dis. Child.* 57:754–57

43. Douglas, G. C., King, B. F. 1990. Uptake and processing of ^{125}I-labelled transferrin and ^{59}Fe-labelled transferrin by isolated human trophoblast cells. *Placenta* 11:41–57

44. Edel, J., Sabboni, E. 1989. Vanadium transport across placenta and milk of rats to the fetus and newborn. *Biol. Trace Elem. Res.* 22:265–75

45. el-Mir, M. Y. A., Eleno, N., Serrano, M. A., Bravo, P., Marin, J. J. G. 1991. Bicarbonate-induced activation of taurocholate transport across the basal plasma membrane of human term trophoblast. *Am. J. Physiol.* 260:G887–94

46. Fisher, G. J., Kelley, L. K., Smith, C. H. 1987. ATP-dependent calcium transport across basal plasma membranes of human placental trophoblast. *Am. J. Physiol.* 252:C38–C46

47. Fisher, S. E., Atkinson, M., Chang, B. 1987. Effect of delta-9-tetrahydrocannabinol on the in vitro uptake of α-amino isobutyric acid by term human placental slices. *Pediatr. Res.* 21:104–7

48. Fletcher, J., Suter, P. E. N. 1969. The transport of iron by the human placenta. *Clin. Sci.* 36:209–20

49. Friedman, P. A., Shia, M. A., Wallace, J. K. 1977. A saturable high affinity site for transcobalamin II-vitamin B_{12} complexes in human placental membrane preparations. *J. Clin. Invest.* 59:51–58

50. Furesz, T. C., Moe, A. J., Smith, C. H. 1991. Two cationic amino acid transport systems in human placental basal plasma membrane. *Am. J. Physiol.* 261:C246–52

51. Ganapathy, M. E., Leibach, F. H., Mahesh, V. B., Devoe, L. D., Ganapathy, V. 1986. Interaction of clonidine with human placental Na^+-H^+ exchanger. *Biochem. Pharmacol.* 35:3989–94

52. Ganapathy, M. E., Leibach, F. H., Mahesh, V. B., Howard, J. C., Devoe, L. D., Ganapathy, V. 1986. Characterization of tryptophan transport in human placental brush-border membrane vesicles. *Biochem. J.* 238:201–8

53. Ganapathy, V., Balkovetz, D. F., Ganapathy, M. E., Mahesh, V. B., Devoe, L. D., Leibach, F. H. 1987. Evidence for histidyl and carboxyl groups at the active site of the human placental Na^+-H^+ exchanger. *Biochem. J.* 245:473–77

54. Ganapathy, V., Balkovetz, D. F., Miyamoto, Y., Ganapathy, M. E., Mahesh, V. B., et al. 1986. Inhibition of human placental Na^+-H^+ exchanger by cimetidine. *J. Pharmacol. Exp. Ther.* 239:192–97

55. Ganapathy, V., Ganapathy, M. E., Tiruppathi, C., Miyamoto, Y., Mahesh, V. B., Leibach, F. H. 1988. Sodium-gradient-driven, high-affinity, uphill transport of succinate in human placental brush border membrane vesicles. *Biochem. J.* 249:179–84

56. Grassl, S. M. 1989. Cl/HCO_3 exchange in human placental brush border membrane vesicles. *J. Biol. Chem.* 264:11103–6

57. Hamulyak, K., de Boer-van den Berg, M. A. G., Thijssen, H. H. W., Hemker, H. C., Vermeer, C. 1987. The placental transport of [^{3}H]vitamin K_1 in rats. *Br. J. Haematol.* 65:335–38

58. Hauguel, S., Desmaizieres, V., Challier, J. C. 1986. Glucose uptake, utilization, and transfer by the human placenta as functions of maternal glucose concentration. *Pediatr. Res.* 20:269–73

59. Hibbard, J. U., Pridjian, G., Whitington, P. F., Moawad, A. H. 1990. Taurine transport in the in vitro perfused human placenta. *Pediatr. Res.* 27:80–84

60. Hoeltzli, S. D., Kelley, L. K., Moe, A. J., Smith, C. H. 1990. Anionic amino acid transport systems in isolated basal plasma membrane of human placenta. *Am. J. Physiol.* 259:C47–C55

61. Hoeltzli, S. D., Smith, C. H. 1989. Alanine transport systems in isolated basal plasma membrane of human placenta. *Am. J. Physiol.* 256:C630–37

62. Iioka, H., Moriyama, I., Itoh, K., Hino,

K., Ichija, M. 1985. Studies on the mechanism of placental transport of L-glutamate (the effects of potassium ion in microvillous vesicles on L-glutamate uptake). *J. Obstet. Gynaecol. Jpn.* 37: 2005–9

63. Illsley, N. P., Glaubensklee, C., Davis, B., Verkman, A. S. 1988. Chloride transport across placental microvillous membranes measured by fluorescence. *Am. J. Physiol.* 255:C789–97

64. Illsley, N. P., Jacobs, M. M. 1990. Control of the sodium-proton antiporter in human placental microvillous membranes by transport substrates. *Biochim. Biophys. Acta* 1029:227–34

65. Illsley, N. P., Wang, Z. Q., Gray, A., Sellers, M. C., Jacobs, M. M. 1990. Simultaneous preparation of paired, syncytial, microvillous and basal membranes from human placenta. *Biochim. Biophys. Acta* 1029:218–26

66. Ingermann, R. L. 1987. Control of placental glucose transfer. *Placenta* 8: 557–71

67. Ingermann, R. L., Bissonnette, J. M. 1983. Effect of temperature on kinetics of hexose uptake by human placental plasma membrane vesicles. *Biochim. Biophys. Acta* 734:329–35

68. Ingermann, R. L., Bissonnette, J. M., Koch, P. L. 1983. D-Glucose-sensitive and -insensitive cytochalasin B binding proteins from microvillous plasma membranes of human placenta. *Biochim. Biophys. Acta* 730:57–63

69. Ingermann, R. L., Stankova, L., Bigley, R. H. 1986. Role of monosaccharide transporter in vitamin C uptake by placental membrane vesicles. *Am. J. Physiol.* 250:C637–41

70. Ingermann, R. L., Stankova, L., Bigley, R. H., Bissonette, J. M. 1988. Effect of monosaccharide on dehydroascorbic acid uptake by placental membrane vesicles. *J. Clin. Endocrinol. Metab.* 67:389–94

71. Johnson, L. W., Smith, C. H. 1980. Monosaccharide transport across microvillous membrane of human placenta. *Am. J. Physiol.* 238:C160–68

72. Johnson, L. W., Smith, C. H. 1982. Identification of the glucose transport protein of the microvillous membrane of human placenta by photoaffinity labelling. *Biochem. Biophys. Res. Commun.* 109:408–13

73. Johnson, L. W., Smith, C. H. 1985. Glucose transport across the basal plasma membrane of human placental syncytiotrophoblast. *Biochim. Biophys. Acta* 815:44–50

74. Johnson, L. W., Smith, C. H. 1987.

Neutral amino acid transport systems of microvillous membrane of human placenta. *Am. J. Physiol.* 254:C773–80

75. Kamath, S. G., Kelley, L. K., Friedman, A. F., Smith, C. H. 1991. Transport and binding in calcium uptake by microvillous membrane of human placenta. *Am. J. Physiol.* 262:C789–94

76. Karl, P. I., Fisher, S. E. 1990. Taurine transport by microvillous membrane vesicles and the perfused cotyledon of the human placenta. *Am. J. Physiol.* 258:C443–51

77. Karl, P. I., Fisher, S. E. 1992. Biotin transport in microvillous membrane vesicles, cultured trophoblasts, and isolated perfused human placenta. *Am. J. Physiol.* 262:C302–8

78. Karl, P. I., Teichberg, S., Fisher, S. E. 1991. Na$^+$-dependent amino acid uptake by human placental microvillous membrane vesicles: Importance of storage conditions and preservation of cytoskeletic elements. *Placenta* 12:239–50

79. Karl, P. I., Tkaczevski, H., Fisher, S. E. 1989. Characteristics of histidine uptake by human placental microvillous membrane vesicles. *Pediatr. Res.* 25: 19–26

80. Kazzi, N. J., Ilagan, N. B., Liang, K. C., Kazzi, G. M., Grietsell, L. A., Brans, Y. W. 1990. Placental transfer of vitamin K_1 in preterm pregnancy. *Obstet. Gynecol.* 75:334–37

81. Kelley, L. K., Borke, J. L., Verma, A. K., Kumar, R., Penniston, J. T., Smith, C. H. 1990. The calcium transporting ATPase and the calcium- or magnesium-dependent nucleotide phosphatase activities of human placental trophoblast basal plasma membrane are separate enzyme activities. *J. Biol. Chem.* 265: 5453–59

82. Kelley, L. K., Smith, C. H., King, B. F. 1983. Isolation and partial characterization of the basal cell membrane of human placental trophoblast. *Biochim. Biophys. Acta* 734:91–98

83. Kilberg, M. S., Han, H.-P., Barber, E. F., Chiles, T. C. 1985. Adaptive regulation of neutral amino acid transport system A in rat h4 hepatoma cells. *J. Cell. Physiol.* 122:290–98

84. King, B. F. 1983. The organization of actin filaments in human placental villi. *J. Ultrastruct. Res.* 85:320–28

85. King, B. F. 1992. Ultrastructural evidence for transtrophoblastic channels in the hemomonochorial placenta of the Degu (*Octodon degus*). *Placenta* 13:35–41

86. Kretsinger, R. H. 1990. Why cells must export calcium. See Ref. 20, pp. 439–57

87. Kudo, Y., Boyd, C. A. R. 1990. Characterization of amino acid transport systems in human placental basal membrane vesicles. *Biochim. Biophys. Acta* 1021:169–74

88. Kudo, Y., Boyd, C. A. R. 1990. Human placental L-tyrosine transport: a comparison of brush-border and basal membrane vesicles. *J. Physiol.* 426:381–95

89. Kudo, Y., Yamada, K., Fujiwara, A., Kawasaki, T. 1987. Characterization of amino acid transport systems in human placental brush-border membrane vesicles. *Biochim. Biophys. Acta* 904:309–18

90. Kulanthaivel, P., Cool, D. R., Ramamoorthy, S., Mahesh, V. B., Leibach, F. H., Ganapathy, V. 1991. Transport of taurine and its regulation by protein kinase C in the JAR human placental choriocarcinoma cell line. *Biochem. J.* 277:53–58

91. Kulanthaivel, P., Leibach, F. H., Mahesh, V. B., Cragoe, E. J. Jr., Ganapathy, V. 1990. The Na$^+$-H$^+$ exchanger of the placental brush border membrane is pharmacologically distinct from that of the renal brush border membrane. *J. Biol. Chem.* 265:1249–52

92. Kulanthaivel, P., Simon, B. J., Burckhardt, G., Mahesh, V. B., Leibach, F. H., Ganapathy, V. 1990. The ATP-binding site of the human placental H$^+$ pump contains essential tyrosyl residues. *Biochemistry* 29:10807–13

93. Kulanthaivel, P., Simon, B. J., Leibach, F. H., Mahesh, V. B., Ganapathy, V. 1990. An essential role for vicinal dithiol groups in the catalytic activity of the human placental Na$^+$-H$^+$ exchanger. *Biochim. Biophys. Acta* 1024:385–89

94. Lajeunesse, D., Brunette, M. G. 1988. Sodium gradient-dependent phosphate transport in placental brush border membrane vesicles. *Placenta* 9:117–28

95. Lemmons, J. A., Adcock, E. W. III, Jones, M. D. Jr., Naughton, M. A., Meschia, G., Battaglia, F. C. 1976. Umbilical uptake of amino acids in the unstressed fetal lamb. *J. Clin. Invest.* 58:1428–34

96. Loh, T. T., Higuchi, D. A., van Bockxmeer, F. M., Smith, C. H., Brown, E. B. 1980. Transferrin receptors on the human placental microvillous membrane. *J. Clin. Invest.* 65:1182–91

97. Marin, J. J. G., Serrano, M. A., el-Mir, M. Y., Eleno, N., Boyd, C. A. R. 1990. Bile acid transport by basal membrane vesicles of human term placental trophoblast. *Gastroenterology* 99:1431–38

98. Mas, A., Sarkar, B. 1991. Binding, uptake and efflux of ^{65}Zn by isolated human trophoblast cells. *Biochim. Biophys. Acta* 1092:35–38

99. Mircheff, A. K., van Os, C. H., Wright, E. M. 1980. Pathways for alanine transport in intestinal basal lateral membrane vesicles. *J. Membr. Biol.* 52:83–92

100. Miyamoto, Y., Balkovetz, D. F., Leibach, F. H., Mahesh, V. B., Ganapathy, V. 1988. Na$^+$ + Cl$^-$-gradient-driven, high-affinity, uphill transport of taurine in human placental brush-border membrane vesicles. *FEBS Lett.* 231:263–67

101. Moe, A. J., Farmer, D. R., Nelson, D. M., Smith, C. H. 1991. Pentose phosphate pathway in cellular trophoblast from term human placentas. *Am. J. Physiol.* 261:C1042–47

102. Moe, A. J., Smith, C. H. 1989. Anionic amino acid uptake by microvillous membrane vesicles from human placenta. *Am. J. Physiol.* 257:C1005–11

103. Morriss, F. H. Jr., Boyd, R. D. H. 1988. Placental transport. In *The Physiology of Reproduction*, ed. E. Knobil, J. D. Neill, pp. 2043–83. New York: Raven

104. Mudd, L. M., Werner, H., Shen-Orr, Z., Roberts, C. T. Jr., LeRoith, D., et al. 1990. Regulation of rat brain/HepG2 glucose transporter gene expression by phorbol esters in primary cultures of neuronal and astrocytic glial cells. *Endocrinology* 126:545–49

105. Ng, W. W., Miller, R. K. 1983. Transport of nutrients in the early human placenta. Amino acid, creatine, vitamin B$_{12}$. *Trophoblast Res.* 1:121–34

106. Ohlsson, R. 1989. Growth factors, protooncogenes and human placental development. *Cell Differ. Dev.* 28:1–15

107. Paterson, P. G., Mas, A., Sarkar, B., Zlotkin, S. H. 1991. The influence of zinc-binding ligands in fetal circulation on zinc clearance across the in situ perfused guinea pig placenta. *J. Nutr.* 121:338–44

108. Philipps, A. F., Holzman, I. R., Teng, C., Battaglia, F. C. 1978. Tissue concentrations of free amino acids in term human placentas. *Am. J. Obstet. Gynecol.* 131:881–87

109. Pitkin, R. M. 1985. Calcium metabolism in pregnancy and the perinatal period: a review. *Am. J. Obstet. Gynecol.* 151:99–109

110. Ratnam, M., Marquardt, H., Duhring, J. L., Freisheim, J. H. 1989. Homologous membrane folate binding proteins in human placenta: Cloning and

sequence of cDNA. *Biochemistry* 28: 8249–54

111. Rindler, M. J., Traber, M. G., Esterman, A. L., Bersinger, N. A., Dancis, J. 1991. Synthesis and secretion of apolipoprotein E by human placenta and choriocarcinoma cell lines. *Placenta* 12:615–24

112. Ron, J., Levitz, M., Chuba, J., Dancis, J. 1984. Transfer of 25-hydroxyvitamin D_3 and 1,25-dihydroxyvitamin D_3 across the perfused human placenta. *Am. J. Obstet. Gynecol.* 148:370–74

113. Schenker, S., Dicke, J. M., Johnson, R. F., Hays, S. E., Henderson, G. I. 1989. Effect of ethanol on human placental transport of model amino acids and glucose. *Alcoholism: Clin. Exp. Res.* 13: 112–19

114. Schenker, S., Johnson, R. F., Hoyumpa, A. M., Henderson, G. I. 1990. Thiamine-transfer by human placenta: Normal transport and effects of ethanol. *J. Lab. Clin. Med.* 116:106–15

115. Schneider, H., Mohlen, K.-H., Challier, J.-C., Dancis, J. 1979. Transfer of glutamic acid across the human placenta perfused in vitro. *Br. J. Obstet. Gynaecol.* 86:299–306

116. Schneider, H., Proegler, M., Sodha, R., Dancis, J. 1987. Asymmetrical transfer of alpha-aminoisobutyric acid (AIB), leucine, and lysine across the in vitro perfused human placenta. *Placenta* 8:141–51

117. Seligman, P. A., Schleicher, R. B., Allen, R. H. 1979. Isolation and characterization of the transferrin receptor from human placenta. *J. Biol. Chem.* 254:9943–46

118. Shaw, A. J., Mughal, M. Z., Mohammed, T., Maresh, M. J. A., Sibley, C. P. 1990. Evidence for active maternofetal transfer of magnesium across the in situ perfused rat placenta. *Pediatr. Res.* 27:622–25

119. Shaw, A. J., Mughal, M. Z., Maresh, M. J. A., Sibley, C. P. 1991. Sodium-dependent magnesium transport across in situ perfused rat placenta. *Am. J. Physiol.* 261:R369–72

120. Shearer, M. J., Rahim, S., Barkhan, P., Stimmler, L. 1982. Plasma vitamin K_1 in mothers and their newborn babies. *Lancet* 2:460–63

121. Shennan, D. B. 1988. Selenium (selenate) transport by human placental brush border membrane vesicles. *Br. J. Nutr.* 59:13–19

122. Shennan, D. B., Boyd, C. A. R. 1987. Ion transport by placenta: a review of membrane transport systems. *Biochim. Biophys. Acta* 906:437–57

123. Shennan, D. B., Davis, B., Boyd, C. A. R. 1986. Chloride transport in human placental microvillus membrane vesicles. I. Evidence for anion exchange. *Pflüger's Arch.* 406:60–64

124. Sideri, M., de Virgiliis, G., Rainoldi, R., Remotti, G. 1983. The ultrastructural basis of the nutritional transfer: evidence of different patterns in the plasma membranes of the multilayered placental barrier. *Trophoblast Res.* 1: 15–26

125. Sivitz, W., DeSautel, S., Walker, P. S., Pessin, J. E. 1989. Regulation of the glucose transporter in developing rat brain. *Endocrinology* 124:1875–80

126. Slegel, P., Kitagawa, H., Maguire, M. H. 1988. Determination of adenosine in fetal perfusates of human placental cotyledons using fluorescence derivatization and reversed-phase high-performance liquid chromatography. *Anal. Biochem.* 171:124–34

127. Smith, C. H., Depper, R. 1974. Placental amino acid uptake. II. Tissue preincubation, fluid distribution, and mechanisms of regulation. *Pediatr. Res.* 8:697–703

128. Steel, R. B., Mosley, J. D., Smith, C. H. 1979. Insulin and placenta: degradation and stabilization, binding to microvillous membrane receptors, and amino acid uptake. *Am. J. Obstet. Gynecol.* 135:522–29

129. Steel, R. B., Smith, C. H., Kelley, L. K. 1982. Placental amino acid uptake. VI. Regulation by intracellular substrate. *Am. J. Physiol.* 243:C46–C51

130. Stegink, L. D., Pitkin, R. M., Reynolds, W. A., Filer, L. J. Jr., Boaz, D. P., Brummel, M. C. 1975. Placental transfer of glutamate and its metabolites in the primate. *Am. J. Obstet. Gynecol.* 122:70–78

131. Stevens, B. R., Kaunitz, J. D., Wright, E. M. 1984. Intestinal transport of amino acids and sugars: advances using membrane vesicles. *Annu. Rev. Physiol.* 46:417–33

132. Stulc, J. 1989. Extracellular transport pathways in the haemochorial placenta. *Placenta* 10:113–19

133. Stulc, J., Stulcova, B. 1984. Transport of inorganic phosphate by the fetal border of the guinea pig placenta perfused in situ. *Placenta* 5:9–20

134. Stulc, J., Stulcova, B., Svihovec, J. 1982. Uptake of inorganic phosphate by the maternal border of the guinea pig placenta. *Pflügers Arch.* 395:326–30

135. Stulc, J., Stulcova, B., Svihovec, J. 1990. Transport of calcium across the

dually perfused placenta of the rat. *J. Physiol.* 420:295–311

136. Sweiry, J. H., Page, K. R., Dacke, C. G., Abramovich, D. R., Yudilevich, D. L. 1986. Evidence of saturable uptake mechanisms at maternal and fetal sides of the perfused human placenta by rapid paired-tracer dilution: studies with calcium and choline. *J. Dev. Physiol.* 8: 435–45

137. Sweiry, J. H., Yudilevich, D. L. 1985. Transport of folates at maternal and fetal sides of the placenta: lack of inhibition by methotrexate. *Biochim. Biophys. Acta* 821:497–501

138. Sweiry, J. H., Yudilevich, D. L. 1988. Characterization of folate uptake in guinea pig placenta. *Am. J. Physiol.* 254:C735–43

139. Takahashi, Y. I., Smith, J. E., Goodman, D. S. 1977. Vitamin A and retinol-binding protein metabolism during fetal development in the rat. *Am. J. Physiol.* 233:E263–72

140. Teasdale, F., Jean-Jacques, G. 1985. Morphometric evaluation of the microvillous surface enlargement factor in the human placenta from mid-gestation to term. *Placenta* 6:375–81

141. Torma, H., Vahlquist, A. 1986. Uptake of vitamin A and retinol-binding protein by human placenta in vitro. *Placenta* 7:295–305

142. Tsunoo, H., Sussman, H. H. 1983. Characterization of transferrin binding and specificity of the placental transferrin receptor. *Arch. Biochem. Biophys.* 225:42–54

143. Tuan, R. S., Cavanaugh, S. T. 1986. Identification and characterization of a calcium-binding protein in the mouse chorioallantoic placenta. *Biochem. J.* 233:41–49

144. Vahlquist, A., Nilsson, S. 1984. Vitamin A transfer to the fetus and to the amniotic fluid in Rhesus monkey *(Macaca mulatta)*. *Ann. Nutr. Metab.* 28:321–33

145. van der Ende, A., du Maine, A., Simmons, C. F., Schwartz, A. L., Strous, G. J. 1987. Iron metabolism in BeWo chorion carcinoma cells. Transferrin-mediated uptake and release of iron. *J. Biol. Chem.* 262:8910–16

146. van Os, C. H. 1987. Transcellular calcium transport in intestinal and renal epithelial cells. *Biochim. Biophys. Acta* 906:195–222

147. Van Winkle, L. J. 1988. Amino acid transport in developing animal oocytes and early conceptuses. *Biochim. Biophys. Acta* 947:173–208

148. Van Winkle, L. J., Campione, A. L., Gorman, J. M., Weimer, B. D. 1990. Changes in the activities of amino acid transport system $b^{0,+}$ and L during development of preimplantation mouse conceptuses. *Biochim. Biophys. Acta* 1021:77–84

149. Vanderpuye, O. A., Kelley, L. K., Smith, C. H. 1988. Transferrin receptors in the basal plasma membrane of the human placental syncytiotrophoblast. *Placenta* 7:391–403

150. Wada, H. G., Hass, P. E., Sussman, H. H. 1979. Transferrin receptor in human placental brush border membrane. *J. Biol. Chem.* 254:12629–35

151. Wheeler, C. P. D., Yudilevich, D. L. 1988. Transport and metabolism of adenosine in the perfused guinea-pig placenta. *J. Physiol.* 405:511–26

152. Wheeler, C. P. D., Yudilevich, D. L. 1989. Lysine and alanine transport in the perfused guinea-pig placenta. *Biochim. Biophys. Acta* 978:257–66

153. White, M. F., Christensen, H. N. 1983. Simultaneous regulation of amino acid influx and efflux by system A in the hepatoma cell HTC. *J. Biol. Chem.* 258:8028–8038

154. Williams, J. W., Cunningham, F. G., MacDonald, P. C., Gant, N. F. 1989. The placenta and fetal membranes. In *Williams Obstetrics,* pp. 39–65. Norwalk, Conn: Appleton & Lange. 18th ed.

155. Williamson, J. R., Monck, J. R. 1989. Hormone effects on cellular Ca^{2+} fluxes. *Annu. Rev. Physiol.* 51:107–24

156. Yudilevich, D. L., Sweiry, J. H. 1985. Transport of amino acids in the placenta. *Biochim. Biophys. Acta* 822:169–201

157. Zafra, F., Gimenez, C. 1989. Characteristics and adaptive regulation of glycine transport in cultured glia cells. *Biochem. J.* 258:403–8

Annu. Rev. Nutr. 1992. 12:207–33

CELLULAR AND MOLECULAR ASPECTS OF ADIPOSE TISSUE DEVELOPMENT

G. Ailhaud, P. Grimaldi, and R. Négrel

Centre de Biochimie (CNRS UMR 134), Université de Nice-Sophia Antipolis, Faculté des Sciences, Parc Valrose, 06108 Nice cédex 2, France

KEY WORDS: adipose cell differentiation, hormonal factors, signaling pathways, gene expression

---

CONTENTS

INTRODUCTION

During the past two decades, growth and development of white adipose tissue (WAT) and brown adipose tissue (BAT) has been extensively studied in

0199-9885/92/0715-0207$02.00

various mammals and humans. Despite a wealth of data, key information is lacking regarding the origin of fat cells and is minimal regarding adipose tissue development during embryogenesis and after birth. However, the relationships between WAT and BAT and the molecular mechanisms leading to adipose cell differentiation are now better understood. Quite recently, the importance of adipose tissue as an endocrine, autocrine, and intracrine organ has been established. In this review we examine the current understanding of the cellular and molecular mechanisms that ultimately determine the hyperplastic development of adipose tissue, and we discuss some of the controversial issues. When needed, references to older but important work are included.

Adipocytes represent between one third and two thirds of the total number of cells in adipose tissue. The remaining cells are various blood cells, endothelial cells, pericytes, adipose precursor cells of varying degree of differentiation, and, most likely, fibroblasts (54). Over the past few years, convincing evidence has accumulated concerning the existence of very small fat cells in addition to mature adipocytes (31, 43, 75). The relationships between a definition of cell types based upon ultrastructural studies and one based upon biological studies are summarized in Figure 1. Most of the cell population of adipose precursor cells corresponds presumably to those cells previously defined in developing and adult rodents as "interstitial cells" (14, 15, 52–54), "nonlipid-filled mesenchymal cells" (95), or "other mesenchymal cells," which includes both "undifferentiated" and "poorly differentiated" mesenchymal cells (100). From a biological perspective, it is generally accepted that adipose precursor cells arise from multipotential stem cells, the origin of which remains undecided. From an embryological perspective, the process of determination from multipotential stem cells leads to the formation of unipotential adipoblasts. Commitment of adipoblasts leads to the formation of preadipocytes, defined as cells that have expressed early but not yet late markers and that have not yet accumulated triacylglycerol stores. Thus, by definition, the population of adipoblasts and preadipocytes should correspond to interstitial cells or nonlipid-filled cells. Although adipoblasts are formed during embryonic development, it is unclear whether some remain postnatally or whether only preadipocytes are present. Clearly, biochemical or molecular markers are urgently needed to carry out decisive studies of the development of adipose tissue.

PROLIFERATION AND DIFFERENTIATION OF WHITE ADIPOSE TISSUE (WAT)

Embryonic Development

White adipose tissue cannot be detected macroscopically during embryonic life and at birth in rodents (rat, mouse), whereas it is present at birth in the

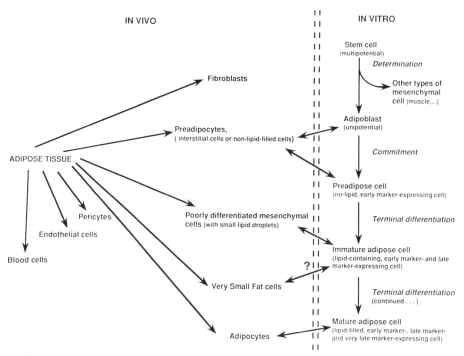

Figure 1 Relationships between morphological types in vivo and stages of cell differentiation in vitro. The adipocyte fraction corresponds to adipocytes and some very small fat cells. The stromal-vascular fraction corresponds to a mixture of the other cell types.

pig, rabbit, guinea pig, and human. The development of white adipose tissue has been studied in pig fetuses (66, 68, 69). Approximately at the beginning of the last third of the gestation period, large and small arterioles can be detected in the vicinity of small developing fat cell clusters that are surrounded by extensive stroma. Cell cluster size increases steadily with fetal age, but the number of clusters does not change significantly before birth. The importance of tissue vascularization is underlined by the fact that fat cell density appears to be positively associated with capillary density and that the largest fat cell clusters are located near the entry points of large blood vessels. The embryonic development of human adipose tissue takes place earlier than that of the pig, i.e. at the beginning of the second third of the gestation period in various sites (buccal, neck, shoulder, gluteal, perirenal) (16, 105–107). Adipose tissue appears and develops in those areas where it remains after birth. Initially, the aggregation of a dense mass of mesenchymal cells (mesenchymal lobules with no lipid accumulation) is associated with the organization of a vascular structure. Primitive fat cell clusters are then formed, composed of densely packed fat cells adjacent to capillaries. In the last stage, growth of adipose tissue is mainly due to an increase in size of fat cell clusters sur-rounded by mesenchyme, which condenses rapidly and then thickens to form septa among the clusters. Both during pig and human embryogenesis, an-giogenesis appears to be tightly coordinated with the formation of fat cell clusters in time and space. Therefore, the characterization of specific growth factors able to trigger and modulate the development of capillaries and fat cell clusters represents an important issue. A wide variety of ubiquitous an-giogenesis factors has been described, some of which (TGF-β, PGE$_2$) are synthesized and secreted by adipocytes. Quite recently, however, monobutyr-in (1-butyrylglycerol) has been described as a novel angiogenesis factor that stimulates motility (35). The biosynthetic pathway has been delineated recent-ly, and monobutyrin appears to be a fat-specific angiogenesis factor (137). Conversely, microvascular endothelial cells from human omental adipose tissue [most likely mesothelial cells (130)] in culture secrete more insulin-like growth factor-I (IGF-I) and IGF-binding protein(s) than do epithelial cell types (79). Clearly, IGF-I mRNAs (and secreted IGF-I) in the human fetus are preferentially localized in fibroblasts and other cells of mesenchymal origin (61). It is now generally accepted that IGF-I acts as an endocrine hormone via the blood, and locally as a paracrine/autocrine factor, and that growth hor-mone (GH) controls its synthesis both in liver cells and in peripheral cells of mesodermal origin (70, 71). Taken together, these observations agree with in vitro data and suggest that the requirement for IGF-I in the adequate hyperplastic development of adipose tissue is met during embryogenesis.

Postnatal Development

Methods of determining fat cell number and size have been used to study the postnatal development of WAT, particularly in rodents. Unfortunately, they are not sufficiently accurate to detect modest changes in cell number and, in any event, they only count lipid-filled cells. In other words, preadipocytes (Figure 1), i.e. cells that have expressed early phenotypes but do not yet contain triacylglycerol droplets, are not counted. This technique excludes a priori a true estimate of the developmental potential of the various adipose depots. The expression of S-100 proteins may help solve this problem, because during the development of rat epididymal fat tissue, nonlipid-filled cells (interstitial cells) stain positively for these proteins while endothelial cells, pericytes, and fibroblasts do not stain (23). Numerous studies using ^{3}H-thymidine have been performed in order to distinguish between cell proliferation and the lipid-filling process. By combining pulse-labeling with ^{3}H-thymidine followed by autoradiography and histochemical visualization of α-naphtyl acetate hydrolase activity, i.e. most likely lipoprotein lipase (LPL) and/or monoglyceride lipase activity, Pilgrim (104) has been able to distin-guish between lipid-free, mesenchymal cells (presumably adipoblasts), lipid-free, esterase-positive cells (presumably preadipocytes), and lipid-filled, es-terase-positive cells (adipocytes). The proliferative activity is highest in preadipocytes, which suggests that committed cells are able to proliferate, whereas fully differentiated cells lose the capacity to divide. Using a similar approach in mice, Cook & Kozak (26) have observed the highest labeling index in cells negative for glycerol-3-phosphate dehydrogenase (GPDH) activity, a late marker of adipose cell differentiation. A dramatic decrease of the labeling index precedes the rise of GPDH activity, which is detected subsequently in all triacylglycerol-containing cells. Thus it appears that early marker-expressing cells undergo mitoses before terminal differentiation takes place.

Sex- and site-related differences in body fat distribution are well known both in human and various animal species. Possibly, part of the explanation regarding the differential hyperplastic growth of adipose tissue lies at the cell level. In any event, the ability of adult rodents and human to increase the number of adipocytes, depending on the localization of the adipose depot, the nature of the diet, and the environmental conditions to which the animals were exposed, has been long known (39). Some controversy exists, however, as to whether the formation of new fat cells takes place during refeeding after a prolonged period of food deprivation. Initial reports indicated that such a regimen caused loss and recovery of endothelial and non-lipid filled mesen-chymal cells only and that no loss or gain of fat cells occurred (95). Quite recently, a similar nutritional study in adult mice that used improved ^{3}H-thymidine autoradiography has shown that poorly differentiated mesenchymal cells can replicate. The labeling index reaches ~ 10% at the sixth day of refeeding, giving rise to labeled adipocytes found near capillaries (100). Some loss of adipocytes appears possible under severe pathological con-ditions: a reduction in the number of fat cells from parametrial and retroperi-toneal sites has been reported in diabetic rats (54), but the turnover of adipocytes clearly is a slow process. Proliferation of mature fat cells has remained a controversial issue. In vivo labeling experiments with ^{3}H-thymidine have shown that mature adipocytes do not incorporate the label to a significant degree. Recently, however, Sugihara and coworkers have reported (129) that approximately 2% of cultured adipocytes from very young rats can undergo mitoses in culture. Possibly, this phenomenon is related to the rearrangement of the extracellular matrix in vitro following collagenase treat-ment of the abdominal subcutaneous adipose tissue; this phenomenon might not take place in vivo to a significant extent.

Late Development

Since the pioneering work of Lemonnier (89), the hyperplastic development of adipose tissue in aging animals fed a high-carbohydrate or high-fat diet has been thoroughly studied. Sprague-Dawley or Osborne-Mendel rats are able to increase the fat cell number in most of their adipose depots (retroperitoneal, perirenal) in response to a high-carbohydrate or a high-fat diet (39). The life-long potential to make new fat cells has been clearly illustrated in rodents. For instance, the perirenal fat depots of very old mice from both sexes contain large amounts of early markers of differentiation, i.e. A2COL6/pOb24 mRNAs, LPL mRNAs, and IGF-I mRNAs, indicating that preadipocytes at stages 2 and 3 are indeed present in these depots (4) (see section on sequential events and Figure 2). A similar conclusion can be indirectly drawn in humans, since a significant proportion of stromal-vascular cells from subcutaneous fat tissue of elderly men and women is able to differentiate in vitro into adipose cells (65). This finding indicates that adipose precursor cells, which are not likely to give rise to new fat cells, are indeed present in those individuals. This observation can also explain the acquisition of new fat cells, which is known to take place at the adult stage in normal subjects and obese patients.

Endocrine Regulation

Pituitary hormones may trigger and/or modulate directly or indirectly the formation of mature fat cells. In humans, GH deficiency and possibly hypothyroidism are accompanied by hypoplasia and hypertrophy of subcutaneous fat tissue (11). However, adipose tissue can be formed in hypophysectomized pigs (67, 109), which suggests that either GH is not obligatory or other hormone(s) act as substitute(s). Hypothyroidism in rats induces a transient hypoplasia, whereas hyperthyroidism induces a transient hyperplasia of retroperitoneal and epididymal fat tissues (90), suggesting a precocious formation of mature fat cells in response to triiodothryronine (T_3). The role of insulin in adipose cell differentiation has been investigated in streptozotocin-diabetic rats, taking advantage of the fact that after one month of diabetes the majority of adipocytes (*i*) have lost the bulk of their triacylglycerol, (*ii*) show a very small diameter, and (*iii*) contain several tiny triacylglycerol droplets (termed pauciadipose cells). After infusion of insulin by mini-osmotic pumps and pulse-labeling for 4 h with ^{3}H-thymidine, the labeling indices at days 1, 4, and 8 have been determined in the various cell types of parametrial adipose tissue (endothelial cells, interstitial cells, pauciadipose cells). A rapid and dramatic hypertrophic effect of insulin on pauciadipose cells transformed into adipocytes is observed, followed by a slower but potent effect of insulin on the proliferation of interstitial cells, which gives rise to neo-formed pauciadipose cells (53).

Glucocorticoids have been long known to increase adipose tissue mass via their hypertrophic effect. In Cushing's syndrome, hypercortisolism, which leads to centrally localized adipose tissue as in abdominal obesity, is likely accompanied by cell hyperplasia (110). A higher density of glucocorticoid

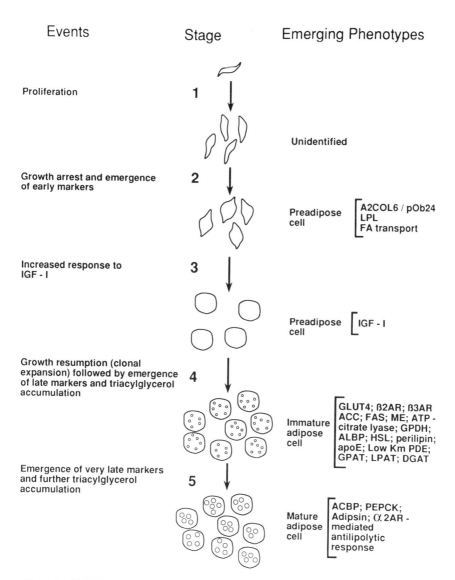

Events   Stage   Emerging Phenotypes

Proliferation   1

            Unidentified

Growth arrest and emergence   2
of early markers

         Preadipose   [A2COL6 / pOb24
         cell   LPL
                  [FA transport

Increased response to   3
IGF - I

         Preadipose   [IGF - I
         cell

Growth resumption (clonal   4
expansion) followed by emergence
of late markers and triacylglycerol
accumulation

                          [GLUT4; ß2AR; ß3AR
                          ACC; FAS; ME; ATP -
         Immature   citrate lyase; GPDH;
         adipose   ALBP; HSL; perilipin;
         cell   apoE; Low Km PDE;
                          [GPAT; LPAT; DGAT

Emergence of very late markers   5
and further triacylglycerol
accumulation

                          [ACBP; PEPCK;
         Mature   Adipsin; α2AR -
         adipose   mediated
         cell   antilipolytic
                          [response

Figure 2 Multiple stages of adipose cell differentiation. The scheme is based upon data obtained with 3T3-L1, 3T3-F442A, and Ob17 cells as well as with rodent adipose precursor cells. The abbreviations are LPL, lipoprotein lipase; FA transport, fatty acid transport; A2COL6/pOb24, α2-chain of collagen VI; IGF-I, insulin-like growth factor I; GLUT-4, insulin-sensitive glucose transporter 4; ME, malic enzyme; ACC, acetyl-CoA carboxylase; FAS, fatty acid synthetase; GPDH, glycerol-3-phosphate dehydrogenase; ALBP, adipocyte lipid-binding protein (aP2); HSL, hormone-sensitive lipase; apoE, apolipoprotein E; low Km PDE, low Km phosphodiesterase; GPAT, glycerophosphate acyltransferase; LPAT, lysophosphatidate acyltransferase; DGAT, diglyceride acyltransferase; ACBP, acyl CoA binding protein; PEPCK, phosphoenolpyruvate carboxykinase; GS, glutamine synthetase; β_2-AR, β_2-adrenoreceptor; β_3-AR, β_3-adrenoreceptor; α_2-AR, α_2-adrenoreceptor.

receptors is found in abdominal fat tissue than in other fat depots, and investigators have proposed that glucocorticoids may regulate this differential development (111). Glucocorticoids such as cortisol reportedly control terminal differentiation of human adipose precursor cells (see below). Thus one can envision that cortisol may play an important role in the development of cell hyperplasia and subsequently cell hypertrophy. The role of steroid sex hormones in the hyperplastic development of adipose tissue has not yet been documented, although low levels of testosterone in men or low levels of female sex hormones with increased levels of testosterone in women have been associated with increased visceral fat mass. This association may not be fortuitous; primates are unique in that a large proportion of androgens in men (40%) and a preponderance of estrogens in women (75% before menopause, $\sim$ 100% after menopause) are synthesized in peripheral tissues (including adipose tissue) from adrenal precursor steroids (85, 86).

PROLIFERATION AND DIFFERENTIATION OF BROWN ADIPOSE TISSUE (BAT)

Considerable evidence now supports the view that BAT is a specific organ distinct from WAT. In most mammalian species, BAT develops during fetal life and is identifiable at birth. Studies on the prenatal development of BAT are scarce and have been hampered by the small size of BAT depots. Developmental changes of BAT were therefore examined in bovine fetuses: the emergence of uncoupling protein (UCP) unique to BAT mitochondria takes place during the last trimester of gestation (18). At birth both in bovine and ovine, the different adipose depots (perirenal, subscapular, retroperitoneal), except for the subcutaneous adipose tissue, can be considered as BAT (19). Taking advantage of the remarkable hyperplasia of adult rat BAT in response to cold acclimation and/or adaptation to hyperphagia, Bukowiecki and coworkers have studied the sequence of events leading to the differentiation of brown fat cells, under conditions where the labeling index increased 60–80 times over control values. After cold exposure for two days, the majority of labeled cells are interstitial and endothelial cells. The labeling index, which remains unchanged in endothelial cells with exposure time, decreases in interstitial cells whereas it increases first in poorly differentiated cells and later in brown adipocytes. Based upon these results and the determination of specific labeling frequency, i.e. the labeling index for a *given* cellular type independent of the other cellular types, the chronology for differentiation appears to be interstitial cells → poorly differentiated brown fat cells → brown adipocytes. Endothelial cells can be excluded as progenitors of brown adipocytes (15). The stimulation of BAT proliferation and differentiation is β-adrenergic-mediated, since continuous infusion of norepinephrine or isoproterenol as β-agonists to warmth-exposed rats mimicks the effect of cold exposure, whereas the α-agonist phenylephrine is ineffective (52, 99, 114). Recent evidence obtained in adult dogs favors a β_3 receptor-mediated process (97). Using a novel β_3 adrenoreceptor agonist (ICI D7114) that has thermogenic and anti-obesity properties in this animal, a potent increase of the UCP content is observed in extracts of perirenal, peribladder, and pericardiac adipose tissue (116).

RELATIONSHIPS BETWEEN BAT AND WAT

Studies on the development of WAT and to a lesser extent of BAT have led to much controversy regarding the relationships between both tissues. The existence of distinct precursor cells (Figure 3) is assumed and is discussed later, but the existence of a single adipoblast giving rise to distinct brown and white preadipocytes cannot be ruled out. A possible transformation of brown adipocyte into white adipocyte, defined by the disappearance of UCP, is strongly suggested: UCP mRNA, which in bovine fetuses at birth reaches its highest level in the various adipose depots, is no longer detectable two days later. Similarly, ovine UCP mRNA disappeared within two days from peritoneal adipose depot and within six days from all adipose depots (18). Since the estimated number of adipocytes remains constant in bovine during the first week following birth, we assume that a true transformation of brown into white adipocytes has been indeed taking place (Figure 3). Despite a dramatic mitochondriogenesis, the reverse phenomenon does not appear to take place in the epididymal fat pad, when rats are exposed to severe cold stress (93). In mice, under similar conditions, inguinal WAT only appears as BAT, expresses UCP mRNA, and the developed mitochondria contain UCP; this expression ceases in rewarmed animals (92).

ADIPOSE CELL DIFFERENTIATION IN VITRO

General Considerations

The process of determination from multipotential stem cells to unipotential adipoblasts cannot be easily studied (see Figure 1). However, a determination-like process can be induced by treatment of mouse 10T1/2 and 3T3 cells (131) and hamster CHEF-18 cells with 5-azacytidine (118), whereas it is spontaneous in T984 cells isolated from a mouse teratocarcinoma (30). In most cases, these treated mesenchymal cells were able to differentiate into adipocytes, chondrocytes, and fibroblasts. Adipose precursor cells or adipoblasts (see Refs. 2, 3, and 4 for reviews) have been cloned from this mixed population of cells (1246 clonal line), from mouse embryo (3T3-L1, 3T3-F442A, A31T, TA1) or hamster embryo (CHEF-18), and from adult mice

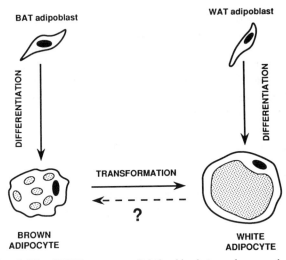

Figure 3   Distinct BAT and WAT precursors. Relationships between brown and white adipocytes.

(Ob17, HFGu, BFC-1, ST13, MS3-2A, MC 3T3-G2/PA6). When dealing with the various "preadipocyte" clonal lines, the process of adipose cell differentiation, which can be analyzed in vitro, corresponds to the phenotypic changes: adipoblast → preadipose cell (usually termed preadipocyte) → immature adipose cell → mature adipose cell (Figure 1). Some investigators have begun to study stromal-vascular cells from adipose tissues of various species, including humans (20, 32, 51, 57, 65, 115, 123); these cells are diploid but have a limited life span. Although the presence of adipoblasts cannot be excluded, the differentiation process of adipose precursor cells isolated from fat tissue corresponds primarily to the sequence: preadipose cell (preadipocyte) → immature adipose cell → mature adipose cell. This tentative conclusion is based upon the fact that rat, mouse, and human stromal-vascular cells contain the bulk of A2COL6/pOb24 mRNA and express LPL and IGF-I mRNAs, i.e. have already expressed early markers (stages 2 and 3 of Figure 2). Note that, when injected into animals at the undifferentiated state, cells from established lines (3T3-F442A, Ob17) or rat stromal-vascular cells develop into mature fat cells (4). When seeded at clonal densities, stromal-vascular cells from rat perirenal and epididymal fat depots show varying capacities for replication and differentiation, irrespective of donor age (34, 80, 135). At any given age, stromal-vascular cells from perirenal fat tissue showed a greater proportion of clones with a high frequency of differentiation than was found in epididymal fat tissue (34). In humans, stromal-vascular cells from abdominal fat tissue show a higher capacity for differentiation than those of the femoral depot (64), in agreement with clinical observations regarding the differential adipose tissue development observed for both sites. Both in rat for the perirenal and epididymal adipose tissues (80) and in human for the subcutaneous adipose tissue (65), aging is associated with a decrease in the proportion of cells undergoing differentiation.

Sequential Events

Morphologically, after reaching confluence, fibroblast-like cells become round, enlarge, and accumulate triacylglycerol droplets in their cytoplasm. Once differentiated, mature adipose cells show most biochemical characteristics and hormonal responses, if not all, of adipocytes. The main events, based upon numerous studies by various investigators, are summarized in Figure 2. Growth arrest at the G_1/S stage of the cell cycle, rather than contact among arrested cells, is necessary to trigger the process of cell commitment (5). This commitment is associated with the emergence of potential regulatory genes such as A2COL6/pOb24 (29, 72) and clone 5 (96), LPL mRNAs and LPL activity (28), as well as the emergence of a selective uptake of long-chain fatty acids (1). The regulation of expression of these early genes takes place primarily at a transcriptional level and appears to be independent of various hormones which, in contrast, are required for terminal differentiation (28, 29). The expression of late and very late genes is associated with limited growth resumption of these committed, early marker-expressing cells. DNA synthesis of preadipose cells precedes terminal differentiation. At least one cell doubling has been consistently observed by various investigators using different cell lines and different culture media, and this process of clonal amplification of committed cells (defined as postconfluent mitoses) is limited both in magnitude and duration. Cell division of preadipose cells appears to be essential for terminal differentiation (defined by the emergence of GPDH activity), providing that cells are exposed to the appropriate hormonal milieu (5, 48, 84, 121). The observations made in vitro are in agreement with those made in vivo by Pilgrim (104) as well as by Cook & Kozak (26) concerning the relationships in rodent adipose tissue between cell proliferation and differentiation. However, a complete dissociation between growth resumption and terminal differentiation takes place in the case of human adipose precursor cells.

The process of terminal differentiation is characterized by the induction of late and very late markers. As shown in Figure 2, the enzymatic machinery required for lipogenesis and triacylglycerol synthesis is then turned on and is responsible for lipid accumulation. Among newly discovered late markers are GLUT-4 (17, 50, 60, 76, 83, 132), β_2- and β_3-adrenoreceptors (40, 41), perilipin (55), and apolipoprotein E (apoE) (142). Very late events include the emergence of acyl-CoA-binding protein (ACBP) (62) and α_2-adrenoreceptor-mediated response (119). In the case of GPDH, adipocyte-lipid-binding protein (ALBP or aP2), and adipsin genes, this emergence is primarily due to increased gene transcription (25, 33). With respect to fatty acid metabolism, the increase in uptake of fatty acid is concomitant with the induction of LPL (Figure 3). Fatty acid entry then becomes sufficient to activate the transcription of ALBP and acyl-CoA synthetase (ACS) genes, whereas cholesterol can activate the expression of apoE gene (142). The time course of appearance of cholesterol ester transfer protein (CETP) mRNA and that of the corresponding protein are not known, but they are abundant in adipose tissue of mammals (74). Investigators report that the emergence of phospholipase A2 activity parallels that of GPDH activity in cells of the 1246 clonal line and in rat adipose precursor cells (49), which suggests an increase in the disposal of fatty acids (at position 2 of glycerolipids), including arachidonic acid required for terminal differentiation (46).

Changes in receptor level and hormone sensitivity during differentiation of 3T3 and Ob17 cells have been extensively described for insulin and lipolytic hormones (2). During the last few years, the characterization of β- and α_2-adrenoreceptors has expanded remarkably, and new strategies regarding the control of adipose cell hypertrophy can be considered (87). Quite recently, using specific ligands, Fève et al (40, 41) investigated the differential regulation of β_1-, β_2-, and β_3-adrenoreceptors in 3T3-F442A cells and that of their respective mRNAs. β_1-adrenoreceptors, detectable in growing, undifferentiated cells, increase their level up to 6-fold at growth arrest, whereas β_2-adrenoreceptors (under glucocorticoid stimulation) and β_3-adrenoreceptors emerge later, at a time when GPDH activity is also emerging. β_3-adrenoreceptors represent 90% of the total population of β-adrenoreceptors in differentiated 3T3-F442A cells. Since both these differentiated cells and mouse adipocytes share the same pharmacological properties with respect to β-adrenergics, β_3-adrenoreceptors likely play an important role in mediating catecholamine-induced lipolysis. Extrapolation of these observations to human adipose precursor cells remains questionable, because no β_3-adrenoreceptor mRNA has been reported in human fat cells (38).

Control of Terminal Differentiation of Preadipose Cells by Adipogenic and Antiadipogenic Factors

The critical role played by various hormones in regard to *differentiating* cells should not be confused with the role of the same hormones in regard to *differentiated* cells. So far, growth arrest at the G_1/S phase of the cell cycle appears sufficient to commit the cells and allow the expression of early markers (5, 28, 29). The combination of hormonal factors that trigger terminal differentiation of preadipose cells (stages 3 to 5 of Figure 3) remains difficult to define owing to the widely different experimental conditions used and the origin of the cells. When obtained in serum-supplemented medium, the data are difficult to interpret, as illustrated recently by the expression of LPL gene in Ob17 cells, since the multiple effectors present in serum regulate its expression either negatively or positively (108). Therefore, serum-free, chemically defined media have been developed for the differentiation of adipose precursor cells from various clonal lines and those derived from various species including rat (32, 123), rabbit (115), porcine (51), ovine, (20) and human (65). For cells from clonal lines, the first hormonal requirement in serum-supplemented medium appears at stage 3 with GH activating the IGF-I gene at a transcriptional level (36). Induction can only occur after cell growth is arrested and early markers are expressed. Zezulak & Green reported that requirement of GH is followed by an increase in the responsiveness of 3T3-F442A cells to IGF-I; IGF-I in turn causes a mitogenic response, leading to clonal expansion, which is defined as post-confluent mitoses (143). This observation is at variance with another showing that exogenous IGF-I is obligatory and sufficient for the differentiation of 3T3-L1 cells (128). Under serum-free conditions, GH is required for differentiation of 3T3-F442A (59), and it has been suggested that GH drives committed cells to a special state of quiescence (the primed state), allowing them to respond to insulin, and then terminates differentiation. These conflicting results may be explained by the fact that cells may represent different developmental stages in the differentiation process and that, under serum-supplemented conditions, different hormones may play the same role while other hormones may merely overcome the inhibition of some serum factors. Data obtained under serum-supplemented and serum-free conditions indicate that GH is not required for the conversion of adipose precursor cells to adipocytes in rodents, domestic animals, and humans. If GH is indeed needed at some stage, a prior exposure of the cells to the hormone in vivo may be sufficient to "prime" the cells for terminal differentiation.

Triiodothyronine appears to be essential for the differentiation of Ob17 cells both in serum-supplemented and serum-free medium (5, 58). Removal of T_3 reduces by 75% the GPDH activity of 3T3-F442A cells under serum-free conditions (59), whereas addition of T_3 to serum-supplemented medium increases the number of differentiated cells, thus suggesting a selective multiplication of committed cells (42). The requirement for T_3 remains partial, if any, for the terminal differentiation of adipose precursor cells isolated from pig, rat, and rabbit adipose tissues. As in the case of GH, prior exposure to T_3 in vivo may explain these results.

Studies of insulin requirements have produced apparently conflicting results: when terminal differentiation of 3T3-L1 cells is induced in the presence of dexamethasone (or corticosterone) and 1-methyl-3-isobutylxanthine (MIX), either supraphysiological concentrations of insulin (active at least in part by binding to IGF-I receptor) or low concentrations of insulin and IGF-I are required, both in serum-supplemented (3) and serum-free medium (63, 121). In the absence of both inducers, insulin appears more potent than IGF-I whereas, in their presence, the reverse is observed (10). Likewise, the optimal expression of GPDH activity and the maximal accumulation of triacylglycerol in rat and rabbit adipose precursor cells require insulin and IGF-I (32, 115). In any event, insulin, by regulating glucose transporters GLUT-1 and GLUT-4, increases the steady-state level of lipogenic enzymes and triacylglycerol stores. These results suggest a subtle interplay between insulin, IGF-I, and their receptors to modulate the maximal expression of terminal differentiation. MIX (added at confluence), which probably acts by raising intracellular cAMP concentrations via phosphodiesterase inhibition, has long been known to accelerate and amplify differentiation of 3T3-L1 cells; it has a similar effect on differentiation of rat and human adipose precursor cells. Because forskolin can substitute for MIX, it is likely that cAMP is a major signal leading to terminal differentiation (121). Arachidonic acid, which behaves as an adipogenic-mitogenic factor, triggers the production of cAMP and 1,2-diacylglycerol from polyphosphoinositide breakdown in Ob17 cells (46). Triggering of the cAMP and inositol phospholipid pathways is followed by at least one round of cell division, and within a few days the whole population of cells becomes differentiated. Indomethacin or aspirin prevents the arachidonic acid-induced differentiation, which would suggest that one or more of the three secreted prostaglandins (PGE_2, PGI_2, $PGF_{2\alpha}$) might be involved. Carbaprostacyclin ($cPGI_2$), a stable analog of prostacyclin, is indeed an efficient activator of cAMP production and terminal differentiation of Ob17 cells; its role in terminal differentiation has been extended to rat and human adipose precursor cells. $PGF_{2\alpha}$, which is able to activate polyphosphoinositide breakdown, dramatically increases the mitogenic-adipogenic effect of $cPGI_2$ that is present at submaximal concentrations (98). Antibodies against both prostaglandins (PGs) are able to counteract the adipogenic-mitogenic effect of arachidonic acid on terminal differentiation (2?), thereby supporting the role of both prostaglandins in this process (4). Like $PGF_{2\alpha}$, GH promotes the formation of 1,2-diacylglycerol (37), but from phosphatidylcholine instead of polyphosphoinositide (21). Both GH and $PGF_{2\alpha}$ are thought to activate the PKC pathway, since 1,2-dioctanoylglycerol and phorbol esters can substitute for GH and $PGF_{2\alpha}$. In the process of terminal differentiation, the cAMP and IGF-I pathways may play a cardinal role, whereas the diacylglycerol and insulin pathways may play a modulating role (4).

The importance of controlling the stage of cell differentiation when examining any hormonal effect is illustrated clearly in the case of glucocorticoids. Some investigators have found that glucocorticoids stimulate the differentiation of 3T3-L1 cells (120, 121), TA-1 cells, and adipose precursor cells derived from rat, rabbit, and human adipose tissues (3). More recent studies have examined the effects of dexamethasone and glucocorticoids on the proliferation and differentiation of rat adipose precursor cells (56, 138): glucocorticoids decrease markedly cell proliferation and enhance terminal differentiation. This stimulation appears to depend mainly on the presence of insulin and an optimal concentration of glucocorticoids, since inhibition of differentiation is observed at high concentrations and is a function of the duration of glucocorticoid treatment. Other studies have shown that cortico-sterone increases the metabolism of arachidonic acid and leads to an increase in the production of prostacyclin, triggering, in turn, cAMP production; this explains a posteriori the effectiveness of the dexamethasone/MIX cocktail used in a large number of studies and also the fact that glucocorticoids can substitute for arachidonic acid or PGs (47). Among steroid hormones, sex steroids such as β-estradiol, testosterone, and progesterone fail to enhance directly the differentiation of 3T3-L1 cells or of human adipose precursor cells, and only steroids with glucocorticoid activity are effective.

Chronic exposure to dehydroepiandrosterone and some structural analogues abolishes the emergence of GPDH activity and blocks differentiation of 3T3-L1 cells (126). The effect of vitamins and their analogues has been also investigated in 3T3-L1 cells: water-soluble vitamins (vitamin B6 group and vitamin C) stimulate GPDH activity and triacylglycerol accumulation, where-as many fat-soluble vitamins (vitamin A, D, E, and K groups) significantly inhibit both parameters (77). Ascorbic acid enhances synthesis and secretion of type IV collagen but not of laminin and entactin (101). Some drug treatments have been reported to enhance terminal differentiation: in 3T3-L1 cells and rat adipose precursor cells, a synergistic effect of various fibrates and different cAMP-elevating agents is observed (13). This requirement for cAMP has led Brandes et al to analyze cAMP-dependent protein phosphoryla-tion: interestingly, both in the absence or presence of bezafibrate, a very early phosphorylation is observed in a 60-kDa acidic protein recovered in the nuclear fraction of dibutyryl cAMP-treated 3T3-L1 cells (12). A recent and intriguing observation (133) is the effect of butyrate in combination with insulin and dexamethasone in Swiss 3T3 cells, which do not usually differen-tiate into adipose cells. After growth arrest at G_1 induced with butyrate, in the absence or presence of MIX, these cells accumulate lipid droplets as well as LPL and ALBP mRNAs, but they do not accumulate adipsin mRNA. This observation is similar to the finding that, at high concentrations of in-domethacin, activation of various differentiation-specific genes can occur by alternate and still unknown pathways, since synthesis of PGs is abolished under these conditions. The search for factors that act via paracrine mech-anisms to trigger or enhance proliferation and/or differentiation of adipose precursor cells has shown that mature adipocytes isolated from human tissue release mitogenic factors (88), whereas those released from rat tissue promote a potent increase in GPDH activity and triacylglycerol accumulation of cultured rat adipose precursor cells (127). Using these parameters as indicators of terminal differentiation, a factor of 63 kDa, distinct from known growth factors and hormones, has been characterized in rat and mouse serum (91).

Various factors that inhibit or abolish differentiation of adipose precursor cells are present in serum: platelet-derived growth factor (PDGF) and transforming growth factor β (TGF-β) have been reported to be antiadipogenic factors for 3T3-L1 and 3T3-F442A cells, as have been epidermal growth factor (EGF) and TGF-β for rat adipose precursor cells (73, 124). The inhibitory effect of TGF-β appears to be independent of its mitogenic property. The situation remains unclear with respect to the effects of FGF and EGF (3). FGF added to serum-containing medium prevents the expression in TA-1 cells of some late differentiation-specific genes. FGF abolishes the expression of those genes in fully differentiated cells independently of its mitogenic properties, but, in contrast, basic or acidic FGF does not inhibit terminal differentiation of rat adipose precursor cells. EGF is required for terminal differentiation of 3T3-L1 cells (121), whereas it is inhibitory for terminal differentiation of rat adipose precursor cells (122 and Ref. 3 for review). In agreement with the latter observation, a recent report notes that subcutaneous administration of EGF to newborn rats results in a large decrease of the weight of inguinal fat pads, which suggests the delayed formation of adipocytes from preadipocytes (125).

The effects of tumor necrosis factor α (TNFα) have been studied extensively, as it can alter energy balance in vivo by a selective action on adipose tissue. In differentiated 3T3-L1, 3T3-F442A, and TA-1 cells, TNFα inhibits the expression of several differentiation-specific genes such as LPL, GPDH, adipsin, and aP2 mRNAS in a dose-dependent manner. Moreover, it stimulates lipolysis and induces a phenotypic "dedifferentiation" on a long-term basis (3). TNFα regulates the expression of procollagen genes in 3T3-L1 cells and of the TGFβ gene in TA-1 cells. The effects of TNFα on procollagen mRNA levels depend upon the stage of differentiation of the cells, as undifferentiated and differentiated 3T3-L1 cells undergo, respectively, a coordinate decrease or increase in the contents of types I, III, and IV procollagen mRNAs (136). On the other hand, TNFα within the same range of concentrations has been shown to kill 3T3-L1 cells or inhibit their differentiation (78); further studies are required to resolve these conflicting results.

Retinoids are able to inhibit terminal differentiation. Studies in 3T3-F442A cells indicate that this action is reversible and takes place after expression of LPL but before expression of GPDH and accumulation of triacylglycerol (102, 103). Moreover, retinoic acid can down-regulate adipsin expression in differentiated 3T3-F442A cells, whereas the expression of other differentiation-specific genes (LPL, ALBP, and GPDH) remains unaffected (8). Thus, in vitro, the differential effects of retinoids depend upon the developmental stage of the cells.

At present, little is known about the intracellular components and mechanisms of the multiple signaling pathways involved in the control of terminal differentiation. The recent observations of Benito et al (9) indicate that Ras proteins are involved in the transduction signals initiated by insulin and IGF-I in differentiating 3T3-L1 cells. The c-Myc protein, whose gene expression is normally abolished at a time when early markers are expressed, prevents 3T3-L1 cells from entering terminal differentiation (44, 141). This blockage of differentiation is accompanied by a large decrease in the expression of genes encoding for the α1- and α2-chains of collagen I and the α3-chain of collagen VI. This observation should be kept in mind because the activation of the gene encoding for the α2-chain of collagen VI (A2COL6/pOb24) is the earliest event of differentiation so far reported (29, 72). These observations suggest that some extracellular matrix components may play a role in initiating events leading to terminal differentiation.

Brown Adipose Precursor Cells

An important observation made independently by Rehnmark et al (112, 113) and Kopecky et al (82) supports the existence of distinct BAT and WAT precursors (Figure 3): stromal-vascular cells isolated from the interscapular BAT of young mice are able after growth arrest to express UCP mRNA and UCP protein and to differentiate into lipid-filled cells. Pharmacological studies of the adrenergic response indicated that the stimulation of UCP gene expression is primarily regulated by β-adrenoreceptors (likely β_3) and that α_1-adrenoreceptors play an additional role. Insulin and T_3 are required for maximal expression. Under the same conditions, stromal-vascular cells isolated from WAT are unable to express UCP mRNA and UCP protein. As in the case of WAT, stromal-vascular cells from adult mouse BAT contain the bulk of A2COL6/pOb24 mRNA and thus should be considered as preadipocytes (Figure 3), although clearly, both for BAT and WAT, the coexistence of adipoblasts and preadipocytes cannot be excluded. In any event, these experiments show that mouse BAT and WAT arise from distinct adipose precursor cells (Figure 3). The possible transformation of BAT precursors into WAT precursors and the fine regulation of the expression of UCP gene have been recently investigated by Casteilla et al (20): in serum-free medium, the expression of UCP mRNA takes place in stromal-vascular cells from perirenal BAT of newborn lambs. Glucocorticoids promote a stimulating effect on this expression. However neither UCP mRNA is expressed nor a glucocorticoid effect is observed with precursor cells from 3-week-old lambs.

Therefore, the extinction of the expression of UCP gene is a rapid process in vivo, and a true conversion of BAT precursors into WAT precursors apparently occurs in ovine during this short period after birth.

ADIPOSE CELLS AS SECRETORY CELLS

Adipocytes are a major source of free fatty acids in mammals. Furthermore, both adipocytes and muscle cells have been long known as the main source of synthesized and secreted LPL. In the last few years, the concept that adipocytes behave as secretory cells for other proteins and metabolites has emerged. In a series of elegant experiments (24, 27, 81, 117), Spiegelman and coworkers were the first to show that adipsin, which is found in vivo in the circulation of both animals and humans, is produced and secreted in vitro by differentiated 3T3-L1 cells (constitutively and, more importantly, upon insulin stimulation) and mouse epididymal fat tissue (81). Both glycosylated forms of adipsin (37 and 44 kDa) are secreted, and considerably more adipsin is found in blood than in adipose tissue. Mouse adipsin is a secreted protease homologous to human complement factor D. Both adipsin and complement factor D are highly specific in clearing complement factor B when it is complexed with activated complement component C_3. In addition to adipsin, factor B and factor C_3 are also secreted by differentiated 3T3-F442A cells. These authors have proposed that adipsin and other factors of the alternative pathway of complement may play a role in the regulation of systemic energy balance in vivo (117).

Of utmost interest is the fact that CETP and apoE are synthesized by various tissues (muscle, heart, liver, adipose) in different species (hamster, rat, rabbit, human) but that, rather unexpectedly, mammalian adipose tissue and muscle appeared as major sources of CETP mRNA. Isolated hamster adipocytes synthesize and secrete active CETP, and analysis by in situ hybridization of adipose tissue reveals coexpression of CETP mRNA and LPL mRNA (74). In many ways, apoE resembles CETP, as it is produced in several peripheral tissues in which LPL is also expressed. During differentiation of 3T3-L1 cells, the expression of apoE gene takes place after that of LPL (142). In contrast to CETP, most of apoE is not secreted and remains cell-associated in differentiated cells. Thus adipocytes are another important extrahepatic source of CETP and possibly of apoE. Adipose tissue is known to contain the major cholesterol pool of the body both in rodents and humans; conceivably, locally synthesized apoE and CETP play a role in the local removal of adipocyte cholesterol. The capacity of adipose tissue as a secretory organ extends to sex steroid synthesis. Among the genes encoding enzymes responsible for the synthesis of androgens and estrogens from adrenal steroid precursors, investigators observed that the expression of P-450 aromatase occurred earlier than that of GPDH during the differentiation of 3T3-L1 cells (140). The P-450 aromatase mRNA, the product of which is required for the synthesis of estrone and estradiol, is also expressed in rat and human adipose tissue. This gene as well as the genes encoding for the 3β-hydroxy-steroid dehydrogenase/Δ4-Δ5 isomerase and the 17β-hydroxysteroid dehydrogenase (required for androgen and estrogen synthesis from dehydroepiandrosterone) are also expressed in a variety of peripheral tissues, including adipose tissue (86, 144). Large amounts of adrenal steroid precursors secreted in human and nonhuman primates are further metabolized by peripheral tissues. This synthesis accounts for 40% of total active androgens in men and for a higher percentage of active estrogens in women. This activity, termed intracrine steroid formation (85), may thus be quantitatively important owing to the weight of fat tissue in normal subjects and obese patients. It is tempting to postulate that intracrine steroid formation may also affect adipose tissue metabolism, as reported recently in the regulation of lipolysis by androgens in rat adipose precursor cells (139). Among metabolites of interest, and as already mentioned, differentiated 3T3-F442A cells have been shown to secrete monobutyrin, which appears as a fat-specific angiogenesis factor (35, 137). Together, these various observations emphasize that not only are adipocytes able to respond to hormonal signals originating from endocrine glands and possibly from other cell types in their immediate vicinity but they are also able to synthesize and secrete peptides and nonpeptide factors that can be recognized and used by other cells.

RESEARCH TRENDS

At the present time, a better estimate of adipose tissue cellularity appears feasible by using in situ hybridization techniques with cDNA or RNA probes corresponding to early markers of adipose cell differentiation. More importantly, the cloning and sequencing of master genes involved in the determination process, leading to unipotential adipoblasts, can be foreseen. We also hope that the cloning and sequencing of the mouse *ob* gene and the rat *fa* gene will shed some light on the initial event(s) leading to adipose tissue hyperplasia (45, 134).

The "reactivation" of BAT, owing to its partial or nearly complete disappearance in many mammals and humans after birth, deserves further study. To date, the increase in rodents and dog of total UCP content of BAT upon adrenergic stimulation appears to be due more to hyperplasia than to an increase in the UCP content per cell. The recent observation that fatty acids can regulate gene expression in adipose cells (6, 7) may introduce a link between the composition of diets and the hyperplastic/hypertrophic response of white adipose tissue; and the characterization of fatty acid-responsive genes may also produce some clues about the development of insulin-resistant states and cell hypertrophy.

From a more general point of view, adipose tissue now appears to perform endocrine, paracrine/autocrine, and intracrine functions. As such, it may play a direct role in some regulatory events related to cholesterol metabolism and energy balance. The potential to analyze in humans the composition of the interstitial fluid surrounding adipocytes in vivo (94) will enable scientists to study the factors that may be involved in adipose tissue hyperplasia and hypertrophy.

SUMMARY

Both in animals and humans, before or after birth, angiogenesis appears to be closely coordinated in time and space with the formation of fat cell clusters. Monobutyrin, a novel fat-specific angiogenesis factor, may play a role in this process. The potential to acquire new fat cells appears to be permanent throughout life in both animals and humans, as revealed by in vitro experiments. Considerable evidence now supports the view that BAT and WAT are distinct organs; in addition, the existence of distinct BAT precursor cells is demonstrated by their unique ability to express the UCP gene. In bovine and ovine, the transformation of BAT into WAT is strongly suggested by the rapid disappearance after birth of UCP from the various BAT depots. Despite the initial cell heterogeneity of the stromal-vascular fraction, cultured stromal-vascular cells of adipose tissue are adipose precursor cells that show varying capacities for replication and differentiation, according to age and fat depot. Studies of adipose cell differentiation in vitro correspond to the sequence: adipoblast (unipotential cells) $\xrightarrow{\text{commitment}}$ preadipose cell (preadipocyte) $\xrightarrow{\text{terminal differentiation}}$ immature adipose cell $\xrightarrow{\text{terminal differentiation}}$ mature adipose cell (adipocyte). Cell commitment is triggered by growth arrest and characterized by the expression of early markers (A2COL6/pOb24; clone 5; LPL), whereas only terminal differentiation of preadipocytes requires the presence of various hormones. Multiple signaling pathways have been characterized and shown to cooperate in the process of terminal differentiation. The concept that adipose cells behave as secretory cells is now emerging from in vitro data, since secretion of various proteins (LPL, adipsin, CETP) and important metabolites (fatty acids, monobutyrin, androgens, estrogens, prostaglandins) takes place both constitutively and upon hormonal stimulation. This suggests that adipose tissue participates more directly than previously thought in metabolic activities and energy balance.

ACKNOWLEDGMENTS

G. Ailhaud and R. Négrel are professors of biochemistry and P. Grimaldi is an INSERM investigator. This research has been supported by funds from

CNRS and grants from INSERM, INRA, and ARC. We thank C. Dani, D. Gaillard, and C. Vannier for their valuable contributions over the years. The important contributions of a large number of other investigators are also gratefully acknowledged. Special thanks are due to G. Oillaux for expert secretarial assistance.

Literature Cited

1. Abumrad, N. A., Forest, C. C., Regen, D. M., Sanders, S. 1991. Increase in membrane uptake of long-chain fatty acids early during preadipocyte differentiation. *Proc. Natl. Acad. Sci. USA* 88:6008–12

2. Ailhaud, G. 1982. Adipose cell differentiation in culture. *Mol. Cell. Biochem.* 49:17–31

3. Ailhaud, G. 1990. Extracellular factors, signalling pathways and differentiation of adipose precursor cells. *Curr. Opin. Cell Biol.* 2:1043–49

4. Ailhaud, G., Amri, E., Bertrand, B., Barcellini-Couget, S., Bardon, S., et al. 1990. Cellular and molecular aspects of adipose tissue growth. In *Obesity: Towards a Molecular Approach,* ed. G. Bray, D. Ricquier, B. Spiegelman, 133:219–236. New York: Liss

5. Ailhaud, G., Dani, C., Amri, E., Djian, P., Vannier, C., et al. 1989. Coupling growth arrest and adipocyte differentiation. *Environ. Health Perspect.* 80:17–23

6. Amri, E., Ailhaud, G., Grimaldi, P. 1991. Regulation of adipose cell differentiation. II. Kinetics of induction of the aP2 gene by fatty acids and modulation by dexamethasone. *J. Lipid Res.* 32:1457–63

7. Amri, E., Bertrand, B., Ailhaud, G., Grimaldi, P. 1991. Regulation of adipose cell differentiation. I. Fatty acids are inducers of the aP2 gene expression. *J. Lipid Res.* 32:1449–56

8. Antras, J., Lasnier, F., Pairault, J. 1991. Adipsin gene expression in 3T3-F442A adipocytes is postranscriptionally down-regulated by retinoic acid. *J. Biol. Chem.* 266:1157–61

9. Benito, M., Porras, A., Nebreda, A. R., Santos, E. 1991. Differentiation of 3T3-L1 fibroblasts to adipocytes induced by transfection of ras oncogenes. *Science* 253:565–68

10. Blake, W. L., Clarke, S. D. 1990. Induction of adipose fatty acid binding protein (a-FABP) by insulin-like growth factor-I (IGF-I) in 3T3-L1 preadipocytes. *Biochem. Biophys. Res. Commun.* 173:87–91

11. Bonnet, F. P., Rocour-Brumioul, D. 1981. Fat cells in leanness, growth retardation, and adipose tissue dystrophic syndromes. In *Adipose Tissue in Childhood,* ed. F. P. Bonnet, pp. 155–63. Boca Raton, Fla: CRC Press

12. Brandes, R., Arad, R., Bar-Tana, J. 1991. The induction of adipose conversion in 3T3-L1 cells is associated with early phosphorylation of a 60 kDa nuclear protein. *FEBS Lett.* 285:63–65

13. Brandes, R., Arad, R., Benvenisty, N., Weil, S., Bar-Tana, J. 1990. The induction of adipose conversion by bezafibrate in 3T3-L1 cells. Synergism with dibutyryl-cAMP. *Biochim. Biophys. Acta* 1054:219–24

14. Bukowiecki, L., Collet, A. J., Follea, N., Guay, G., Jahjah, L. 1982. Brown adipose tissue hyperplasia: a fundamental mechanism of adaptation to cold and hyperphagia. *Am. J. Physiol.* 242:E353–59

15. Bukowiecki, L. J., Géloën, A., Collet, A. J. 1986. Proliferation and differentiation of brown adipocytes from interstitial cells during cold acclimation. *Am. J. Physiol.* 250:C880–87

16. Burdi, A. R., Poissonnet, C. M., Garn, S. M., Lavelle, M., Sabet, M. D., et al. 1985. Adipose tissue growth patterns during human gestation: a histometric comparison of buccal and gluteal fat depots. *Int. J. Obes.* 9:247–56

17. Calderhead, D. M., Kitagawa, K., Tanner, L. I., Holman, G. D., Lienhard, G. E. 1990. Insulin regulation of the two glucose transporters in 3T3-L1 adipocytes. *J. Biol. Chem.* 265:13800–6

18. Casteilla, L., Champigny, O., Bouillaud, F., Robelin, J., Ricquier, D. 1989. Sequential changes in the expression of mitochondrial protein mRNA during the development of brown adipose tissue in bovine and ovine species. Sudden occurrence of uncoupling protein mRNA during embryogenesis and its disappearance after birth. *Biochem. J.* 257:665–71

19. Casteilla, L., Forest, C., Robelin, J., Ricquier, D., Lombet, A., et al. 1987. Characterization of mitochondrial-uncoupling protein in bovine fetus and

newborn calf. Disappearance in lamb during aging. *Am. J. Physiol.* 252: E627–36

20. Casteilla, L., Nouguès, J., Reyne, Y., Ricquier, D. 1991. Differentiation of ovine brown adipocyte precursor cells in a chemically defined serum-free medium. Importance of glucocorticoids and age of animals. *Eur. J. Biochem.* 198:195–99

21. Catalioto, R. M., Ailhaud, G., Négrel, R. 1990. Diacylglycerol production induced by growth hormone in Ob1771 preadipocytes arises from phosphatidylcholine breakdown. *Biochem. Biophys. Res. Commun.* 173:840–48

22. Catalioto, R. M., Gaillard, D., Maclouf, J., Ailhaud, G., Négrel, R. 1991. Autocrine control of adipose cell differentiation by prostacyclin and PGF$_{2\alpha}$. *Biochim. Biophys. Acta* 1091:364–69

23. Cinti, S., Cigolini, M., Morroni, M., Zingaretti, M. C. 1989. S-100 protein in white preadipocytes: an immunoelectronmicroscopy study. *Anat. Rec.* 224:466–72

24. Cook, K. S., Groves, D. L., Min, H. Y., Spiegelman, B. M. 1985. A developmentally regulated mRNA from 3T3 adipocytes encodes a novel serum protease homologue. *Proc. Natl. Acad. Sci. USA* 82:6480–84

25. Cook, K. S., Hunt, C. R., Spiegelman, B. M. 1985. Developmentally regulated mRNAs in 3T3-adipocytes: analysis of transcriptional control. *J. Cell Biol.* 100:514–20

26. Cook, J. R., Kozak, L. P. 1982. sn-Glycerol-3-phosphate dehydrogenase gene expression during mouse adipocyte development in vivo. *Dev. Biol.* 92: 440–48

27. Cook, K. S., Min, H. Y., Johnson, D., Chaplinski, R. J., Flier, J. S., et al. 1987. Adipsin: a circulating serine protease homolog secreted by adipose tissue and sciatic nerve. *Science* 237:402–5

28. Dani, C., Amri, E., Bertrand, B., Enerback, S., Bjursell, G., et al. 1990. Expression and regulation of pOb24 and lipoprotein lipase genes during adipose conversion. *J. Cell. Biochem.* 43:103–10

29. Dani, C., Doglio, A., Amri, E., Bardon, S., Fort, P., et al. 1989. Cloning and regulation of a mRNA specifically expressed in the preadipose state. *J. Biol. Chem.* 264:10119–25

30. Darmon, M., Serrero, G., Rizzino, A., Sato, G. 1981. Isolation of myoblastic, fibro-adipogenic, and fibroblastic clonal cell lines from a common precursor and study of their requirements for growth

and differentiation. *Exp. Cell Res.* 132:313–27

31. DeMartinis, F. D., Francendese, A. 1982. Very small fat cell populations: mammalian occurrence and effect of age. *J. Lipid Res.* 23:1107–19

32. Deslex, S., Négrel, R., Ailhaud, G. 1987. Development of chemically defined serum-free medium for differentiation of rat adipose precursor cells. *Exp. Cell Res.* 168:15–30

33. Djian, P., Phillips, M., Green, H. 1985. The activation of specific gene transcription in the adipose conversion of 3T3 cells. *J. Cell. Physiol.* 124:554–56

34. Djian, P., Roncari, D. A. K., Hollenberg, C. H. 1983. Influence of anatomic site and age on the replication and differentiation of rat adipocyte precursors in culture. *J. Clin. Invest.* 72: 1200–8

35. Dobson, D. E., Kambe, A., Block, E., Dion, T., Lu, H., et al. 1990. 1-butyrylglycerol: a novel angiogenesis factor secreted by differentiating adipocytes. *Cell* 61:223–30

36. Doglio, A., Dani, C., Fredrikson, G., Grimaldi, P., Ailhaud, G. 1987. Acute regulation of insulin-like growth factor-I gene expression by growth hormone during adipose cell differentiation. *EMBO J.* 6:4011–16

37. Doglio, A., Dani, C., Grimaldi, P., Ailhaud, G. 1989. Growth hormone stimulates c-fos gene expression by means of protein kinase C without increasing inositol lipid turnover. *Proc. Natl. Acad. Sci. USA* 86:1148–52

38. Emorine, L. J., Marullo, S., Briend-Sutren, M. M., Patey, G., Tate, K., et al. 1989. Molecular characterization of the human β_3-adrenergic receptor. *Science* 245:1118–21

39. Faust, I. M., Miller, W. H. Jr. 1983. Hyperplastic growth of adipose tissue in obesity. In *The Adipocyte and Obesity: Cellular and Molecular Mechanisms*, ed. A. Angel, C. H. Hollenberg, D. A. K. Roncari, pp. 41–51. New York: Raven

40. Fève, B., Emorine, L. J., Briend-Sutren, M. M., Lasnier, F., Strosberg, A. D., et al. 1990. Differential regulation of β_1- and β_2-adrenergic receptor protein and mRNA levels by glucocorticoids during 3T3-F442A adipose differentiation. *J. Biol. Chem.* 265:16343–49

41. Fève, B., Emorine, L. J., Lasnier, F., Blin, N., Baude, B., et al. 1991. Atypical β-adrenergic receptor in 3T3-F442A adipocytes: pharmacological and molecular relationship with the human

β_3-adrenergic receptor. *J. Biol. Chem.* In press

42. Flores-Delgado, G., Marsh-Moreno, M., Kuri-Harcuch, W. 1987. Thyroid hormone stimulates adipocyte differentiation of 3T3 cells. *Mol. Cell. Biochem.* 76:35–43

43. Francendes, A., DeMartinis, F. D. 1985. Very small fat cells. II. Initial observations on basal and hormone-stimulated metabolism. *J. Lipid Res.* 26:149–57

44. Freytag, S. O. 1988. Enforced expression of the *c-myc* oncogene inhibits cell differentiation by precluding entry into a distinct predifferentiation state in G_0/G_1. *Mol. Cell Biol.* 8:1614–24

45. Friedman, J. M., Leibel, R. L., Bahary, N., Chua, S., Leib, A., et al. 1988. Molecular mapping of the mouse *ob* and *db* genes. In *Mouse News Letter,* ed. M. Festing, p. 110. Dorchester: Henry Ling

46. Gaillard, D., Négrel, R., Lagarde, M., Ailhaud, G. 1989. Requirement and role of arachidonic acid in the differentiation of pre-adipose cells. *Biochem. J.* 257:389–97

47. Gaillard, D., Wabitsch, M., Pipy, B., Négrel, R. 1991. Control of terminal differentiation of adipose precursor cells by glucocorticoids. *J. Lipid Res.* 32:569–79

48. Gamou, S., Shimizu, N. 1986. Adipocyte differentiation of 3T3-L1 cells in serum-free hormone-supplemented media: effects of insulin and dihydroteleocidin B. *Cell Struct. Funct.* 11:21–30

49. Gao, G., Serrero, G. 1990. Phospholipase A2 is a differentiation-dependent enzymatic activity for adipogenic cell line and adipocyte precursors in primary culture. *J. Biol. Chem.* 265:2431–34

50. Garcia de Herreros, A., Birnbaum, M. J. 1989. The regulation by insulin of glucose transporter gene expression in 3T3 adipocytes. *J. Biol. Chem.* 264:9885–90

51. Gaskins, H. R., Kim, J. W., Wright, J. T., Rund, L. A., Hausman, G. J. 1990. Regulation of insulin-like growth factor-I: ribonucleic acid expression, polypeptide secretion, and binding protein activity by growth hormone in porcine pre-adipocyte cultures. *Endocrinology* 126:622–30

52. Géloen, A., Collet, A. J., Guay, G., Bukowiecki, L. J. 1988. β-adrenergic stimulation of brown adipocyte proliferation. *Am. J. Physiol.* 254:C175–82

53. Géloen, A., Collet, A. J., Guay, G., Bukowiecki, L. J. 1989. Insulin stimulates in vivo cell proliferation in white

adipose tissue. *Am. J. Physiol.* 256: C190–96

54. Géloen, A., Roy, P. E., Bukowecki, L. J. 1989. Regression of white adipose tissue in diabetic rats. *Am. J. Physiol.* 257:E547–53

55. Greenberg, A. S., Egan, J. J., Wek, S. A., Garty, N. B., Blanchette-Mackie, E. J., et al. 1991. Perilipin, a major hormonally regulated adipocyte-specific phosphoprotein associated with the periphery of lipid storage droplets. *J. Biol. Chem.* 256:11341–46

56. Grégoire, F., Genart, C., Hauser, N., Remacle, C. (1991. Glucocorticoids induce a drastic inhibition of proliferation and stimulate differentiation of adult rat fat cell precursors. *Exp. Cell Res.* 196:270–78

57. Grégoire, F., Todoroff, G., Hauser, N., Remacle, C. 1990. The stromavascular fraction of rat inguinal and epididymal adipose tissue and the adipoconversion of fat cell precursors in primary culture. *Biol. Cell.* 69:215–22

58. Grimaldi, P., Djian, P., Négrel, R., Ailhaud, G. 1982. Differentiation of Ob17 preadipocytes to adipocytes: requirement of adipose conversion factor(s) for fat cell cluster formation. *EMBO J.* 1:687–92

59. Guller, S., Corin, R. E., Mynarcik, D. C., London, B. M., Sonenberg, M. 1988. Role of insulin in growth hormone-stimulated 3T3 cells adipogenesis. *Endocrinology* 122:2084–89

60. Hainque, B., Guerre-Millo, M., Hainault, I., Moustaid, N., Wardzala, L. J., et al. 1990. Long-term regulation of glucose transporters by insulin in mature 3T3-F442A adipose cells. *J. Biol. Chem.* 265:7982–86

61. Han, V. K., D'Ercole, A. J., Lund, P. K. 1987. Cellular localization of somatomedin (insulin-like growth factor) messenger RNA in the human fetus. *Science* 236:193–97

62. Hansen, H. O., Andreasen, P. H., Mandrup, S., Kristiansen, K., Knudsen, J. 1991. Induction of acyl-CoA-binding protein and its mRNA in 3T3-L1 cells by insulin during preadipocyte-to-adipocyte differentiation. *Biochem. J.* 277:341–44

63. Hauner, H. 1990. Complete adipose differentiation of 3T3-L1 cells in a chemically defined medium: comparison to serum-containing culture conditions. *Endocrinology* 127:865–72

64. Hauner, H., Entenmann, G. 1991. Regional variation of adipose differentiation in cultured stromal-vascular cells from the abdominal and femoral adipose

tissue of obese women. *Int. J. Obes.* 15:121–26

65. Hauner, H., Entenmann, G., Wabitsch, M., Gaillard, D., Ailhaud, G., et al. 1989. Promoting effect of glucocorticoids on the differentiation of human adipocyte precursor cells cultured in chemically defined medium. *J. Clin. Invest.* 84:1663–70

66. Hausman, G. J. 1987. Identification of adipose tissue primordia in perirenal tissues of pig fetuses: utility of phosphatase histochemistry. *Acta Anat.* 128:236–42

67. Hausman, G. J., Hentges, E. J., Thomas, G. B. 1987. Differentiation of adipose tissue and muscle in hypophysectomized pig fetuses. *J. Anim. Sci.* 64:1255–61

68. Hausman, G. J., Richardson, R. L. 1987. Adrenergic innervation of fetal pig adipose tissue. Histochemical and ultrastructural studies. *Acta Anat.* 130:291–97

69. Hausman, G. J., Thomas, G. B. 1986. Structural and histochemical aspects of perirenal adipose tissue in fetal pigs: relationships between stromal-vascular characteristics and fat cell concentration and enzyme activity. *J. Morphol.* 190:271–83

70. Humbel, R. E. 1990. Insulin-like growth factors I and II. *Eur. J. Biochem.* 190:445–62

71. Hynes, M. A., Van Wik, J. J., Brooks, P. J., D'Ercole, A. J., Jansen, M., et al. 1987. Growth hormone dependence of somatomedin-C/Insulin-like growth factor-I and insulin-like growth factor-II messenger ribonucleic acids. *Mol. Endocrinol.* 1:233–42

72. Ibrahimi, A., Bardon, S., Bertrand, B., Ailhaud, G., Dani, C. 1991. In *Obesity in Europe 1991*, ed. G. Ailhaud, B. Guy-Grand, M. Lafontan, D. Ricquier. London: John Libbey. In press

73. Ignotz, R. A., Massagué, J. 1985. Type β transforming growth factor controls the adipogenic differentiation of 3T3 fibroblasts. *Proc. Natl. Acad. Sci. USA* 82:8530–34

74. Jiang, X. C., Moulin, P., Quinet, E., Goldberg, I. J., Yacoub, L. K., et al. 1991. Mammalian adipose tissue and muscle are major sources of lipid transfer protein mRNA. *J. Biol. Chem.* 266:4631–39

75. Julien, P., Despres, J. P., Angel, A. 1989. Scanning electron microscopy of very small fat cells and mature fat cells in human obesity. *J. Lipid Res.* 30:293–99

76. Kaestner, K. H., Christy, R. J., McLenithan, J. C., Braiterman, L. T., Cornelius, P., et al. 1989. Sequence, tissue distribution, and differential expression of mRNA for a putative insulin-responsive glucose transporter in mouse 3T3-L1 adipocytes. *Proc. Natl. Acad. Sci. USA* 86:3150–54

77. Kawada, T., Aoki, N., Kamei, Y., Maeshige, K., Nishiu, S., et al. 1990. Comparative investigation of vitamins and their analogues on terminal differentiation, from preadipocytes to adipocytes, of 3T3-L1 cells. *Comp. Biochem. Physiol.* 96:323–26

78. Kawakami, M., Watanabe, N., Ogawa, H., Kato, A., Sando, H., et al. 1989. Cachectin/TNF kills or inhibits the differentiation of 3T3-L1 cells according to developmental stage. *J. Cell. Physiol.* 138:1–7

79. Kern, P. A., Svoboda, M. E., Eckel, R. H., Van Wyk, J. J. 1989. Insulin-like growth factor action and production in adipocytes and endothelial cells from human adipose tissue. *Diabetes* 38:710–16

80. Kirkland, J. L., Hollenberg, C. H., Gillon, W. S. 1990. Age, anatomic site, and the replication and differentiation of adipocyte precursors. *Am. J. Physiol.* 258:C206–10

81. Kitagawa, K., Rosen, B. S., Spiegelman, B. M., Lienhard, G. E., Tanner, L. I. 1989. Insulin stimulates the acute release of adipsin from 3T3-L1 adipocytes. *Biochim. Biophys. Acta* 1014:83–89

82. Kopecky, J., Baudysova, M., Zanotti, F., Janikova, D., Pavelka, S., et al. 1990. Synthesis of mitochondrial uncoupling protein in brown adipocytes differentiated cell culture. *J. Biol. Chem.* 265:22204–9

83. Kozka, I. J., Clark, A. E., Holman, G. D. 1991. Chronic treatment with insulin selectively down-regulates cell-surface GLUT4 glucose transporters in 3T3-L1 adipocytes. *J. Biol. Chem.* 266:11726–31

84. Kuri-Harcuch, W., Marsch-Moreno, M. 1983. DNA synthesis and cell division related to adipose differentiation of 3T3 cells. *J. Cell. Physiol.* 114:39–44

85. Labrie, F. 1991. At the cutting edge: Intracrinology. *Mol. Cell. Endocrinol.* 78:C113–18

86. Labrie, F., Simard, J., Luu-The, V., Trudel, C., Martel, C., et al. 1991. Expression of 3β-hydroxysteroid dehydrogenase/$\Delta4$–$\Delta5$ isomerase (3β-HSD) and 17β-hydroxysteroid dehydrogenase (17β-HSD) in adipose tissue. *Int. J. Obes.* 15:91–99

87. Lafontan, M., Saulnier-Blache, J. S.,

Carpene, C., Langin, D., Galitzky, J., et al. 1991. See Ref. 72. In press
88. Lau, D. C. W., Roncari, D. A. K., Hollenberg, C. H. 1987. Release of mitogenic factors by cultured pre-adipocytes from massively obese human subjects. J. Clin. Invest. 79:632–36
89. Lemonnier, D. 1972. Effect of age, sex and site on the cellularity of the adipose tissue in mice and rats rendered obese by a high-fat diet. J. Clin. Invest. 51:2907–12
90. Levacher, C., Sztalryd, C., Kinebanyan, M. F., Picon, L. 1984. Effects of thyroid hormones on adipose tissue development in Sherman and Zucker rats. Am. J. Physiol. 246:C50–C56
91. Li, Z. H., Lu, Z., Kirkland, J. L., Gregarman, R. I. 1989. Preadipocyte stimulating factor in rat serum: evidence for a discrete 63 kDa protein that promotes cell differentiation of rat preadipocytes in primary cultures. J. Cell. Physiol. 141:543–57
92. Loncar, D. 1991. Convertible adipose tissue in mice. Cell Tissue Res. 266: 149–61
93. Loncar, D., Afzelius, B. A., Cannon, B. 1988. Epididymal white adipose tissue after cold stress in rats. II. Mitochondrial changes. J. Ultrastruct. Mol. Struct. Res. 101:199–209
94. Lönroth, P., Jansson, P. A., Smith, U. 1987. A microdialysis method allowing characterization of intercellular water space in humans. Am. J. Physiol. 253: E228–31
95. Miller, W. H. Jr., Faust, I. M., Goldberger, A. C., Hirsch, J. 1983. Effects of severe long-term food deprivation and refeeding on adipose tissue cells in the rat. Am. J. Physiol. 245: E74–E80
96. Navre, M., Ringold, G. M. 1988. A growth factor repressible gene associated with protein kinase C-mediated inhibition of adipocyte differentiation. J. Cell Biol. 107:279–86
97. Nedergaard, J., Néchad, M., Rehnmark, S., Herron, D., Jacobsson, A., et al. 1991. See Ref. 72. In press
98. Négrel, R., Gaillard, D., Ailhaud, G. 1989. Prostacyclin as a potent effector of adipose-cell differentiation. Biochem. J. 257:399–405
99. Obregon, M. J., Jacobsson, A., Kirchgessner, T., Schotz, M. C., Cannon, B., Nedergaard, J. 1989. Postnatal recruitment of brown adipose tissue is induced by the cold stress experienced by the pups. An analysis of mRNA levels for thermogenin and lipoprotein lipase. Biochem. J. 259:341–46
100. Ochi, M., Yoshioka, H., Sawada, T., Kusunoki, T., Hattori, T. 1991. New adipocyte formation in mice during refeeding after long-term deprivation. Am. J. Physiol. 260:R468–74
101. Ono, M., Aratani, Y., Kitagawa, I., Kitagawa, Y. 1987. Ascorbic acid phosphate stimulates type IV collagen synthesis and accelerates adipose conversion of 3T3-L1 cells. Exp. Cell Res. 187:309–14
102. Pairault, J., Lasnier, F. 1987. Control of adipogenic differentiation of 3T3-F442A cells by retinoic acid, dexamethasone, and insulin: a topographic analysis. J. Cell. Physiol. 132:279–86
103. Pairault, J., Quignard-Boulangé, A., Dugail, I., Lasnier, F. 1988. Differential effects of retinoic acid upon early and late events in adipose conversion of 3T3 preadipocytes. Exp. Cell Res. 177:27–36
104. Pilgrim, C. 1971. DNA synthesis and differentiation in developing white adipose tissue. Dev. Biol. 26:69–76
105. Poissonnet, C. M., Burdi, A. R., Bookstein, F. L. 1984. Critical periods of human adipogenesis: the buccal fat pad model. In Human Growth and Development, ed. J. Borms, R. Hauspie, A. Sand, C. Suzanne, M. Hebbelinck, pp. 243–52. New York: Plenum
106. Poissonnet, C. M., Burdi, A. R., Garn, S. M. 1984. The chronology of adipose tissue appearance and distribution in human fetus. Early Hum. Dev. 10:1–11
107. Poissonnet, C. M., Lavelle, M., Burdi, A. R. 1988. Growth and development of adipose tissue. J. Pediatr. 113:1–9
108. Pradines-Figuères, A., Barcellini-Couget, S., Dani, C., Baudoin, C., Ailhaud, G. 1990. Inhibition by serum components of the expression of lipoprotein lipase gene upon stimulation by growth hormone. Biochem. Biophys. Res. Commun. 166:1118–25
109. Ramsay, T. G., Hausmann, G. J., Martin, R. J. 1987. Pre-adipocyte proliferation and differentiation in response to hormone supplementation of decapitated fetal pig sera. J. Anim. Sci. 64:735–44
110. Rebuffé-Scrive, M., Krotkiewski, M., Elfverson, J., Björntorp, P. 1988. Muscle and adipose tissue morphology and metabolism in Cushing's syndrome. J. Clin. Endocrinol. Metab. 67:1122–28
111. Rebuffé-Scrive, M., Lundholm, K., Björntorp, P. 1985. Glucocorticoid hormone binding to human adipose tissue. Eur. J. Clin. Invest. 15:267–71
112. Rehnmark, S., Kopecky, J., Jacobsson, A., Néchad, M., Herron, D., et al. 1989. Brown adipocytes differentiated

in vitro can express the gene for the uncoupling protein thermogenin: effects of hypothyroidism and norepinephrine. *Exp. Cell Res.* 182:75–83

113. Rehnmark, S., Néchad, M., Herron, D., Cannon, B., Nedergaard, J. 1990. α- and β-adrenergic induction of the expression of the uncoupling protein thermogenin in brown adipocytes differentiated in culture. *J. Biol. Chem.* 265:16464–71

114. Rehnmark, S., Nedergaard, J. 1989. DNA synthesis in mouse brown adipose tissue is under β-adrenergic control. *Exp. Cell Res.* 180:574–79

115. Reyne, Y., Nouguès, J., Dulor, J. P. 1989. Differentiation of rabbit adipocyte precursor cells in a serum-free medium. *In Vitro Cell. Dev. Biol.* 25:747–52

116. Ricquier, D., Bouillaud, F., Casteilla, L., Cassard, A. M., Champigny, O., et al. 1990. Molecular studies on the uncoupling protein of brown adipose tissue. In *Progress in Obesity Research 1990*, ed. Y. Oomura, S. Tarui, S. Inoue, T. Shimazu, pp. 119–26. London: John Libbey

117. Rosen, B. S., Cook, K. S., Yaglom, J., Groves, D. L., Volanakis, J. E., et al. 1989. Adipsin and complement factor D activity: an immune-related defect in obesity. *Science* 244:1483–87

118. Sager, R., Kovac, P. 1982. Preadipocyte determination either by insulin or by 5-azacytidine. *Proc. Natl. Acad. Sci. USA* 79:480–84

119. Saulnier-Blache, J. S., Dauzats, M., Daviaud, D., Gaillard, D., Ailhaud, G., et al. 1991. Late expression of α2-adrenergic-mediated antilipolysis during differentiation of hamster preadipocytes. *J. Lipid Res.* 32:1489–99

120. Schiwek, D. R., Löffler, G. 1987. Glucocorticoid hormones contribute to the adipogenic activity of human serum. *Endocrinology* 120:469–74

121. Schmidt, W., Pöll-Jordan, G., Löffler, G. 1990. Adipose conversion of 3T3-L1 cells in a serum-free culture system depends on epidermal growth factor, insulin-like growth factor I, corticosterone, and cyclic AMP. *J. Biol. Chem.* 265:15489–95

122. Serrero, G. 1987. EGF inhibits the differentiation of adipocyte precursors in primary cultures. *Biochem. Biophys. Res. Commun.* 146:194–202

123. Serrero, G., Mills, D. 1987. Differentiation of newborn rat adipocyte precursors in defined serum-free medium. *In Vitro Cell. Dev. Biol.* 23:63–66

124. Serrero, G., Mills, D. 1991. Decrease in transforming growth factor β1 binding

during differentiation of rat adipocyte precursors in primary culture. *Cell Growth Differ.* 2:173–78

125. Serrero, G., Mills, D. 1991. Physiological role of epidermal growth factor on adipose tissue development in vivo. *Proc. Natl. Acad. Sci. USA* 88:3912–16

126. Shantz, L. M., Talalay, P., Gordon, G. B. 1989. Mechanism of inhibition of growth of 3T3-L1 fibroblasts and their differentiation to adipocytes by dehydroepiandrosterone and related steroids: Role of glucose-6-phosphate dehydrogenase. *Proc. Natl. Acad. Sci. USA* 86:3852–56

127. Shillabeer, G., Forden, J. M., Lau, D. C. W. 1989. Induction of preadipocyte differentiation by mature fat cells in the rat. *J. Clin. Invest.* 84:381–87

128. Smith, P. J., Wise, L. S., Berkowitz, R., Wan, C., Rubin, C. S. 1988. Insulin-like growth factor-I is an essential regulator of the differentiation of 3T3-L1 adipocytes. *J. Biol. Chem.* 263:9402–8

129. Sugihara, H., Yonemitsu, N., Toda, S., Miyabara, S., Funatsumaru, S., et al. 1988. Unilocular fat cells in three-dimensional collagen gel matrix culture. *J. Lipid Res.* 29:691–98

130. Takahashi, K., Goto, T., Mukai, K., Sawasaki, Y., Hata, J. 1989. Cobblestone monolayer cells from human omental adipose tissue are possibly mesothelial, not endothelial. *In Vitro Cell. Dev. Biol.* 25:109–10

131. Taylor, S. M., Jones, P. A. 1979. Multiple new phenotypes induced in 10T1/2 and 3T3 cells treated with 5-azacytidine. *Cell* 17:771–79

132. Tordjman, K. M., Leingang, K. A., Mueckler, M. M. 1990. Differential regulation of the HepG2 and adipocyte/muscle glucose transporters in 3T3-L1 adipocytes. *Biochem. J.* 271:201–7

133. Toscani, A., Soprano, D. R., Soprano, K. J. 1990. Sodium butyrate in combination with insulin or dexamethasone can terminally differentiate actively proliferating Swiss 3T3 cells into adipocytes. *J. Biol. Chem.* 265:5722–30

134. Truett, G. E., Bahary, N., Friedman, J. M., Leibel, R. L. 1991. Rat obesity gene fatty (fa) maps to chromosome 5: evidence for homology with the mouse gene diabetes (db). *Proc. Natl. Acad. Sci. USA* 88:7806–9

135. Wang, H., Kirkland, J. L., Hollenberg, C. H. 1989. Varying capacities for replication of rat adipocyte precursor clones and adipose tissue growth. *J. Clin. Invest.* 83:1741–46

136. Weiner, F. R., Shah, A., Smith, P. J.,

Rubin, C. S. 1989. Regulation of collagene gene expression in 3T3-L1 cells. Effects of adipocyte differentiation and tumor necrosis factor α. *Biochemistry* 28:4094–99

137. Wilkison, W. O., Choy, L., Spiegelman, B. M. 1991. Biosynthetic regulation of monobutyrin, an adipocyte-secreted lipid with angiogenic activity. *J. Biol. Chem.* 266:16886–91

138. Xu, X., Björntorp, P. 1990. Effects of dexamethasone on multiplication and differentiation of rat adipose precursor cells. *Exp. Cell Res.* 189:247–52

139. Xu, X., De Pergola, G., Björntorp, P. 1990. The effects of androgens on the regulation of lipolysis in adipose precursor cells. *Endocrinology* 126:1229–34

140. Yamada, K., Harada, N. 1990. Expression of estrogen synthetase (P-450 aromatase) during adipose differentiation of 3T3-L1 cells. *Biochem. Biophys. Res. Commun.* 169:531–36

141. Yang, B. S., Geddes, T. J., Pogulis, R. J., De Crombrugghe, B., Freytag, S. O. 1991. Transcriptional suppression of cellular gene expression by c-Myc. *Mol. Cell. Biol.* 11:2991–95

142. Zechner, R., Moser, R., Newman, T. C., Fried, S. K., Breslow, J. L. 1991. Apolipoprotein E gene expression in mouse 3T3-L1 adipocytes and human adipose tissue and its regulation by differentiation and lipid content. *J. Biol. Chem.* 266:10583–88

143. Zezulak, K. M., Green, H. 1986. The generation of insulin-like growth factor-I-sensitive cells by growth hormone action. *Science* 233:551–53

144. Zhao, H. F., Labrie, C., Simard, J., de Launoit, Y., Trudel, C., et al. 1991. Characterization of rat 3β-hydroxysteroid dehydrogenase/$\Delta 4$-$\Delta 5$ isomerase cDNAs and differential tissue-specific expression of the corresponding mRNAs in steroidogenic and peripheral tissues. *J. Biol. Chem.* 266:583–93

Annu. Rev. Nutr. 1992. 12:235-56

THE EOSINOPHILIA-MYALGIA SYNDROME AND TRYPTOPHAN

Edward A. Belongia,[1] Arthur N. Mayeno,[2] and Michael T. Osterholm[1]

[1]Acute Disease Epidemiology Section, Minnesota Department of Health, Minneapolis, Minnesota 55440; [2]Departments of Immunology and Medicine, Mayo Clinic and Foundation, Rochester, Minnesota 55905

KEY WORDS: eosinophilia, myositis, fasciitis, eosinophilic fasciitis, toxic oil syndrome, amino acids, tryptophan, food supplements

CONTENTS

INTRODUCTION

During the summer and fall of 1989, an epidemic of a new, multisystem illness occurred in the United States. The disease was characterized by severe

235

muscle pain and profound eosinophilia. It was initially recognized in October when physicians in New Mexico treated three women with similar clinical findings. They suspected an association with tryptophan consumption after observing that all three had consumed the food supplement prior to onset of illness (35). This finding was publicized by the local news media, and additional cases were reported by other New Mexico physicians. Shortly thereafter, cases were also recognized in other regions of North America and Europe. The major clinical features formed the basis for the name of the new disease: eosinophilia-myalgia syndrome, or EMS.

In early November, case-control studies were initiated in Minnesota and New Mexico to determine if there was an epidemiologic association between tryptophan use and EMS, or if tryptophan use was a surrogate for another unidentified risk factor. Both investigations demonstrated a significant association between antecedent tryptophan consumption and EMS (9, 23). A national surveillance program was initiated by the Centers for Disease Control (CDC). The case definition was developed based on review of the clinical findings of the initial cases. It included (a) eosinophil count greater than $1,000/mm^3$, (b) generalized debilitating myalgia, and (c) no evidence of infection or neoplasm that would explain the clinical findings.

On November 11, 1989, the US Food and Drug Administration issued a warning that advised consumers to discontinue use of tryptophan food supplements. The agency subsequently requested a nationwide recall of all over-the-counter food supplements that contained at least 100 milligrams of tryptophan in a daily dose. Although over 1500 EMS cases were identified, the epidemic was essentially halted by the removal of tryptophan from the consumer market.

THERAPEUTIC USE OF TRYPTOPHAN

Medical research in the 1970s and early 1980s suggested that tryptophan might be useful for the treatment of depression (17, 32, 84). Since then a number of investigations have examined its efficacy for a variety of other conditions, including insomnia, chronic pain, schizophrenia, premenstrual syndrome, affective disorders, and behavioral disorders (6, 12, 16, 20, 25, 31, 33, 49, 52–55, 64). The emphasis on treatment of psychiatric and behavioral disorders stemmed in part from the observation that brain serotonin content could be altered by changes in plasma tryptophan levels (24).

During the 1980s, reports in the popular press encouraged consumers to use tryptophan for therapeutic purposes. The product was widely available without a prescription, and it was promoted as an over-the-counter remedy for a variety of problems (51). A 1990 survey of tryptophan use in the Minneapolis-St. Paul area found that 4% of households had at least one person who had used tryptophan between 1980 and 1989 (3). The prevalence of use increased markedly between 1985 and 1989 and was highest in women (Figure 1). Although many consumers purchased tryptophan for therapeutic use, it was marketed as a food supplement. The manufacturers made no claims regarding therapeutic efficacy, and the product was not regulated or approved by the US Food and Drug Administration (5a).

BIOCHEMISTRY AND METABOLISM OF TRYPTOPHAN

Tryptophan is an essential amino acid. It is catabolized in mammals along two main pathways, resulting in the formation of kynurenine and serotonin (Figure 2). Most ingested tryptophan is degraded via the kynurenine pathway and provides precursors for the biosynthesis of niacin (nicotinic acid) and nicotinamide adenine dinucleotide. In this pathway, tryptophan is first oxidized to N-formylkynurenine by tryptophan 2,3-dioxygenase (TDO) or indoleamine 2,3-dioxygenase (IDO). This first enzymatic step is the rate-limiting step in the degradation of tryptophan. TDO (also called tryptophan pyrrolase) is localized to the liver while IDO is distributed throughout various tissues. Induction of IDO or TDO increases tryptophan catabolism and the formation of kynurenine and its metabolites. TDO activity is partially regulated by the hypothalamic-pituitary-adrenal axis. The enzyme is induced by glucocorticoids and adrenocorticotropic hormones; it is down-regulated by growth hormone. TDO activity is also increased by tryptophan loading. IDO activity is induced by gamma interferon (IFN-γ), and administration of this cytokine leads to increased levels of tryptophan metabolites in vitro and in vivo (8).

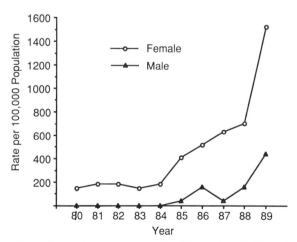

Figure 1   Prevalence of tryptophan use among male and female household members in a survey of 2012 randomly selected households in metropolitan Minneapolis-St. Paul (from Reference 3).

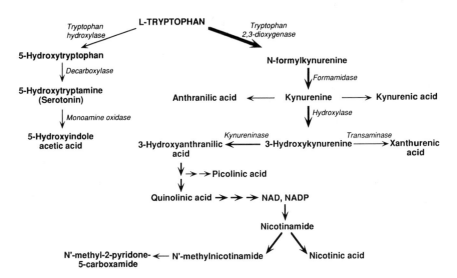

Figure 2   Major metabolic pathways of tryptophan degradation in humans.

Administration of interleukin 2 also induces IDO (7). Unlike TDO, IDO is not induced by either glucocorticoids or tryptophan loading.

Kynurenine is catabolized through several routes. The major pathway involves hydroxylation to 3-hydroxykynurenine, followed by degradation to 3-hydroxyanthranilic acid. Transaminases convert small portions of kynurenine and 3-hydroxykynurenine to kynurenic acid and xanthurenic acid, respectively.

A small portion of ingested tryptophan is converted to serotonin, a neurotransmitter. This metabolic pathway is found primarily in the central nervous system. Serotonin is degraded by monoamine oxidase and is excreted as 5-hydroxyindoleacetic acid. In the carcinoid syndrome, excessive amounts of serotonin and 5-hydroxyindoleacetic acid are produced by this route. A small proportion of ingested tryptophan is also metabolized by bacteria in the large intestine to indole, skatole, and other indole derivatives. Indole is converted to indican, which is excreted in the urine along with other indole compounds such as tryptamine, indole pyruvic acid, and indole acetic acid.

EOSINOPHILIA-MYALGIA SYNDROME

National Surveillance Data

By mid-1990, a total of 1531 EMS cases had been reported to the Centers for Disease Control, including 27 deaths (74). Eighty-four percent of patients were female, 97% were non-Hispanic white, and 86% were over 34 years old (median age, 49 years). Surveillance data demonstrated a dramatic increase in the incidence of EMS during the summer and fall of 1989 (Figure 3).

The prevalence of EMS was higher in the western United States than in other parts of the country, possibly because of a higher rate of tryptophan consumption in those states (74). High prevalence rates were also found in states that carried out investigations of EMS, including Minnesota, South Carolina, New Mexico, and Oregon. The high prevalence in these states may be partly attributed to more active surveillance and case identification.

The true prevalence of EMS is underestimated by surveillance reports. Persons with mild disease were excluded by the surveillance case definition even if the clinical diagnosis was consistent with EMS. In addition, surveillance data were compiled from reports submitted by physicians, and it is likely that a number of cases were diagnosed but not reported to state or federal health agencies.

Epidemiologic Studies

After initial case-control studies implicated tryptophan consumption as a major risk factor for EMS, additional investigations sought to determine the basis for this association. Two hypotheses were initially advanced to explain the association. According to one hypothesis, tryptophan itself triggered EMS in susceptible individuals, possibly owing to abnormalities of tryptophan

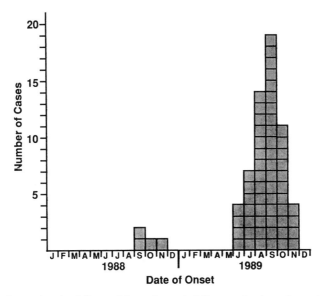

Figure 3 Cases of eosinophilia-myalgia syndrome in Minnesota by date of symptom onset (from Reference 3).

metabolism (15). According to the other hypothesis, EMS was triggered by a contaminant that was present in some lots of manufactured tryptophan. The latter hypothesis was consistent with the sudden appearance of the outbreak after tryptophan had been marketed for several years with no apparent ill effects. Epidemiologic investigations subsequently demonstrated that EMS was not triggered by tryptophan per se, but rather by exposure to a contaminant in tryptophan manufactured by one company.

Evidence for a tryptophan contaminant was provided by two case-control studies. In Minnesota, EMS patients and two control groups were evaluated to determine the manufacturer of their tryptophan and to assess potential risk factors (3). One control group consisted of self-referred, asymptomatic tryptophan users; the other group included asymptomatic tryptophan users who were randomly selected during a telephone survey. Retail lot numbers were obtained and traced back to determine the tryptophan source, and lots of bulk tryptophan were analyzed using high performance liquid chromatography.

Analysis of the tryptophan source for case patients and controls demonstrated a strong association between EMS and consumption of tryptophan manufactured by Showa Denko, K.K. (Tokyo, Japan). Twenty-nine (97%) of 30 case-patients consumed tryptophan (during the month before onset) that was manufactured by this company, compared to 21 (60%) of 35 in the combined control groups (odds ratio 19.3; 95% confidence interval 2.5 to 844.9). The tryptophan consumed by the 29 case-patients was manufactured by Showa Denko between October 1988 and June 1989. To assess the role of manufacturing changes during this time period, additional analyses were carried out utilizing the 29 case-patients and 21 controls who consumed tryptophan that was manufactured by Showa Denko. Product lots that were used prior to illness onset were considered "case lots", whereas lots that were not consumed by any patients were considered "control lots."

The company utilized a fermentation process to manufacture tryptophan. A strain of *Bacillus amyloliquefaciens* was used to synthesize tryptophan from precursors. Following fermentation, the tryptophan was extracted from the broth and purified using a series of filtration, crystallization, and separation processes. In December 1988, the company introduced a new strain of *B. amyloliquefaciens* (strain V) that had been genetically modified to increase the synthesis of intermediates in the tryptophan biosynthetic pathway. In 1989 the company also processed some fermentation batches with a reduced amount of powdered activated carbon (10 kg) in one of the purification steps. According to the company, these changes did not significantly alter the purity of the tryptophan produced, which was maintained at 99.6% or greater.

In the comparison of "case lots" and "control lots," both a reduction in the amount of powdered activated carbon and use of *B. amyloliquefaciens* (strain V) were significant manufacturing changes according to univariate analysis.

The independent contribution of each manufacturing change could not be assessed because there was a high correlation between them. Studies carried out by the company suggested that the biochemical and physiologic characteristics of *B. amyloliquefaciens* (strain V) did not differ from those of earlier strains. At this time it is not known whether the production of contaminant varied with different bacterial strains or if the efficiency of removal varied with changes in the separation and purification processes. Possibly both factors contributed to the presence of the etiologic agent.

The association between EMS and consumption of Showa Denko tryptophan was also found in an investigation of EMS patients and asymptomatic tryptophan users in Oregon (67). Ninety-eight percent of case patients had consumed tryptophan manufactured by Showa Denko, compared to 44% of controls. Eighty-five percent of the case lots were manufactured from January through May, 1989. An analysis of manufacturing conditions for case lots and control lots was not reported.

Chemical analyses of bulk tryptophan lots provided additional support for the epidemiologic findings (3). High performance liquid chromatography demonstrated a unique pattern, or "fingerprint," for tryptophan manufactured by different companies. The chromatographic pattern consisted of multiple peaks, with each peak representing a trace chemical constituent other than tryptophan. The chromatogram for Showa Denko tryptophan was distinctive and included 5 "signature" peaks that were present in all tryptophan manufactured by this company. Comparison of individual peaks in case and control lots demonstrated one peak ("peak E") that was significantly associated with case lots. This peak was present in 9 (75%) of 12 case lots and 3 (27%) of 11 control lots (odds ratio, 8.0; 95 percent confidence interval, 0.9 to 76.6; $P = 0.022$). The presence of peak E was also associated with the manufacturing changes described earlier. The chemical structure of peak E was subsequently determined to be 1,1'-ethylidenebis[tryptophan] (EBT) (Figure 4) (50, 68). The chemical is hydrolyzed under acidic conditions, and its biologic activity is uncertain. However, preliminary results from animal studies suggest that EBT may cause abnormalities of the fascia and microvasculature (46).

In addition to EBT, two other contaminants have been reported to be associated with case lots of tryptophan manufactured by Showa Denko (77). Both were found by using high performance liquid chromatography. One of the peaks, labeled UV-5 (also called FL-7), eluted before tryptophan and may be a low molecular weight aromatic compound. The mean concentration of UV-5 was lower than that of EBT. The other peak (UV-28) eluted much later than EBT and was present in even lower concentrations. The chemical structure of these peaks had not been reported as of January 1992.

There is no convincing evidence that any EMS cases were caused by

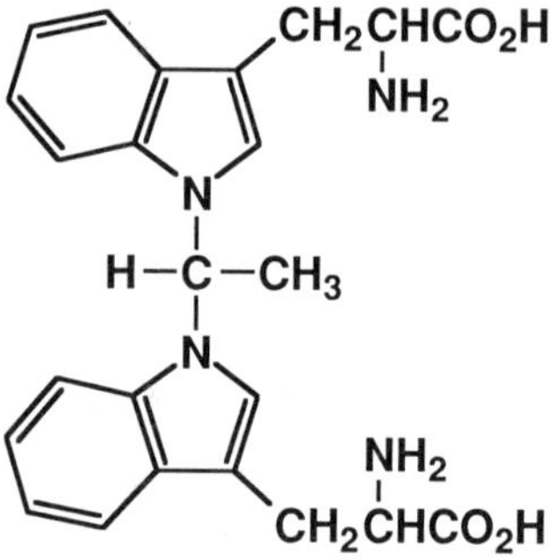

Figure 4   Structure of 1,1'-ethylidenebis[tryptophan], a tryptophan contaminant that has been associated with EMS in epidemiologic studies.

consumption of tryptophan manufactured by companies other than Showa Denko. One patient in the Minnesota series consumed tryptophan that was traced back to another manufacturer. This person had reportedly consumed a lifetime total of 20 tablets from one bottle during a two-month period. The tablets from this bottle yielded a chromatogram that was characteristic of tryptophan manufactured by Showa Denko, including the presence of EBT. These data suggest that her tryptophan was manufactured by Showa Denko rather than by the company identified by the product traceback. The explanation for the discrepant results from the product traceback is unclear. In the Oregon study, one EMS patient reportedly took tryptophan manufactured by another company, but this patient had also used another unknown brand of tryptophan before onset of her symptoms. Thus, exposure to tryptophan manufactured by Showa Denko could not be ruled out.

Data from a cohort of tryptophan users in a South Carolina psychiatric practice provides an estimate of the EMS attack rate for persons exposed to the etiologic agent (37). In this group, 157 persons consumed a single brand of tryptophan that was manufactured by Showa Denko. The product traceback indicated that only three manufacturer lots (all from Showa Denko) were represented in the tablets. The EMS attack rate was 29% in this group. An additional 23% were classified as "possible cases" because they had some clinical findings of EMS with suggestive symptoms, but without incapacitating myalgia. Thus, the pooled attack rate may have been as high as 52% among persons exposed to the etiologic agent.

RISK FACTORS

Few risk factors for EMS have been identified other than consumption of implicated tryptophan lots. The amount of tryptophan consumed has been identified as a risk factor in three investigations. In Minnesota, EMS patients consumed a median of 40.5 grams of tryptophan per month, compared to 6.0 grams for random controls and 15.0 grams for self-referred controls ($P < .001$) (3). In Oregon, EMS patients consumed an average of 1.3 grams of tryptophan per day, while the average in the two control groups was less than half that amount (67). In the South Carolina cohort, the risk of EMS was 3.5 times higher for persons taking more than 4 grams of implicated tryptophan per day than for those using 0.5 to 1.5 grams daily (37). This demonstrates a dose-response relationship between the amount of implicated tryptophan ingested and the risk of EMS.

There are several possible explanations for the association between tryptophan dose and risk of EMS. First, persons who consume larger doses of tryptophan may be exposed to greater quantities of the etiologic agent. Alternatively, persons consuming large doses (i.e. many tablets) may have a greater probability of encountering a contaminated tablet that causes illness. Evidence for a dose-response relationship in the South Carolina cohort suggests that the former hypothesis is correct; tryptophan dose is probably a surrogate for dose of ingested contaminant.

Age was also found to be a risk factor for EMS in two investigations. In the Minnesota study, the median age of case patients (45 years) was significantly older than the median age of randomly selected tryptophan users (39 years) after controlling for dose. In the South Carolina cohort, the risk of EMS was increased in older patients and was independent of dose (37). In Oregon, no significant difference in age was found between EMS patients and controls who used tryptophan (67). It is plausible that increasing age could increase the risk of EMS owing to physiologic changes in renal or hepatic function that alter the metabolism or delay the clearance of a toxic substance. Age-related changes in gastrointestinal absorption could also play a role.

There is no evidence that other host factors significantly alter the risk of developing EMS. A variety of factors were examined in both the Minnesota and South Carolina studies, including preexisting illnesses, asthma, smoking, alcohol consumption, and use of specific food supplements or prescription medications (e.g. nonsteroidal antiinflammatory drugs, tricyclic antidepressants, benzodiazepines, pyridoxine). None of these factors was significant when EMS patients were compared with controls who consumed tryptophan manufactured by Showa Denko. However, it is possible that other unknown genetic or environmental factors increase the risk of EMS after exposure to the etiologic agent.

Investigators have speculated that persons with impaired hypothalamic-pituitary-adrenal (HPA) function may be at increased risk of developing EMS when exposed to an inflammatory trigger (65). In the normal host, the presence of inflammatory mediators will signal the HPA axis to restrain the intensity of the inflammatory response through the release of glucocorticoids.

In Lewis rats, arthritis can be induced by exposure to streptococcal cell wall antigen, and susceptibility is increased if the HPA response to an inflammatory stimulus is impaired (70). However, at present no evidence indicates that EMS patients have impairment of the HPA axis.

CLINICAL AND PATHOLOGIC FEATURES

EMS is a syndrome with multiple clinical presentations and variable severity. Early clinical reports indicated that most patients developed profound eosinophilia and severe myalgias; these features provided the basis for the CDC case definition. In addition to myalgia, the most commonly reported early symptoms included arthralgias, weakness or fatigue, dyspnea or cough, rash, headache, peripheral edema, fever, and paresthesia (3, 10, 58, 74).

According to the CDC surveillance case definition for eosinophilia, all patients had at least 1,000 cells/mm.^{3} In different groups of EMS patients, the median eosinophil count has been reported to be 4,000 to 6,000 cells/mm^3 (3, 10, 67). The majority of patients also had an elevated leukocyte count with abnormally high levels of serum aldolase, a marker for muscle injury. Serum creatine phosphokinase, another indicator of muscle injury, was normal in most patients. Approximately one half of patients had abnormal liver function tests. The erythrocyte sedimentation rate, rheumatoid factor, and levels of IgE, complement, cryoglobulin, and thyroid stimulating hormone were normal in most patients tested (10, 29, 38, 58, 74).

Pathologic studies have demonstrated a perivascular, lymphocytic infiltrate in the dermis, fascia, and skeletal muscle, with variable numbers of eosinophils (34, 69). The perivascular infiltrate was accompanied by thickening of the capillary and arteriolar endothelium in dermal, fascial, and muscle vessels. The frequent occurrence of microangiopathy in biopsy specimens suggests that ischemia may contribute to tissue injury (69). Cytotoxic eosinophil degranulation products can be found in affected tissue, and these may play an important role in pathogenesis (35, 47).

The cutaneous and subcutaneous induration in EMS patients may be attributed to excessive accumulation of connective tissue in the affected fascia and lower dermis (79). Immunohistochemical studies have demonstrated increased deposition of transforming growth factor-β_1 (TGF-β_1), type VI collagen, and fibronectin in the extracellular matrix of the affected fascia (57). Fibroblasts in affected fascia demonstrate increased expression of several genes by in situ hybridization, including type I procollagen, type VI collagen, and TGF-β_1 (79). Since TGF-β_1 has been shown to play a prominent role in the regulation of fibroblast connective tissue production (61), this cytokine may play a role in the pathogenesis of EMS.

Most patients with EMS reported paresthesias, and in some patients peripheral neuropathy was the most prominent clinical feature. In some patients, persistent paresthesias have been accompanied by axonal and demyelinating abnormalities revealed by electrophysiologic testing (78). Perineural inflammation and type II fiber atrophy with denervation features have been observed, but muscle fiber necrosis was uncommon (80). The severe myalgias may be related to inflammation of nerves in the fascia or muscle, peripheral nerve injury caused by eosinophil-derived neurotoxin, or ischemia of nerves caused by occlusive microangiopathy (47).

Respiratory symptoms have been reported frequently by EMS patients, but the proportion with significant pulmonary disease is unknown. In the South Carolina cohort, 22% of hospitalized patients with EMS had radiographic evidence of pulmonary infiltrates (37). Lung biopsies performed in a small number of patients have demonstrated a vasculitis and perivasculitis with a chronic interstitial pneumonitis (73, 76). Disturbances of cardiac rhythm and conduction have also been documented. Examination of autopsy specimens has demonstrated neural lesions throughout the conduction system, similar to the neuropathology seen in skeletal muscle (36). Inflammatory lesions of the small coronary arteries were also present. The prevalence of cardiac abnormalities among all patients with EMS is unknown, although life-threatening rhythm disturbances appear to be rare.

The clinical and histopathologic findings of EMS overlap those of eosinophilic fasciitis (22). The latter is a scleroderma-like syndrome characterized by tender swelling and induration of the subcutaneous tissue, primarily in the arms and legs. Eosinophilia and hypergammaglobulinemia are characteristic findings, along with inflammatory infiltrates in the fascia and dermis. Eosinophilic fasciitis is distinguished from systemic sclerosis (scleroderma) by the relative absence of visceral involvement, digital ulcerations, and Raynaud's phenomenon (44), although the two diseases share many common features and may be variants of the same pathologic process.

The relationship between EMS, eosinophilic fasciitis, and systemic sclerosis was investigated in a retrospective review of patients diagnosed with eosinophilic fasciitis between 1977 and 1989 (48). Eight (24%) of 34 patients with onset from 1986 to 1989 had taken tryptophan before onset of symptoms. However, none of 25 patients with onset from 1977 to 1984 had consumed tryptophan ($P < .001$). In addition, none of 11 patients with systemic sclerosis had used tryptophan before the onset of illness. Review of biopsy material demonstrated no histopathologic differences between tryptophan-associated eosinophilic fasciitis and idiopathic eosinophilic fasciitis.

These results indicate that the histopathologic features of EMS and eosinophilic fasciitis are identical and that some cases of eosinophilic fasciitis were caused by tryptophan consumption. However, eosinophilic fasciitis may also be triggered by factors other than tryptophan consumption. Few, if any, eosinophilic fasciitis cases with onset before 1986 can be attributed to tryptophan use. This finding could be explained by manufacturing changes that caused sporadic contamination of tryptophan manufactured from 1985 to 1988, with an increase in the quantity or concentration of this contaminant in 1989.

Abnormalities of tryptophan metabolism have been reported in patients with EMS, the toxic oil syndrome, and other scleroderma-like conditions, leading to speculation that one or more metabolites may play a role in the pathogenesis of these diseases (5, 59, 65, 71). In patients with EMS, both kynurenine and quinolinic acid levels are elevated compared to those of controls (66), and quinolinic acid is a potential neurotoxin (26). However, the same abnormalities of tryptophan metabolism are found in unrelated conditions that involve chronic immune system activation, including HIV infection (27). These abnormalities appear to be mediated by IFN-γ, which induces IDO, the major rate-limiting enzyme in the kynurenine pathway. Administration of IFN-γ to cancer patients causes increased serum levels of kynurenine and quinolinic acid and decreased levels of tryptophan. However, it does not produce scleroderma-like changes or eosinophilia. Overall, studies of tryptophan metabolism suggest that the observed abnormalities are secondary to immune system activation; there is no evidence that they contribute to the specific pathologic changes seen in EMS.

NATURAL HISTORY AND TREATMENT

Only limited information is available regarding the clinical progression of EMS and the response to treatment. In Washington, 45 patients were followed up with serial telephone interviews and review of physical exam findings for up to 15 months after illness onset (19). The initial symptoms were myalgia and fatigue in most patients, followed by pulmonary symptoms (cough or dyspnea) within several weeks. Pruritic rash, peripheral edema, and/or paresthesias also developed in over 75% of patients, typically during the first two months after onset of myalgia. Some patients had relapses of myalgia after nearly complete symptom resolution. Neurologic symptoms generally consisted of diffuse paresthesias, but 15% of patients also described neurocognitive symptoms such as memory loss and difficulty concentrating. Scleroderma-like skin changes developed in 42%, but this finding was not observed until later in the disease course (median, 80 days after onset). The changes included dry, leathery, thickened skin, usually accompanied by changes in pigmentation. After 6 months, there was a steady decrease in the number of patients reporting severe myalgia, pulmonary symptoms, rash, and edema. One year after onset, myalgias had resolved in 42% of patients and had improved by an average of 72% in the remainder.

In New York, a follow-up questionnaire was completed by 91 patients (11); the median interval from symptom onset was 16 months (range, 11 to 40 months). At follow-up, 64% of patients reported persistent EMS symptoms that were "moderate" or "extreme." The most commonly reported persistent symptoms included fatigue (64%), muscle weakness (60%), muscle cramps (57%), myalgia (55%), and arthralgia (48%). Only 10% of patients reported complete resolution of symptoms, although the majority had experienced some reduction of severity.

Both of the previous investigations relied primarily on subjective reports of symptoms and severity to assess the progression of EMS. Objective measurements, such as serial muscle biopsies or laboratory results, were not available. The findings suggest that EMS symptoms improve gradually in most patients, but complete recovery is uncommon during the first one to two years after onset. Additional studies are needed to determine if tissue histopathology correlates with symptoms over time.

The response to therapy has been disappointing. Multiple therapeutic interventions have been suggested, but no clearly effective treatment has been identified (14). Because of the nature of the outbreak, it has not been possible to assess treatment regimens in a randomized, prospective study. Glucocorticoid treatment (usually prednisone) has been reported to cause symptomatic improvement in some patients, with a reduction of the eosinophil count (42). However, some patients have not responded to high doses of prednisone, and others have had an exacerbation of symptoms when the dose was tapered (29, 48). There is no evidence that prednisone therapy alters the natural history of the disease or the risk of neuropathy (19). Other treatments that have been utilized include nonsteroidal antiinflammatory drugs, cyclosporin A, cyclophosphamide, hydroxychloroquine, D-penicillamine, methotrexate, octreotide (a somatostatin analogue), and plasmapheresis (13, 14, 29, 38, 48, 72). Many of these therapies have been tried in patients with severe illness, and insufficient information is available to assess efficacy.

EMS AND THE TOXIC OIL SYNDROME

The clinical and pathologic findings of EMS bear a striking resemblance to those of the toxic oil syndrome (TOS). The latter outbreak occurred in Spain during 1981. Nearly 20,000 persons were affected, including 315 who died (41). Unlike EMS, respiratory symptoms (cough or dyspnea) were prominent and severe in TOS during the first week of illness. Bilateral pulmonary infiltrates due to noncardiogenic pulmonary edema were present in over 90% of patients with chest radiographs (2). The respiratory symptoms usually resolved and the chest radiographs returned to normal, although some patients developed pulmonary hypertension (30). Other early symptoms included fever, malaise, headache, nausea, splenomegaly, diffuse adenopathy, and pruritic rash (41, 75). In some patients, the disease progressed to an intermediate and chronic phase that more closely resembled EMS. The intermediate phase (2 to 8 weeks after onset) was characterized by eosinophilia and leukocytosis. Patients who progressed to the late phase developed muscle cramps and severe myalgias, peripheral edema, scleroderma-like skin changes, and polyneuropathy. The histopathology of skin, nerve, and skeletal muscle is remarkably similar in EMS and TOS (34, 63).

Epidemiologic investigations of TOS implicated consumption of denatured industrial rapeseed oil that was illegally sold by itinerant salesmen (60). Chemical analyses of implicated oil samples and "control" oil samples demonstrated that free aniline and fatty acid anilides were significantly associated with case-related samples (4, 40). However, efforts to evaluate the biologic activity of these substances have been limited by the absence of an animal model.

The strong similarities between EMS and TOS suggest that they may share the same final common pathway that leads to neuromuscular damage. However, they may not be triggered by the same etiologic agent. The vehicle of transmission was clearly different (oil versus manufactured tryptophan), and no contaminants common to both vehicles have been reported. In addition, the type and level of exposure to the etiologic agent may have been different for TOS compared to EMS. In the TOS epidemic, inhalation of oil vapor (during or immediately after cooking) could have been an alternate source of exposure and might account for the more severe pulmonary pathology.

A population-based follow-up evaluation of patients with TOS demonstrated that the majority reported an improvement of symptoms over 8 years (1). Overall, 49% had complete resolution. Of those with residual symptoms, only 7% had significant functional impairment. If the natural history of EMS is similar, severe long-term disability may be uncommon, and complete recovery may be expected in approximately half of affected patients.

PROGRESS TOWARD UNDERSTANDING ETIOLOGY AND PATHOGENESIS

Research to elucidate the etiology and pathogenesis of EMS is ongoing. Development of an appropriate animal model has been a priority. Ideally, such a model would be based on an inexpensive, readily-available species that develops eosinophilia and typical pathologic changes after exposure to implicated tryptophan. Investigators have also attempted to develop an in vitro system to study the mechanism of disease at the cellular level.

Animal Models

Lewis rats have been proposed as an animal model for EMS (18). This species is known to be susceptible to several inflammatory diseases in response to inflammatory stimuli. To assess the utility of the model for EMS, Lewis rats were given either implicated tryptophan or pharmaceutical grade tryptophan at a dose of 1,600 mg/kg per day. Eosinophil counts were obtained weekly and histopathologic changes were assessed after 38 days. Muscle biopsy specimens demonstrated perimysial inflammation in 7 of 9 animals receiving implicated tryptophan, compared to 0 of 10 receiving USP grade tryptophan ($P < .001$, Fisher exact test). A significant increase in fascial thickening was also observed in rats receiving implicated tryptophan. However, leukocyte counts and eosinophil counts remained normal in both groups. Gastrointestinal changes were also noted with an increased number of degranulating inflammatory cells in the lamina propria of the rats that received case-implicated tryptophan (21).

Subsequent studies have demonstrated that Lewis rats treated with EBT or case-associated tryptophan developed significant fascial thickening compared to rats that received pure (nonimplicated) tryptophan or methyl cellulose vehicle (46). Although the animals did not develop eosinophilia, this is the first evidence that EBT, a contaminant that is epidemiologically associated with EMS, may cause pathologic changes in an animal model. These findings have not yet been confirmed by other investigators, and no other animal model for EMS has been proposed.

In Vitro Studies

In vitro investigations have attempted to clarify the mechanism of immune activation. One study has reported that EBT induces interleukin 5 (IL-5) production from human T cells in a dose-dependent manner (83). In addition, data from a small number of EMS patients suggests that IL-5 is elevated in the serum (56). Levels of granulocyte-macrophage colony stimulating factor (GM-CSF) and interleukin 3 were normal. IL-5 can increase eosinophil production, degranulation, and in vitro survival (28, 39). It can also convert eosinophils to a hypodense phenotype with enhanced antibody mediated cytotoxicity (62). Thus, the immunologic effects of IL-5 may mediate some of the end-organ damage in EMS. Additional investigations are needed to confirm this hypothesis.

Peripheral blood mononuclear cells were used by Gleich and coworkers to evaluate an in vitro bioassay for EMS. The cells were exposed to case- and control-lots of tryptophan and incubated for 24 h. GM-CSF and eosinophil viability enhancing activity were measured in the supernatant (manuscript in preparation). Case-associated tryptophan did not stimulate release of GM-CSF from mononuclear cells, and the supernatant did not enhance eosinophil viability. Endotoxin was found in random lots of tryptophan and was associated with release of GM-CSF; endotoxin was not associated with case lots.

Possible Pathogenetic Mechanisms

The sequence of events leading to the pathologic changes of EMS is undoubtedly complex. However, a preliminary framework for possible mechanisms can be advanced based on current knowledge. One scenario involves a direct effect of the etiologic agent on mononuclear cells, leading to production of IL-5 (83). This cytokine could then activate tissue eosinophils and convert them to a hypodense phenotype. Effector functions would be augmented with release of cytotoxic cationic molecules, including major basic protein, eosinophil-derived neurotoxin, and eosinophil peroxidase. In addition, eosinophils may release cytokines such as interleukin 3, GM-CSF, and leukotriene C4 (43, 82). A cascade of interacting cytokines, including those produced by eosinophils themselves, could then lead to recruitment of additional inflammatory cells and increased collagen synthesis by fibroblasts. Eosinophils might also interact collaboratively with lymphocytes and other cells through expression of CD4 and the major histocompatibility complex HLA-DR (81). Associated microvascular changes such as vasculitis and endothelial cell swelling could contribute to ischemia and peripheral neuropathy.

The predominance of inflammatory changes in fascia suggests that mediators produced by mesenchymal cells (fibroblasts and endothelial cells) may also play a role in the pathogenesis. For example, fibroblasts have been shown to augment IL-5-dependent eosinophil survival and stimulate conversion to the hypodense phenotype (62). Fibroblasts can also produce interleukin 8, which recruits neutrophils and lymphocytes when injected in vivo (45). Thus, one can speculate that the etiologic agent interacts with these cells to stimulate release of inflammatory mediators and increase collagen synthesis.

This framework is consistent with epidemiologic evidence for a dose-response relationship between tryptophan consumption (a surrogate for exposure to the etiologic agent) and risk of EMS. It is also consistent with recent reports that EBT may stimulate IL-5 production in a dose-dependent manner and that IL-5 is elevated in the serum of some patients with EMS. In this model, the degree of eosinophil activation would be correlated with the amount of etiologic agent consumed. Clinical signs and symptoms of EMS would presumably develop in persons who were exposed above a critical threshold level; the latter could vary depending on individual susceptibility. However, it is unclear why the inflammatory process can become self-sustaining in individuals after exposure to the etiologic agent is stopped (i.e. tryptophan use discontinued).

Another general hypothesis involves incorporation of the etiologic agent into metabolic or biosynthetic pathways that utilize tryptophan. EBT and tryptophan have obvious structural similarities. If EBT is the etiologic agent, it might function as a tryptophan analogue with adverse immunologic effects.

For example, if EBT is recognized by the transfer RNA that is specific for tryptophan, it might be incorporated into a nascent protein molecule, stimulating an autoimmune response. Alternatively, EBT might be incorporated into either the serotonin or kynurenine pathway of tryptophan degradation, leading to production of one or more toxic metabolites. However, the hypothesis that EBT acts as a tryptophan analogue does not explain the eosinophilia or the predilection for involvement of the fascia.

Additional work is needed to determine if EMS is triggered by exposure to EBT or the chemicals represented by peaks UV-5 and UV-28. It would be useful to determine if any of these substances can stimulate release of cytokines from different cell lines, particularly fibroblasts and endothelial cells. In addition, the reported IL-5-producing effect of EBT on mononuclear cells requires confirmation by other investigators. The hypothesis that EBT functions as a tryptophan analogue can be further evaluated by studies to determine if EBT (*a*) is recognized by transfer RNA and incorporated into protein, or (*b*) interacts with enzymes or intermediates in the tryptophan metabolic pathways.

CONCLUSION

EMS is a chronic inflammatory disease that occurred in epidemic proportions during 1989 and was associated with tryptophan consumption. It preferentially affected fascia with variable involvement of other tissues; typical findings included striking eosinophilia with severe myalgias. Pathologic changes included perimyositis, microangiopathy, and increased collagen deposition in fascia. Response to treatment has been poor, and over 50% of patients remain symptomatic after one year of follow-up. EMS is clinically and pathologically similar to eosinophilic fasciitis and the toxic oil syndrome, and some cases of the former have been attributed to tryptophan consumption.

Epidemiologic studies indicate that EMS is triggered by one or more contaminants in tryptophan that was manufactured by one company. The chemical that has been most strongly implicated is 1,1'-ethylidenebis[tryptophan] (EBT), a molecule that is structurally similar to tryptophan. Results from animal studies suggest that EBT may cause pathologic changes in fascia that resemble EMS. The mechanism of immune activation is unknown, but preliminary investigations suggest that EBT can stimulate production of IL-5, a potent eosinophil-activating cytokine. Two other contaminants (peaks UV-5 and UV-28) have also been reported to be associated with case lots of tryptophan, but their structures and biologic activity have not been described. Epidemiologic investigations have shown that the presence of EBT was associated with changes in the tryptophan manufacturing process. However, the specific manufacturing source of the contaminant has not been determined. Consumption of high tryptophan doses and increased age have been identified as possible risk factors in persons who consumed implicated tryptophan; tryptophan dose may be a surrogate for the amount of etiologic agent that is ingested.

Ongoing research in animals and in vitro systems is underway to elucidate the pathogenesis of the syndrome and identify the etiologic agent. Additional research is needed to determine how this substance was synthesized during the manufacturing process and to develop effective monitoring procedures to assure that it does not occur in other products. Success in these endeavors will greatly increase our understanding of eosinophilic and sclerosing diseases and may lead to more effective therapies.

ACKNOWLEDGMENTS

We thank Gerald Gleich, Craig Hedberg, and Mary Kamb for their useful comments and suggestions.

Literature Cited

1. Alonso-Ruiz, A., Calabozo, M., Perez-Ruiz, F., Mancebo, L., Rodriguez-Morua, J. 1991. Results of long term follow-up of the toxic oil syndrome (abstract). *Arthritis Rheum.* 34:S131 (Suppl.)
2. Alonso-Ruiz, A., Zea-Mendoza, A. C., Salazar, M. A., Vallinas, J. M., Roca-Moraripoll, A., Beltran-Gutierrez, J. 1986. Toxic oil syndrome: a syndrome with features overlapping those of various forms of scleroderma. *Semin. Arthritis Rheum.* 15:200–12
3. Belongia, E. A., Hedberg, C. W., Gleich, G. J., White, K. E., Mayeno, A. N., et al. 1990. An investigation of the cause of the eosinophilia-myalgia syndrome associated with tryptophan use. *New Engl. J. Med.* 323:357–65
4. Bernert, J. T., Kilbourne, E. M., Akins, J. R., Posada de la Paz, M., Meredith, N. K., et al. 1987. Compositional analysis of oil samples implicated in the Spanish toxic oil syndrome. *J. Food Sci.* 52:1562–69
5. Binazzi, M., Calandra, P. 1973. Tryptophan-niacin pathway in scleroderma and in dermatomyositis. *Arch. Dermatol. Forsch.* 246:142–45
5a. Bluhm, R. E. 1992. The Food and Drug Administration and its problems (letter). *New Engl. J. Med.* 326:70
6. Brewerton, T. D., Reus, V. I. 1983. Lithium carbonate and L-tryptophan in the treatment of bipolar and schizoaffective disorders. *Am. J. Psychiatry* 140:757–60
7. Brown, R. R., Lee, C. M., Kohler, P. C., Hank, J. A., Storer, B. E., et al. 1989. Altered tryptophan and neopterin metabolism in cancer patients treated with recombinant interleukin 2. *Cancer Res.* 49:4941–44
8. Byrne, G. I., Lehmann, L. K., Kirschbaum, J. G., Borden, E. C., Lee, C. M., et al. 1986. Induction of tryptophan degradation in vitro and in vivo: A gamma-interferon-stimulated activity. *J. Interferon Res.* 6:389–96
9. Centers for Disease Control. 1989. Eosinophilia-myalgia syndrome and L-tryptophan-containing products—New Mexico, Minnesota, Oregon, and New York, 1989. *Morbid. Mortal. Wkly. Rep.* 38:785–88
10. Centers for Disease Control. 1990. Clinical spectrum of eosinophilia-myalgia syndrome—California. *Morbid. Mortal. Wkly. Rep.* 39:89–91
11. Centers for Disease Control. 1991. Eosinophilia-myalgia syndrome: Follow-up survey of patients—New York, 1990–1991. *Morbid. Mortal. Wkly. Rep.* 40:401–3
12. Chouinard, G., Young, S. N., Annable, L. 1985. A controlled clinical trial of L-tryptophan in acute mania. *Biol. Psychiatry* 20:546–57
13. Clauw, D. J., Alloway, J. A., Read, C., Katz, P. 1991. The use of cyclosporin A in the eosinophilia myalgia syndrome. (Abstract). *Arthritis Rheum.* 34:S194 (Suppl.)
14. Clauw, D. J., Katz, P. 1990. Treatment

of the eosinophilia-myalgia syndrome. (Letter). *New Engl. J. Med.* 323:417

15. Clauw, D. J., Nashel, D. J., Umhau, A., Katz, P. 1990. Tryptophan-associated eosinophilic connective-tissue disease. A new clinical entity? *J. Am. Med. Assoc.* 263:1502–6

16. Cole, W., Lapierre, Y. D. 1986. The use of tryptophan in normal-weight bulimia. *Can. J. Psychiatry* 31:755–56

17. Coppen, A., Shaw, D. M., Herzberg, B., Maggs, R. 1967. Tryptophan in the treatment of depression. *Lancet* 2:1178–80

18. Crofford, L. J., Rader, J. I., Dalakas, M. C., Hill, R. J., Page, S. W., et al. 1990. L-tryptophan implicated in human eosinophilia-myalgia syndrome causes fasciitis and perimyositis in the Lewis rat. *J. Clin. Invest.* 86:1757–63

19. Culpepper, R. C., Williams, R. G., Mease, P. J., Koepsell, T. D., Kobayashi, J. M. 1991. Natural history of the eosinophilia-myalgia syndrome. *Ann. Intern. Med.* 115:437–42

20. Demisch, K., Bauer, J., Georgi, K., Demisch, L. 1987. Treatment of severe chronic insomnia with L-tryptophan: results of a double-blind cross-over study. *Pharmacopsychiatry* 20:242–44

21. DeSchryver-Kecskemeti, K., Gramlich, T. L., Crofford, L. J., Rader, J. I., Page, S. W., et al. 1991. Mast cell and eosinophil infiltration in intestinal mucosa of Lewis rats treated with L-tryptophan implicated in human eosinophilia-myalgia syndrome. *Mod. Pathol.* 4:354–57

22. Doyle, J. A., Ginsburg, W. W. 1989. Eosinophilic fasciitis. *Med. Clin. North Am.* 73:1157–66

23. Eidson, M., Philen, R. M., Sewell, C. M., Voorhees, R., Kilbourne, E. M. 1990. L-tryptophan and eosinophilia-myalgia syndrome in New Mexico. *Lancet* 335:645–48

24. Fernstrom, J. D., Wurtman, R. J. 1971. Brain serotonin content: Physiological dependence on plasma tryptophan levels. *Science* 173:149–52

25. Fitten, L. J., Profita, J., Bidder, T. G. 1985. L-tryptophan as a hypnotic in special patients. *J. Am. Geriatr. Soc.* 33:294–97

26. Freese, A., Schwartz, K. J., During, M. 1988. Potential neurotoxicity of tryptophan. *Ann. Intern. Med.* 108:312–13

27. Fuchs, D., Moller, A. A., Reibnegger, G., Werner, E. R., Werner-Felmayer, G., et al. 1991. Increased endogenous interferon-gamma and neopterin correlate with increased degradation of tryptophan in human immunodeficiency virus type 1 infection. *Immunol. Lett.* 28:207–11

28. Fujisawa, T., Abu-Ghazaleh, R., Kita, H., Sanderson, C. J., Gleich, G. J. 1990. Regulatory effect of cytokines on eosinophil degranulation. *J. Immunol.* 144:642–46

29. Glickstein, S. L., Gertner, E., Smith, S. A., Roelofs, R. I., Hathaway, D. E., et al. 1990. Eosinophilia-myalgia syndrome associated with L-tryptophan use. *J. Rheumatol.* 17:1534–43

30. Gomez-Sanchez, M. A., Mestre de Juan, M. J., Gomez-Pajuelo, C. G., Lopez, J. I., Diaz de Atauri, M. J., et al. 1989. Pulmonary hypertension due to toxic oil syndrome: A clinicopathologic study. *Chest* 95:325–31

31. Harrison, W. M., Endicott, J., Rabkin, J. G., Nee, J. 1984. Treatment of premenstrual dysphoric changes: clinical outcome and methodological implications. *Psychopharmacol. Bull.* 20:118–22

32. Hartmann, E., Chung, R., Chien, C. P. 1971. Tryptophan and an MAOI (nialamide) in the treatment of depression. A double-blind study. *Int. Pharmacopsychiatry* 6:92–97

33. Hartmann, E., Lindsley, J. G., Spinweber, C. 1983. Chronic insomnia: effects of tryptophan, flurazepam, secobarbital, and placebo. *Psychopharmacology (Berlin)* 80:138–42

34. Herrick, M. K., Chang, Y., Horoupian, D. S., Lombard, C. M., Adornato, B. T. 1991. L-tryptophan and the eosinophilia-myalgia syndrome: pathologic findings in eight patients. *Hum. Pathol.* 22:12–21

35. Hertzman, P. A., Blevins, W. L., Mayer, J., Greenfield, B., Ting, M., et al. 1990. Association of the eosinophilia-myalgia syndrome with the ingestion of tryptophan. *New Engl. J. Med.* 322:869–73

36. James, T. N., Kamb, M. L., Sandberg, G. A., Silver, R. M., Kilbourne, E. M. 1991. Postmortem studies of the heart in three fatal cases of the eosinophilia-myalgia syndrome. *Ann. Intern. Med.* 115:102–10

37. Kamb, M. L., Murphy, J. J., Jones, J. I.., Caston, J. C., Nederlof, K., et al. 1991. Eosinophilia-myalgia syndrome in L-tryptophan exposed patients in a South Carolina psychiatric practice. *J. Am. Med. Assoc.* 267:77–82

38. Kaufman, L. D., Seidman, R. J., Gruber, B. L. 1990. L-tryptophan-asso-

ciated eosinophilic perimyositis, neuritis, and fasciitis. A clinicopathologic and laboratory study of 25 patients. *Medicine* 69:187–99
39. Kelso, A., Metcalf, D. 1990. T-lymphocyte-derived colony-stimulating factors. *Adv. Immunol.* 48:69–105
40. Kilbourne, E. M., Bernert, J. T., Posada, M., Hill, R. H., Abaitua Borda, I., et al. 1988. Chemical correlates of pathogenicity of oils related to the toxic oil syndrome epidemic in Spain. *Am. J. Epidemiol.* 127:1210–26
41. Kilbourne, E. M., Rigau, P. J., Heath, C. J., Zack, M. M., Falk, H., et al. 1983. Clinical epidemiology of toxic-oil syndrome: Manifestations of a new illness. *New Engl. J. Med.* 309:1408–14
42. Kilbourne, E. M., Swygert, L. A., Philen, R. M., Sun, R. K., Auerbach, S. B., et al. 1990. Interim guidance on the eosinophilia-myalgia syndrome. *Ann. Intern. Med.* 112:85–87
43. Kita, H., Ohnishi, T., Okubo, Y., Weiler, D., Abrams, J. S., et al. 1991. Granulocyte/macrophage colony-stimulating factor and interleukin 3 release from human peripheral blood eosinophils and neutrophils. *J. Exp. Med.* 174:745–48
44. Lakhanpal, S., Ginsburg, W. W., Michet, C. J., Doyle, J. A., Moore, S. B. 1988. Eosinophilic fasciitis: clinical spectrum and therapeutic response in 52 cases. *Semin. Arthritis Rheum.* 17:221–31
45. Larsen, C. G., Anderson, A. O., Oppenheim, J. J., Matsushima, K. 1989. Production of interleukin-8 by human dermal fibroblasts and keratinocytes in response to interleukin-1 or tumour necrosis factor. *Immunology* 68:31–36
46. Love, L. A., Rader, J. I., Crofford, L. J., Page, S. W., Hill, R. H., et al. 1991. L-tryptophan (L-TRP) and 1,1'-ethylidenebis[tryptophan] (EBT), a contaminant in eosinophilia myalgia syndrome (EMS) case-associated L-TRP, cause myofascial thickening and pancreatic fibrosis in Lewis rats (abstract). *Arthritis Rheum.* 34:S131 (Suppl.)
47. Martin, R. W., Duffy, J., Engel, A. G., Lie, J. T., Bowles, C. A., et al. 1990. The clinical spectrum of the eosinophilia-myalgia syndrome associated with L-tryptophan ingestion. Clinical features in 20 patients and aspects of pathophysiology. *Ann. Intern. Med.* 113:124–34
48. Martin, R. W., Duffy, J., Lie, J. T. 1991. Eosinophilic fasciitis associated with use of L-tryptophan: A case-control

study and comparison of clinical and histopathologic features. *Mayo Clin. Proc.* 66:892–98
49. Mattes, J. A. 1986. A pilot study of combined trazodone and tryptophan in obsessive-compulsive disorder. *Int. Clin. Psychopharmacol.* 1:170–73
50. Mayeno, A. N., Lin, F., Foote, C. S., Loegering, D. A., Ames, M. M., et al. 1990. Characterization of "peak E": a novel amino acid associated with eosinophilia-myalgia syndrome. *Science* 250:1707–8
51. Mazer, E. 1983. Tryptophan: The three-way misery reliever. *Prevention* May: 135–39
52. McGrath, R. E., Buckwald, B., Resnick, E. V. 1990. The effect of L-tryptophan on seasonal affective disorder. *J. Clin. Psychiatry* 51:162–63
53. Millinger, G. S. 1986. Neutral amino acid therapy for the management of chronic pain. *Cranio* 4:157–63
54. Morand, C., Young, S. N., Ervin, F. R. 1983. Clinical response of aggressive schizophrenics to oral tryptophan. *Biol. Psychiatry* 18:575–78
55. Nemzer, E. D., Arnold, L. E., Votolato, N. A., McConnell, H. 1986. Amino acid supplementation as therapy for attention deficit disorder. *J. Am. Acad. Child Psychiatry* 25:509–13
56. Owen, W. F., Petersen, J., Sheff, D. M., Folkerth, R. D., Anderson, R. J., et al. 1990. Hypodense eosinophils and interleukin 5 activity in the blood of patients with the eosinophilia-myalgia syndrome. *Proc. Natl. Acad. Sci. USA* 87:8647–51
57. Peltonen, J., Varga, J., Sollberg, S., Uitto, J., Jimenez, S. A. 1991. Elevated expression of the genes for transforming growth factor-beta 1 and type VI collagen in diffuse fasciitis associated with the eosinophilia-myalgia syndrome. *J. Invest. Dermatol.* 96:20–25
58. Philen, R. M., Eidson, M., Kilbourne, E. M., Sewell, C. M., Voorhees, R. 1991. Eosinophilia-myalgia syndrome. A clinical case series of 21 patients. *Arch. Intern. Med.* 151:533–37
59. Price, J. M., Yess, N., Brown, R. R., Johnson, S. A. 1967. Tryptophan metabolism: A hitherto unreported abnormality occurring in a family. *Arch. Dermatol.* 95:462–72
60. Rigau-Perez, J. G., Perez-Alvarez, L., Duenas-Castro, S., Choi, K., Thacker, S. B., et al. 1984. Epidemiologic investigation of an oil-associated pneumonic paralytic eosinophilic syn-

drome in Spain. *Am. J. Epidemiol.* 119:250–60

61. Roberts, A. B., Sporn, M. B., Assoian, R. K., Smith, J. M., Roche, N. S., et al. 1986. Transforming growth factor type-beta: rapid induction of fibrosis and angiogenesis in vivo and stimulation of collagen formation in vitro. *Proc. Natl. Acad. Sci. USA* 83:4167–71

62. Rothenberg, M. E., Petersen, J., Stevens, R. L., Silberstein, D. S., McKenzie, D. T., et al. 1989. IL-5-dependent conversion of normodense human eosinophils to the hypodense phenotype uses 3T3 fibroblasts for enhanced viability, accelerated hypodensity, and sustained antibody-dependent cytotoxicity. *J. Immunol.* 143:2311–16

63. Seidman, R. J., Kaufman, L. D., Sokoloff, L., Miller, F., Iliya, A., et al. 1991. The neuromuscular pathology of the eosinophilia-myalgia syndrome. *J. Neuropathol. Exp. Neurol.* 50:49–62

64. Seltzer, S. 1985. Pain relief by dietary manipulation and tryptophan supplements. *J. Endodontics* 11:449–53

65. Silver, R. M., Heyes, M. P., Maize, J. C., Quearry, B., Vionnet, F. M., et al. 1990. Scleroderma, fasciitis, and eosinophilia associated with the ingestion of tryptophan. *New Engl. J. Med.* 322: 874–81

66. Silver, R. M., Sutherland, S. E., Carreira, P. E., Heyes, M. P. 1992. Alteration of tryptophan metabolism in the toxic oil syndrome and in the eosinophilia-myalgia syndrome. *J. Rheumatol. In press*

67. Slutsker, L., Hoesly, F. C., Miller, L., Williams, L. P., Watson, J. C., et al. 1990. Eosinophilia-myalgia syndrome associated with exposure to tryptophan from a single manufacturer. *J. Am. Med. Assoc.* 264:213–17

68. Smith, M. J., Mazzola, E. P., Farrell, T. J., Sphon, J. A., Page, S. W., et al. 1991. 1,1'-Ethylidenebis(L-tryptophan), structure determination of contaminant "97" implicated in the eosinophilia myalgia syndrome (EMS). *Tetrahedron Lett.* 32:991–94

69. Smith, S. A., Roelofs, R. I., Gertner, E. 1990. Microangiopathy in the eosinophilia-myalgia syndrome. *J. Rheumatol.* 17:1544–50

70. Sternberg, E. M., Hill, J. M., Chrousos, G. P., Kamilaris, T., Listwak, S., et al. 1989. Inflammatory mediator-induced hypothalamic-pituitary-adrenal axis activation is defective in streptococcal cell wall arthritis-susceptible Lewis rats.

Proc. Natl. Acad. Sci. USA 86:21374–78

71. Sternberg, E. M., Van Woert, M. H., Young, S. N., Magnussen, I., Baker, H., et al. 1980. Development of a scleroderma-like illness during therapy with L-5-hydroxytryptophan and carbidopa. *New Engl. J. Med.* 303:782–87

72. Strongwater, S. L., Woda, B. A., Yood, R. A., Rybak, M. E., Sargent, J., et al. 1990. Eosinophilia-myalgia syndrome associated with L-tryptophan ingestion. Analysis of four patients and implications for differential diagnosis and pathogenesis. *Arch. Intern. Med.* 150:2178–86

73. Strumpf, I. J., Drucker, R. D., Anders, K. H., Cohen, S., Fajolu, O. 1991. Acute eosinophilic pulmonary disease associated with the ingestion of L-tryptophan-containing products. *Chest* 99:8–12

74. Swygert, L. A., Maes, E. F., Sewell, L. E., Miller, L., Falk, H., et al. 1990. Eosinophilia-myalgia syndrome. Results of national surveillance. *J. Am. Med. Assoc.* 264:1698–1703

75. Tabuenca, J. M. 1981. Toxic-allergic syndrome caused by ingestion of rapeseed oil denatured with aniline. *Lancet* 2:567–68

76. Tazelaar, H. D., Myers, J. L., Drage, C. W., King, T. J., Aguayo, S., et al. 1990. Pulmonary disease associated with L-tryptophan-induced eosinophilic myalgia syndrome. Clinical and pathologic features. *Chest* 97:1032–36

77. Toyo'oka, T., Yamazaki, T., Tanimoto, T., Sato, K., Sato, M., et al. 1991. Characterization of contaminants in EMS-associated L-tryptophan samples by high performance liquid chromatography. *Chem. Pharm. Bull.* 39:820–22

78. Varga, J., Heiman-Patterson, T. D., Emery, D. L., Griffin, R., Lally, E. V., et al. 1990. Clinical spectrum of the systemic manifestations of the eosinophilia-myalgia syndrome. *Semin. Arthritis Rheum.* 19:313–28

79. Varga, J., Peltonen, J., Uitto, J., Jimenez, S. 1990. Development of diffuse fasciitis with eosinophilia during L-tryptophan treatment: demonstration of elevated type I collagen gene expression in affected tissues. A clinicopathologic study of four patients. *Ann. Intern. Med.* 112:344–51

80. Verity, M. A., Bulpitt, K. J., Paulus, H. E. 1991. Neuromuscular manifestations of L-tryptophan-associated eosinophilia-myalgia syndrome: a histomorphologic

analysis of 14 patients. *Hum. Pathol.* 22:3–11

81. Weller, P. F. 1989. Eosinophils and fibroblasts: The medium in the mesenchyme. *Am. J. Respir. Cell Mol. Biol.* 1:267–68

82. Weller, P. F. 1991. The immunobiology of eosinophils. *New Engl. J. Med.* 324:1110–18

83. Yamaoka, K., Miyasaka, N., Kashiwazaki, S. 1991. L-tryptophan (TRP) containing contaminant "peak E" induces interleukin-5 from human T lymphocytes. Its possible role in the pathogenesis of eosinophilia-myalgia syndrome (abstract). *Arthritis Rheum.* 34:S131 (Suppl.)

84. Young, S. N., Chouinard, G., Annable, L. 1981. Tryptophan in the treatment of depression. *Adv. Exp. Med. Biol.* 133: 727–37

Annu. Rev. Nutr. 1992. 12:257–78

NUTRITIONAL QUESTIONS RELEVANT TO SPACE FLIGHT[1]

Helen W. Lane

Biomedical Operations and Research Branch, NASA–Johnson Space Center, Houston, Texas 77058

Leslie O. Schulz

Department of Health Sciences, University of Wisconsin at Milwaukee, Wisconsin 53201

KEY WORDS: body composition, energy requirements, nutrient requirements

CONTENTS

INTRODUCTION

Space exploration represents a new frontier in the nutritional sciences, where the research questions are evolving with the duration and complexity of the

---

missions. Humans have been eating in space since cosmonaut Yuri Gagarin's 108-minute flight in 1961. Americans were first propelled into orbit in the Mercury Program (1963), but given its relatively short duration (34 hours for the longest flight), nutritional concerns were minor and nutrients provided as dry, bite-sized, and tubed foods sufficed (37). The Gemini Program (1965–1966) brought forth new issues as mission durations extended to two weeks and the heavy physical demands of extravehicular activities were introduced. In addition to the need for oxygen and water, innovative technology was required to provide for the packaging, preserving, and storing of food. (See Table 1 for summary of American space program.)

The longer-duration flights of the Apollo Program (1968–1973) provided an opportunity to study in greater depth a number of physiological changes, including cardiovascular deconditioning and bone demineralization, that had been identified in the Gemini Program (15, 43). Also, vestibular disturbances were added to the inventory of significant physiological findings incident to space flight; previously, no American astronaut had reported symptoms of space motion sickness (7). During the Apollo 15, 16, and 17 flights, balance studies including energy intake and fecal losses, body weight changes, biochemical analyses, and body volume measurements were conducted. Although meal menus and food selections were expanded in terms of variety and food acceptability was evaluated, crude analysis of the food indicated that energy intake in flight was less than optimal, as expressed by loss of body fat and body mass (40, 44).

The three Project Skylab missions (1973–1974) of 28, 59, and 84 days duration offered further opportunities to study the energy and nutrient requirements of astronauts. Detailed metabolic studies in Skylab provided important additional information regarding the cardiovascular, musculoskeletal, hematologic, vestibular, and endocrine alterations experienced in flight. These balance experiments were coordinated with the Skylab food system consisting of 70 foods from which the crew could select their in-flight diets. Food types included frozen, thermostabilized, and freeze-dried foods, and menus were planned in 6-day cycles. The extensive food needs of the 84-day mission led to the inclusion of a high-calorie food bar stowed in the command module to augment energy intake (18).

Table 1   History of manned US space flights

Year	Flight program	Flight length
1961–1963	Mercury	15 min–34 hours
1965–1966	Gemini	5 hours–14 days
1968–1972	Apollo	5–13 days
1973–1974	Skylab	28, 59, and 84 days
1981–present	Space Shuttle	4–10 days

The space shuttle, the principal component of the Space Transportation System, had its inaugural orbital flight in 1981. Flights of the world's first reusable spacecraft thus far have been of limited duration (4–10 days) but have included studies related to nutrition and metabolism. From the time of the Mercury flights to the present shuttle, there has been steady progress in the development of space food systems. Shuttle foods comprise menus geared to the personal preferences of each crewmember and consist primarily of commercially available food items.

Plans for the immediate future include the US space station *Freedom,* where optimal nutrition will be essential for maintaining crew health, productivity, and morale. It is estimated that space station crews will consist of four to six individuals whose tours of duty will be 90 to 180 days. Data from bed rest studies (45) suggest that the number of 90-day tours an astronaut would be permitted is influenced by the possibility that bone and muscle losses may be irreversible, as well as by the danger of exposure to total radiation doses of up to 10–15 rems over a 90-day period.

And the challenge is only beginning. A lunar base, and eventually manned flights to Mars, will require missions lasting up to three years in duration. Thus, the negative physiological effects of microgravity must be overcome by appropriate countermeasures. In addition, the logistical problems associated with providing sufficient food and fluids during long missions will be addressed by the use of regenerative life support systems, which will include small hydroponic gardens. Human nutrient requirements in space therefore must be quantified and the best means of providing those nutrients identified. Overall, to ensure the safety of the crew and the success of longer space missions, the following nutritional questions must be satisfactorily addressed: (*a*) What role does nutrition play in counteracting the major physiological effects of microgravity? (*b*) How does the space environment influence total energy needs and the requirements for specific macro- and micronutrients? and (*c*) What is the most effective system to meet nutritional needs in space?

NUTRITION AS A COUNTERMEASURE TO EFFECTS OF MICROGRAVITY

Methods of Investigation

To determine whether nutritional modifications can play a role in countering some of the major physiological effects of microgravity (see Table 2), it will be necessary to understand the mechanisms involved as humans adapt to the space environment. Technically, to do so would involve observing the process of change in its entirety, without intervention. However, it is not possible to avoid intervention in studies of human responses to the stresses of space flight because medical personnel are responsible for ensuring the health, well-being, and optimum performance of the crews under observation. As a

Table 2 Microgravity effects on the body

Space motion sickness	Experienced by 60–70% of astronauts and cosmonauts; produces malaise, headache, anorexia, nausea, and vomiting. Symptoms appear early in flight and last about 2–7 days.
Cardiovascular deconditioning	Cephalad shift of fluid estimated at 1.5 to 2.0 liters from the lower extremities. Decreased orthostatic tolerance; increased heart rate, decreased pulse pressure, tendency toward spontaneous syncope.
Hematological changes	Reduction in plasma volume and red blood cell mass.
Bone mineral loss	Skeletal changes and loss of total body calcium have been observed in both humans and animals who have flown from one week to more than 237 days in space.
Muscular deconditioning	Loss of lean tissue and decreased muscle strength.

result, some type of intervention is invariably required. In addition, data collection in flight is extremely limited with respect to sample size—only a small number of individuals have flown in space. Plus, the extensive demands on crew time and, in most instances, the relatively short duration of missions limit opportunities for scientific observation.

Because of these limitations, changes in vestibular and cardiovascular function, hematological indices, and the effects of space radiation and microgravity on basic biological processes are being studied through complemental in-flight and ground-based research. A number of simulations and analogs for microgravity have been developed for both human and animal studies on ground (see Table 3). By allowing tighter control over conditions and longer observation periods, ground-based models have provided a great deal of information regarding the adaptation of humans to microgravity.

Body Composition

Several of the physiological changes associated with space flight manifest themselves in changes in body composition. Body composition is classically compartmentalized into fat mass and fat-free mass; the latter includes muscle, organs, blood and bones. Space flight presents a unique challenge for quantifying body composition changes since fluid, bone, muscle, and adipose tissue levels all vary independently of one another in space, and body weight loss does not follow classical patterns (see below).

Body mass has been measured during space flight using the device depicted in Figure 1. The first of these body-mass measurements, taken during 28- to

Table 3 Ground-based research models

Method	Explanation	Selected references
Humans		
Bed rest	The most widely used analog for microgravity in human studies—although not a true simulation, since gravity is still present—does induce many physiological changes similar to space flight.	35, 46
Head-down tilt (−4° to −12°)	To elicit some of the early physiological effects of microgravity with greater fidelity.	19, 21, 35
Water immersion	Used mostly for studies of acute renal and circulatory changes associated with space flight. Other methods are usually preferred because problems with maintaining hygiene and precise thermal control, and skin maceration impair feasibility of long-term studies.	9, 11
Dry immersion	Used by Soviet scientists to address limitations of water immersion; subjects are protected from water contact by a thin, plastic sheet.	10
Animals		
Partial or whole-body casts—confinement in small cages	Hypokinesia is produced by immobilization; particularly useful for studies of circulatory dynamics and the musculoskeletal system.	6, 23
Hind-limb suspension	To more accurately reproduce events in space while minimizing problems associated with immobilization.	4
In vitro	Study of cells by means of rotation.	5
Mathematical models	Can be used to generate computer simulations.	28

84-day Skylab missions, revealed 0.91 kg to 3.64 kg losses of preflight body weight ($n = 9$) (see Figure 2). Analysis of the components of that weight loss was based on both direct whole-body measurements and on indirect metabolic balance data. A conclusion from that analysis was that more than half of the weight loss was derived from fat-free mass and the remainder from fat stores (25, 27). About half the total weight loss that occurred within the first two days of flight was due to water loss. The remaining, later loss occurred more gradually over the duration of the missions and was attributed to both fat and protein depletion. All studies of fluid balance during microgravity have indicated a decrease in total body fluids of approximately 500 to 900 ml (26).

Figure 1   Measuring body mass in space during a Space Shuttle mission (NASA photo #S40-205001).

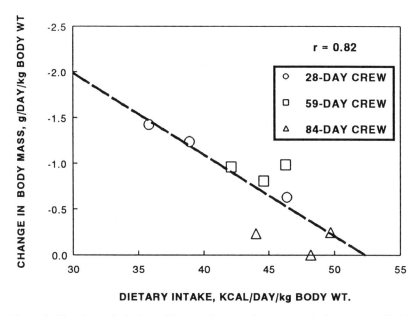

Figure 2   The change in body weight as a function of energy intake for the nine Skylab crewmembers.

This decrease is attributed in part to a 6 to 13% decrease in plasma volume found at landing (16), but it also involves a flight-induced loss of lean body mass (25).

Atrophy of skeletal muscles, especially those used for locomotion, maintaining posture, and counteracting gravity on Earth, occurs during space flight (14). This is reflected by reductions in muscle volume, mass, strength, exercise capacity, and neuromuscular coordination. Disturbances in protein turnover are suggested by the combined observations of skeletal-muscle atrophy, marginal or negative balances of nitrogen and potassium, and a persistent rise in urinary excretion of nitrogen and 3-methylhistidine (14).

Bone loss appears to progress in rough proportion to mission length; the greatest loss determined to date is about 20% (45). In the Skylab 4 mission, the negative calcium balance reached nearly 300 mg per day by flight day 84 (42). Both compact and trabecular bone are lost from the os calcis (heel); the level of loss and rate of recovery vary between sites, from complete recovery of bone mass in the calcaneus to continued deficits in the spine measured up to 6 months after landing. The documented negative calcium balances due to increased urinary and fecal calcium suggest bone resorption during microgravity exposure. Similar losses of bone calcium have been observed during horizontal bed rest (47). In one 19-week bed rest study (13), calcitonin therapy neither prevented negative calcium and phosphorus balances nor increased serum levels of parathyroid hormone (PTH). However, supplementing calcium and phosphorus intake (total intakes 1800 mg per day and 3000 mg per day, respectively) resulted in calcium and phosphorus balances significantly less negative than those in control subjects. Further, these mineral supplements reduced urinary excretion of hydroxyproline, which is typically elevated during bed rest. The apparent lack of hormonal control over the bone demineralization process during bed rest in this study agrees with other observations that neither PTH nor 1,25-dihydroxyvitamin D_3 are activated during bed rest (H. Lane, unpublished data). Because the artificial lighting systems on the US space shuttle and the Soviet space station *Mir* do not activate vitamin D, this vitamin will probably have to be provided through the diet in order to maintain appropriate levels in crewmembers.

The problem of reduced plasma volume upon reentry may be partially alleviated by water and electrolyte replenishment shortly before landing. Both the US and USSR space programs recommend that crewmembers prepare for landing by ingesting approximately one liter of drinking water made isotonic after ingestion by simultaneously swallowing coated tablets of sodium chloride. Postflight studies have indicated that this practice significantly decreases the degree of cardiovascular deconditioning as shown by the postflight blood pressure and heart rate response to a passive standing test (3). Additional fluids and electrolytes are given to returning crewmembers within the first 2 hours after landing in order to further reduce the body fluid volume deficit (34).

Nutrition and Exercise

Appropriate nutrition combined with exercise is likely to be the most effective countermeasure for the changes associated with body composition. In-flight exercise combined with sufficient energy intake is the primary countermeasure being tested against muscle atrophy. An increase in exercise level was imposed during each of the Skylab missions (0.5, 1.0, and 1.5 hours per day), and successive improvements were seen after flight with respect to indicators of muscular deconditioning (50). During the first manned Skylab mission only the bicycle ergometer was used (Figure 3). During the second mission, isokinetic devices were added, and during the last and longest mission daily use of a treadmill was included. Leg-muscle strength declined after both the first and second Skylab missions, but significant improvements were noted in all crewmembers after the last mission. The effect of space flight on precise measures of muscle strength, endurance, and exercise capacity is being

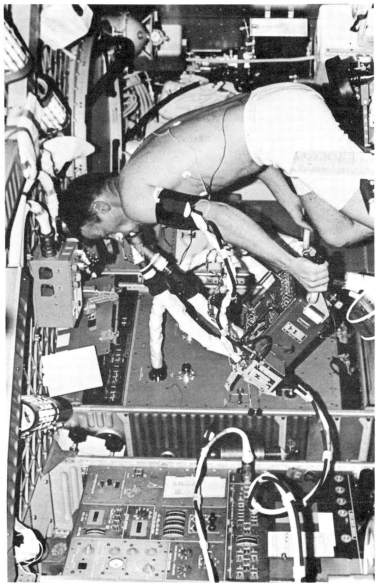

Figure 3 A Skylab crewmember exercising on the cycle-ergometer during flight (NASA photo #SL2-3-227).

studied actively at this time in the US Space Shuttle Program in order to protect crews' ability to leave the vehicle promptly, without help, after landing.

Weight-loading exercise and artificial gravity conferred by centrifugation designed to counteract the loss of gravitational and muscular stress are being investigated for their potential to limit the negative effects of space flight on the skeleton as well. However, Skylab results and preliminary data from the Soviet space program suggest that bungee cords and other devices simulating weight-bearing are not effective in preventing skeletal mineral losses during space flight (45).

Because aerobic exercise can also prevent cardiovascular deconditioning, customized in-flight aerobic exercise protocols are also being developed for use on shuttle flights. Maintenance of adequate balances of energy, protein, fluids and electrolytes assumes increasing importance because of mandatory strenuous exercise during flight.

NUTRITIONAL REQUIREMENTS IN SPACE

Energy

Energy deficits and body weight losses that were easily sustained on short missions cannot be tolerated on missions of long duration. The original hypothesis that life in microgravity would significantly decrease energy requirements has not held true (29). Whereas it was logical to assume that physical activity in a weightless environment would require less energy than at 1 g, total energy requirements are determined by numerous factors other than those associated with counteracting the force of gravity. Space flight can be expected to influence several of these factors. For example, the loss of lean body mass associated with weightlessness would be expected to decrease basal metabolic rate—yet energy utilization does not appear to decrease. The stress of changes in environmental factors such as humidity, pressure, and temperature may also play a role. In addition, the extent of physical activity may well be altered by the restricted area available or by the demands of extravehicular activity.

Weight loss is one of the most consistent findings in astronauts returning from space flight. However, it is important to differentiate whether weight losses related to space travel occur as a result of decreased energy intake, increased energy expenditure, or weightlessness per se. Energy intake for all Apollo astronauts averaged about 25 kcal/day per kg body weight (40). Weight changes following recovery indicated that each man lost about 0.15 kg of fat each day in flight, for an average deficit of about 19 kcal/day per kg body weight. The energy intake of astronauts under ground-based conditions and during hypobaric exposure indicated an energy requirement that was not significantly different from the in-flight requirement when adjusted for weight loss (40).

Measurements of food consumption on Gemini (2), Apollo (40), and the Russian flights of Vostok, Voskhod, and Soyuz (52) have all indicated considerably diminished energy intake in flight. Since it has generally been the case that more food was available in flight than was actually consumed, low intake cannot be explained by lack of availability. Space motion sickness, with its associated symptoms such as stomach awareness, nausea, and vomiting, undoubtedly plays a role. Yet, while these symptoms are common reactions to weightlessness, they are not experienced universally and some crewmen who reported no problems lost weight while assuming that their energy intake was adequate. If these symptoms are not the only reason crewmembers restrict their food intake, possibly the mechanisms by which the body regulates energy intake in proportion to energy expenditure are affected by weightlessness. Energy utilization should therefore be quantified in space through the use of techniques such as excretion of doubly labeled water, calorimetry, and computer-controlled diet records.

A detailed examination of energy balance was conducted during Apollo 17 (17). The urine and fecal excretions of the crewmembers were collected and returned to Earth for analysis, with records of fluid and dietary intake to allow estimation of the energy and fluid balance of each crewmember during the mission. The liquid intake-output data were considered with other methods of measuring body fluid volumes to determine whether body weight loss was a result of negative water balance or negative energy balance. The results indicated that water loss could not be the only cause of the weight loss; body tissue had been lost as well. Unlike bed rest, loss of adipose tissue predominated over loss of lean body mass (27).

The detailed nutritional studies performed during the three Skylab missions included analysis of detailed dietary records in combination with metabolic balance studies (32). The energy levels of Skylab menus were determined directly using bomb calorimetry and indirectly by calculating from crewmembers' records the amounts of carbohydrate, protein, and fat consumed. Energy levels in feces were determined by bomb calorimetry, and these data were used to calculate energy availability. Apparent metabolizable energy intake was determined by energy in the diet minus energy found in urine and feces.

Less body weight was lost as mission duration—and energy intake—increased from the first through the third Skylab mission (see Figure 2). Weight loss and body composition data from Skylab support the notion that body fat losses result mainly from negative energy balance while lean body mass and its fluid, muscle, and electrolyte constituents are reduced during space flight primarily as a result of weightlessness. Since the in-flight diet was intentionally increased on each Skylab mission, along with the time devoted to exercise, energy intake varied dramatically among the nine crewmen (35.8 to 49.7 kcal/day per kg body weight). Five of the subjects decreased their mean energy intake during flight (compared to preflight intake), and these subjects lost 75% more weight than the other four crewmen who increased or maintained their diet at preflight levels. Body fat losses were greatest for those crewmen who had the lowest energy intake. Changes in dietary sodium, potassium, and nitrogen between preflight and in-flight phases were insignificant and bore no relationship to the major tissue losses (24). However, in-flight intake of carbohydrate was higher (and fat intake lower) than preflight levels (in-flight carbohydrate 412 ± 60 g per day versus 363 ± 52 before flight; in-flight fat 83 ± 14 g per day versus 110 ± 10 before flight). These changes in intake, which probably resulted from differences in food choice among crewmembers, may have influenced energy utilization (24).

Protein

The only direct measures of nitrogen balance during flight were those taken during the Skylab missions. Like energy balance, nitrogen balance became progressively more negative during Skylab flights (24, 29). Urinary excretion of nitrogen, potassium, and 3-methylhistidine during flight indicated that muscle degradation may have been partially responsible for loss of lean body mass. Excretion of nitrogen in the feces, however, was no different in flight than before flight (24, 29). Muscle atrophy may account for some of the loss of body nitrogen, particularly in crewmembers who do not exercise. However, Skylab crewmembers who did exercise as well as consume high levels of energy and protein during flight still remained in negative nitrogen balance (24).

In attempts to clarify the causes of protein loss during flight, Soviet scientists are examining plasma levels of free essential and nonessential amino acids in cosmonauts after lengthy space flights (38, 39, 51). Plasma samples taken from male cosmonauts before flight, immediately upon landing, and later after flight showed decreased levels of plasma amino acids after flights of at least 75 days, with greater decreases found after longer flights. The plasma amino-acid levels in one crewmember who flew for 185 days reportedly dropped below normal levels. These decreases took place despite consumption of large amounts of proteins and essential amino acids (20) and regular in-flight exercise.

Biochemical studies in rats exposed to microgravity on shuttle flights and on the Soviet Kosmos missions (30, 31) have revealed increased blood levels of glucose, blood-urea nitrogen (BUN), creatinine, and cholesterol compared to levels observed in ground-based control animals. In addition, hepatic glycogen stores in the exposed rats showed a threefold increase.

However, access to animals after landing is frequently delayed for hours or days, and thus readaptation to normal gravity may be confounding these results. Elevated BUN and creatinine levels possibly may be related to a suspected decrease in renal clearance in flight (16). Also, high glycogen stores may stem from an abundance of free amino acid precursors released through protein turnover. In sum, some evidence, albeit indirect, exists to support the supposition that protein turnover may be altered in microgravity.

Stuart and co-workers (49) evaluated the effect of horizontal bed rest on whole-body protein synthesis using the [1-^{13}C]leucine constant-infusion method. In this method, [1-^{13}C]leucine is infused through an artery, and blood samples are collected at a vein. Blood levels of $^{13}CO_2$ (with ^{13}C from leucine) and [α-^{13}C]ketoisocaproic acid (KIC), the transamination product of leucine, can be used to calculate the rates of leucine oxidation and turnover (53). Thus, the calculated nonoxidative [1-^{13}C]leucine disappearance is equated with protein synthesis. Stuart found that bed rest, like microgravity, induced negative nitrogen balance despite consumption of 30 kcal and 0.6 g protein per kg of body weight per day, respectively. Bed rest increased leucine oxidation and decreased nonoxidative disappearance of plasma leucine, suggesting that bed rest may increase protein catabolism and suppress protein synthesis. However, when subjects were fed protein at levels comparable to those consumed during space flight (1.0 g protein per kg per day) (48), bed rest had no effect on leucine oxidation, nonoxidative leucine disappearance, or nitrogen status. However, astronauts still show evidence of negative nitrogen balance despite consuming up to 1.5 g of protein per kg of body weight and 40 to 50 kcal per kg of body weight during flight. It would be interesting to see whether bed rest with head-down tilt, which simulates fluid shifts and cardiovascular changes similar to those seen in microgravity, also affects measures of protein metabolism.

Fluids, Electrolytes, and Minerals

One of the most well-described effects of microgravity upon humans is the redistribution of body fluids from the lower extremities to the central torso and head. Head-down bed rest induces similar fluid shifts (8, 11, 12, 36). These fluid shifts induce diuresis within 12 to 24 hours: During the first day of space flight, total body water decreases up to 3%, total blood volume decreases 3 to 11%, and plasma volume decreases 2 to 9% (16). Neurohormonal responses to signals from the atrial and arterial pressure receptors have been reported to produce negative fluid balance in head-down bed rest (36). Fluid intake has been reported to decrease during both US and Soviet space flights, suggesting that the normal thirst mechanisms may not be operating. The combined effects of fluid shifts, diuresis, decreased thirst, and resultant increases in urine concentration all suggest that requirements for fluid and electrolyte intake in space deserve careful study.

After 24 hours in flight, levels of sodium, phosphate, and calcium salts in the urine are elevated in a manner similar to that in people with a history of renal-stone formation (41). Sodium intake tends to be high in flight, since most space foods are processed (48). Maintaining adequate hydration in flight thus becomes important for this reason as well. Crewmembers in the US space program are encouraged to consume from 1500 to 2000 ml of fluids daily, despite lack of thirst; consumption of foods high in fluid content is also encouraged.

The potential for cardiac arrhythmias associated with low potassium levels has prompted careful control of potassium in the space diet; adequate intake has been documented since the second Skylab mission (48). Other minerals of interest include those associated with bone demineralization and erthyropoiesis; the two most-studied have been zinc and iron. A 5-week period of horizontal bed rest induced a significant negative zinc balance, with increasing excretion of zinc in the urine and feces over time; interestingly, calcium balance was not affected in this study (22). The increasing excretion of zinc in the presence of consistent zinc intake is suggestive of bone demineralization and implies that long space flight may also compromise zinc status and balance.

As for iron, space flight consistently induces a 15 to 25% reduction in red blood cell mass (34). As might be expected, serum ferritin levels increased, presumably because less iron is needed for erythrocyte synthesis (unpublished data); possibly, the iron stored as ferritin during flight is available for red blood cell formation upon landing. Apparently, iron intake required during space flight may be lower than that required on Earth; this supposition needs to be explored further.

Vitamins

Although space diets generally are believed to provide adequate macronutrients, little is known about the vitamin content of space foods, although efforts are made to meet minimal nutrient requirements (see Table 4). Ongoing attempts to quantify vitamin levels in the highly processed US space diets suggest that vitamin A intake in space may exceed the usual recommended dietary allowances. On the other hand, space foods contain very little vitamin D; foods higher in this essential vitamin are being developed for longer flights. Maintaining sufficient folic acid and riboflavin levels is also a concern given the lack of fresh vegetables, milk, or most milk products in present-day space diets.

Still less is known of how the physiological manifestations of space flight in humans influence vitamin requirements. The stresses associated with space

Table 4 Minimum nutrient levels supplied by Space
Shuttle menus[a]

Nutrient	Amount	Nutrient	Amount
Kcal	2800	Vitamin A	5000 IU
Protein	56 gm	Vitamin D	400 IU
Calcium	800 mg	Vitamin E	15 IU
Phosphorus	800 mg	Ascorbic acid	45 mg
Sodium	150 mEq	Folacin	400 μg
Potassium	70 mEq	Niacin	18 mg
Iron	18 mg	Riboflavin	1.6 mg
Magnesium	350 mg	Thiamin	1.4 mg
Zinc	15 mg	Vitamin B_6	2.0 mg
		Vitamin B_{12}	3.0 μg

[a] Adapted from Ref. 44.

flight, including increased potential exposure to radiation, are expected to affect the turnover and thus the daily requirements for some water-soluble, and perhaps some fat-soluble, vitamins. For example, the Food and Nutrition Board of the National Research Council recently amended their recommendation of 60 mg of ascorbic acid per day to include at least 100 mg per day for people who smoke (33). Cosmonauts were shown to increase their urinary output of ascorbic acid during preflight training in hypobaric chambers (20). NASA is using these and other available data as a base for establishing micronutrient requirements for humans during space missions, with the present goal of protecting micronutrient status during 90- to 180-day tours.

Although many micronutrients can be provided by vitamin-mineral supplements, NASA encourages the use of food as the primary source of nutrients, and uses supplements only when absolutely necessary. Natural foods contain essential nonnutritive substances such as fiber and carotenoids; in addition, consumption of natural foods also provides a sense of psychological well-being that will be important during long space missions. A means of providing low-fat dairy products, which are rich in vitamins A and D, riboflavin, zinc, and calcium, is also being sought for long flights. Provision and consumption of fresh fruits and vegetables is also encouraged.

SYSTEMS FOR NUTRIENT DELIVERY IN SPACE

After 30 years of space flight, it is clear that microgravity is not a limiting factor for the consumption of prepared foods. In both the US and USSR space food systems, the emphasis is on foods with the following characteristics: minimal in-flight preparation time, minimal waste, microbial safety under ambient storage conditions, and good taste as well as nutritional soundness

(Table 4). All foods are provided in individual portions in order to meet sanitation requirements as well as tight crew schedules. An example of the food system used in space is shown in Figure 4. A typical 5-day Space Shuttle menu is shown in Table 5. Most foods in both the US and Soviet programs are preserved through canning, freeze-drying, or thermovacuum packing. A great many types of foods are flown in both space programs, with the exception of alcohol (US), carbonated beverages (US and USSR), and foods that produce excessive crumbs or do not store well at ambient temperatures. Although crews' food preferences reflect cultural differences, crews of all nationalities like fresh foods; fresh fruits and breads are supplied when possible on brief missions or during resupply to the Soviet space station. The major limiting factors in space food systems are the lack of electrical power for storage and cooking (refrigeration and freezers are usually not available), and constraints on stowage space and weight. These constraints, as well as the difficulties associated with manipulating objects in microgravity, preclude preparing food in bulk during flight for future consumption. Nonetheless, a great many foods can be provided for consumption in space despite these operational constraints.

As the space program extends the human presence farther from Earth, economic, logistic, and safety considerations will demand the increasing use of bioregenerative processes in the environmental control systems. Water and oxygen, abundant on Earth, become precious in space and must be recycled. Although high-quality air and water can be produced and recycled effectively using present-day technology, efficient chemosynthesis of food still lies ahead. During long flights and extended habitation in space, crewmembers will depend increasingly upon the biosynthesis of food; early diets will be primarily vegetarian. Providing nutritious, palatable foods within an overall regenerative life support system presents an exciting challenge.

CONCLUSIONS

The major nutritional questions relevant to space flight include identifying the role of nutrition in countering the deleterious physiological effects of microgravity, quantifying the requirements for macro- and micronutrients, and establishing the most effective system to meet nutritional needs in space (1). With respect to countermeasures designed to ensure crew safety and health, the first step is the determination of energy requirements. These investigations must take into consideration the working and living conditions in space plus the design of exercise protocols effective in preventing cardiovascular and muscular deconditioning. Studies combining different exercise regimens with various energy intakes will be needed to identify optimal

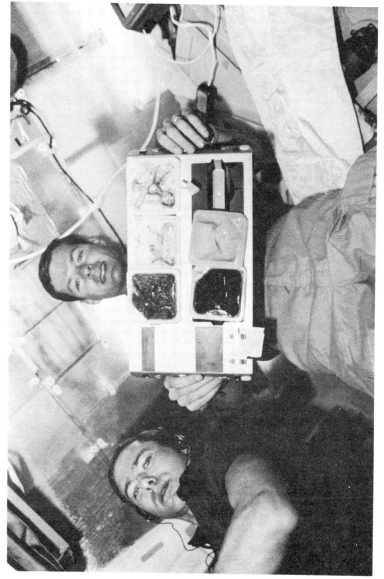

Figure 4   The Space Shuttle food system in use (NASA photo #61B-09-029).

Table 5 A sample 5-day menu for a Space Shuttle astronaut[a]

Day 1	Day 2	Day 3	Day 4	Day 5
Dried apricots (IM)	Strawberries (R)	Dried pears (IM)	Instant Breakfast (FF)	Bran flakes (R)
Sausage pattie (R)	Granola bar (NF)	Oatmeal w/brown sugar (R)	Dried apricots (IM)	Granola w/blueberries (R)
Scrambled eggs (R)	Breakfast roll (FF)	Cherry drink w/A/S (×2) (B)	Mexican scrambled eggs (R)	Orange drink (B)
Breakfast roll (FF)	Orange juice (B)		Cornflakes (R)	
Orange Drink (B)	Decaf coffee		Orange drink w/A/S (B)	
Decaf coffee	w/cream & sugar (B)		Cocoa (B)	
w/cream & sugar (B)				
Mushroom soup (R)	Shrimp cocktail	Macaroni & cheese (R)	Mushroom soup (R)	Tuna creole (T)
Cheddar cheese spread (T)	Rice & chicken (R)	Tomatoes & eggplant (T)	Peanut butter (T)	Tortilla (×2) (FF)
Rye bread (FF)	Tortilla (×2) (FF)	Pineapple (T)	Grape jelly (T)	Diced pears (T)
Vanilla pudding (T)	Crunchy peanut butter (FF)	Cashews (NF)	White bread (×2) (FF)	Candy-coated peanuts (F)
Macadamia nuts (NF)	Trail mix (IM)	Graham crackers (NF)	Fruit cocktail (T)	Orange-mango drink (×2)
Lemonade (×2) (B)	Tropical punch (×2) (B)	Grape drink w/A/S (×2) (B)	Candy-coated peanuts (NF)	(B)
Banana (FF)	Crackers, plain (NF)	Apple (FF)	Almonds (NF)	
			Lemonade w/A/S (×2) (B)	
Chicken consomme (R)	Spaghetti w/meat sauce (R)	Turkey tetrazzini (R)	Chicken a la king (T)	Chicken consomme (R)
Beef pattie (R)	Potatoes au gratin (R)	Rice pilaf (R)	Cauliflower w/cheese (R)	Ham (T)
Potato pattie (R)	Italian vegetables (R)	Italian vegetables (R)	Asparagus (R)	Asparagus (R)
Cauliflower w/cheese (R)	Tortilla (×2) (FF)	Blueberry yogurt (T)	Chocolate pudding (T)	Tortilla (×2) (FF)
Tortilla (FF)	Diced peaches (T)	Candy-coated chocolates	Grape drink w/A/S (×2) (B)	Diced peaches (T)
Brownie (NF)	Candy-coated chocolates	(NF)		Tapioca pudding (T)
Strawberry drink (B)	(NF)	Instant Breakfast (FF)		Shortbread cookies (NF)
	Tea/lemon w/A/S (×2) (B)	Tropical punch w/A/S (×3)		Orange-grapefruit drink (B)
		(B)		

[a] A/S = artificial sweetener, B = beverage, FF = fresh food, IM = intermediate moisture, NF = natural form, R = rehydratable, T = thermostabilized.

combinations. Additional studies must quantify energy expenditure during extravehicular activity.

Whether alternatives in the proportions of dietary carbohydrate, protein, or fat play a role either in metabolic efficiency in space or in preservation of optimal body composition remains to be determined. Little doubt, however, exists about the potential seriousness of muscular and skeletal changes associated with long-term exposure to microgravity. Given the complex nutrient interactions involved in maintaining these body compartments, this area of investigation demands a high priority. Additional studies are needed to determine whether protein synthesis or oxidation are affected by microgravity, and if so, the utility of interventions such as administering branch-chain amino acids must be investigated. Quantifying vitamin and mineral requirements in space will be especially important, since the nature of foods available to support missions of years in duration will probably depend heavily on genetically engineered and synthetic food sources.

The means of providing nutrients under these conditions will necessitate imagination and inventiveness on the part of nutrition scientists. Whereas nutrient needs for current missions are being met satisfactorily, the extended needs of the future will be far more challenging. Limitations on weight and storage capacity negate the possibility of sufficient provision of ready-to-eat foods. Regenerative food systems developed for closed or partially closed ecological support systems will provide novel sources of nutrients.

Many of the nutritional issues to be investigated in space will provide models for similar research on Earth. Space flight-induced changes in body composition, for example, may be useful in understanding the changes that occur with aging. Both aging and space flight invoke declines in muscle capabilities, muscle mass, and bone-mineral structure. If these losses can be countered effectively in space, similar methods might be explored for the elderly. Ample evidence exists of the similarities between bone demineralization in space and the early development of osteoporosis; collaborative efforts would strengthen research in both programs. Clinical studies of disease processes suggest that diet plays a central role in preventing muscle losses; ground-based and in-flight studies can "fill in the blanks" as to how nutritional interventions work. The role of iron and other nutrients in the formation of red blood cells, which is to be studied in space, will provide another model for elucidating the interactions among exercise, body composition, erythropoiesis, and menstruation status in women. Studies of how space flight affects the physiological requirement for vitamin C will be applicable to Earth-based situations. Task-specific studies of energy utilization planned for space may be useful in determining the energy requirements in paraplegia and quadriplegia. The nutritional information gleaned from spacecraft regenerative life-support systems will be useful in meeting the expanding nutritional needs of the world population. In summary, the space environment provides a unique opportunity to study perturbations in nutrient utilization and requirements in healthy people; the nutritional questions to be answered therein may also provide surprising answers for use at home.

ACKNOWLEDGMENTS

The editorial assistance of Christine Wogan of KRUG Life Sciences is acknowledged with thanks. Partial support for L. O. Schulz was provided by the NASA/American Society for Engineering Education Faculty Fellowship Program.

Literature Cited

1. Altman, P. L., Fisher, K. D., eds. 1986. Research opportunities in nutrition and metabolism in space. Prepared for NASA, Washington, D.C., under contract NASW 3924 by the Life Sci. Res. Off., Fed. Am. Soc. Exp. Biol., Bethesda, MD
2. Altman, P. L., Talbot, J. M. 1987. Nutrition and metabolism in space flight. *J. Nutr.* 117:421–27
3. Bungo, M. W., Charles, J. B., Johnson, P. C. 1985. Cardiovascular deconditioning during space flight and the use of saline as a countermeasure to orthostatic intolerance. *Aviat. Space Environ. Med.* 56:985–90
4. Caren, L. D., Mandel, A. D., Nunes, J. A. 1980. Effect of simulated weightlessness on the immune system in rats. *Aviat. Space Environ. Med.* 51:251–55
5. Cogoli, A., Valluchi-Morf, M., Mueller, M., Briegler, W. 1980. Effect of hypogravity on human lymphocyte activation. *Aviat. Space Environ. Med.* 51:29–34
6. Dickey, D. T., Billman, G. E., Teoh, K., Sandler, H., Stone, H. L. 1982. The effects of horizontal body casting on blood volume, drug responsiveness and +Gz tolerance in the rhesus monkey. *Aviat. Space Environ. Med.* 53:142–46
7. Dietlein, L. F. 1977. Skylab: a beginning. See Ref. 18, pp. 408–20
8. Fortney, S. M., Hyatt, K. H., Davis, J. E., Vogel, J. M. 1991. Changes in body fluid compartments during a 28-day bed rest. *Aviat. Space Environ. Med.* 62:97–104
9. Gauer, D. H. 1975. Recent advances in the study of whole body immersion. *Acta Astronaut.* 2:39
10. Gogolev, K. I., Aleksandrova, Y. A., Shul'zhenko, Ye. B. 1980. Comparative evaluation of changes in the human body during orthostatic (head-down) hypokinesia and immersion. NBD 311. *Fiziol. Cheloveka* 6:978–83 (From Russian)
11. Greenleaf, J. E. 1984. Physiological responses to prolonged bed rest and fluid immersion in humans. *J. Appl. Physiol: Respir. Environ. Exer. Physiol.* 57:615–33
12. Greenleaf, J. E., Shvartz, E., Keil, L. C. 1981. Hemodilution, vasopressin suppression, and diuresis during water immersion in man. *Aviat. Space Environ. Med.* 52:329–36
13. Hantman, D. A., Vogel, J. M., Donaldson, C. L., Friedman, R., Goldsmith, R. S., Hulley, S. B. 1973. Attempts to prevent disuse osteoporosis by treatment with calcitonin, longitudinal compression, and supplementary calcium and phosphate. *J. Clin. Endocrinol. Metab.* 36:845–58
14. Herbison, G. J., Talbot, J. M. 1984. Final Report Phase IV: Research opportunities in muscle atrophy. *NASA Contractor Report 3796.* Washington, DC: Nat. Aeronaut. Space Admin. Sci. Tech. Inf. Off. 95 pp.
15. Hoffler, G. W., Johnson, R. L. 1975. Apollo flight crew cardiovascular evaluations. See Ref. 18a, pp. 227–64
16. Huntoon, C. L., Johnson, P. C., Cintron, N. M. 1989. Hematology, immunology, endocrinology, and biochemistry. See Ref. 34, pp. 227–39
17. Johnson, P. C., Leach, C. S., Rambaut, P. C. 1973. Estimates of fluid and energy balances of Apollo 17. *Aerospace Med.* 44:1227–30
18. Johnston, R. S. 1977. Skylab medical program overview. In *Biomedical Results from Skylab, NASA SP-377*, ed. R. S. Johnston, L. F. Dietlein, pp. 3–19. Washington, DC: US Gov. Print. Off.

18a. Johnston, R. S., Dietlein, L. F., Berry, C. A., eds. 1975. *Biomedical Results of Apollo.* Washington, DC: Natl. Aeronaut. Space Admin.

19. Kakurin, L. I., Lobachik, V. I., Mikhailov, V. M., Senkevich, Yu. A. 1976. Antiorthostatic hypokinesia as a method of weightlessness simulation. *Aviat. Space Environ. Med.* 47:1083–86

20. Kalandarov, S., Markaryan, M. V., Sedova, Ye. A., Sivuk, A. K., Khokhlova, O. S. 1982. Diet of crew in Salyut-6 orbital station. *Kosm. Biol. Aviakosm. Med.* 16:10–13

21. Kotovskaya, A. R., Gavrilova, L. N., Galle, R. R. 1981. Effect of hypokinesia in head-down position on man's equilibrium function. *Space Biol. Aerospace Med.* 15:34–38

22. Krebs, J. M., Schneider, V. S., LeBlanc, A. D. 1988. Zinc, copper, and nitrogen balances during bed rest and fluoride supplementation in healthy adult males. *Am. J. Clin. Nutr.* 47:509–14

23. Kurash, S., Andzheyevska, A., Gurski, Ya. 1981. Morphological changes in different types of rat muscle fibers during long-term hypokinesia. *Space Biol. Aerospace Med.* 14:45–52

24. Lane, H. W. 1992. Critical review: metabolic energy requirements for space flight. *J. Nutr.* 102:13–18

25. Leach, C. S., Leonard, J. I., Rambaut, P. C. 1982. Dynamic of weight loss during prolonged space flight. *Physiologist* 22:S61–S62

26. Leach, C. S., Leonard, J. I., Rambaut, P. C., Johnson, P. C. 1978. Evaporative water loss in man in a gravity-free environment. *J. Appl. Physiol.* 45:430–36

27. Leonard, J. I. 1982. Energy balance and the composition of weight loss during prolonged space flight. General Electric Co. Tech. Rep. TIR-2114-MED-2012, Houston, Tex.

28. Leonard, J. I. 1985. Fluid-electrolyte responses during prolonged space flight: A review and interpretation of significant findings. General Electric Co. Tech. Rep. TIR-2114-MED-5008, Houston, Tex.

29. Leonard, J. I., Leach, C. S., Rambaut, P. C. 1983. Quantitation of tissue loss during prolonged space flight. *Am. J. Clin. Nutr.* 38:667–79

30. Merrill, A. H., Wang, E., Jones, D. P., Hargrove, J. L. 1987. Hepatic function in rats after space flight: effects on lipids, glycogen, and enzymes. *Am. J. Physiol.* 252: R222–26

31. Merrill, A. H., Wang, E., Mullins, R. F., Grindeland, R. E., Popova, I. A.

32. Michel, E. L., Rummel, J. A., Sawin, C. F., Buderer, M. C., Lem, J. D. 1977. Results of Skylab medical experiment M171-metabolic activity. See Ref. 18, pp. 372–87

33. National Research Council. 1989. *Recommended Dietary Allowances.* Washington, DC: Natl. Acad. Press. 285 pp. 10th ed.

34. Nicogossian, A. E. 1989. Countermeasures to space deconditioning. In *Space Physiology and Medicine,* ed. A. E. Nicogossian, C. L. Huntoon, S. L. Pool, pp. 294–311. Philadelphia: Lea & Febiger

35. Nicogossian, A. E., Dietlein, L. F. 1989. Microgravity: simulations and analogs. See Ref. 34, pp. 240–48

36. Nixon, J. V., Murray, R. G., Bryant, C., Johnson, R. L., Mitchell, J. H., et al. 1979. Early cardiovascular adaptation to simulated zero gravity. *J. Appl. Physiol: Respir. Environ. Exerc. Physiol.* 46:541–48

37. Popov, I. G. 1975. Food and water supply. In *Foundations of Space Biology and Medicine,* ed. M. Calvin, O. G. Gazenko, 3:22–55. Washington, DC: Natl. Aeronaut. Space Admin.

38. Popov, I. G., Latskevich, A. A. 1983. Effect of 140-day flight on blood amino acid levels in cosmonauts. *Kosm. Biol. Aviakosm. Med.* 17:23–30

39. Popov, I. G., Latskevich, A. A. 1984. Blood amino acids of cosmonauts before and after 211-day spaceflight. *Kosm. Biol. Aviakosm. Med.* 18:10–15

40. Rambaut, P. C., Heidelbaugh, N. D., Reid, J. M., Smith, M. C. 1973. Caloric balance during simulated and actual space flight. *Aerospace Med.* 44:1264–69

41. Rambaut, P. C., Johnson, P. C. 1989. Nutrition. See Ref. 34, pp. 202–13

42. Rambaut, P. C., Johnston, R. S. 1979. Prolonged weightless and calcium loss in man. *Acta Astronaut.* 6:1113–22

43. Rambaut, P. C., Smith, M. C. Jr., Mack, P. B., Vogel, J. M. 1975. Skeletal response. See Ref. 18a, pp. 303–22

44. Rambaut, P. C., Smith, M. C. Jr., Wheeler, H. O. 1975. Nutritional studies. See Ref. 18a, pp. 277–302

45. Schneider, V. S., LeBlanc, A., Rambaut, P. C. 1989. Bone and mineral metabolism. See Ref. 34, pp. 214–21

46. Shangraw, R. E., Stuart, C. A., Price, M. J., Peters, E. J., Wolfe, R. R. 1988. Insulin responsiveness of protein metabolism in vivo following bed rest in humans. *Am. J. Physiol.* 255:E548–58

47. Smith, M. C., Rambaut, P. C., Vogel, J. M., Whittle, M. W. 1977. Bone mineral measurement—Experiment M078. See Ref. 18, pp. 372–87
48. Stadler, C. R., Rapp, R. M., Bourland, C. T., Fohey, M. F. 1988. Space Shuttle food-system summary, 1981–1986. NASA Tech. Memo. 100469, Johnson Space Center, Houston, Tex.
49. Stuart, C. A., Shangraw, R. E., Peters, E. J., Wolfe, R. R. 1990. Effect of dietary protein on bed-rest-related changes in whole-body-protein synthesis. Am. J. Clin. Nutr. 52:509–14
50. Thornton, W. E., Rummel, J. A. 1977. Muscular deconditioning and its prevention in space flight. See Ref. 18, pp. 191–97
51. Ushakov, A. S., Vlasova, T. F., Miroshnikova, Ye. B., Mikhaylovs, V. M., Biryukov, Ye. N. 1983. Effect of long-term space flights on human amino acid metabolism. Kosm. Biol. Aviakosm. Med. 17:10–15
52. Yegorov, A. D. 1980. Results of medical studies during long-term manned flights on the orbital Salyut-6 and Soyuz complex. Presented during the 11th meeting of the Joint Soviet-American Working group on Space Biology and Medicine, Moscow. NASA Tech. Memo. 76014. Washington, DC: Natl. Aeronaut. Space. Admin.
53. Young, V. R., Yu, Y. M., Krempf, M. 1991. Protein and amino acid turnover using stable isotopes ^{15}N, ^{13}C, and ^{2}H as probes. In New Techniques in Nutrition Research, ed. R. G. Whitehead, A. Prentice, pp. 17–72. San Diego: Academic

Annu. Rev. Nutr. 1992. 12:279–98

HYPERHOMOCYST(E)INEMIA AS A RISK FACTOR FOR OCCLUSIVE VASCULAR DISEASE

Soo-Sang Kang, Paul W. K. Wong, and M. Rene Malinow

Department of Pediatrics, Rush Medical College and Presbyterian-St. Luke's Medical Center, Chicago, Illinois, 60612; Laboratory of Cardiovascular Diseases, Oregon Regional Primate Research Center, Beaverton, Oregon 97006

KEY WORDS: homocysteine, coronary artery disease, cerebrovascular disease, thromboembolism, arteriosclerosis, atherosclerosis

CONTENTS

Introduction

Homocysteine is metabolized by transsulfuration to cysteine via cystathionine or by remethylation to methionine. An excessive accumulation of plasma homocysteine and its derivatives is found in homocystinuria, which is caused by inborn metabolic defects in the transsulfuration or remethylation of homocysteine (18, 56). Patients with untreated homocystinuria develop premature occlusive vascular disease (18, 56). The major cause of morbidity and early mortality among these patients, irrespective of the simultaneous occur-

279

0199-9885/92/0715-0279$02.00

rence of hyper- or hypomethioninemia, is the development of thrombosis and thromboembolism, with resultant strokes, coronary occlusions, and other complications.

Pathological changes similar to those observed in arteriosclerosis have been described in homocystinuric patients (3, 43, 49), and the accumulation of homocyst(e)ine and its derivatives in tissue fluids appears to be the primary factor associated with pathological changes in the intimal cells (49, 56). Thus, it is important to determine whether or not a correlation exists between moderate hyperhomocyst(e)inemia and the development of arterial changes leading to vascular diseases in nonhomocystinuric subjects. Quantification of plasma cysteine-homocysteine disulfide after methionine loading was used initially to investigate this correlation. Subsequently, the determination of total homocyst(e)ine, including protein-bound homocysteine, introduced new insight into this area of research.

In this chapter we review the mechanisms that characterize moderate hyperhomocyst(e)inemia and the results of studies demonstrating the association of elevated plasma concentrations of homocyst(e)ine with the occurrence of occlusive vascular disease. We define the terminology used and describe the etiologies of various types of hyperhomocyst(e)inemias. The role of homocysteine as a risk factor in occlusive vascular disease is discussed based on three lines of evidence: premature thromboembolism and arteriosclerosis in severe hyperhomocyst(e)inemia; experimental production of arterial lesions in hyperhomocyst(e)inemic animals; and epidemiologic studies in patients with occlusive arterial diseases. In addition, homocysteine toxicity to endothelial cells is discussed.

Several reviews addressing the issue of moderate hyperhomocyst(e)inemia in occlusive vascular disease have been published within the last several years (6, 55, 56, 83, 85).

Total Homocyst(e)ine and Classification of Hyperhomocyst(e)inemia

In normal circulating plasma, most free homocysteine molecules are present as oxidized forms in the supernatant after acid precipitation. However, neither homocysteine nor homocystine is detectable in the plasma of nonhomocystinuric subjects by conventional amino acid analysis because concentrations are low. Sardharwalla et al (72) first observed the presence of cysteine-homocysteine disulfide after methionine loading in normal subjects and in subjects heterozygous for cystathionine β-synthase deficiency. Subsequently, the presence of cysteine-homocysteine disulfide without methionine loading was reported (28), but its quantification in nonhomocystinuric subjects is difficult owing to its low concentration.

In 1979, Kang et al (37) demonstrated that protein-bound homocysteine is the major form of plasma homocyst(e)ine in nonhomocystinuric subjects and that unbound forms convert to protein-bound homocysteine during storage of plasma, even at $-70°C$. Refsum and co-workers (66) have confirmed these findings by using radioenzymatic methods. Hence, plasma homocysteine and its derivatives are currently classified into free and protein-bound homocysteine. The latter accounts for 70–85% of total homocyst(e)ine in normal individuals. Free homocyst(e)ine includes homocysteine, homocystine, and cysteine-homocysteine disulfide. In plasma, albumin is the major protein that binds homocysteine, mostly through disulfide bonds between homocysteine and protein molecules. However, there is evidence that some homocystine and cysteine-homocysteine disulfide molecules are bound noncovalently to proteins (37).

Since homocyst(e)ine is a normal constituent of tissues and tissue fluids, the condition of increased homocyst(e)ine concentration is designated as hyperhomocyst(e)inemia in this review. In addition, moderate, intermediate, and severe hyperhomocyst(e)inemia are defined as basal values for plasma homocyst(e)ine less than 30, between 31 and 100, and above 100 nmol/ml, respectively. All values are expressed as homocysteine; thus homocystine values are converted to homocysteine, and cysteine-homocysteine disulfide is treated as equivalent to homocysteine. Finally, the condition previously known as homocystinuria is called severe hyperhomocyst(e)inemia.

Homocysteine Metabolism and Etiology of Hyperhomocyst(e)inemia

Homocysteine is formed by the cleavage of S-adenosylhomocysteine, which is produced from the versatile methyl donor S-adenosylmethionine. Normal rat liver contains more than 50 nmol per gram of wet tissue of S-adenosylhomocysteine and an approximately similar amount of S-adenosylmethionine (2, 71). In contrast, the concentration of homocysteine is usually less than one-tenth of S-adenosylmethionine or S-adenosylhomocysteine (i.e. < 5 μM). Homocysteine is readily transsulfurated to cysteine via cystathionine or is remethylated to methionine. In humans on a normal diet, approximately 50% of the available homocysteine is remethylated (57). The amount of methionine intake influences homocysteine synthesis and controls the ratio between transsulfuration and remethylation of homocysteine. The methyl group of methionine in humans on high protein diets disappears twice as fast as when they are on normal diets; and 70% of homocysteine is converted to cystathionine (21, 22). In contrast, only 10% of homocysteine is transsulfurated to cysteine in humans on low protein diets (21, 22).

The remethylation of homocysteine is carried out either by betaine-homocysteine methyltransferase or 5-methyltetrahydrofolate-homocysteine methyltransferase (methionine synthase). The former is found mainly in the liver, whereas the latter is distributed ubiquitously in all tissues. In humans, the rates of homocysteine remethylation by each of these two enzymes are approximately equal. S-adenosylmethionine and methionine are important modulators for the transsulfuration and remethylation of homocysteine. S-adenosylmethionine inhibits the activities of both betaine-homocysteine methyltransferase and methylenetetrahydrofolate reductase; it also activates cystathionine β-synthase (21, 22). Excess methionine intake increases the concentration of S-adenosylmethionine and also inhibits the activity of methionine synthase. Therefore one can speculate that methionine loading may help to evaluate the pathway of transsulfuration but may hinder the evaluation of homocysteine remethylation.

The interruption of transsulfuration or remethylation of homocysteine produces hyperhomocyst(e)inemia. Genetic defects, inadequate availability of cofactors or substrates, or other mechanisms can interfere with both processes. Severe hyperhomocyst(e)inemia, with an incidence of 1:200,000, is found in individuals that are homozygous for cystathionine β-synthase deficiency. Although some individuals that are heterozygous for cystathionine β-synthase deficiency have moderate hyperhomocyst(e)inemia, approximately 30–50% have normal plasma homocyst(e)ine concentrations (60, 73). Since pyridoxal 5'-phosphate is a coenzyme of cystathionine β-synthase, nutritional deficiency of pyridoxine may also cause hyperhomocyst(e)inemia. Hyperhomocyst(e)inemia has been demonstrated in experimental animals fed diets deficient in pyridoxine (75).

A severe genetic defect of methylenetetrahydrofolate reductase has been documented (18, 59). 5-Methyltetrahydrofolate and homocysteine are substrates for methionine synthesis, and the lack of 5-methyltetrahydrofolate produces severe hyperhomocyst(e)inemia. Homozygotes for a new type of mutation, thermolabile methylenetetrahydrofolate reductase, have recently been found in 5% of the general population (38, 41). Subjects with thermolabile methylenetetrahydrofolate reductase have a 50% reduction in enzyme activity and a tendency toward moderate hyperhomocyst(e)inemia. Some of these subjects exhibit intermediate hyperhomocyst(e)inemia resulting from compound heterozygosity of severe and thermolabile mutations (38).

A deficiency of 5-methyltetrahydrofolate is expected to result from any impairment of folate metabolism. Moderate and intermediate hyperhomocyst(e)inemia is most commonly seen in subjects with low concentrations of serum folate that are due to either inadequate nutrition or to certain medications (40, 66). Plasma homocyst(e)ine greater than 2 SD above the normal mean was observed in 61% of subjects with subnormal serum folate; 19% of those had intermediate hyperhomocyst(e)inemia. Moreover, 53% of the subjects with low normal serum folate (between 2.1 and 4.0 ng/ml) showed moderate hyperhomocyst(e)inemia (40).

Impaired methionine synthase activity is also caused by a deficiency of the cofactor methylcobalamin. Known genetic defects include abnormalities in intestinal absorption of cobalamin, transcobalamin II, cbl C, D, and cbl E, F, G (14, 19). Kang et al (39) have demonstrated the presence of intermediate hyperhomocyst(e)inemia in subjects with subnormal serum levels of B_{12}. A negative correlation is found between serum cobalamin levels and plasma total homocyst(e)ine. Elevation of homocyst(e)ine as a result of cobalamin and/or folate deficiency has been also reported by Brattström et al (8) and Stabler et al (77).

Normal plasma homocyst(e)ine has been observed in pregnant women who had subnormal folate and/or cobalamin levels (39), which suggests that increased betaine-homocysteine methyltransferase activity prevents the occurrence of hyperhomocyst(e)inemia. Plasma homocyst(e)ine levels in pregnant women are less than 60% of the normal mean (42), and this reduction appears to be associated with increased cortisol level (S.-S. Kang, unpublished results). Cortisol is a potent activator of betaine-homocysteine methyltransferase (20). Reduced homocyst(e)ine levels in premenopausal women have previously been interpreted as a result of the direct interaction of homocysteine metabolism with estrogen (7, 83). However, it is plausible that augmented betaine-homocysteine methyltransferase activity is responsible for maintaining low levels of plasma homocyst(e)ine in premenopausal women, as well as during pregnancy.

Hyperhomocyst(e)inemia may occur also through an interference of betaine-homocysteine methyltransferase. Although a genetic defect of this enzyme has not been reported, rats fed choline-deficient diets for 10 days exhibit increased homocyst(e)ine concentration in the spleen (80). The severity of hyperhomocyst(e)inemia that is due to folate deficiency is amplified by the concomitant restriction of choline in the diet (S.-S. Kang, unpublished data). Betaine supplementation effectively corrects most types of hyperhomocyst(e)inemia, irrespective of their etiology (76). Hyperhomocyst(e)inemia that is due to other diseases such as renal failure and certain medications has been discussed in the comprehensive review by Ueland & Refsum (83).

In summary, hyperhomocyst(e)inemia is caused by genetic and/or nutritional defects. Since plasma homocyst(e)ine levels are controlled by various genetic and nongenetic factors, the evaluation of moderate and intermediate hyperhomocyst(e)inemia requires a thorough assessment of parameters involved in homocysteine metabolism. Hence, identification of a genetic defect(s) plays an important role in detecting those subjects susceptible to the development of hyperhomocyst(e)inemia.

Vascular Disease in Severe Hyperhomocyst(e)inemia

Gibson et al (26) published the first pathological description of children with homocystinuria that was due to cystathionine β-synthase deficiency. Subsequent reports by other investigators confirmed the pathological findings (9, 10). Thromboembolism is a frequent cause of death. Thrombosis is observed in arteries of various sizes. The consequences of thrombosis depend on the size and locations of the arteries affected; infarctions may result. Venous thrombosis is also common and may occur in the vena cava, cerebral sinuses, and/or peripheral veins. Emboli from these thrombi may be found in various organs. Some thrombi may be organized and recanalized.

The arterial walls may be dilated, stretched thin, or thickened; and the arterial lumen may be narrowed. Aneurysms may also occur. Intimal hyperplasia and fibrosis are evident. The arteries are often affected in a patchy fashion. The internal elastic lamina is prominent and often thick or frayed. The most obvious pathological change is intimal fibrosis, which may be concentric, resulting in ridges and causing severe narrowing, or eccentric in the form of pads. In these intimal pads, layers of elastic fibrils may have split off from the main elastic lamina. Areas of atheromatous plaques may be found. Medial hyperplasia is often found. The muscle fibers may be enlarged and separated from each other. Deposits of interstitial material that have the staining properties of collagen are found. The external elastic lamina is often swollen and frayed. Generally, the adventitia is little affected.

Kanwar et al (43) were the first to provide a detailed description of the pathological findings in a 10-year-old child with methylenetetrahydrofolate reductase deficiency. A later study was reported by Baumgartner et al (3). The pathological changes are very similar to those observed in cystathionine β-synthase deficiency.

Pathologic findings in patients with inherited metabolic defects of cobalamin have been reported by McCully (49, 51) and Baumgartner et al (3). The vascular lesions are very similar to those observed in cystathionine β-synthase deficiency and methylenetetrahydrofolate reductase deficiency. McCully (49) implicated homocysteine as the etiologic factor for the premature vascular disease in these patients. He also suggested that hyperhomocyst(e)inemia might be a factor in the pathogenesis of arteriosclerosis (49, 51).

Experimental Hyperhomocyst(e)inemia and Vascular Damage

Early experiments were conducted by subcutaneous or oral administration of dl-homocysteine thiolactone in rabbits for 7 to 12 weeks (50). Although arteriosclerotic plaques were reported in the animals, these results should not be interpreted as the direct action of homocyst(e)ine. Exogenous homocysteine is readily metabolized or excreted through the kidney in a short period of time. Thus, this model would be unable to maintain an abnormal plasma homocysteine level.

Harker et al (31) used a continuous infusion method in baboons for three months. The ensuing endothelial desquamation and early proliferative lesions observed in certain arteries and the shortening of platelet survival supported the "reaction to injury" hypothesis of atherogenesis (31, 32). The arterial lesions were prevented by the antiplatelet agent dipyridamole, suggesting that platelets mediated intimal proliferation of smooth muscle cells (31). Reddy & Wilcken (64) found no arterial changes in young pigs infused with *dl*-homocysteine thiolactone over a two-month period. Because patients with severe hyperhomocyst(e)inemia usually exhibit 0.2 mM *l*-homocystine levels, the finding of less than 0.04 mM concentration of *l*-homocystine in their study was probably too low to produce vascular damage within two months. Species difference may also account for the discrepancy.

Hyperhomocyst(e)inemia has been induced in several species of animals fed diets deficient in folate (36), pyridoxine (75), or choline (80). Hence, the vascular pathology found in animals fed diets deficient in one of these nutrients may be interpreted as the result of nutritional deficiency or hyperhomocyst(e)inemia.

Prior to the discovery of hyperhomocyst(e)inemia, several investigators studied the development of arterial lesions in animals fed diets deficient in choline, pyridoxine, or folate. Hartcroft et al (33) observed pathological changes resembling those of atheroma in the major arterial trunks and coronary arteries of rats maintained up to seven months on diets low in choline. The initial lesions consisted of microscopic deposits of lipid in the endothelial cells. In later stages, proliferation of intimal cells took place and they appeared as plaques. In rhesus monkeys, pyridoxine deficiency caused lesions resembling arteriosclerosis in humans; similar lesions were also produced by feeding monkeys a "suboptimal" pyridoxine diet (67). Recently, hyperhomocyst(e)inemia and arteriosclerotic changes were observed in rats fed folic acid–deficient diets (88).

In summary, certain nutritional hyperhomocyst(e)inemias may be associated with the development of arteriosclerotic plaques, irrespective of which nutrient is involved. Hyperhomocyst(e)inemia is most likely the common denominator, although the contribution of the nutritional deficiency to the vascular pathology can not be excluded in these models. Since betaine supplement largely corrects any type of hyperhomocyst(e)inemia, deficiency of pyridoxine, folate, or cobalamin without hyperhomocyst(e)inemia can be established by concomitant betaine supplementation. Whether hyperhomocyst(e)inemia is an indispensable condition or nutritional deficiency is sufficient for the production of vascular damage may be clarified in these models.

Moderate Hyperhomocyst(e)inemia and Occlusive Arterial Disease

Only within the last 15 years have abnormalities in the metabolism of homocysteine been investigated in adults with occlusive arterial disease.

Wilcken & Wilcken (86) first demonstrated that abnormally elevated plasma cysteine-homocysteine disulfide levels after methionine loading were three times more common in patients with coronary artery disease than in controls. Similar observations were made by many investigators using various methods. These studies are summarized in Tables 1 and 2.

Before 1986, all studies were conducted by the determination of unbound homocysteine. Because of its low concentration in plasma, most studies during this period were performed after methionine loading. Brattström et al (5) reported, for the first time, comparisons of pre- and post-methionine loading plasma cysteine-homocysteine disulfide levels in patients with cerebrovascular disease. An approximately threefold greater level of plasma cysteine-homocysteine disulfide was noted in the patient group, both before and after methionine loading. Murphy-Chutorian et al (60) reported a similar positive association between plasma homocyst(e)ine values and coronary artery disease. In contrast, an elevated cysteine-homocysteine disulfide level was found by Boers et al (4) in patients with occlusive peripheral artery and cerebrovascular diseases but not in patients with myocardial infarctions. A recent report by Clarke et al (13) demonstrated abnormally high plasma cysteine-homocysteine disulfide and homocystine values after methionine loading in 30–40% of patients with cerebrovascular, peripheral, or coronary artery diseases.

After the discovery of protein-bound homocysteine, most investigators determined plasma total homocyst(e)ine without methionine loading as shown in Tables 1 and 2. Methionine loading does not increase the ratio of the patient homocysteine value to the control value. This is probably due to the fact that the inhibition of homocysteine remethylation is balanced by the activation of cystathione β-synthase following methionine loading. Although values of plasma total homocyst(e)ine are slightly different depending on the methods used, these studies arrived at the same conclusion (Table 1 and 2). In coronary artery disease, the ratio for mean homocyst(e)ine of patients to controls varied from 1.2 to 1.3, whereas in peripheral and cerebrovascular disease the ratio ranged from 1.5 to 1.8 (Table 1). The ratio of the incidence of hyperhomocyst(e)inemia in patients versus controls ranged from 2.6 to 5.3 in coronary artery disease and from 2.2 to 9 in cerebrovascular disease (Table 2). These results suggest a more pronounced correlation of hyperhomocyst(e)inemia with peripheral and cerebrovascular arterial disease than with coronary artery disease. When all available data on plasma total homocyst(e)ine from various investigators were taken together, hyperhomocyst(e)inemia was found in 41.8% of patients with peripheral and cerebrovascular arterial diseases and in 11.9% of those with coronary artery disease (44).

Since Sardharwalla et al (72) first observed elevated concentrations of plasma cysteine-homocysteine disulfide in subjects heterozygous for cystathionine β-synthase deficiency, investigators have speculated that hyperhomocyst(e)inemia found in patients with vascular diseases is exclusively due to heterozygosity for cystathionine β-synthase deficiency. Moreover, Boers et al (4) reported that 13 out of 14 patients with abnormal accumulation of cysteine-homocysteine disulfide had decreased enzyme activity that was comparable to the range of activity found in individuals that were obligate heterozygotes for cystathionine β-synthase deficiency. This proposition has been strengthened by the recent observations of Clarke et al (13), who confirmed the presence of cystathionine β-synthase deficiency in 18 of 23 hyperhomocyst(e)inemic patients with vascular disease. However, defining heterozygosity by enzyme activity determination is not a simple matter in many genetic disorders (38, 52, 73). It has been well established that several genetic and nongenetic factors, singly or in combination, may control plasma homocyst(e)ine concentration.

One significant challenge is posed by the results from a survey of families with cystathionine β-synthase deficiency. Mudd et al (58) found no increase of vascular disease in individuals heterozygous for cystathionine β-synthase deficiency. Since there is considerable overlap of homocyst(e)ine values between these obligate heterozygotes and controls, one can speculate that a positive correlation is confined to hyperhomocyst(e)inemic heterozygotes for cystathionine β-synthase deficiency.

Note that the distribution pattern of plasma homocyst(e)ine values in patients with coronary artery disease is similar to that of controls, except for the values above the 95th percentile (39, 87). This pattern is consistent with a multifactorial disorder. Reed et al (65) found a significantly greater intraclass correlation coefficient of homocyst(e)ine levels in monozygotic twins than in dizygotic twins, which suggests that the level of plasma homocyst(e)ine is influenced by genetic factor(s). A strong family correlation in homocyst(e)ine values was also observed by Williams et al (87) and Genest et al (24). High homocyst(e)ine values may be the result of a combination of a major genetic defect and other minor genetic and nongenetic factors.

Evidence indicates that moderately elevated plasma homocyst(e)ine level is an independent risk factor in the development of occlusive arterial disease (13, 24, 60). Other major risk factors including smoking, hypertension, diabetes mellitus, hyperlipidemia, and elevated levels of fibrinogen or apolipoprotein AI or B showed no positive correlation with hyperhomocyst(e)inemia in patients with vascular disease. However, Swift & Shultz (81) reported a higher plasma homocyst(e)ine value in a high risk group of patients than in a low risk group. Genest et al (23) also found that patients with four or more risk factors had higher homocyst(e)ine levels than patients with two risk factors.

The association of hyperhomocyst(e)inemia and coronary heart disease was

Table 1 Mean homocysteine levels (nmol/ml) in human plasma or serum of patients with vascular disease and controls[a]

Disease studied	Homocysteine determined as	Basal levels			Postmethionine loading levels			Refs.
		Controls	Patients	Ratio	Controls	Patients	Ratio	
Coronary artery disease	Cysteine-homo-cysteine disulfide				6 ± 2	18 ± 8	3.0	86
Arteriosclerotic cere-brovascular disease	Cysteine-homo-cysteine disulfide	3.5 ± 0.2	5.1 ± 0.6	1.5	11.9 ± 1.0	16.0 ± 3.6	1.3	5
Premature occlusive artery disease (male)	Homocystine plus cysteine-homo-cysteine disulfide	3.3 ± 1.3	4.7 ± 1.0	1.4	13.3 ± 3.6	23.8 ± 1.4	1.9	4
Coronary artery disease	Homocystine	0.06 ± 0.12	0.03 ± 0.11	0.5	0.59 ± 0.37	0.70 ± 0.68	1.6	60
	Cysteine-homo-cysteine disulfide	2.73 ± 1.11	3.22 ± 2.07	1.2	9.39 ± 3.67	10.47 ± 5.17	1.1	60
Coronary artery disease	Total homocyst(e)ine	8.50 ± 2.80	10.96 ± 3.44	1.3				39
Myocardial infarction	Total homocyst(e)ine	13.5 ± 3.6	16.4 ± 6.9	1.2				35
Cerebral infarction	Total homocyst(e)ine	7.3 ± 2.9	13.1 ± 5.6	1.8				1

								n
Peripheral occlusive artery	Total homocyst(e)ine	9.80 ± 3.44	16.6 ± 6.94	1.7				45
Cerebral infarction	Total homocyst(e)ine	10.7 ± 3.2	15.8 ± 5.4	1.5				15
Coronary heart disease	Total homocyst(e)ine							46
Male		11.3 ± 3.7	13.1 ± 4.3	1.2				
Female		10.1 ± 5.0	13.0 ± 7.4	1.3				
Coronary artery disease	Total homocyst(e)ine	10.9 ± 4.9	13.7 ± 6.4	1.3				23
Familial Coronary artery disease	Total homocyst(e)ine							87
Male		11.1 ± 2.6	14.3 ± 7.1	1.3				
Aortoiliac disease	Total homocyst(e)ine	11.0 ± 3.4	18.7 ± 14.9	1.7	21.1 ± 7.3	32.9 ± 2.03	1.6	6
Cerebral thrombosis	Total homocyst(e)ine	11.0 ± 3.4	12.5 ± 7.8	1.1	21.1 ± 7.3	26.0 ± 14.2	1.2	
Coronary artery disease	Total homocyst(e)ine	13.7 ± 4.8	18.1 ± 7.5	1.3				82

[a]Differences between patients and controls are statistically significant ($p < 0.05$) except cysteine-homocysteine disulfide studies by Murphy-Chutorian (60).

Table 2 Hyperhomocyst(e)inemia in controls and patients with occlusive vascular disease

Disease studied	Homocysteine determined as	Basal levels			Postmethionine loading			Refs.
		Controls	Patients	Ratio	Controls	Patients	Ratio	
Coronary artery disease	Cysteine-homocysteine disulfide				5/22 (23%)	17/25 (68%)	3.1	86
Arteriosclerotic cerebrovascular disease	Cysteine-homocysteine disulfide	2/17 (12%)	5/19 (26%)	2.2	1/16 (6%)	4/19 (21%)	3.5	5
Occlusive peripheral artery	Homocystine and cysteine-homocysteine disulfide				0/40 (0%)	7/25 (28%)	(5.6)[a]	4
Occlusive cerebrovascular disease						7/25 (28%)	(5.6)[a]	4
Myocardial infarction						0/25 (0%)		4
Coronary artery disease	Homocystine and cysteine-homocysteine disulfide				1/39 (3%)	16/99 (16%)	5.3	60
					3/39 (8%)	12/99 (12%)	1.5	60

Disease	Measurement	Controls	Patients	Odds ratio	Ref.
Coronary artery disease	Total homocyst(e)ine	8/202 (4%)	25/240 (10%)	2.6	39
Myocardial infarction	Total homocyst(e)ine	2/36 (6%)	5/21 (24%)	4.0	35
Peripheral artery disease	Total homocyst(e)ine	0/29 (0%)	22/47 (47%)	(9.4)[a]	45
Cerebral infarction	Total homocyst(e)ine	1/31 (3%)	11/41 (27%)	9.0	15
Coronary artery disease	Total homocyst(e)ine				
Male			12/64 (19%)	(3.8)[a]	46
Female			3/35 (9%)	(1.8)[a]	46
Cerebrovascular disease	Total homocyst(e)ine	2/46 (4%)	20/72 (28%)	7.0	6
Cerebrovascular disease	Homocystine and cysteine-homocysteine disulfide	0/27 (0%)	16/38 (42%)	(8.4)[a]	13
Peripheral artery disease			7/25 (28%)	(5.6)[a]	13
Coronary artery disease			18/60 (30%)	(6.0)[a]	13

[a] Frequency of hyperhomocyst(e)inemic is 5% in general population; hyperhomocyst(e)inemia was defined as total homocyst(e)ine level above 2 SD of normal mean.

investigated using a prospective design in 14,916 participants in the Physician's Health Study (78). In a five-year follow-up, 271 cases of acute myocardial infarction were reported and were matched to 271 controls. Results showed that the patients had a higher mean level of homocyst(e)ine than did the controls (47). Moreover, hyperhomocyst(e)inemia was observed in 31 patients, but only in 13 controls (47).

In summary, whether cysteine-homocysteine disulfide or total homocyst(e)ine was determined with or without methionine loading, the results have been consistent. In every series (with one exception, Ref. 4), patients with coronary artery disease, cerebrovascular disease, or peripheral arterial disease had mean values significantly higher than those found in controls. The difference between the mean values of basal homocyst(e)ine in patients with vascular disease and in controls was small. The patients usually had mean values 30–50% above those of the controls. Moreover, hyperhomocyst(e)inemia was demonstrated to be independent of other risk factors.

Pathogenic Mechanisms of Vascular Damage and Thromboembolism in Hyperhomocyst(e)inemia: The in vitro Studies

At least three lines of in vitro evidence suggest that excessive homocyst(e)ine is associated with the development of vascular disease. These include endothelial injury, platelet adhesion to endothelium, and release of mitogenic factors leading to intimal smooth muscle cell proliferation.

Chemical endothelial injury has been proposed as the initiating event, followed by patchy desquamation and focal proliferation of intimal smooth muscle cells, similar in appearance to early atherosclerotic lesions in humans (31). This hypothesis was further examined by the evaluation of adhesion of human umbilical vein endothelial cells to glass or plastic surfaces. Wall et al (84) found that 0.1–1.0 mM dl-homocysteine thiolactone induced endothelial cell detachment in direct proportion to its concentration. The cytotoxicity of homocysteine on cultured endothelial cells from an individual who was an obligate heterozygote for cystathionine β-synthase deficiency was also explored by this method (16). Cultured cells from the heterozygote were more susceptible to injury induced by sulfur-containing amino acids than were normal endothelial cells. These observations support the hypothesis that cystathionine β-synthase is expressed in normal endothelium.

Homocysteine-induced endothelial cell detachment was prevented by catalase but not by superoxide dismutase, which would suggest that hydrogen peroxide plays a role in mediating injury (62, 70, 84). It seems that homocysteine participates in the generation of hydrogen peroxide, which interacts with the synthesis of arterial prostacyclin (PGI_2). Thus the effect of homocysteine on PGI_2 synthesis may be dependent on the oxidation of homocysteine and the subsequent generation of hydrogen peroxide.

Normal hemostasis depends in part on the balance achieved between platelet thromboxane A_2 and endothelial PGI_2 production. Graeber et al (27) examined the effect of homocysteine on platelet arachidonic acid metabolism by studying thromboxane B_2, which is the stable end-product of thromboxane A_2. In their studies, they were unable to find any stimulatory or inhibitory effect of homocysteine on PGI_2 synthesis, but they found increased platelet thromboxane production in the presence of 1 mM homocysteine. In contrast, cystine, cysteine, or methionine did not have a similar effect. The conflicting observations of the two studies in regard to PGI_2 syntheses were thought to result from the employment of different assay systems (62).

Because copper-catalyzed oxidation of thiol compounds can lead to the reduction of oxygen and the generation of hydrogen peroxide (12, 89), Starkebaum & Harlan (79) have further studied the effect of homocysteine and copper on endothelial cells. Cultured human and bovine endothelial cells were lysed in the presence of 66–500 μM homocysteine and copper ions, in a dose-dependent manner. Although the generation of hydrogen peroxide was not measured directly, these results suggested a mechanism whereby elevated levels of homocysteine could injure endothelial cells through copper-catalyzed generation of hydrogen peroxide.

Several investigators have demonstrated superoxide (54), hydrogen peroxide (89), and hydroxyl radical generation (70) during autooxidation of thiol compounds, including protein-bound thiols. A wide spectrum of both oxygen- and sulfur-derived free radicals can be produced in the presence of thiols and trace metals, thus initiating lipid peroxidation. Parthasarathy (63) showed that homocysteine promoted the oxidation of LDL. Other thiol compounds including cysteine and reduced glutathione also showed these effects. In contrast, oxidized thiols were ineffective in oxidizing LDL. Homocystine, therefore, is unlikely to produce these effects by a superoxide-mediated mechanism (34). However, this does not exclude the possibility that smooth muscle cells are able to reduce homocystine, since homocysteine autooxidizes by a two-electron transfer reaction that results in the production of hydrogen peroxide (34).

Features of blood coagulation in homocystinuric patients have been recently reviewed by Palareti & Coccheri (61). Reduced antithrombin III and factor VII levels were observed in patients with severe hyperhomocyst(e)inemia (12, 25, 48, 53). An increased homocyst(e)ine level appeared to be directly linked to low antithrombin III and factor VII activity levels, which were correctable by the administration of pyridoxine and folate. Rodgers & Kane (69) demonstrated that 0.5–10 mM homocysteine increased factor V activity and prothrombin activation in bovine aortic endothelial cell cultures. These results indicated that homocysteine-treated vascular endothelium induced the activation of factor V. Their studies explored further the mechanism of factor V activation by which homocysteine reduced endothelial cell protein C activation (68). These data suggested that coagulation abnormalities induced by homocysteine may have contributed to the thrombotic tendency seen in patients with severe hyperhomocyst(e)inemia.

In summary, homocysteine-induced chemical injury to cultured endothelial cells is probably due to the modification of arachidonic acid metabolism. Oxidation of homocysteine causes copper-catalyzed generation of hydrogen peroxide, leading to the inhibition or stimulation of PGI_2 synthesis. An increased platelet thromboxane production is also involved. Oxidation of low density lipoprotein is another effect of thiol compounds. Coagulation abnormalities, such as reduced antithrombin III and factor VII, and increased factor V and prothrombin activities are observed in the presence of homocysteine, suggesting its association with thrombotic tendency. These in vitro effects are initiated by homocysteine, whereas the major portion of plasma homocyst(e)ine is present as disulfides. Hence, investigators have speculated that cells take up homocystine and reduce it to homocysteine intracellularly; subsequently, reoxidation of homocysteine leads to the generation of hydrogen peroxide (34, 79).

Concluding Remarks

In the past several years, significant progress has been achieved in our understanding of various hyperhomocyst(e)inemias and their association with the development of occlusive arterial diseases. Homocysteine and its derivatives are reliably measured as total homocyst(e)ine in nonhomocystinuric individuals. Not only is plasma homocyst(e)ine level affected by genetic defects of homocysteine metabolism but it is also controlled by various nongenetic factors such as nutritional inadequacy of folate, cobalamin, pyridoxine, or choline, endocrinological status, other diseases, and some medications.

Many retrospective studies have substantiated the positive correlation between moderate hyperhomocyst(e)inemia and occlusive arterial diseases. Recently, Malinow et al (47) reported a prospective study in which moderately high levels of homocyst(e)ine were associated with increased risk of myocardial infarction. Although definite proof is still lacking, it is reasonable to assume that the accumulation of excessive homocyst(e)ine causes damage to endothelial and smooth muscle cells and alters the activity of coagulation factor(s). No ideal animal models are available to demonstrate homocysteine-induced vascular pathology. Chronic infusion studies are technically difficult because of the high metabolic turnover of homocysteine molecules. In nutritionally induced hyperhomocyst(e)inemias, it has not been possible to rule out the direct effect of nutritional deficiency itself on the vascular system.

Irrespective of the etiology, moderate and intermediate hyperhomocyst(e)inemia is readily correctable by supplementation with folate and/or betaine. Whether or not normalization of hyperhomocyst(e)inemia will improve morbidity and mortality is a question that must await the outcome of clinical trials.

ACKNOWLEDGMENTS

We express our appreciation to S. Harvey Mudd, J. B. Ubbink, and W. J. H. Vermack for communicating papers prior to publication. We also thank Delilah Delgado for her excellent typing of the manuscript. Work in the authors' laboratories was supported by grants from the National Heart, Lung and Blood Institute (HL 36135) and the National Institute of Health (RR00163-32).

Literature Cited

1. Araki, A., Sako, Y., Fukushima, Y., Matsumoato, M., Asada, T., Kita, T. 1989. Plasma sulfhydryl-containing amino acids in patients with cerebral infarction and in hypertensive subjects. *Atherosclerosis* 79:139–46

2. Baldessarini, R. J., Kepin, I. J. 1963. Assay of tissue levels of S-adenosylmethionine. *Anal. Biochem.* 6:289–92

3. Baumgartner, E. R., Wick, H., Linnell, J. C., Gaull, G. E., Bachmann, C., Steinmann, B. 1979. Congenital defect in intracellular cobalamin metabolism resulting in homocystinuria and methylmalonic aciduria. *Helv. Paediatr. Acta* 34:483–96

4. Boers, G. H. J., Smals, A. G. H., Trijbels, F. J. M., Fowler, B., Bakkeren, J. A., et al. 1985. Heterozygosity for homocystinuria in premature peripheral and cerebral occlusive arterial disease. *New Engl. J. Med.* 313:709–15

5. Brattström, L. E., Hardebo, J. E., Hultberg, B. L. 1984. Moderate homocysteinemia—a possible risk factor for arteriosclerotic cerebrovascular disease. *Stroke* 14:1012–16

6. Brattström, L., Israelsson, B., Hultberg, B. 1988. Impaired homocysteine metabolism—a possible risk factor for arteriosclerotic vascular disease. In *Genetic Susceptibility to Environmental Factors: A Challenge for Public Intervention*, ed. U. Smith, S. Ericksson, F. Lindgarde, pp. 25–34. Stockholm: Almqvist & Wiksell Int.

7. Brattström, L., Israelsson, B., Jeppsson, J. O. 1986. Heterozygosity for homocystinuria in premature arterial disease. *New Engl. J. Med.* 314:849–50

8. Brattström, L., Israelsson, B., Lind-

garde, F., Hultberg, B. 1988. Higher total plasma homocysteine in vitamin B_{12} deficiency than in heterozygosity for homocystinuria due to cystathionine B-synthase deficiency. *Metabolism* 37:175–78

9. Carey, M. C., Donovan, D. E., Fitz-Gerald, O., McAuley, F. D. 1968. Homocystinuria. I. A clinical and pathological study of nine subjects in six families. *Am. J. Med.* 45:7–25

10. Carson, N. A. J., Neill, D. W. 1962. Metabolic abnormalities detected in a survey of mentally backward individuals in Northern Ireland. *Arch. Dis. Child.* 37:505–13

11. Deleted in proof

12. Charlot, J. C., Haye, C., Chaumien, J. P. 1982. Homocystinuric et deficit en facteur VII. *Bull. Soc. Ophthalmol. Fr.* 82:787–89

13. Clarke, R., Daly, L., Robinson, K., Naughten, E., Cahalane, S., et al. 1991. Hyperhomocysteinemia: An independent risk factor for vascular disease. *New Engl. J. Med.* 324:1149–55

14. Cooper, B. A., Rosenblatt, D. S. 1987. Inherited defects of vitamin B_{12} metabolism. *Annu. Rev. Nutr.* 7:291–320

15. Coull, B. M., Malinow, M. R., Beamer, N., Saxton, G., Nordt, F., deGarmo, P. 1990. Elevated plasma homocyst(e)ine concentration as a possible independent risk factor for stroke. *Stroke* 21:572–76

16. De Groot, P. G., Willems, C., Boers, G. H. J., Gonsalves, M. D., Van Aken, W. G., Van Mourik, J. A. 1983. Endothelial cell dysfunction in homocystinuria. *Eur. J. Clin. Invest.* 13:405–10

17. Deleted in proof

18. Erbe, R. W. 1986. Inborn errors of folate metabolism. In *Folates and Pterins*, ed. R. L. Blakely, pp. 413–65. Boston: Wiley
19. Fenton, W. A., Rosenberg, L. E. 1989. Inherited disorders of cobalamin transport and metabolism. In *The Metabolic Basis of Inherited Disease*, ed. C. R. Scriver, A. L. Beadet, W. S. Sly, D. Valle, pp. 2065–82. New York: McGraw-Hill
20. Finkelstein, J. D., Kyle, W. E., Harris, B. J. 1971. Methionine metabolism in mammals. Regulation of homocysteine methyltransferases in rat tissue. *Arch. Biochem. Biophys.* 146:84–92
21. Finkelstein, J. D., Martin, J. J. 1984. Methionine metabolism in mammals. Distribution of homocysteine between competing pathways. *J. Biol. Chem.* 259:9508–13
22. Finkelstein, J. D., Martin, J. J. 1986. Methionine metabolism in mammals. Adaptation to methionine excess. *J. Biol. Chem.* 261:1582–87
23. Genest, J. J., McNamara, J. R., Salem, D. N., Wilson, P. W. F., Schaefer, E. J., Malinow, M. R. 1990. Plasma homocyst(e)ine levels in men with premature coronary artery disease. *J. Am. Coll. Cardiol.* 16:1114–19
24. Genest, J. J., McNamara, J. R., Upson, B., Salem, D. N., Ordovas, J. M., et al. 1991. Prevalence of familial hyperhomocyst(e)inemia in men with premature coronary artery disease. *Arterioscler. Thromb.* 11:1129–36
25. Giannini, M. J., Coleman, M., Innerfield, I. 1975. Antithrombin activity in homocystinuria. *Lancet* 1:1094
26. Gibson, J. B., Carson, N. A. J., Neill, D. W. 1964. Pathological findings in homocystinuria. *J. Clin. Pathol.* 17:427–37
27. Graeber, J. E., Slott, J. H., Ulane, R. E., Schulman, J. D., Stuart, J. J. 1982. Effect of homocysteine and homocystine on platelet and vascular arachidonic acid metabolism. *Pediatr. Res.* 16:490–93
28. Gupta, V. J., Wilcken, D. E. L. 1978. The detection of cysteine-homocysteine mixed disulphide in plasma of normal fasting man. *Eur. J. Clin. Invest.* 8:205–7
29. Deleted in proof
30. Deleted in proof
31. Harker, L. A., Ross, R., Slichter, S. J., Scott, C. R. 1976. Homocysteine-induced arteriosclerosis. The role of endothelial cell injury and platelet response in its genesis. *J. Clin. Invest.* 58:731–41
32. Harker, L. A., Slichter, S. J., Scott, C. R., Ross, R. 1974. Homocystinemia:

Vascular injury and arterial thrombosis. *New Engl. J. Med.* 291:537–43
33. Hartcroft, W. S., Ridout, J. H., Sellers, E. A., Best, C. H. 1952. Atheromatous changes in aorta, carotid and coronary arteries of choline-deficient rats. *Proc. Soc. Exp. Biol. Med.* 81:384–93
34. Heinecke, J. W., Rosen, H., Suzuki, L. A., Chait, A. 1987. The role of sulfur-containing amino acids in superoxide production and modification of low density lipoprotein by arterial smooth muscle cells. *J. Biol. Chem.* 262:10098–10103
35. Israelsson, B., Brattstrom, L. E., Hultberg, B. L. 1988. Homocysteine and myocardial infarction. *Atherosclerosis* 71:227–33
36. Lin, J. Y., Kang, S. S., Zhou, J., Wong, P. W. K. 1989. Homocysteinemia in rats induced by folic acid deficiency. *Life Sci.* 44:319–25
37. Kang, S. S., Wong, P. W. K., Becker, N. 1979. Protein-bound homocyst(e)ine in normal subjects and in patients with homocystinuria. *Pediatr. Res.* 13:1141–43
38. Kang, S. S., Wong, P. W. K., Bock, H. G., Horwitz, A., Grix, A. 1991. Intermediate hyperhomocysteinemia resulting from compound heterozygosity of methylenetetrahydrofolate reductase mutation. *Am. J. Hum. Genet.* 48:546–51
39. Kang, S. S., Wong, P. W. K., Cook, H. Y., Norusis, M., Messer, J. V. 1986. Protein-bound homocyst(e)ine. A possible risk factor for coronary artery disease. *J. Clin. Invest.* 77:1482–86
40. Kang, S. S., Wong, P. W. K., Norusis, M. 1987. Homocysteinemia due to folate deficiency. *Metabolism* 36:458–62
41. Kang, S. S., Wong, P. W. K., Susmano, A., Sora, J., Norusis, M., Ruggie, N. 1991. Thermolabile methylenetetrahydrofolate reductase: an inherited risk factor for coronary artery disease. *Am. J. Hum. Genet.* 48:536–45
42. Kang, S. S., Wong, P. W. K., Zhou, J., Cook, H. Y. 1986. Preliminary report: total homocyst(e)ine in plasma and amniotic fluid of pregnant women. *Metabolism* 35:889–91
43. Kanwar, Y. S., Manaligod, J. R., Wong, P. W. K. 1976. Morphologic studies in a patient with homocystinuria due to 5,10-methylenetetrahydrofolate reductase deficiency. *Pediatr. Res.* 10:598–609
44. Malinow, M. R. 1991. Homocyst(e)ine and vascular occlusive disease. *Nutr. Metab. Cardiovasc. Dis.* 1:166–69
45. Malinow, M. R., Kang, S. S., Taylor,

L. M., Wong, P. W. K., Coull, B., et al. 1989. Prevalence of hyperhomocyst(e)inemia in patients with peripheral arterial occlusive disease. *Circulation* 79:1180–88

46. Malinow, M. R., Sexton, G., Averbuch, M., Grossman, M., Wilson, D., Upson, B. 1990. Homocyst(e)inemia in daily practice: levels in coronary artery disease. *Coron. Artery Dis.* 1:215–20

47. Malinow, M. R., Stampfer, M. J., Willett, W. C., Newcomer, L., Upson, B., et al. 1991. A prospective study of plasma homocyst(e)ine and risk of myocardial infarction. *Arterioscler. Thromb.* 11:1480A

48. Maruyama, I., Fukuda, R., Kazama, M. 1977. A case of homocystinuria with low antithrombin activity. *Acta Haematol. Jpn.* 40:267–71

49. McCully, K. S. 1969. Vascular pathology of homocysteinemia: Implications for the pathogenesis of arteriosclerosis. *Am. J. Pathol.* 56:111–28

50. McCully, K. S., Ragsdale, B. D. 1970. Production of arteriosclerosis by homocysteinemia. *Am. J. Pathol.* 61:1–8

51. McCully, K. S., Wilson, R. B. 1975. Homocysteine theory of arteriosclerosis. *Atherosclerosis* 22:215–27

52. McGill, J. J., Mettler, G., Rosenblatt, D. S., Scriver, C. R. 1990. Detection of heterozygotes for recessive alleles. Homocyst(e)inemia: Paradigm of pitfalls in phenotypes. *Am. J. Med. Genet.* 36:45–52

53. Mercky, J., Kuntz, F. 1981. Deficit en facteur VII et homocytinurie. Association fortuite ou syndrome? *Nouv. Presse Med.* 10:3796

54. Misra, H. P. 1974. Generation of superoxide free radical during the autoxidation of thiols. *J. Biol. Chem.* 249:2151–55

55. Mudd, S. H. 1988. Vascular disease and homocysteine metabolism. See Ref. 6, pp. 11–24

56. Mudd, S. H., Levy, H. L., Skovby, F. 1989. Disorders of transsulfuration. See Ref. 19, pp. 693–734

57. Mudd, S. H., Poole, J. R. 1975. Labile methyl balances for normal humans on various dietary regimens. *Metabolism* 24:721–35

58. Mudd, S. H., Skovby, F., Levy, H. L., Pettigrew, K. D., Wilcken, B., et al. 1985. The natural history of homocystinuria due to cystathionine β-synthase deficiency. *Am. J. Hum. Genet.* 37:1–31

59. Mudd, S. H., Uhlendorf, B. W., Freeman, J. M., Finkelstein, J. D., Shih, V.

E. 1972. Homocystinuria associated with decreased methylenetetrahydrofolate reductase activity. *Biochem. Biophy. Res. Commun.* 46:905–12

60. Murphy-Chutorian, D. R., Wexman, M. P., Grieco, A. J., Heiringer, J. E., Glassman, E., et al. 1985. Methionine intolerance: a possible risk factor for coronary artery disease. *J. Am. Coll. Cardiol.* 6:725–30

61. Palareti, G., Coccheri, S. 1989. Lowered antithrombin III activity and other clotting changes in homocystinuria: Effects of a pyridoxine-folate regimen. *Haemostasis* 19(Suppl. 1):24–28

62. Panganamala, R. V., Karpen, C. W., Merola, A. J. 1986. Peroxide-mediated effects of homocysteine on arterial prostacyclin synthesis. *Prostaglandins Leukot. Med.* 22:349–56

63. Parthasarathy, S. 1987. Oxidation of low density lipoprotein by thiol compounds leads to its recognition by the acetyl LDL receptor. *Biochim. Biophys. Acta* 917:337–40

64. Reddy, G. S. R., Wilcken, D. E. L. 1982. Experimental homocysteinemia in pigs: comparison with studies in sixteen homocystinuric patients. *Metabolism* 31:778–83

65. Reed, T., Malinow, M. R., Christian, J. C., Upson, B. 1991. Estimates of heritability for plasma homocyst(e)ine levels in aging adult male twins. *Clin. Genet.* 39:425–28

66. Refsum, H., Helland, S., Ueland, P. M. 1985. Radioenzymic determination of homocysteine in plasma and urine. *Clin. Chem.* 31:624–28

67. Rinehart, J. F., Greenberg, L. D. 1956. Vitamin B_6 deficiency in the rhesus monkey with particular reference to the occurrence of atherosclerosis, dental caries and hepatic cirrhosis. *Am. J. Clin. Nutr.* 4:318–25

68. Rodgers, G. M., Conn, M. T. 1990. Homocysteine, an atherogenic stimulus reduces protein C activation by arterial and venous endothelial cells. *Blood* 75:895–901

69. Rodgers, G. M., Kane, W. H. 1986. Activation of endogenous factor V by a homocysteine-induced vascular endothelial cell activator. *J. Clin. Invest.* 77:1909–16

70. Saez, G., Thornalley, P. J., Hill, H. A. O., Hems, R., Bannister, J. V. 1982. The production of free radicals during autoxidation of cysteine and their effect on isolated rat hepatocytes. *Biochim. Biophys. Acta* 719:24–31

71. Salvatore, F., Zappia, V, Shapiro, S. K. 1968. Quantitative analysis of S-

adenosylhomocystein in liver. *Biophys. Biochim. Acta* 158:461–64

72. Sardharwalla, I. B., Fowler, B., Robins, A. J., Komrower, G. M. 1974. Detection of heterozygotes for homocystinuria. Study of sulphur-containing amino acids in plasma and urine after L-methionine loading. *Arch. Dis. Child.* 49:553–59

73. Sartorio, R., Carrozzo, R., Corbo, L., Andria, G. 1986. Protein-bound plasma homocyst(e)ine and identification of heterozygotes for cystathionine-synthase deficiency. *J. Inherit. Metab. Dis.* 9:25–29

74. Deleted in proof

75. Smolin, L. A., Benevenga, N. J. 1982. Accumulation of homocyst(e)ine in Vitamin B-6 deficiency: A model for the study of cystathionine β-synthase deficiency. *J. Nutr.* 112:1264–72

76. Smolin, L. A., Benevenga, N. J., Berlow, S. 1981. The use of betaine for the treatment of homocystinuria. *J. Pediatr.* 99:467–72

77. Stabler, S. P., Marcell, P. D., Podell, E. R., Allen, R. H., Savage, D. G., Lindenbaum, J. 1988. Elevation of total homocysteine in the serum of patients with cobalamin or folate deficiency detected by capillary gas chromatography-mass spectrometry. *J. Clin. Invest.* 81:466–74

78. Stampfer, M. R., Buring, J., Willett, W., Rosner, B., Eberlein, K., et al. 1985. The 2 × 2 factorial design: its application to a randomized trial of aspirin and caretone among US physicians. *Stat. Med.* 4:111–16

79. Starkebaum, G., Harlan, J. M. 1986. Endothelial cell injury due to copper-catalyzed hydrogen peroxide generation from homocysteine. *J. Clin. Invest.* 77:1370–76

80. Svardal, A., Ueland, P. M., Berge, R. K., Aarsland, A., Aarsaether, N., et al. 1988. Effect of methotrexate on homocysteine and other sulfur compounds in tissues of rats fed a normal or a defined, choline-deficient diet. *Cancer Chemother. Pharmacol.* 21:313–18

81. Swift, M. E., Shultz, T. D. 1986. Relationship of vitamin B_6 and B_{12} to homocysteine levels: Risk for coronary artery disease. *Nutr. Rep. Int.* 34:1–14

82. Ubbink, J. G., Vermack, W. J. H., Bennett, J. M., Becker, P. G., VanStaden, D. A., Bissbort, S. 1991. The prevalence of homocysteinemia and hypercholesterolemia in angiographically defined coronary heart disease. *Klin. Wochenschr.* 69:527–34

83. Ueland, P. M., Refsum, H. 1989. Plasma homocysteine, a risk factor for vascular disease: Plasma levels in health, disease, and drug therapy. *J. Lab. Clin. Med.* 114:473–501

84. Wall, R. T., Harlan, J. M., Harker, L. A., Striker, G. E. 1980. Homocysteine-induced endothelial cell injury in vitro: a model for the study of vascular injury. *Thromb. Res.* 18:113–21

85. Wilcken, D. E. L., Dudman, N. P. B. 1989. Mechanism of thrombogenesis and accelerated atherogenesis in homocysteinemia. *Haemostasis* 19(Suppl. 1):14–23

86. Wilcken, D. E. L., Wilcken, B. 1976. The pathogenesis of coronary artery disease. A possible role for methionine metabolism. *J. Clin. Invest.* 57:1079–82

87. Williams, R. R., Malinow, M. R., Hunt, S. C., Upson, B., Wu, L. L., et al. 1990. Hyperhomocyst(e)inemia in Utah siblings with early coronary disease. *Coron. Artery Dis.* 1:681–85

88. Wong, P. W. K., Kang, S. S., Kanwar, Y. W. 1990. Folic acid induced homocysteinemia and atherosclerosis. *Clin. Res.* 38:806A

89. Zwart, J., Van Wolput, H. M. C., Van der Cammen, J. C. M., Konigsberger, D. C. 1981. Accumulation and reactions of hydrogen peroxide during the copper ion catalyzed auto oxidation of cysteine in alkaline medium. *J. Mol. Catal.* 11:69–82

Annu. Rev. Nutr. 1992. 12:299–326

DIETARY IMPACT ON BILIARY LIPIDS AND GALLSTONES[1]

K. C. Hayes, Anne Livingston, and Elke A. Trautwein

Foster Biomedical Research Laboratory, Brandeis University, Waltham, Massachusetts 02254

KEY WORDS: lipoprotein profile, dietary fiber, energy intake, dietary fat, animal models

CONTENTS

[1]Abbreviations: LI, lithogenic index; FC, free cholesterol; TGSR, triglyceride secretion rate; CE, cholesteryl ester; PUFA, polyunsaturated fatty acid; BA, bile acid; EFAD, essential fatty acid deficient; TG, triglyceride; LRP, LDL receptor-related protein; LDLr, LDL receptor; CSI, cholesterol saturation index; IgA, immunoglobin A.

INTRODUCTION

This review summarizes current information concerning the dietary modulation of gallstone formation. The emphasis is on cholesterol gallstones associated with Western diets as opposed to less frequently encountered pigment
gallstones. Although the genesis of cholesterol gallstones is multifactorial
(19), basic physiological correlates associated with dietary modulation of
biliary lipid metabolism (bile acids, phospholipids, cholesterol) are discussed
in relation to gallstones. Several reviews on the subject of gallstones are
available (24a, 27, 34, 58), and some include nutritional aspects of the
problem (15, 36, 37, 85, 87, 122).

GENERAL PHYSIOLOGY OF ENTEROHEPATIC CIRCULATION

The liver and gut interact via the enterohepatic circulation to transport important lipid factors in a relatively closed circuit. Absorptive processes by the
intestine and secretory processes by the liver are involved. Bile acids and
lecithin secreted by the liver reach the intestine via the bile duct to solubilize
fat and fat-soluble nutrients, thereby making their absorption possible. To that
end, the liver must synthesize bile acids from free cholesterol as well as
phospholipids (lecithin) from glycerol, choline, and fatty acids, including an
important contribution from specific polyunsaturated fatty acids. In addition
to these two lipids, bile contains a variable amount of free cholesterol.
Lecithin is secreted in the form of lamellae (140) or together with free
cholesterol as vesicles, via the smooth endoplasmic reticulum and Golgi
apparatus of hepatocytes (29, 44, 94). Bile acids are secreted separately in
much greater volume into the bile canaliculus, whereupon they disperse the
lamellae and vesicles to form the micelles necessary for solubilizing and
absorbing fat in the small intestine (59).

The ratio of bile salts to lecithin relative to the total lipid concentration is
the overriding factor modulating the lithogenicity of bile (27). In other words,
if cholesterol solubilization is inadequate, gallstones are apt to form. The
relationship between the secretion of bile acids and either phospholipids or
free cholesterol has been referred to in "linkage coefficients" representing
important ratios of the secreted molecules that generally become distorted in
lithogenic bile (59). The linkage coefficient between bile acid and free
cholesterol secretion is highly species dependent (relative biliary cholesterol
secretion is low in rats, high in humans) and, within certain species, readily
affected by diet.

Bile Acids

Bile acids are formed exclusively in the liver via the rate-limiting initial hydroxylation of free cholesterol at the 7α position (161). The enzyme, cholesterol 7α-hydroxylase, forms the basic dihydroxy steroid molecule that gives rise to the primary bile acid known as (3,7-OH)chenodeoxycholic acid (cheno). A second hydroxylation at the twelfth position of the steroid nucleus (12C) leads to the other primary bile acid, cholic acid (3,7,12-OH). These bile acids rapidly conjugate with glycine or taurine by an amide linkage to form the respective conjugates of cholic or chenodeoxycholic acids, which are immediately secreted into bile. Taurine conjugation predominates if taurine is readily available from synthesis, but especially if the diet is rich in taurine (52). Since taurine is not available in plants, herbivorous animals (e.g. rabbits) tend to be glycine conjugators, whereas carnivores (e.g. cats) are mainly taurine conjugators. Species also vary substantially in their ability to synthesize taurine (54).

Once in the intestine, the distinct polarity of the two conjugates differentially impacts fat absorption. Thus, the taurine conjugates are more stable at low pH and less apt to be deconjugated during intestinal transit. They remain intraluminally and sustain fat absorption throughout the small intestine until they are actively absorbed in the terminal ileum. By contrast, glycine conjugates are more apt to be passively absorbed along the entire gut. Taurocholate is most hydrophilic and least able to solubilize cholesterol, whereas glycodeoxycholate and glycocheno (both dihydroxy bile acids) are least hydrophilic and solubilize fat and cholesterol more effectively (8).

The 3–5% of bile acids reaching the large bowel during each of the 5–7 daily circulations of the bile acid pool can be modified by bacterial flora that remove the 7-OH group, forming the secondary bile acids, deoxycholic (3,12-OH) and lithocholic acid (3-OH). About one third of these secondary bile acids are passively absorbed from the colon to be reincorporated into the bile acid pool. The amount depends, in part, on the mass of primary bile acids reaching the colon, colonic microflora activity, the concentration and type of dietary fiber, and fecal transit time. The bile acids in the gallbladder, gut, and liver constitute the bulk of the pool size, which is an important variable in lithogenesis, since gallstones often develop in individuals with a bile acid pool size insufficient to solubilize the biliary cholesterol present (19).

Bile acid synthesis and secretion follows a circadian rhythm that is closely associated with the feeding cycle and reflux of bile acids via the portal vein (113, 161). The nadir of 7α-hydroxylase activity occurs during fasting with peak activity postprandially. Although the ratio between primary bile acids may vary in normal individuals, an association between specific bile acids and cholesterol gallstones can be inferred from the literature. Healthy human bile tends to have slightly more cholic acid than chenodeoxycholic acid, but this distribution is often reversed in patients with cholesterol gallstones (23, 59, 60). Hamsters (30) and prairie dogs (31) develop an exaggerated cheno profile as they develop cholesterol-induced gallstones, but cheno-supplemented hamsters fed a fat-free diet develop gallstones at an accelerated rate (37). On the other hand, cheno supplementation in humans desaturates bile and dissolves gallstones, in part, by decreasing hepatic cholesterol synthesis (2, 68, 124, 163).

The primary precursor pool for bile acid synthesis is thought to derive from lipoprotein cholesterol (51, 125, 127), but the exact mechanism of cholesterol delivery is unclear (Figures 1 and 2). Empirical evidence suggests that the two primary bile acids may receive substantial free cholesterol from different lipoproteins. Free cholesterol enters via the LDL receptor (LDLr), arriving with LDL under normal circumstances, and appears to be associated with cholate synthesis and glycine conjugation. By comparison, cheno synthesis appears to be enhanced by free cholesterol entering via HDL_2 or the apoE receptor (LDL receptor-related protein or LRP receptor), presumably during periods of lipoprotein cholesterol overload (HDL_2 postprandially) or from apoE-rich lipoproteins (HDL_2, IDL, or VLDL remnants) during dietary saturated fat feeding or cholesterol loading (9, 20, 48, 86, 88, 143). Cheno seems to prefer taurine conjugation. When taurine is available (high synthesis in rat or high diet source for carnivorous cat), it preferentially conjugates most bile acids, i.e. taurocholate predominates in normal cats and rats. In herbivorous animals such as rabbits, with a large cecum, minimal lipoprotein lipid pool and good LDLr activity, the main bile acid is glycodeoxycholate.

Humans and hamsters (the most widely studied model of cholelithiasis) have slightly more cholate than cheno, and glycine conjugation predominates under normolipemic conditions. A cheno profile typically develops as they become hyperlipemic from dietary excesses, in part because their relatively limited LDLr activity under normal circumstances is quickly down-regulated owing to their inability to compensate cholesterol balance by depressing an already marginal hepatic cholesterol synthesis (158). Furthermore, both hamsters and humans (women) have relatively generous HDL_2 profiles. The biological implications of the bile acid profile are unclear, but elucidation of the point may be revealing in terms of gallstone risk and comparative aspects of cholesterol flux across the liver (see below).

Lecithin

Lecithin (diacylphosphatidylcholine) is the predominant phospholipid in bile of humans and other animals, representing 15 to 25% of the total biliary lipids. In bile lecithin the sn-1 fatty acid tends to be the least hydrophobic, palmitic acid (16:0). The sn-2 position is unsaturated, generally with oleic

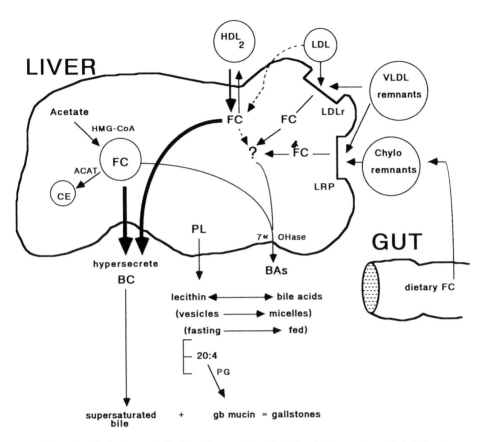

Figure 1 The basic metabolic determinants of hepatic biliary lipid secretion and cholesterol gallstone formation are depicted. The key variable is hypersecretion of biliary cholesterol (BC), which depends on the size and flux of the hepatic free cholesterol (FC) pool. The latter derives from free cholesterol returning to the liver via lipoproteins (LPs) plus newly synthesized cholesterol via HMG-CoA reductase activity. BC output can be modulated by diverting FC into cholesteryl esters (CE) via ACAT or into bile acids (BAs) via 7α-hydroxylase activity (7α-OHase). Lipoprotein-FC can enter the liver at least four ways; via the LDL receptor (LDLr), via apoE-rich LPs through the LDL receptor-like protein (LRP), from HDL$_2$-FC "exchange," or via nonreceptor-mediated LDL uptake *(dotted arrow)*. The exact distribution of these FC sources is uncertain, but HDL$_2$-FC is presumed to strongly influence BC secretion. Gallstones develop when BC is excreted disproportionally to phospholipids (PL) and bile acids (BAs), thus causing supersaturation of bile. Increasing arachidonate in biliary lecithin enhances gallbladder mucin production through increased prostaglandin (PG). Mucin acts as a nucleating matrix for gallstone formation. Diet manipulation impacts most of these variables (see text).

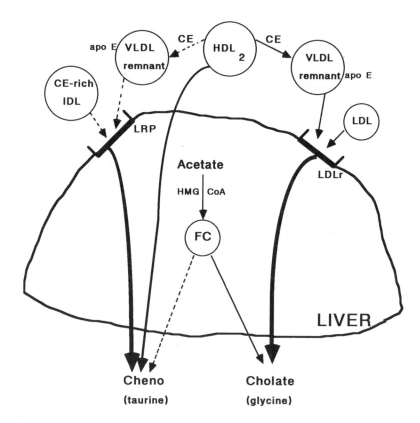

Figure 2 A scheme for the putative disposition of lipoprotein-free cholesterol (LP-FC) and newly synthesized cholesterol in the genesis of primary bile acids is depicted. The literature suggests that LP-FC, especially HDL_2-FC, represents the major precursor pool for bile acid synthesis. Under normolipemic conditions in most species, LDL receptors provide the route for LDL and VLDL remnant uptake that feeds preferentially into the cholate pool, which tends to be mostly glycine-conjugated. During enhanced (or increased) VLDL catabolism, HDL_2 increases and CE overflow into VLDL (IDL) occurs via transfer from HDL_2 *(broken arrow)*. Cholesterol uptake by the liver is increased by $IIDL_2$ directly (process unclear) or indirectly via the apo E receptor (LRP). The LP-FC entering by this route putatively favors cheno synthesis, which tends to be taurine-conjugated. Gallstones develop in the second case, in part, because much of the HDL_2-FC appears to be secreted directly into bile (see Figure 1).

(18:1), linoleic (18:2), or arachidonic acid (20:4). The amount and exact fatty acid profile varies with species and diet. Lecithin composition is important because fatty acid saturation can greatly modify the micelle-vesicle equilibrium in bile to influence biliary cholesterol solubilization (33). In fact, nascent lecithin lamellae may be an important cholesterol carrier in bile (140). In addition, elevated arachidonic acid content can stimulate mucin production by the gallbladder mucosa, which enhances cholesterol crystal formation (28).

Biliary Cholesterol

Free cholesterol represents the least concentrated lipid in bile (1–5 mol% in healthy human bile, but generally < 1.0 mol% in most mammalian biles). Like bile acids, the principle source of biliary cholesterol appears to be lipoprotein free cholesterol coupled with a modest contribution from the newly synthesized pool of cholesterol. A growing body of evidence suggests that in humans (125, 126), rats (20, 116), and monkeys (131, 143), HDL_2-FC is the primary donor to the biliary free cholesterol pool. This contribution may be rather direct, bypassing general mixing with the hepatic intracellular pool of free cholesterol by lateral transfer through the cell membrane for direct secretion into the bile canaliculus (116, 131). Plasma cholesteryl esters, on the other hand, appear to contribute minimally to biliary sterols (17).

Biliary Proteins

In addition to lipids, a number of proteins, the character and importance of which are increasingly appreciated (24, 28, 50, 114, 139), are secreted in bile. Albumin is the most prevalent, constituting over 70% of bile protein. IgA, secretory protein, and apolipoproteins AI, AII, CIII, and E are also present, but their concentration is low and their role unknown (75, 136). Apo E and apo AI are the most prevalent apolipoproteins in nascent bile of hamsters (145). Although apolipoproteins are potential antinucleating proteins in bile, their functional role in vivo as a factor in the solubilization of biliary cholesterol is relatively unexplored (95). In vitro apo AI in low concentrations can delay the shift from micelles to vesicles, thereby enhancing the cholesterol-solubilizing capacity of bile acids (24a, 43). Immunoglobins (IgM, IgA) in bile can serve as pronucleating agents (53). Similarly, mucins and other glycoproteins in gallbladder bile are thought to provide a nucleating matrix for cholesterol crystal formation (50, 139), whereas other proteins inhibit crystalization (24). No evidence has been presented to indicate that the concentration of biliary proteins is affected by diet except indirectly by the fact that dihydroxy bile acids stimulate mucin secretion (105).

Bile Secretion

Bile secretion is stimulated by food intake. The presence of food in the duodenum (especially fat) induces secretion of cholecystokinin (CCK) by mucosal endocrine cells which, in turn, causes contraction and release of bile by the gallbladder (117). The failure of gallbladder contraction can result in stasis of gallbladder bile and can enhance the formation of biliary "sludge," characterized by the accumulation of a lipid-mucin-proteinaceous debris with calcium bilirubinate salts that generally precedes gallstone formation (82). Physiological events associated with digestion and absorption stimulate hepatic bile synthesis and secretion to effectively increase the concentration of biliary micelles, i.e. phospholipid, bile acid, and free cholesterol, and decrease the relative biliary cholesterol concentration. As the postabsorptive condition develops, micelle formation gives way to biliary vesicles, which are phospholipid and cholesterol-rich but bile acid-poor (106).

Because vesicles do not solubilize cholesterol as effectively as micelles, they increase the chance of cholesterol crystalization during the fasting phase of the diurnal feeding cycle (44, 62). This is especially true in cholesterol gallstone disease, as bile acid and lecithin secretion rates tend to be depressed and cholesterol secretion rates elevated (29).

Gallstones

The pathogenesis of cholesterol gallstones involves a complex set of variables affecting the relative concentration of bile acid, phospholipid, free cholesterol, and protein during their collective flux through the biliary system. In essence, the supersaturation of bile with cholesterol results from increased biliary cholesterol secretion (28) and/or decreased secretion of bile acids and, to a lesser extent, lecithin. In addition to lipid changes, nucleating and/or antinucleating factors, coupled with stasis of gallbladder bile, are important considerations, since supersaturated bile does not always form crystals or gallstones (62).

BILIARY LIPID HOMEOSTASIS AND DIET

Since the above discourse indicates that the relative flux of cholesterol through biliary lipids is the critical issue in cholesterol gallstone disease, it is relevant to review nutritional factors that impact this relationship. In fact, diet affects most of the secreted components in bile.

Caloric Intake and Lipoprotein Profile

The most critical variable in lithogenic bile formation is the relative mass of free cholesterol fluxing through bile. In humans this correlates with the absolute load of calories consumed (89, 121, 122) and the resulting lipoprotein cholesterol that must eventually return to the liver and pass into bile as bile acids or, failing that, as free cholesterol itself (63). Not surprisingly, obesity represents the greatest risk factor for gallstones. Females whose body mass index (BMI) exceeded 32 have 6-times the risk of those with a BMI <22. Equally remarkable was the increased risk of gallstones associated with increasing caloric consumption in nonobese women in this cohort (89). However, short-term overconsumption by clinically normal, nonobese persons does not seem to be associated with increased cholesterol secretion (15), in contrast to the situation for persons afflicted with gallstones and fed a high level of calories and protein (121).

The association of gallstones with obesity (and caloric overconsumption) presumably reflects cholesterol production, or flux, which is substantially elevated in the obese and gallstone patients (15, 119). This is further complicated by high insulin levels in obesity (63). Hyperinsulinemia itself is a risk factor for gallstones (133), possibly because insulin is known to stimulate hepatic cholesterol synthesis (99). Increased hepatic cholesterol synthesis and food intake have been linked to biliary cholesterol secretion in humans in certain circumstances but not in others (19, 112, 119). On the other hand, inhibition of cholesterol synthesis by lovastatin in humans depressed bile acid synthesis and secretion, but not biliary cholesterol secretion (98). However, in cholesterol-fed prairie dogs lovastatin actually prevented gallstone formation (123).

In essence, the liver of obese persons must cope with an increased flux of cholesterol derived both from newly synthesized cholesterol associated with elevated hepatic lipoprotein secretion (VLDL) as well as from an increment of exogenous cholesterol entering the lipoprotein cholesterol pool from the diet. Unlike rats, the ability of humans to dispose of this expanded cholesterol pool is limited by our ability to synthesize bile acids. As a consequence, free biliary cholesterol rises to dangerous concentrations, i.e. the cholesterol solubility index (CSI) or lithogenic index (LI) of gallbladder bile becomes supersaturated (>1.0) and readily exceeds the solubilization threshold, setting the stage for cholesterol nucleation and stone formation. By contrast, the rat (and most other species) readily convert any excess cholesterol to bile acids so that the biliary cholesterol concentration seldom approaches the saturation point.

Since the absolute mass of cholesterol fluxing through the liver into bile acids or biliary cholesterol depends, in part, upon energy intake and the lipoprotein cholesterol returning to the liver, dietary modification of the lipoproteins, both in terms of their relative distribution (VLDL, IDL, LDL, HDL_2, HDL_3) as well as their absolute mass, would appear to have an important bearing on lithogenesis. Unfortunately, minimal definitive information of a clinical nature is available on this subject (143).

The lipoprotein profile in persons with cholesterol gallstones is variable and probably complicated by gender differences. Correlations between gallstones and both depressed HDL (high VLDL) (41, 107, 134, 148a, 150, 151) and slightly elevated HDL (93) profiles have been reported in women and men, respectively. Although the LDL-C concentration appears to be related to the biliary cholesterol concentration in normal people (151), no correlation seems to exist between high LDL-C (type IIa) and gallstones. On the other hand, type IV individuals with elevated VLDL-TG and LDL-C have a high incidence of gallstones (41, 46, 63). This concurs with human data correlating elevated VLDL-C with the CSI of bile (4, 107, 134) as well as with recent findings in cholesterol-fed hamsters that a VLDL-C/HDL-C ratio greater than 1.0 was highly associated with cholesterol gallstone formation (56). In both humans and hamsters under these conditions a cheno profile predominates, suggesting enhanced flow of lipoprotein cholesterol into the liver via non LDLr pathways (see Figures 1 and 2).

The association between gallstones and excessive caloric intake may also reflect increased cholesterol delivery to bile via VLDL remnants and HDL_2 because remnants and the HDL_2/HDL_3 ratio increase postprandially (105a, 143), which could represent a significant portion of the day in individuals with excessive energy consumption.

Rapid weight loss in the obese presents a high risk for lithogenic bile formation, but the lipoprotein dynamics of this situation have not been adequately characterized. It may seem ironic that severe caloric restriction and precipitous weight loss (the metabolic antithesis of obesity) should also induce a high degree of bile supersaturation and rapid gallstone induction (21, 84, 135). Presumably reverse transport of cholesterol as HDL_2 from the shrinking adipose pool (49) is coupled with enhanced biliary vesicle formation; and reduced micelle formation associated with decreased bile acid secretion is known to occur during prolonged periods of fasting, frequent dieting (137), or extended parenteral nutrition (96, 106). In addition, a reduced rate of enterohepatic circulation is associated with decreased gallbladder contraction and increased bile stasis (15, 84, 135), which are thought to increase cholesterol precipitation and gallstone nucleation.

The relevance of the lipoprotein profile to lithogenic bile is also supported by experiments with animals having HDL_2 as a prominent lipoprotein. The Brazilian squirrel monkey, but not the Bolivian strain, is prone to cholesterol gallstones when fed cholesterol. The susceptible Brazilian strain exhibits an HDL_2 profile (like women) while the resistant strain presents an HDL_3 profile (like men) (109, 110). In African green monkeys fed cholesterol-enriched diets, the total HDL-C also correlated well with bile lithogencity (132). Furthermore, a PUFA- and cholesterol-rich diet in these monkeys yielded a higher lithogenic index than the comparable saturated fat diet, a response similar to that of the gallstone-susceptible squirrel monkey but unlike the response of the resistant cebus monkey (9). In African green monkeys, PUFA plus cholesterol reduced the total HDL-C but increased the relative percentage of the HDL_{2a} subfraction compared to expansion of the HDL_{2b} subfraction by saturated fat and cholesterol (11). The hepatic CE pool in African green monkeys also increased with the PUFA-cholesterol diet (70), and in cebus monkeys the circulating cholesterol redistributed from the plasma (coconut oil-fed) into the enterohepatic circulation (corn oil-fed) (143) because polyunsaturated fatty acids up-regulate the hepatic LDL receptors, thereby shrinking the plasma LDL cholesterol pool (103).

The relationship between HDL_2 and gallstones is further supported by studies in the Syrian hamster, which develops cholesterol gallstones under two experimental conditions, i.e. during cholesterol supplementation of a low-fat diet or, less predictably, when all fat and cholesterol are removed from the diet to induce essential fatty acid (EFA) deficiency (6, 30, 37, 55, 72). Although these two nutritional paradigms would seem diametrically opposed, they accomplish the same end by increasing the net flux of cholesterol across the liver and the relative concentration (mol%) of biliary cholesterol. In the EFAD hamster model this is the consequence of exaggerated fatty acid synthesis and a 2–3 fold increase in the hepatic triglyceride secretion rate (55), with the attendant increases in cholesterol synthesis needed for VLDL production (66) and HDL_2 reflux to the liver. In the cholesterol-supplemented model the dietary overload reaches the lipoprotein pool, dramatically expanding HDL_2 initially, followed by expansion of the VLDL cholesterol pool. Eventually, the hepatic FC and CE pools are increased and lithogenic bile and gallstones ensue. A minimum cholesterol concentration of 0.85 mg/kcal of diet appears to be the threshold at which biliary cholesterol saturation (CSI > 1.0) is evidenced in most hamsters (56).

EFAD, obesity, and abrupt weight loss have a common metabolic theme in that all three include extreme reflux of free fatty acids through the liver accompanied by elevated triglyceride and VLDL output. VLDL secretion requires free cholesterol synthesis, which is often associated with bile supersaturation. At least two scenarios for lithogenic bile formation could result from such metabolism. First, as a consequence of increased cholesterol synthesis expanding hepatocyte pools, more HDL-FC may be diverted into bile. A second possibility might be increased secretion of the newly synthesized pool directly into bile. The first possibility seems more likely because HDL_2-FC is considered the preferred substrate for biliary cholesterol, and newly synthesized cholesterol favors bile acid synthesis, not the biliary cholesterol pool.

Whereas expanded pools of CE and FC in the liver are generally associated with gallstone development in most animal models (6), this is not particularly true for humans, and a high concentration of hepatic cholesterol can be maintained without gallstone formation in animals (155) if substantial sequestration and removal of bile acids and neutral sterols is achieved by dietary components (e.g. soluble fiber or sequestering agents). On the other hand, the hepatic cholesterol concentration in EFAD hamsters remains low even though hepatic synthesis of cholesterol is greatly elevated. Gallstones often form because biliary cholesterol saturation increases as phospholipid output declines and nucleating factors in the gallbladder are optimized (80,

115). Depressed lecithin output presumably reflects the lack of available polyunsaturated fatty acids needed for lecithin synthesis, since a minimal amount of polyunsaturated fat (2% w/w) added to the hamster diet greatly reduces the incidence of stones (37) and normalizes hepatic triglyceride secretion (55) along with cholesterol synthesis.

The relative depletion of essential fatty acids (18:2) may pertain to certain cases of human gallstone disease as well, since the unfavorable biliary lipid balance that ensues when cholesterol increases and phospholipids decrease in bile has been noted in humans (100). Collectively, the data suggest that a substantial increase in cholesterol transport to the liver (as apoE rich remnant particles and HDL_2, see Figure 1) may accentuate the flux of free cholesterol into bile. This possibility can be inferred from experiments in humans as well as in animals (116, 125, 126, 141, 143).

Dietary Fat

The overall impact of dietary fat on bile lipids in humans is unclear, possibly because gender and the inherent lipid metabolism, e.g., the lipoprotein profile, of the individual (or population) at the time of dietary challenge with polyenes or saturates may affect the outcome. In addition to the substantial energy load contributed by fat to caloric flux through the liver, the degree of fat saturation itself has been examined for its influence on lithogenesis with equivocal results. In one study, autopsied male patients who consumed a high PUFA diet to reduce their high risk for coronary heart disease (and who were also more obese!) had more gallstones than the leaner control population consuming the usual house diet containing more saturated fat (146). The tendency for PUFA to increase risk in men, but decrease risk in women, was noted by others (42). However, large-scale intervention or epidemiological studies either found no effect of dietary fat in men or women (97, 137) or a protective effect among women (89,000 nurses after 4 years) that was associated with vegetable oil consumption and could not be distinguished from the associated intake of vegetable protein (90). Also, female vegetarians were found to have one-half the incidence of gallstones of nonvegetarian controls (108). Thus, a gender difference may exist in the response to PUFA consumption, with a possible increased risk for men and decreased risk for women.

Metabolic studies of the impact of dietary fat on bile lipids are also equivocal, possibly because of inherent differences in lipid metabolism of the host at the time of dietary intervention. Both normal and hyperlipidemic subjects with and without gallstones have been studied. No consistent pattern has evolved (38, 77, 78), but in some cases (mostly men) PUFA has been found to increase the excretion of biliary cholesterol (15) while decreasing the moles percent of bile acids. A more consistent finding is the PUFA-induced increase in the moles percent of cholate (increased LDLr activity) and decrease in the moles percent of cheno associated with enhanced glycine conjugation (83, 86). A complicating possibility with most of these studies is that dietary PUFA may transiently increase the flux of biliary cholesterol or bile acids in the acute phase until the body has adjusted to a new steady-state of sterol distribution when the difference may no longer exist for one or both biliary sterols.

The protective role of adding PUFA to the EFAD-hamster gallstone model has been noted for its contribution to biliary phospholipid synthesis and secretion (79, 80, 115). Feeding humans diacyl linolyl lecithin enriched bile lecithin with 18:2 (120). Ironically, high PUFA intake reduces the conversion of linoleic acid to arachidonic acid as well as eicosanoid metabolism (44a), in part by limiting δ-6 desaturase activity. In fact, the substitution of butter for 18:2-rich margarine was found to increase the 20:4 content of bile lecithin in humans (38). Since the availability of arachidonate (20:4) for prostaglandin production and mucin secretion by gallbladder mucosa has been preferred as a risk factor for gallstones (28), the butter results would argue against the hypothesis that dietary PUFA (lower 20:4 content of bile lecithin) enhances lithogenesis, at least from the point of view of gallbladder mucin production. A comprehensive experiment examining a wide range of PUFA effects on gallbladder metabolism (prostaglandins) is not yet available.

On the other hand, fish oil, which interferes with normal prostaglandin metabolism, exerts a protective effect against gallstones in hamsters (36, 37) as well as in cholesterol-fed prairie dogs (18) and African green monkeys (129). This protective action was associated with increased 20:5 (n3) in biliary lecithin, which reduced prostaglandin-initiated mucin production by the gallbladder mucosa. Increased phospholipid and reduced biliary cholesterol saturation were noted as well. Fish oil also greatly reduces hepatic and intestinal triglyceride and cholesterol secretion in man and animals (101), which would ultimately reduce the return of lipoprotein cholesterol for excretion into bile. The n-6 polyenes may also reduce triglyceride synthesis and VLDL output relative to saturated fatty acids (70), and thereby act in a fashion similar to n-3-rich oils. In this sense, in the absence of an extreme dietary cholesterol load, polyenes might reduce lithogenesis both by decreasing lipoprotein cholesterol flux into bile and by favoring synthesis of biliary lecithin low in 20:4. On the negative side, fish oil has been reported to accelerate cholesterol nucleation in bile from healthy subjects to a rate comparable to that in gallstone patients, even though the saturation index decreased slightly (67). In rats fish oil increased biliary cholesterol concentration (12), whereas it prevented gallstones in the mouse model (13).

Whether or not fat saturation exerts an influence on bile acid synthesis and pool size has been examined in monkeys (9, 25, 111, 130). Although fat saturation (aside from n-3 fatty acids) does not seem to alter the lithogenic index appreciably, polyunsaturates without added cholesterol stimulated bile acid synthesis and secretion in rhesus monkeys (25, 111). Adding cholesterol to squirrel monkey diets depressed the relative distribution of taurocheno by at least 30% relative to taurocholate (9), producing a shift in primary bile acid composition opposite to that found in hamsters (30), prairie dogs (31), and humans with gallstones (23, 59, 60). Polyenes plus cholesterol fed to African green monkeys increased the lithogenic index and gallstones relative to saturated fat plus cholesterol in association with a marked decrease in bile acid secretion (130). The most consistent production of gallstones in monkeys has been in Brazilian squirrel monkeys fed butter and cholesterol (0.9 mg/kcal), although the type of fat did not appear to have an appreciable effect when sufficient cholesterol was present (109).

Dietary Cholesterol

In epidemiological studies, high intake of cholesterol ironically appears to have a protective effect against gallstones (42, 108, 134). In contrast, the experimental consumption of cholesterol by men and women tends to increase the relative concentration of biliary cholesterol and decrease the moles percent of bile acids in both healthy persons (40, 81) and subjects with gallstones (81). Women have not responded to a cholesterol challenge in other studies (5, 37).

Feeding excessive cholesterol (> 0.85 mg/kcal or 2 g/day human equivalent) is essentially the only way to regularly induce gallstones in most animal models (see section on models below). Under these conditions bile is typically enriched with cholesterol and reduced in bile acids (e.g. see 56). A possible mechanism would be that inhibition of hepatic cholesterol synthesis impairs bile acid production in a manner similar to the coordinated inhibition of both cholesterol and bile acid synthesis by lovastatin in humans (98), while apoE-rich lipoproteins deliver the excess absorbed cholesterol directly to bile (Figure 1).

Ironically, adding dietary cholesterol to the EFAD hamster model decreases the incidence of cholesterol gallstones while enhancing pigment stone formation (37). This presumably reflects absorbed cholesterol causing feedback inhibition of hepatic cholesterol synthesis.

Dietary cholesterol also affects the amount and composition of lecithin in bile (38, 115). When prairie dogs are fed cholesterol, arachidonic acid in bile lecithin increases while linoleic acid decreases. In hamsters and patients with untreated gallstones, palmitic and arachidonic acids increase while linoleic acid decreases (1, 26, 27, 159, 160). The more hydrophobic bile acids (e.g. deoxycholate, cheno) are associated with biliary excretion of 20:4- and 18:0-rich lecithins, and arachidonic acid in bile might be expected to induce the production of prostaglandins and mucin (105, 159).

Dietary Protein

In humans no direct evidence indicates that the amount or type of protein consumed affects cholesterol gallstone incidence; however, epidemiological studies reveal a lower incidence of gallstones in vegetarians than in omnivores (108), and vegetable protein consumption may have a protective effect against gallstone occurrence in women (90). Distinguishing vegetable protein from the possible contribution by vegetable oil or fiber is difficult in such cases.

In hamster studies when dietary casein was replaced by soy protein or other vegetable proteins a "decreased incidence of gallstones" resulted (147), but the character and diversity of stones (cholesterol or pigment) has not been sufficiently documented to determine the protein impact on pure cholesterol lithogenesis.

Theoretically, it is possible that dietary sulfur amino acids, including taurine, could influence bile acid conjugation and secretion and/or lipoprotein clearance rates (144), but a relationship between taurine and cholesterol stones in humans is not evident from the literature. Although feeding taurine tends to increase the taurine conjugation of bile acids (14, 52, 76, 138, 141, 148), neither the biliary lipid profile nor bile lithogenicity appears to be affected in taurine-supplemented humans (52, 54, 156). A possible exception is the ameliorating effect on gallstones of a high taurine intake (5%) in mice consuming a lithogenic diet (162).

In infant monkeys the depletion of available taurine exerted a substantial influence on the bile acid profile without affecting the lithogenic index in either cebus or cynomolgus monkeys (141). Specifically, taurine-depleted cynomolgus reduced taurine conjugation of bile salts while increasing glycine conjugates, but the loss of taurine (gain in glycine) was almost exclusively observed in the cholate, and not cheno pool; the latter seemed to be preferentially conserved as taurocheno. Taurine depletion was linked to a relative increase in biliary phospholipid and a 4-fold increase in the taurocheno/taurocholate ratio. These data suggest that distortions in the bile acid profile (including conjugation pattern) of gallstone-bearing individuals may not be causative but rather may be indicative of abnormal hepatic sterol metabolism.

Why taurine appears closely linked to cheno metabolism and glycine to cholate, and whether these associations have any relevance to human gallstones, is unclear. Nevertheless, these relationships have been observed during manipulation of human (86) and monkey (9) bile lipids by dietary fat and of hamster bile lipids by dietary fiber (157). Specifically, humans and Brazilian squirrel monkeys fed corn oil lowered their plasma cholesterol, presumably increasing LDL receptor activity (103), which led to a rise in glycocholate. Coconut oil, which increases LDL, VLDL, and HDL (by down-regulating the LDL receptors), resulted in an expanded taurocheno pool. Thus, a $VLDL/HDL_2$-cheno-taurine precursor-product linkage appears to be in contrast to an LDL-cholate-glycine connection. These observations reaffirm the association between gallstones, an HDL_2 profile (women), and an expanded cheno pool, whereas a low HDL_2 profile (males) relies on LDL-C delivery to yield more glycocholate and less gallstones, but more atherosclerosis, from the expanded plasma LDL pool.

Taurine depletion might decrease LDLr activity, divert lipoprotein clearance to HDL_2 for return to the liver, and favor the taurocheno profile observed in monkeys (141). In addition, taurine or cysteine supplementation of HEP-G2 cells increased LDLr activity and bile acid secretion independent of any effect of taurine conjugation, since HEP-G2 cells secrete unconjugated bile acids (144). Stimulation of bile acid synthesis by sulfur amino acids apparently is related to enhanced availability of a sulfhydryl protein (39), which may pertain to the whole animal as well, since hamsters fed a taurine supplement reportedly increased their bile acid output and reduced the biliary moles percent of cholesterol concentration (14). One potential advantage of taurine is that increased conversion of cholesterol to bile acids by taurine (especially taurine conjugates) would increase the opportunity for bile acid removal by naturally occurring dietary fibers, thus facilitating the steady removal of bile acids and diverting hepatic cholesterol from direct secretion into bile (see section on fiber below).

Dietary Carbohydrate

Certain epidemiological evidence suggests that persons consuming highly refined carbohydrates are more apt to have cholesterol gallstones than populations eating more complex forms of carbohydrate (42, 134, 149). However, when confounding factors were controlled in a large epidemiologic study, a carbohydrate effect was not apparent (90) but a protective influence of vegetable protein was noted that could not be separated from vegetable oil.

In contrast to the limited data on humans, the literature describing the impact of carbohydrate on gallstones in hamsters is extensive. Dam (36, 37) was the first to report that the development of cholesterol stones in Syrian hamsters fed the EFAD diet depended on all the carbohydrate being fed as glucose or sucrose, as opposed to other more complex sugars and starches. Lactose was also protective whereas a galactose and glucose mix was not, indicating that the protection afforded by lactose reflected its ability to partially bypass digestion and absorption in the small intestine, to reach the large bowel, and to stimulate bacterial flora activity. Lactose feeding results in lower plasma lipids, especially triglycerides, and increased bile acid turnover and secretion with a favorable impact on the lithogenic index, i.e. decreasing bile cholesterol supersaturation and preventing cholesterol gallstones in the EFAD model or pigment stones in hamsters fed a purified diet with fat added (55, 57). Cholesterol supplementation of the latter diet can eventually exceed the protection against cholesterol stones afforded by lactose, even though lactose may increase large bowel size 3–4-fold (55). To a certain extent various complex carbohydrates (rice flour, cornstarch) exert a similar protective effect depending on the amount of digesta that reaches the cecum.

In the absence of sufficient nutrients reaching the large bowel, cecal flora activity wanes and the cecum atrophies, eventually resulting in diarrhea and death. This "wet tail" syndrome appears in a variety of forms, but is widely recognized as a deterrent to sustained nutritional studies in hamsters (57).

In fact, the cecum is larger than the liver in the suckling hamster at the time of weaning, presumably due to the lactose in suckled milk (K. C. Hayes, unpublished data). Thus lactose may assure the early, rapid development of normal large bowel flora to act as a counterbalance [via bile acid catabolism and generation of short-chain fatty acids to modulate hepatic lipid metabolism (153)] to an expanding lipoprotein cholesterol pool in the newborn. If lactose is continued as the major source of carbohydrate in the post-weaning diet, the cecum continues to outweigh the liver, plasma triglycerides and cholesterol concentrations are reduced, and cholesterol gallstones are prevented or reduced in number (37, 57). Ordinarily, a small amount of complex carbohydrate and an increment of dietary fiber would reach the cecum in mature hamsters to sustain the cecum. Note that once the cecum "reaches maturity" it becomes extremely difficult to induce cholesterol gallstones in the EFAD hamster model (37). Based on relative cecal weight, the "stressed" cecal flora would appear to fare better with dietary PUFA, possibly delivered by the fiber component of the diet (57). Previous reports also demonstrate the differential effect of polyenes and saturates on cultures of large bowel bacteria (104). The protective role of age on gallstone susceptibility in EFAD hamsters may reflect the accumulation of PUFAs in adipose reserves that protect against the metabolic consequences of EFAD. These PUFA stores also may impact the metabolism of the large bowel flora.

Dietary Fiber

From the above discussion on carbohydrates it should be apparent that dietary fiber might influence gallstone incidence via its impact on large bowel metabolism. This hypothesis was first expounded by Burkitt & Trowell (22) for humans when they noted the relative absence of chronic diseases (including diabetes, gallstones, coronary heart disease, appendicitis, diverticulitis) in Third World populations and attributed this to the consumption of complex carbohydrates and fiber. The fiber hypothesis is supported by some epidemiologic data (134), but not by others (100). Dietary fiber includes a wide array of complex substances commonly divided into soluble and insoluble components.

A number of metabolic studies have attempted to establish the fiber-gallstone link in humans (71, 73). Insoluble and soluble fibers exert different effects. It is known that slow fecal transit time (constipation) is associated with elevated deoxycholate (a secondary bile acid) and gallstones (91). In general, insoluble fibers are good bulking agents and increase fecal transit rate and bile acid removal. Simply doubling the dietary mixed-fiber load (12 to 25 g/per day) increased fecal bile acid output 13% (74). Several investigators have examined the effects of wheat bran with varied results. Humans with gallstones or highly lithogenic bile tend to improve when fed wheat bran, associated with a decrease in deoxycholate, whereas normal persons are not affected (71, 73). However, normal individuals fed high mixed-fiber diets or pectin tend to increase deoxycholate while decreasing cheno (71, 73). Experiments designed to vary bowel transit time (92) indicate that rapid transit decreases the opportunity for colonic flora to generate secondary bile acids. Deoxycholate can be considered lithogenic, in part, because it is associated with increased arachidonic acid in biliary lecithin and because it can transport the most cholesterol into bile (8, 105, 159).

Western diets high in sucrose and low in fiber may also increase gallstones because such a diet is highly glycemic and insulinemic, the latter stimulating cholesterol synthesis (99) in conjunction with elevated triglyceride (VLDL) secretion by the liver. Fiber, especially soluble fibers like pectin and psyllium, lowers the glycemic index and reduce insulin secretion (69), which decreases an important risk factor for gallstones (133).

Soluble fibers (64, 155, 157) also exert an interesting selective removal of taurine-conjugated bile acids by virtue of their gelling action. Because taurine conjugates are not passively absorbed by the small intestine and thus remain longer intraluminally than glycine conjugates, taurine-conjugated bile acids (especially cheno) become entrapped and bulk-removed by fiber gels. In hamsters the net effect was to reduce lithogenic bile and gallstone formation (155). Whether this is applicable to humans is unexplored, although psyllium fed to hypercholesteremic humans does increase LDL receptor activity and decreases plasma LDL without increasing cholesterol synthesis or absorption (7, 47).

Colonic fermentation of nonstarch polysaccharides (including fibers) is a rapidly evolving area sure to expand our knowledge of bile lipid metabolism and lithogenesis. Currently it is unclear how volatile short chain fatty acids, released during bacterial colonic fermentation, impact cholesterol metabolism, but research is actively exploring this aspect of lipid metabolism (153).

Experiments in hamsters have demonstrated the protective role of both soluble (psyllium) and insoluble (lignin, "lactulose") fibers as well as artificial resins (cholestyramine) against cholesterol gallstone formation (16, 118, 155). The mechanism of this protection is not fully appreciated, but fibers fed to hamsters have revealed a variety of effects on the bile acid profile, and cholestyramine is known for its efficient binding of bile acids in the gut (45). Hamsters fed 1% cholestyramine revealed dramatic lowering of chenodeoxycholate, which appeared to interfere with cholesterol absorption. The result was lower plasma cholesterol, prevention of hepatic cholesterol accumulation, an increase in the cholate bile acid profile, and lower LI while preventing gallstones. By comparison, cholesterol-fed (1.12 mg/kcal) control hamsters experienced elevated lipids, hepatic cholesterol deposition, a high cheno profile, bile supersaturated with cholesterol, and a high incidence of cholesterol stones (155). Psyllium, on the other hand, prevented cholesterol gallstones by the less efficient, but selective removal of taurine-conjugated bile acids. Psyllium was associated with moderately lower blood lipids without prevention of hepatic cholesterol accumulation, but the LI was reduced to the point of minimal supersaturation.

In rats citrus pectin mediated an increase in bile acid secretion that depleted the hepatic taurine pool available for bile acid conjugation and increased glycine conjugation of bile acids (64). Rats fed methionine and taurine in addition to pectin experienced a striking increase in hepatic taurine concentration, and glycine conjugation of bile acids was almost totally replaced by taurine conjugates (65).

Thus, it appears that the fiber effect is multifaceted and potentially extremely complex in its influence on lipoproteins, bile lipids, and intestinal sterols, including metabolism by the large bowel flora.

Alcohol

Although a protective effect of alcohol consumption has been noted in an epidemiologic study (134) as well as experimentally in cholesterol-fed prairie dogs (128) and in humans in terms of the bile LI (152), other experimental results are inconsistent as to possible mechanisms. An acute increase in bile acid synthesis (10) and output with a minimal change in total sterol excretion has been noted in some hyperlipidemic patients (102) and pigs (154), but not in other individuals (35). Again, the metabolic status of the host may be an important variable.

ANIMAL MODELS

Selection of an animal model for gallstone study is limited by the fact that relatively few species readily develop cholesterol gallstones under any circumstance, and none, other than humans, are thought to spontaneously develop supersaturated gallbladder bile (58). The reason for this has not been fully delineated, but a key limitation is the failure of most species to secrete supersaturated bile, i.e. conversion of hepatic free cholesterol to cholesteryl esters and bile acids is sufficient to preclude excess free cholesterol from entering bile (3).

The fact that none of the models (with the exception of the EFAD hamster) develop gallstones without extraordinary cholesterol loading is probably a serious limitation of their relevance to the actual human experience unless careful monitoring of the prevailing lipoprotein dynamics elicited by the total dietary regimen is included for comparison. As indicated earlier, the influence of PUFA on bile acid secretion may be totally reversed by the presence of excess dietary cholesterol. Since humans are not "cholesterol loaded" in the way animals models tend to be, results from animal studies must be interpreted with care.

Among species whose bile becomes supersaturated with cholesterol feeding (hamster, prairie dog, ground squirrel, and certain monkeys), not all are equally endowed with the "other" prerequisites of sufficient gallbladder mucin production such as sludge formation, bile stasis, nucleating factors, etc, to allow cholesterol nucleation and stone formation once cholesterol supersaturation is achieved (61).

Consequently, only two species are commonly used for gallstone modeling based on their size, availability, and relative cost effectiveness. These are the Syrian hamster (30, 37, 56, 85) and the prairie dog (31), both of which develop lithogenic bile and stones in 3 to 8 weeks depending primarily on the amount of cholesterol fed (0.3–2.0%). Although of limited availability, the cholesterol-fed ground squirrel responds to cholesterol feeding as well.

The Brazilian squirrel monkey fed cholesterol (0.3% or more) is also an excellent model (109, 110), but these monkeys are no longer available for general study. Other monkeys such as the African green monkey (129, 130) and baboon develop stones albeit at a reduced incidence, whereas others such as rhesus, cynomolgus, Bolivian squirrel monkey, and cebus monkeys, often used in cardiovascular research, do not readily form cholesterol stones even when fed cholesterol. The reason for this comparative resistance or susceptibility has not been studied in detail, but the cholesterol-fed cebus (HDL_3 predominates) has a relatively greater bile acid pool size than the Brazilian squirrel monkey (HDL_2 profile), which increases the moles percent of biliary cholesterol under comparable conditions of cholesterol feeding (9). The fasting cebus also has three times the bile acid secretion rate of the cynomolgus, another resistant species (142). On the other hand, Brazilian squirrel monkeys, but not the Bolivian strain, reportedly expand their HDL and VLDL cholesterol pools like the cholesterol-fed hamster (109, 110).

Syrian hamsters have the decided advantage of availability, cost effectiveness, and relative uniformity, but their size (adult 125–160 gm) and inconsistent induction of gallstones can limit their general applicability in certain experiments. The complicating factor of the high incidence of pigment gallstones in hamsters has been noted (37) and described in detail (32, 57). The pathogenesis of pigment stones is relatively unexplored, but the high incidence of these stones serves to underscore the facility with which hamsters develop them. Feeding hamsters purified diets, per se, without added cholesterol, seems to enhance pigment stone formation (32, 55, 57). Many reports in hamsters are sufficiently vague about the composition of the induced stones to render their results of questionable value in a discussion focusing specifically on cholesterol gallstones.

Until recently (57) the inability to prevent the lethal enteritis "wet tail" in hamsters fed purified diets has limited long-term (several weeks) study of nutritional factors, which necessitates the use of purified diets. Prairie dogs are larger (1–2 kg), but their nutritional requirements and the application of purified diets in this model are poorly defined. Mice hold great promise for study of the genetic variables involved in lithogenesis but they only develop cholesterol stones when fed high (0.4–2.0%) levels of dietary cholesterol, usually with cholic acid added to the diet to enhance cholesterol absorption and depress bile acid synthesis (3).

SUMMARY

Although dietary factors influence bile lithogenicity and gallstone formation, the main dietary effect appears to be indirect, depending on an interaction between caloric consumption and gender-specific aspects of lipoprotein metabolism. Excessive energy intake elicits its detrimental effect by altering lipoprotein and hepatic cholesterol metabolism in association with hyperinsulinemia.

Factors, dietary and genetic, that favor elevated hepatic cholesterol synthesis and production of a bile acid profile in which chenodeoxycholic acid predominates appear to be associated with lithogenic bile. An inconsistent effect of dietary fat saturation on gallstones is that polyunsaturates possibly increase risk in men and decrease risk in women. Vegetable protein may reduce the risk of cholelithiasis. Whereas both the amount and type of dietary fiber influence cholesterol and bile lipid metabolism, specific associations between fiber and gallstones in humans remain elusive.

ACKNOWLEDGMENTS

This work was supported in part by NIH grant DK 35375 and by a fellowship (Elke A. Trautwein) from Deutsche Forschungsgemeinschaft, Bonn, Germany.

Literature Cited

1. Ahlberg, J., Curstedt, T., Einarsson, K., Sjovall, J. 1981. Molecular species of biliary phosphatidylcholines in gallstone patients: the influence of treatment with cholic acid and chenodeoxycholic acid. *J. Lipid Res.* 22:404–8
2. Albers, J. J., Grundy, S. M., Cleary, P. A., Small, D. M., Lachin, J. M., et al. 1982. National cooperative gallstone study, the effect of chenodeoxycholic acid on lipoproteins and apoproteins. *Gastroenterology* 82:638–46
3. Alexander, M., Portman, O. W. 1987. Different susceptibilities to the formation of cholesterol gallstones in mice. *Hepatology* 7:257–65
4. Alvaro, D., Angelico, F., Attili, A. F., Antonini, R., Mazzarella, B., et al.

1986. Plasma lipoproteins and biliary lipid composition in female gallstone patients. *Biomed. Biochim. Acta* 54: 761–68

5. Andersen, E., Hellstrom, K. 1979. The effect of cholesterol feeding on bile acid kinetics and biliary lipids in normolipidemic and hypertriglyceridemic subjects. *J. Lipid Res.* 20:1020–27

6. Anderson, J. M., Cook, L. R. 1986. Regulation of gallbladder cholesterol concentration in the hamster. Role of hepatic cholesterol level. *Biochim. Biophys. Acta* 875:582–92

7. Anderson, J. W., Zettwoch, N., Feldman, T., Tietyen-Clark, J., Oeltgen, P., Bishop, W. 1988. Cholesterol-lowering effects of psyllium hydrophilic mucilloid for hypercholesterolemic men. *Arch. Intern. Med.* 148:292–96

8. Armstrong, M. J., Carey, M. C. 1982. The hydrophobic-hydrophilic balance of bile salts. Inverse correlation between reverse-phase high performance liquid chromatographic mobilities and micellar cholesterol-solubilizing capacities. *J. Lipid Res.* 23:70–80

9. Armstrong, M. J., Stephan, Z., Hayes, K. C. 1982. Biliary lipids in new world monkeys: dietary cholesterol, fat, and species interactions. *Am. J. Clin. Nutr.* 36:592–601

10. Axelson, M., Mork, B., Sjovall, J. 1991. Ethanol has an acute effect on bile acid biosynthesis in man. *FEBS Lett.* 281:155–59

11. Babiak, J., Lindgren, F. T., Rudel, L. L. 1988. Effects of saturated and polyunsaturated dietary fat on the concentrations of HDL subpopulations in African green monkeys. *Arteriosclerosis* 8:22–32

12. Balasubramaniam, S., Simons, L. A., Chang, S., Hickie, J. B. 1985. Reduction in plasma cholesterol and increase in biliary cholesterol by a diet rich in n-3 fatty acids in the rat. *J. Lipid Res.* 26:684–89

13. Banerjee, B., Singh, E. 1989. Fish oil prevents cholesterol crystal formation and development of gall stones in mice. *Gut* 30:A1459

14. Bellentani, S., Pecorari, M., Cordoma, P., Marchegiano, F., Manenti, F., et al. 1987. Taurine increases bile acid pool size and reduces bile saturation index in the hamster. *J. Lipid Res.* 28:1021–27

15. Bennion, L. J., Grundy, S. M. 1978. Risk factors for the development of cholelithiasis in man. *New Engl. J. Med.* 299:1161–67, 1221–27

16. Bergman, F., van der Linden, W. 1975. Effect of dietary fibre on gallstone

formation in hamsters. *Z. Ernährungswiss.* 14:217–23

17. Bhattacharya, S., Balasubramaniam, S., Simons, L. A. 1986. Quantification of LDL cholesteryl ester contribution to biliary steroids in the rat. *Biochim. Biophys. Acta* 876:413–16

18. Booker, M. L., Scott, T. E., La Morte, W. W. 1990. Effects of dietary fish oil on biliary phospholipids and prostaglandin synthesis in the cholesterol-fed prairie dog. *Lipids* 25:27–32

19. Bouchier, I. A. D. 1984. Debits and credits: a current account of cholesterol gallstone disease. *Gut* 25:1021–28

20. Bravo, E., Cantafora, A. 1990. Hepatic uptake and processing of free cholesterol from different lipoproteins with and without sodium taurocholate administration. An in vivo study in the rat. *Biochim. Biophys. Acta* 1045:74–80

21. Broomfield, P. H., Chopra, R., Scheinbaum, R. C., Bonorris, G. G., Silverman, A., et al. 1988. Effects of ursodeoxylcholic acid and aspirin on the formation of lithogenic bile and gallstones during loss of weight. *New Engl. J. Med.* 319:1567–72

22. Burkitt, D. P., Trowell, H. C. 1977. Dietary fibre and Western diseases. *Ir. Med. J.* 70:272–77

23. Burnett, W. 1965. The pathogenesis of gall stones. In *The Biliary System*, ed. W. Taylor, pp. 601–11. Philadelphia: F. A. Davis

24. Busch, N. B., Holzbach, R. T. 1990. Crystal growth-inhibiting proteins in bile. *Hepatology* 12:195S–99S

24a. Busch, N., Matern, S. 1991. Current concepts in cholesterol gallstone pathogenesis. *Eur. J. Clin. Invest.* 21:453–60

25. Campbell, C. B., Cowley, D. J., Dowling, R. H. 1972. Dietary factors affecting biliary lipid secretion in the rhesus monkey. *Eur. J. Clin. Invest.* 2:332–41

26. Cantafora, A., Angelico, M., DiBiase, A., Pieche, U., Bracci, F., et al. 1981. Structure of biliary phosphatidylcholine in cholesterol gallstone patients. *Lipids* 16:589–92

27. Carey, M. C. 1989. Formation of cholesterol gallstones: the new paradigms. In *Trends in Bile Acid Research,* ed. G. Paumgartner, A. Stiehl, W. Gerde, pp. 259–81. Boston: Kluwer Acad. Publ.

28. Carey, M. C., Cahalane, M. J. 1988. Whither biliary sludge? *Gastroenterology* 95:508–23

29. Carey, M. C., Mazer, N. A. 1984. Biliary lipid secretion in health and in

cholesterol gallstone disease. *Hepatology* 4:31S–37S

30. Cohen, B. I., Matoba, N. M., Mosbach, E. H., McSherry, C. K. 1989. Dietary induction of cholesterol gallstones in hamsters from three different sources. *Lipids* 24:151–56

31. Cohen, B. I., Mosbach, E. H., McSherry, C. K., Stenger, R. J., Kuroki, S., et al. 1986. Gallstone prevention in prairie dogs: Comparison of chow vs semisynthetic diets. *Hepatology* 6:874–80

32. Cohen, B. I., Setoguchi, T., Mosbach, E. H., McSherry, C. K., Stenger, R. J., et al. 1987. An animal model of pigment cholelithiasis. *Am. J. Surg.* 153:130–38

33. Cohen, D. E., Carey, M. C. 1991. Acyl chain unsaturation modulates distribution of lecithin molecular species between mixed micelles and vesicles in model bile. Implications for particle structure and metastable cholesterol solubilities. *J. Lipid Res.* 32:1291–1302

34. Cooper, A. D. 1991. Metabolic basis of cholesterol gallstone disease. *Gastroenterol. Clin. N. Am.* 20:21–46

35. Crouse, J. R., Grundy, S. M. 1984. Effects of alcohol on plasma lipoproteins and cholesterol and triglyceride metabolism in man. *J. Lipid Res.* 25:486–96

36. Dam, H. 1969. Nutritional aspects of gallstone formation with particular reference to alimentary production of gallstones in laboratory animals. *World Rev. Nutr. Diet.* 11:199–239

37. Dam, H. 1971. Determinates of cholesterol cholelithisis in man and animals. *Am. J. Med.* 51:596–613

38. Dam, H., Kruse, I., Krough-Jensen, M., Kallehauge, H. E. 1966. Studies in human bile: II. Influence of two different fats on the composition of human bile. *Scand. J. Clin. Lab. Invest.* 19:367–78

39. Danielson, H., Kalles, I., Wikvall, K. 1984. Regulation of hydroxylations in biosynthesis of bile acids. *J. Biol. Chem.* 259:4258–62

40. DenBesten, L., Connor, W. E., Bell, S. 1973. The effect of dietary cholesterol on the composition of human bile. *Surgery* 73:266–73

41. Diehl, A. K., Haffner, S. M., Hazuda, H. P., Stern, M. P. 1987. Coronary risk factors and clinical gallbladder disease: An approach to the prevention of gallstones? *Am. J. Public Health* 77:841–45

42. Diehl, A. K., Haffner, S. M., Knapp, J. A., Hazuda, H. P., Stern, M. P. 1989. Dietary intake and the prevalence of gallbladder disease in Mexican Americans. *Gastroenterology* 97:1527–33

43. Donovan, J. M., Benedek, G. B.,

Carey, M. C. 1987. Formation of mixed micelles and vesicles of human apolipoproteins A-I and A-II with synthetic and natural lecithins and the bile salt sodium taurocholate: Quasi-elastic light scattering studies. *Biochemistry* 26: 8125–33

44. Donovan, J. M., Carey, M. C. 1990. Separation and quantitation of cholesterol "carriers" in bile. *Hepatology* 12: 94S–105S

44a. Dupont, J., Dowd, M. K. 1990. Icosanoid synthesis as a functional measurement of essential fatty acid requirement. *J. Am. Coll. Nutr.* 9:272–76

45. Einarsson, K., Ericsson, S., Ewerth, S., Reihner, E., Rudling, E., et al. 1991. Bile acid sequestrants: mechanisms of action on bile acid and cholesterol metabolism. *Eur. J. Clin. Pharmacol.* 40:S53–S58

46. Einarsson, K., Hellstrom, K., Kallner, M. 1975. Gallbladder disease in hyperlipoproteinaemia. *Lancet* 1:484–87

47. Everson, G. T., Daggy, B. P., McKinley, C., Story, J. A. 1991. Dietary fiber (psyllium) lowers LDL cholesterol by stimulating bile acid synthesis in man. *Hepatology* 14:148a

48. Ford, R. P., Botham, K. M., Suckling, K. E., Boyd, G. S. 1985. The effect of a rat plasma high-density lipoprotein subfraction on the synthesis of bile salts by rat hepatocyte monolayers. *FEBS Lett.* 179:177–80

49. Galbraith, W. B., Connor, W. E., Stone, E. D. 1966. Weight loss and serum lipid changes in obese subjects given low calorie diets of varied cholesterol content. *Ann. Intern. Med.* 64:268–76

50. Groen, A. K. 1990. Nonmucus glycoproteins as pronucleating agents. *Hepatology* 12:189S–94S

51. Halloran, L. G., Schwartz, C. C., Vlahcevic, Z. R., Nisman, R. M., Swell, L. 1978. Evidence for high-density lipoprotein-free cholesterol as the primary precursor for bile acid synthesis in man. *Surgery* 84:1–7

52. Hardison, W. G. M., Grundy, S. M. 1983. Effect of bile acid conjugation pattern on bile acid metabolism in normal humans. *Gastroenterology* 84:617–20

53. Harvey, P. R. C., Upadhya, G. A., Strasberg, S. M. 1991. Immunoglobulins as nucleating proteins in gallbladder bile of patients with cholesterol gallstones. *J. Biol. Chem.* 266:13996–14003

54. Hayes, K. C. 1988. Taurine nutrition. *Nutr. Res. Rev.* 1:99–113

55. Hayes, K. C., Khosla, P., Kaiser, A., Yeghiazarians, V., Pronczuk, A. 1992. Dietary fat and cholesterol modulate the plasma lipoprotein distribution and production of pigment or cholesterol gallstones in hamsters. *J. Nutr.* 122:374–84

56. Hayes, K. C., Khosla, P., Pronczuk, A. 1991. Diet-induced type IV-like hyperlipidemia and increased body weight are associated with cholesterol gallstones in hamsters. *Lipids* 26:729–35

57. Hayes, K. C., Stephan, Z. F., Pronczuk, A., Lindsey, S., Verdon, C. 1989. Lactose protects against estrogen-induced pigment gallstones in hamsters fed nutritionally adequate purified diets. *J. Nutr.* 119:1726–36

58. Hofmann, A. F. 1988. Pathogenesis of cholesterol gallstones. *J. Clin. Gastroenterol.* 10:S1–S11

59. Hofmann, A. F. 1990. Bile acid secretion, bile flow and biliary lipid secretion in humans. *Hepatology* 12:17S–22S

60. Hofmann, A. F., Grundy, S. M., Lachin, J. M., Lan, S., Baum, R. A., et al. 1982. Pretreatment biliary lipid composition in white patients with radiolucent gallstones in the National Cooperative Gallstone Study. *Gastroenterology* 83:738–52

61. Holzbach, R. T. 1984. Animal models of cholesterol gallstone disease. *Hepatology* 4:191S–98S

62. Holzbach, R. T. 1990. Current concepts of cholesterol transport and crystal formation in human bile. *Hepatology* 12:26S–32S

63. Howard, B. 1986. Obesity, cholelithiasis, and lipoprotein metabolism in man. *Atheroscler. Rev.* 15:169–86

64. Ide, T., Horii, M. 1989. Predominant conjugation with glycine of biliary and lumen bile acids in rats fed on pectin. *Br. J. Nutr.* 61:545–57

65. Ide, T., Horii, M., Kawashima, K., Yamamoto, T. 1989. Bile acid and hepatic taurine concentration in rats fed on pectin. *Br. J. Nutr.* 62:539–50

66. Iijima, Y., Yamazaki, M., Maruyama, M. 1979. Effects of dietary fatty acids on hepatic HMG-CoA reductase activity in hamsters fed a high-glucose diet. *Arch. Biochim. Biophys.* 196:265–69

67. Janowitz, P., Swobodnik, W., Wechsler, J. G., Janowitz, A., Saal, D., et al. 1991. Fish oil enriched with polyunsaturated fatty acids of the omega-3-type accelerates the nucleation time in healthy subjects. *Klin. Wochenschr.* 69:289–93

68. Janowitz, P., Wechsler, J. G., Janowitz, A., Kuhn, K., Swobodnik, W., et al. 1991. Nucleation time, cholesterol saturation index and biliary bile acid pattern. *Scand. J. Gastroenterol.* 26:367–73

69. Jenkins, D. J. A., Jenkins, A. L. 1987. The glycemic index, fiber, and the dietary treatment of hypertriglyceridemia and diabetes. *J. Am. Coll. Nutr.* 6:11–17

70. Johnson, F. L., St. Clair, R. W., Rudel, L. L. 1985. Effects of the degree of saturation of dietary fat on the hepatic production of lipoproteins in the African green monkey. *J. Lipid Res.* 26:403–17

71. Judd, P. A. 1985. Dietary fibre and gallstones. In *Dietary Fibre Perspectives: Reviews and Bibliography*, ed. A. R. Leeds, 1:40–46. London: John Libbey

72. Kajiyama, G., Kubota, S., Sasaki, H., Kawamoto, T., Miyoshi, A. 1980. Lipid metabolism in the development of cholesterol gallstones in hamsters. I. Study on the relationship between serum and biliary lipids. *Hiroshima J. Med. Sci.* 29:133–41

73. Kay, R. M. 1982. Dietary fiber. *J. Lipid Res.* 23:221–42

74. Kesäniemi, Y. A., Tarpila, A., Miettinen, T. A. 1990. Low vs high dietary fiber and serum, biliary, and fecal lipids in middle-aged men. *Am. J. Clin. Nutr.* 51:1007–12

75. Kibe, A., Holzbach, R. T., LaRusso, N. F., Mao, S. J. T. 1984. Inhibition of cholesterol crystal formation by apolipoproteins in supersaturated model bile. *Science* 225:514–16

76. Kibe, A., Wake, C., Kuramoto, T., Hoshita, T. 1980. Effect of dietary taurine on bile acid metabolism in guinea pigs. *Lipids* 15:224–29

77. Kohlmeier, M., Schlierf, G., Stiehl, A. 1988. Metabolic changes in healthy men using fat-modified diets. II. Composition of biliary lipids. *Ann. Nutr. Metab.* 32:10–14

78. Kohlmeier, M., Stricker, G., Schlierf, G. 1985. Influences of "normal" and "prudent" diets on biliary and serum lipids in healthy women. *Am. J. Clin. Nutr.* 42:1201–5

79. Kubota, S., Kajiyama, G., Sasaki, H., Horiuchi, I., Miyoshi, A. 1981. Lipid metabolism in the development of cholesterol gallstones in hamsters. IV. The effect of essential phospholipids and plant sterols on the biliary lipids. *Hiroshima J. Med. Sci.* 30:301–9

80. Kubota, S., Kajiyama, G., Sasaki, H., Kawamoto, T., Miyoshi, A. 1980. Lipid metabolism in the development of cholesterol gallstones in hamsters. II. The effect of dietary cholesterol on biliary phospholipids and gallstone formation. *Hiroshima J. Med. Sci.* 29:143–53

81. Lee, D. W. T., Gilmore, C. J., Bonorris, G., Cohen, H., Marks, J. W., et al. 1985. Effect of dietary cholesterol on biliary lipids in patients with gallstones and normal subjects. *Am. J. Clin. Nutr.* 42:414–20

82. Lee, S. P. 1990. Pathogenesis of biliary sludge. *Hepatology* 12:200S–5S

83. Lewis, B. 1958. Effect of certain dietary oils on bile-acid secretion and serum-cholesterol. *Lancet* 1:1090–92

84. Liddle, R. A., Goldstein, R. B., Saxton, J. 1989. Gallstone formation during weight-reduction dieting. *Arch. Intern. Med.* 149:1750–53

85. Liepa, G. U., Gorman, M. A., Duffy, A. M. 1988. The use of animals in studying the effects of diet on gallstone formation. *Comp. Anim. Nutr.* 6:149–73

86. Lindstedt, S., Avigan, J, Goodman, D. S., Sjovall, J., Steinberg, D. 1965. The effect of dietary fat on the turnover of cholic acid and on the composition of the biliary bile acids in man. *J. Clin. Invest.* 44:1754–65

87. Low-Beer, T. S. 1985. Nutrition and cholesterol gallstones. *Proc. Nutr. Soc.* 44:127–34

88. Mackinnon, A. M., Drevon, C. A., Sand, T. M., Davis, R. A. 1987. Regulation of bile acid synthesis in cultured rat hepatocytes: stimulation by apoE-rich high density lipoproteins. *J. Lipid Res.* 28:847–55

89. Maclure, K. M., Hayes, K. C., Colditz, G. A., Stampfer, M. J., Speizer, F. E., et al. 1989. Weight, diet, and the risk of symptomatic gallstones in middle-aged women. *New Engl. J. Med.* 321:563–69

90. Maclure, K. M., Hayes, K. C., Colditz, G. A., Stampfer, M. J., Willet, W. C. 1990. Dietary predictors of symptom-associated gallstones in middle-aged women. *Am. J. Clin. Nutr.* 52:916–22

91. Marcus, S. N., Wheaton, K. W. 1986. Intestinal transit rate, deoxycholic acid and the cholesterol saturation of bile—three interrelated factors. *Gut* 27:550–58

92. Marcus, S. N., Wheaton, K. W. 1986. Effects of a new, concentrated wheat fibre preparation on intestinal transit, deoxycholic acid metabolism and the composition of bile. *Gut* 27:893–900

93. Marks, J. W., Cleary, P. A., Albers, J. J. 1984. Lack of correlation between serum lipoproteins and biliary cholesterol saturation in patients with gallstones. *Dig. Dis. Sci.* 29:1118–22

94. Marzalo, M. P., Rigotti, A., Nervi, F. 1990. Secretion of biliary lipids from the hepatocyte. *Hepatology* 12:134S–42S

95. Mendez-Sanchez, N., Panduro, A., Gonzalez, L., Castrillon, L., Poo, J. L., et al. 1991. Low doses of ursodeoxycholic acid induces apolipoprotein A-1 gene expression and prolongates cholesterol nucleation time in patients with cholesterol gallstones. *Hepatology* 14:148a

96. Messing, B., Bories, C., Kunstlinger, F., Bernier, J. J. 1983. Does total parenteral nutrition induce gallbladder sludge formation and lithiasis? *Gastroenterology* 84:1012–29

97. Miettinen, M., Turpeinen, O., Karvonen, M. J., Paavilainen, E., Elosuo, R. 1976. Prevalence of cholelithiasis in men and women ingesting a serum-cholesterol-lowering diet. *Ann. Clin. Res.* 8:111–16

98. Mitchell, J. C., Logan, G. M., Stone, B. G., Duane, W. C. 1991. Effects of lovastatin on biliary lipid secretion and bile acid metabolism in humans. *J. Lipid Res.* 32:71–78

99. Nepokroeff, C. M., Lakshmanan, M. R., Ness, G. C., Dugan, R. E., Porter, J. W. 1974. Regulation of the diurnal rhythm of rat liver beta-hydroxy-beta-methylglutaryl coenzyme A reductase activity by insulin, glucagon, cyclic AMP, and hydrocortisone. *Arch. Biochim. Biophys.* 160:387–93

100. Nervi, F., Covarrubias, C., Bravo, P., Velasco, N., Ulloa, N., et al. 1989. Influence of legume intake on biliary lipids and cholesterol saturation in young Chilean men. *Gastroenterology* 96:825–30

101. Nestel, P. J. 1990. Effects of n-3 fatty acids on lipid metabolism. *Annu. Rev. Nutr.* 10:149–67

102. Nestel, P. J., Simons, L. A., Homma, Y. 1976. Effects of ethanol on bile acid and cholesterol metabolism. *Am. J. Clin. Nutr.* 29:1007–15

103. Nicolosi, R. J., Stucchi, A. F., Kowala, M. C., Hennessy, L. K., Hegsted, D. M., et al. 1990. Effect of dietary fat saturation and cholesterol on LDL composition and metabolism. *Arteriosclerosis* 10:119–28

104. Nieman, C. 1954. Influence of trace amounts of fatty acids on the growth of microorganisms. *Bacteriol. Rev.* 18:147–63

105. O'Leary, D. P., Murray, F. E., Turner, B. S., LaMont, J. T. 1991. Bile salts stimulate glycoprotein release by guinea pig gallbladder in vitro. *Hepatology* 13:957–61

105a. Patsch, J. R., Prasad, S., Gotto, A. M., Patsch, W. 1987. Relationship of the plasma levels of HDL$_2$ to its composition, postprandial lipemia, and activities of lipoprotein lipase and hepatic lipase. *J. Clin. Invest.* 80:241–47

106. Pattinson, N. R., Chapman, B. A. 1986. Distribution of cholesterol between mixed micelles and nonmicelles in relation to fasting and feeding in humans. *Gastroenterology* 91:697–702

107. Petitti, D. B., Friedman, G. D., Klatsky, A. L. 1981. Association of a history of gallbladder disease with a reduced concentration of high density lipoprotein cholesterol. *New Engl. J. Med.* 304:1396–98

108. Pixley, F., Wilson, D., McPherson, K., Mann, J. 1985. The effect of vegetarianism on development of gallstones in women. *Br. Med. J.* 91:11–12

109. Portman, O. W., Alexander, M., Tanaka, N., Osuga, T. 1979. Role of diet in normal biliary physiology and gallstone formation. In *Primates in Nutritional Research*, ed. K. C. Hayes, pp. 140–75. New York: Academic

110. Portman, O. W., Alexander, M., Tanaka, N., Osuga, T. 1980. Relationships between cholesterol gallstones, biliary function, and plasma lipoproteins in squirrel monkeys. *J. Lab. Clin. Med.* 96:90–101

111. Redinger, R. N., Hermann, A. H., Small, D. M. 1973. Primate biliary physiology. X. Effects of diet and fasting on biliary lipid secretion and relative composition and bile salt metabolism in the rhesus monkey. *Gastroenterology* 65:610–21

112. Reihner, E., Angelin, B., Björkhem, I., Einarsson, K. 1991. Hepatic cholesterol metabolism in cholesterol gallstone disease. *J. Lipid Res.* 32:469–76

113. Reihner, E., Björkheim, I., Angelin, B., Ewerth, S., Einarsson, K. 1989. Bile acid synthesis in humans: regulation of hepatic microsomal cholesterol 7α-hydroxylase activity. *Gastroenterology* 97.1498–1505

114. Reuben, A. 1984. Biliary proteins. *Hepatology* 4:46S–50S

115. Robins, S. J., Fasulo, J. M. 1973. Mechanism of lithogenic bile production: Studies in the hamster fed an essential fatty acid-deficient diet. *Gastroenterology* 65:104–14

116. Robins, S. J., Fasulo, J. M., Collins, M. A., Patton, G. M. 1985. Evidence for separate pathways of transport of newly synthesized and preformed cholesterol into bile. *J. Biol. Chem.* 260:6511–13

117. Roslyn, J. J., DenBesten, L., Pitt, H. A., Kuckenbecker, S., Polarek, J. W. 1981. Effects of cholecystokinin on gall-

bladder stasis and cholesterol gallstone formation. *J. Surg. Res.* 30:200–4

118. Rotstein, O. D., Kay, R. M., Wayman, M., Strasberg, S. M. 1981. Prevention of cholesterol gallstones by lignin and lactulose in the hamster. *Gastroenterology* 81:1098–1103

119. Salen, G. S., Nicolau, G., Shefer, S., Mosbach, E. H. 1975. Hepatic cholesterol metabolism in patients with gallstones. *Gastroenterology* 69:676–84

120. Salvioli, G., Lugli, R. 1989. Polyunsaturated phosphatidylcholine and bile lipid composition. (Fosfatidilcolina polinsatura e composizione lipidica della bile.) *Clin. Ter.* 131:219–24

121. Sarles, H., Crotte, C., Gerolami, A., Mule, A., Domingo, N., Hauton, J. 1970. The influence of calorie intake and of dietary protein on the bile lipids. *Scand. J. Gastroenterol.* 6:189–91

122. Sarles, H., Hauton, J., Planche, N. E., Hafont, H., Gerolami, A. 1970. Diet, cholesterol gallstones, and composition of the bile. *Am. J. Dig. Dis.* 15:251–60

123. Saunders, K. D., Cates, J. A., Abedin, M. Z., Rege, S., Festekdjian, S. F., et al. 1991. Lovastatin inhibits gallstone formation in the cholesterol-fed prairie dog. *Ann. Surg.* 214:149–54

124. Schoenfield, L. J., Lachin, J. M., Baum, R. A., Habig, R. L., Hanson, R. F., et al. 1981. Chenodiol (chenodeoxycholic acid) for dissolution of gallstones: The National Cooperative Gallstone Study. *Ann. Intern. Med.* 95:257–82

125. Schwartz, C. C., Berman, M., Vlahcevic, Z. R., Halloran, L. G., Gregory, D. H., et al. 1978. Multicompartmental analysis of cholesterol metabolism in man. *J. Clin. Invest.* 61:408–23

126. Schwartz, C. C., Halloran, L. G., Vlahcevic, Z. R., Gregory, D. H., Swell, L. 1978. Preferential utilization of free cholesterol from high-density lipoproteins for biliary cholesterol secretion in man. *Science* 200:62–64

127. Schwartz, C. C., Vlahcevic, Z. R., Halloran, L. G., Swell, L. 1981. An in vivo evaluation in man of the transfer of esterified cholesterol between lipoproteins and into the liver and bile. *Biochim. Biophys. Acta* 663:143–62

128. Schwesinger, W. H., Kurtin, W. E., Johnson, R. 1988. Alcohol protects against cholesterol gallstone formation. *Ann. Surg.* 207:641–47

129. Scobey, M. W., Johnson, F. L., Parks, J. S., Rudel, L. L. 1991. Dietary fish oil effects on biliary lipid secretion and cholesterol gallstone formation in the

African green monkey. *Hepatology* 14: 679–84

130. Scobey, M. W., Johnson, F. L., Rudel, L. L. 1985. Dietary polyunsaturated fat effects on bile secretion in a primate model of cholelithiasis. *Gastroenterology* 88:1693a

131. Scobey, M. W., Johnson, F. L., Rudel, L. L. 1989. Delivery of high-density lipoprotein free and esterfied cholesterol to bile by the perfused monkey liver. *Am. J. Physiol.* 257:G644–52

132. Scobey, M. W., Johnson, F. L., Rudel, L. L. 1991. Plasma HDL cholesterol concentrations are correlated to bile cholesterol saturation index in the African green monkey. *Am. J. Med. Sci.* 301:97–101

133. Scragg, R. K. R., Calvert, G. D., Oliver, J. R. 1984. Plasma lipids and insulin in gall stone disease: a case control study. *Br. Med. J.* 289:521–25

134. Scragg, R. K. R., McMichael, A. J., Baghurst, P. A. 1984. Diet, alcohol, and relative weight in gallstone disease: a case-control study. *Br. Med. J.* 288:1113–19

135. Seaton, T. B., Heymsfield, S., Rosenthal, W. 1989. Ursodeoxycholic acid and gallstones during weight loss. (Letter). *New Engl. J. Med.* 320:1351–52

136. Sewell, R. B., Mao, S. J. T., Kawamoto, T., LaRusso, N. F. 1983. Apolipoproteins of high, low and very low density lipoproteins in human bile. *J. Lipid Res.* 24:391–401

137. Sichieri, R., Everhardt, J. E., Roth, H. 1991. A prospective study of hospitalization with gallstone disease among women: Role of dietary factors, fasting period, and dieting. *Am. J. Public Health* 81:880–84

138. Sjovall, J. 1959. Dietary glycine and taurine on bile acid conjugation in man. Bile acids and steroids. *Proc. Soc. Exp. Biol. Med.* 100:676–78

139. Smith, B. F. 1990. Gallbladder mucin as a pronucleating agent for cholesterol monohydrate crystals in bile. *Hepatology* 12:183S–88S

140. Sömjen, G. J., Marikovsky, Y., Wachtel, E., Harvey, P. R. C., Rosenberg, R., et al. 1990. Phospholipid lamellae are cholesterol carriers in human bile. *Biochim. Biophys. Acta* 1042:28–35

141. Stephan, Z. F., Armstrong, M. J., Hayes, K. C. 1981. Bile lipid alterations in taurine-depleted monkeys. *Am. J. Clin. Nutr.* 34:204–10

142. Stephan, Z. F., Hayes, K. C. 1985. Evidence for distinct precursor pools for

biliary cholesterol and primary bile acids in cebus and cynomolgus monkeys. *Lipids* 20:343–49

143. Stephan, Z. F., Hayes, K. C. 1986. Mechanisms underlying the dietary fat impact on lipoproteins, atherogenesis and gallstones. *Pathol. Immunopathol. Res.* 5:46

144. Stephan, Z. F., Lindsey, S., Hayes, K. C. 1987. Taurine enhances low density lipoprotein binding: Internalization and degradation by cultured HepG2 cells. *J. Biol. Chem.* 262:6069–75

145. Stephan, Z. F., Royal, D., Hayes, K. C. 1985. Fat effect on plasma lipid and biliary lipid and protein in the hamster. *Fed. Proc.* 44:1456

146. Sturdevant, R. A. L., Pearce, M. L., Dayton, S. 1973. Increased prevalence of cholelithiasis in men ingesting a serum-cholesterol-lowering diet. *New Engl. J. Med.* 288:24–27

147. Sullivan-Gorman, M. A., Anderson, J. M., DiMarco, N. M., Johnson, J., Chin, I., et al. 1987. Dietary protein effects on cholelithiasis in hamsters: Interaction with amino acids and bile acids. *J. Am. Oil Chem. Soc.* 64:1196–99

148. Tanno, N., Oikawa, S., Koizumi, M., Fuji, Y., Hori, S., et al. 1989. Effect of taurine administration on serum lipid and biliary lipid composition in man. *Tohoku J. Exp. Med.* 159:91–99

148a. Thijs, C., Knipschild, P., Brombacher, P. 1990. Serum lipids and gallstones: A case-control study. *Gastroenterology* 99:843–49

149. Thornton, J. R., Emmett, P. M., Heaton, K. W. 1983. Diet and gall stones: effects of refined and unrefined carbohydrate diets on bile cholesterol saturation and bile acid metabolism. *Gut* 24:2–6

150. Thornton, J. R., Heaton, K. W. 1986. Bile cholesterol saturation and serum lipoproteins. (Letter). *Dig. Dis. Sci.* 31:109

151. Thornton, J. R., Heaton, K. W., MacFarlane, D. J. 1981. A relationship between high density lipoprotein cholesterol and bile cholesterol saturation. *Br. Med. J.* 2:1352–54

152. Thornton, J. R., Symes, C., Heaton, K. W. 1983. Moderate alcohol intake reduces bile cholesterol saturation and raises HDL cholesterol. *Lancet* 2:819–21

153. Topping, D. L. 1991. Soluble fiber polysaccharides: Effects on plasma cholesterol and colonic fermentation. *Nutr. Rev.* 49:195–203

154. Topping, D. L., Weller, R. A., Nader,

C. J., Calvert, G. D., Illman, R. J. 1982. Adaptive effects of dietary ethanol in the pig: Changes in plasma high-density lipoproteins and fecal sterol excretion and mutagenicity. *Am. J. Clin. Nutr.* 36:245–50

155. Trautwein, E. A., Siddiqui, A., Hayes, K. C. 1991. Feeding psyllium or cholestyramine reduces cholesterol gallstone formation in hamsters by differentially modulating chenodeoxycholic acids. *FASEB J.* 5:A1082

156. Truswell, A. S., McVeigh, S., Mitchell, W. D., Bronte-Stewart, B. 1965. Effect in man of feeding taurine on bile acid conjugation and serum cholesterol levels. *J. Atheroscler. Res.* 5:526–32

157. Turley, S. D., Daggy, B. S., Dietschy, J. M. 1991. Cholesterol-lowering action of psyllium mucilloid in the hamster. Sites and possible mechanisms of action. *Metabolism* 40:1063–73

158. Turley, S. D., Spady, D. K., Dietschy, J. M. 1983. Alteration of the degree of biliary cholesterol saturation in the hamster and rat by manipulation of the pools of preformed and newly synthesized cholesterol. *Gastroenterology* 84:253–65

159. van Berge, Henegouwen, G. P., van der Werf, S. D., Ruben, A. T. 1987. Fatty acid composition of phospholipids in bile in man: promoting effect of deoxycholate on arachidonate. *Clin. Chim. Acta* 165:27–37

160. van Erpecum, K. J., van Berge Henegouwen, G. P., Stolk, M. F. J. 1991. Bile acid and phospholipid fatty acid composition in bile of patients with cholesterol and pigment gallstones. *Clin. Chim. Acta* 199:295–304

161. Vlahcevic, Z. R., Heuman, D. M., Hylemon, P. B. 1991. Regulation of bile acid synthesis. *Hepatology* 13:590–600

162. Yamanaka, Y., Tsuji, K., Ichikawa, T., Nakagawa, Y., Kawamura, M. 1985. Effect of dietary taurine on cholesterol gallstone formation and tissue cholesterol contents in mice. *J. Nutr. Sci. Vitaminol.* 31:225–32

163. Zuin, M., Petroni, M. L., Grandinetti, G., Crosignani, I., Bertolini, E., et al. 1991. Comparison of effects of chenodeoxycholic and ursodeoxycholic acid and their combination on biliary lipids in obese patients with gallstones. *Scand. J. Gastroenterol.* 26:257–62

Annu. Rev. Nutr. 1992. 12:327–43

COORDINATED MULTISITE REGULATION OF CELLULAR ENERGY METABOLISM

Dean P. Jones [1,2,3], *Xiaoqin Shan*[1,2], *and Youngja Park*[1]

[1]Department of Biochemistry, [2]Department of Pediatrics, and [3]Winship Cancer Center, Emory University, Atlanta, Georgia 30322

KEY WORDS: respiratory control, mitochondria, oxygen, ATP, calcium

CONTENTS

PERSPECTIVES AND OVERVIEW

The ability to extract energy from the environment and utilize that energy to organize cell functions and support reproduction is fundamental to all living systems. Animals have evolved mechanisms to ensure adequate energy supply and regulate its delivery to cells according to cellular and organismic demands. In addition, cells have the ability to respond to variations in nutrient supply to optimize utilization of available nutrients. Thus, a dynamic interac-

0199-9885/92/0715-0327$02.00

tion exists between the supply of foods and the metabolic, transport, and mechanical functions of cells.

This interaction directly or indirectly affects nearly all aspects of biology and medicine. More than 120 scientific reviews have been published on the regulation of mitochondrial function and energy metabolism during the past 5 years. We do not intend to duplicate these, and readers are referred to more comprehensive considerations of the biochemical (8, 9, 29, 33, 50), biophysical (60), genetic (56), pathophysiological (18, 40), and mechanistic (39, 51) aspects of energy supply and utilization. Additional reviews are available on respiratory control and exercise (34), thyroid hormone action (15, 21), nutrient deficiency (57), hyperthermia (53), fatty acid oxidation (23), obesity (49), alcohol metabolism (37), aging (43), and in vivo analysis (24, 46).

As is apparent from this wealth of information, the subject of respiratory control has a rich history. However, a clear understanding of regulation of cellular respiration is not available. Heineman & Balaban (20) concluded in their 1990 review of myocardial respiratory control, "After reviewing the controversies in the literature surrounding the regulation of oxidative phosphorylation, a unifying theory to integrate the disparate results would be welcome." The present review focuses on a new description of the regulation of oxidative phosphorylation, termed coordinated multisite regulation. This regulatory scheme evolved from our studies of mitochondrial regulation during anoxia (26) and expands the concept that control is shared among different control sites, perhaps orchestrated by second messengers (20). It reconciles previously contradictory conclusions and provides a rational basis for interpreting whole body nutritional, physiological, and pathological effects, such as reflected in the diverse list of reviews, cited above, on cellular energy metabolism.

PRINCIPLES OF ENERGY REGULATION

The regulatory scheme is developed in the context of three general principles. Firstly, in terms of general descriptors of energy metabolism, mammalian cell and organ systems function largely as "regulators" rather than "conformers." In higher organisms, cell function does not simply conform to the availability of calories; instead, the utilization of calories is regulated by the physiological demands upon the system. Under most conditions, preparedness is required for any function beyond basal activity. This preparedness has an energy cost and can be affected by the timing and sources of dietary calories.

Secondly, the preparedness for physiological demand is expressed at the cellular level in terms of expression and activities of specific metabolic, transport, and mechanical machinery. Studies with modern molecular techniques provide new insight into the diversity and expression of this machinery (7). Of considerable importance, data from in vivo and cellular studies suggest that "basal" cellular energy metabolism is variable and can be regulated by mediators of and differential expression of enzyme, transport, and mechanical machinery. This variability allows energy metabolism to be optimized between the degree of preparedness and a more energy-efficient pattern of limited reserve capacity.

Thirdly, a hierarchy of supply of energy to support functions is based upon the immediacy of needs by the cell. This molecular triage provides a flexible yet economical way to optimize energy utilization. However, in addition to specialized functions, such as contraction, absorption, etc, cells are exposed to diverse challenges such as chemical intoxication, oxidative damage, viral infection, and hormonal and development signals to alter cell structure or function. Optimal energy supply for one function may not equate with optimal supply for other functions. Because cells differ in their specialized functions and utilization of oxidizable substrates, adequacy for one cell type may not reflect adequacy for other cell types. Thus, nutritional energy supply to support the hierarchy of nutritional needs must be matched to the specific cellular demands of the physiological and/or pathological state.

ENERGY METABOLISM IN CONFORMERS AND REGULATORS

Contemporary interpretations of the regulation of energy metabolism are derived from early studies of pathways in prokaryotes and yeast. These organisms provided glimpses of elegantly simple feedback inhibition and protein induction mechanisms that allowed self-regulation of pathways and efficient responses to altered nutritional conditions. Although these serve as examples for regulation in higher organisms, there is a fundamental difference in energy metabolism between lower and higher organisms. In lower organisms, cell functions at any given time largely conform to concurrently available nutrient supply (Figure 1). Thus, simple feedback regulation of energy metabolism provides a good description of how the energy metabolism is actually regulated. This regulation is exemplified by the energy charge concept introduced by Atkinson (4) in which activities of energy-producing and energy-consuming systems respond directly to a ratio of high-energy phosphate bond content to total adenine nucleotide content.

In contrast to this elegantly simple means of allowing cell functions to conform to energy status, higher organisms evolved tissue specializations that

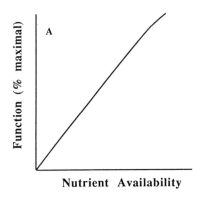 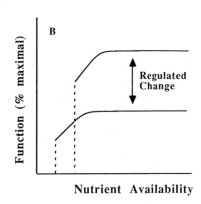

Figure 1 Comparison of functional responses of conformers and regulators to changes in nutrient availability. *A*. In conformers, function is proportional to nutrient availability. For example, growth rate in prokaryotes is proportional to nutrient availability. *B*. In regulators, function is regulated independently of nutrient availability over a wide range. As nutrient availability becomes limiting, only a small change occurs prior to collapse of the function.

allow them to regulate cell functions almost independently of concurrent nutritional energy supply. This concept has been developed in detail for O_2 metabolism and is also applicable to energy metabolism (22, 44). In conformers, energy-related function increases with increasing nutrient availability under essentially all conditions (Figure 1*A*). In contrast, energy-related function in regulators is determined by other factors and is independent of nutrient availability except under conditions of extreme limitation, when the function fails (Figure 1*B*).

Understanding this difference between lower and higher life forms is essential to understanding the regulation of cellular energy metabolism in humans and other higher animals and is illustrated by the energy needs of the mammalian heart. Supply of energy to support cardiac function is of primary importance for organismic survival. This need is so great that during starvation, macromolecules in other tissues are degraded to supply oxidizable substrates to the heart. Clearly, the function of the heart cannot conform to the nutritional energy supply of the organism but must be supplied and regulated independently. Simply put, a hungry animal needs to be able to exert itself to acquire food; it can not slow down and wait until food happens to enter its mouth!

Many strategies have been used by complex multicellular organisms to maintain a reserve capacity to rapidly accommodate altered work demands. A striking feature of these responses is that in many tissues in higher organisms, changes in work load can be accomplished with little, if any, change in adenylate energy change or in ATP/ADP ratio. Thus, earlier studies demonstrating energy-charge-dependent regulation in higher animals obscured the fact that this is not the primary or most immediate type of response in most cells in these organisms. Moreover, the reliance upon the paradigm of regulation of energy metabolism by high energy phosphates suggested that the regulation was remarkably sensitive to changes in high energy phosphate or was related to other nonmitochondrial O_2-dependent reactions because functional changes occurred under conditions in which changes in high energy phosphates were very small or could not be detected (14).

Heineman & Balaban (20) clearly illustrated this point in their review of the control of mitochondrial respiration in heart with ^{31}P-NMR studies showing that substantial changes in cardiac work load (e.g. from 2 to 10 μmol O_2 consumed per gram per minute) occur with little or no change in ATP, ADP, or inorganic phosphate (P_i) concentrations. They concluded that the two most commonly considered mechanisms of respiratory control, namely, kinetic limitation by ADP or P_i (12) and near-equilibrium (thermodynamic) control by ATP, ADP, and P_i concentrations [kinetically limited at cytochrome oxidase (17)], cannot account for the changes in energy utilization and supply. The unavoidable conclusion from these studies is that ATP, ADP, and P_i are not the most important regulators of energy metabolism in heart and other mammalian tissues.[1] This conclusion is particularly liberating because it allows an unbiased consideration of possible effectors of respiratory control and eliminates constraints upon thermodynamic analyses that have led to conflicts between observation and rational interpretation.

COORDINATED MULTISITE REGULATION

Sites of Regulation

For decades, the prevailing interpretation has been that the redox state, high energy phosphate metabolites, and O_2 are the primary determinants of respiratory control (12, 17, 54). Conditions can be readily established for mitochondrial studies in which each of these could limit (or control) O_2

---

[1]A wealth of data show that ATP, ADP, AMP, and P_i are effectors of many mammalian enzymes, and thus, this conclusion does not rule out a role for these molecules in regulation. Instead, the conclusion is that the adenylates and phosphate normally are not the primary regulators in controlling energy supply and utilization in response to altered work loads in mammalian cells.

consumption rate and/or ATP production rate. Because no single site could be identified to provide an adequate explanation for respiratory control in cells, the concept of multiple-site regulation was introduced (27, 54). This regulatory scheme provides a mathematical description of respiration in terms of fractional regulation at different sites in the overall process of oxidative phosphorylation (27).

Potential sites for regulation include the adenine nucleotide transporter, NAD^+-linked dehydrogenases, ATP synthase, and cytochrome oxidase (Figure 2). The phosphate transporter is also regulated under anoxic conditions (5). Before considering details of the regulation at each of these sites, it is important to examine their positions in the overall process of oxidative phosphorylation. The coupling mechanism of oxidative phosphorylation involves capture of energy from oxidation of NADH in the form of an electrochemical proton gradient (Δp) across the mitochondrial inner membrane (36, 51). This Δp is composed of two parts, the energy available from the proton gradient (ΔpH) across the membrane and the energy available from moving charges across the membrane in response to the potential difference ($\Delta \Psi$). Mathematically, it is usually expressed in millivolts by the equation $\Delta p = \Delta \Psi - Z\Delta pH$, where Z is a factor to convert ΔpH to millivolts. The expression can be converted to kilocalories per mole or kilojoules per mole of ATP synthesized by the equation

$$\Delta G = -nF\Delta p$$

where n is the stoichiometry of H^+ transported per ATP synthesized and F is Faraday constant. From such calculations, the energy available from the Δp is often not measurably different from that available from the free energy of hydrolysis of ATP to ADP and P_i, which suggests that the oxidative reactions are in equilibrium with the phosphorylation reaction, i.e. a thermodynamic control condition (17).

A usual interpretation is that ATP synthase and phosphate transporter have relatively high rates and contribute little to rate determination. Thus, the amount of energy available for Δp and ATP synthesis is determined by the rate of supply of NADH through the dehydrogenases and cytochrome oxidase to O_2, and the rate of ATP synthesis is limited by either the rate of hydrolysis of ATP in the cytoplasm (equilibrium or thermodynamic control) or the return of ADP by the adenine nucleotide transporter (kinetic control). If one assumes that the oxidative reactions (generating Δp) are in equilibrium with the phosphorylation state, one need not even consider the regulation of dehydrogenases to be important, because these dehydrogenase-catalyzed reactions also could be relatively fast and, hence, in equilibrium with the rest of the electron transport chain and ATP synthase.

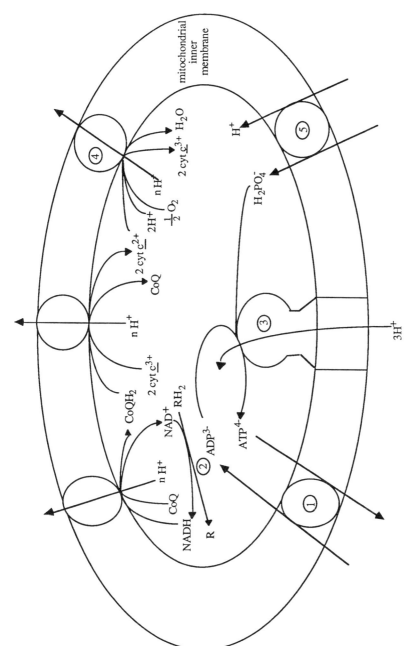

Figure 2 Potential sites for regulation of oxidative phosphorylation. 1. Adenine nucleotide transporter; 2. NAD$^+$-linked dehydrogenases; 3. ATP synthase; 4. cytochrome oxidase; 5. electroneutral phosphate carrier.

These descriptions are suitable for conformers because when nutrients are ample the system provides energy as rapidly as can be used, whereas when nutrients are limiting, the system limits function to the amount of energy available. However, such an interpretation provides no mechanism to anticipate energy demands or alter work load without changes in ATP/ADP. Mammalian tissues, which regulate metabolism, anticipate work load, and change work load without changes in ATP/ADP, must have additional sites of regulation and regulatory signals.

Inclusion of the five sites listed above allows for regulation of oxidative phosphorylation that can behave like a conformer under some conditions but also can be modulated to anticipate increased work load and to change work load without changes in ATP/ADP (Figure 3). This modulation without changes in ATP/ADP can be accomplished with the following coordinated, multisite regulation mechanism.

The basics of this mechanism are as follows. (*a*) The generation of Δp is regulated by control of the NAD^+-linked dehydrogenases and cytochrome oxidase. Coordinated regulation at these two sites provides the ability to alter the energy available in Δp for chemical, osmotic, and electrical work. (*b*) The utilization of Δp for ATP synthesis is regulated by control of the ATP synthase. (*c*) ATP synthesis in the mitochondria can also be regulated by kinetic control of the adenine nucleotide carrier and phosphate carrier. Coordinated regulation at these sites provides the ability to control the utilization of Δp for ATP synthesis and also to control the extent to which this ATP is maintained within the mitochondria or delivered to the cytoplasm. (*d*) Regulation of the use of Δp for other functions, e.g. Ca^{2+} transport, also provides the ability to control the distribution of utilization of Δp between ATP synthesis and other functions.

Evidence for Multisite Regulation

Two specific issues are important in considering multisite regulation, namely, whether physiological effectors are known for these sites and whether regulation actually occurs at these sites in cells under physiological conditions. During the past decade considerable evidence has accumulated to show that certain NAD^+-linked dehydrogenases are regulated by Ca^{2+} (16, 32). Although earlier studies indicated that mitochondria buffer cytosolic Ca^{2+} (10), we now know that cytosolic Ca^{2+} is an effector of mitochondrial function (16, 38). An increase in cytosolic Ca^{2+} serves as an activating signal for diverse cellular functions (39) and also activates mitochondrial supply of NADH by increasing the activities of pyruvate dehydrogenase, 2-oxoglutarate dehydrogenase, and isocitrate dehydrogenase (16). Variations in O_2 consumption rate in response to changes in intra- and extra-mitochondrial Ca^{2+} have been found in isolated mitochondria (38) and intact cells (60). Measure-

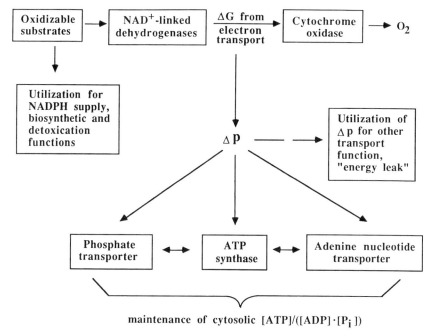

Figure 3 Coordinated multisite regulation of oxidative phosphorylation. Oxidizable substrates can be used for NADPH supply and biosynthetic functions or can be used to support oxidative phosphorylation. Regulation of NAD$^+$-linked dehydrogenases provides control between these two processes. Coordinated regulation of cytochrome oxidase with NAD$^+$-linked de-hydrogenases provides a means to regulate generation of Δp. Higher Δp is less energy efficient because it increases nonproductive ion leak across the membrane. However, higher Δp also increases the energy available for ATP synthesis and allows for increased ATP/ADP to increase work capacity. Utilization of Δp for ATP synthesis can be controlled by regulating the ATP synthase or by regulating the adenine nucleotide transporter or phosphate carrier. Coordinated regulation of these allows control over cytosolic and mitochondrial [ATP]/([ADP]·[P$_i$]) and the relative proportion of energy from Δp that is used for ATP synthesis as opposed to other mitochondrial functions, e.g. Ca^{2+} cycling or energy-dependent transhydrogenase function.

ments of mitochondrial Ca^{2+} showed that increased Ca^{2+} was associated with positive ionotropic stimulation of heart (13, 59). These results are consistent with regulation of dehydrogenase activity in cells.

Cytochrome oxidase is not usually considered a regulatory enzyme, yet Kadenbach and co-workers have obtained substantial evidence that noncatalytic subunits alter the enzyme kinetics (2, 28). Only two out of the 9 to 13 subunits are involved in the oxidation-reduction reaction and one in proton transport. The functions of the other subunits have not been resolved, but some of these affect the kinetics of the electron transport rate, suggesting that these subunits function in regulation of the oxidase (28). In addition, tissue-specific isozymes of some subunits have been identified (2); at present, except for regulation, no apparent function for the different isozymes is known. Direct spectral analyses have indicated that a substantial fraction of cytochrome oxidase is reduced in intact tissues (35, 52) even though this extent is not observed in isolated mitochondria (25). This difference may be due to regulatory mechanisms that are lost in vitro.

Electron flow between NADH and O_2 provides the energy available for establishing Δp. In the usual model of respiratory control, the rate of electron flow is thought to be determined by changes in the Δp that are due to alterations in its utilization for ATP synthesis. Coordinated regulation of NADH dehydrogenases and cytochrome oxidase provides a mechanism to increase the rate of O_2 consumption without a decreased Δp.

An endogenous inhibitor of F_1-ATPase (ATP synthase) was described in 1963 (45), and a Ca^{2+}-antagonized inhibitor was subsequently described (62). These inhibitors are usually considered to be safety valves to prevent hydrolysis of ATP by the ATP synthase working in reverse. On the other hand, variations of mitochondrial Ca^{2+} following Ca^{2+} depletion or anoxia in hepatocytes have been associated with altered O_2 consumption rate and altered oligomycin-inhibitable ATPase activity (6). Under anoxic conditions, the energy available from the Δp was substantially more than that necessary for ATP synthesis at prevailing mitochondrial concentrations of ATP, ADP, and P_i (1). Thus, at least under anoxic conditions, the ATP synthase is regulated in intact cells. Similar data for aerobic conditions are consistent with kinetic limitation at this site, but the combined errors from measurement of the various components do not allow certainty in this conclusion. Thus, the data show that regulation of utilization of Δp by the ATP synthase occurs, but they do not clearly establish that this regulation contributes to respiratory control under normal conditions.

The adenine nucleotide transporter is important in limiting oxidative phosphorylation in many in vitro experiments (19, 20, 54). Different isozymic forms are present in different tissues, further suggesting that this is an important control site (48). The activity appears to be inhibited by anoxia (5) and is inhibitable by acyl CoA's (41). The adenine nucleotide transporter is associated with an inner membrane-outer membrane complex that includes porin and other associated proteins (11). Doubtlessly, it is a central component in the regulation of oxidative phosphorylation. The primary effectors of the transporter remain unknown, however, and the ways in which regulation at this site are integrated with other sites of regulation also remain undefined.

Regulation of the phosphate transporter has not been considered extensively in respiratory control, perhaps because its activity is so great that it does not appear to be a likely site of regulation. However, under anoxic conditions, cytosolic phosphate accumulates dramatically because of the hydrolysis of ATP (5). Continued activity of the phosphate transporter would result in accumulations of up to 100 mM matrix phosphate and would be expected to cause extensive mitochondrial swelling. This phosphate loading and mitochondrial swelling does not occur, thus indicating that the transporter is dramatically inhibited under anoxic conditions (1). In principle, regulation of the phosphate transporter could regulate the production of ATP by limiting mitochondrial phosphate concentration. It could also function to regulate matrix Ca^{2+} or availability of matrix anions, such as dicarboxylates. Thus, evidence suggests that the phosphate carrier is regulated, but the conditions (other than anoxia) and mechanisms involved remain unknown.

Coordination of Multisite Regulation

While there is little doubt that multisite regulation of oxidative phosphorylation occurs, little is known about how this multisite regulation is coordinated, especially since it does not involved high energy phosphates. One of the most important challenges ahead is to identify the signalling agents and their molecular mechanisms of action. Two specific examples provide a framework for more detailed consideration, effects of Ca^{2+} and of di-calciphor, a synthetic prostaglandin B_1 analogue.

As discussed above, increased cytosolic Ca^{2+} is an intracellular second messenger that signals activation of diverse, specialized cell responses. Most of these responses either directly or indirectly require increased energy production, e.g. for cytoskeletal changes, altered gene expression, altered ion transport. Thus, it is reasonable to assume that mitochondria respond to the same activating signal. Indeed, mechanisms are known whereby increased Ca^{2+} increases NAD^+-linked dehydrogenase (16), ATP synthase (62), and adenine nucleotide carrier (6) activities. Increased Ca^{2+} also increases mitochondrial P_i uptake (47) and stimulates mitochondrial O_2 consumption (6, 38). Thus, Ca^{2+} appears to be a likely agonist for coordinated activation of mitochondrial respiration.

Studies of anoxia provide evidence for a second type of coordinated regulation, one involving simultaneous inhibition of ATP synthase, P_i uptake, and adenine nucleotide exchange (1, 5, 6). Recent studies suggest that di-calciphor, the dimer of 16,16-dimethyl-15-dehydroprostaglandin B_1, inhibits each of these activities in cyanide-treated liver cells (42). The mechanism of this inhibition is not known but may occur through a fatty acid binding site that appears to be common to several of the mitochondrial inner membrane transport systems (3, 31). Such a pattern of coordinated regulation provides a means to adjust oxidative phosphorylation to a reduced functional state in opposition to activation provided by Ca^{2+} or other positive effectors.

NUTRITIONAL RELEVANCE OF THE COORDINATED MULTISITE REGULATION OF MITOCHONDRIAL RESPIRATION

Coordinated regulation of mitochondrial respiration is of fundamental importance to nutritional support of higher organisms. Cells in higher organisms are usually regulators, not conformers. Thus, provision of oxidizable substrates without consideration of the regulatory factors may be of little or no benefit.

The most general implication of the coordinated multisite regulation is that cells and organisms can be adjusted between relatively energy-inefficient, high work capacity machines and highly efficient, low capacity machines. This concept is easiest to apply to muscle, but it is also applicable to all cells and tissues that can vary work load, e.g. kidney, lungs, etc. A high work capacity can be achieved through two distinct mechanisms—expression of more metabolic, transport, and mechanical machinery (61) and operation of that machinery at higher rates of oxidation, higher Δp, and higher $[ATP]/([ADP] \cdot [P_i])$. Increases such as these reduce energy efficiency at a low work load because of the energy required to maintain the machinery and metabolic and ionic homeostasis. For example, at higher Δp, more ion leak occurs. At higher ATP/ADP, more spontaneous hydrolysis occurs. Coordinated multisite activation of cellular respiration allows changes in Δp and ATP/ADP to anticipate work load and/or respond rapidly and efficiently to an imposed work load.

In contrast, chronic lack of work load allows acclimatization to a more highly efficient, reduced capacity state. This physiological state not only has decreased work machinery but also has oxidative phosphorylation machinery in a lower state of readiness for function. Substrate oxidation is less, Δp is lower, and $[ATP]/([ADP] \cdot [P_i])$ is lower; overall utilization of energy is more energy efficient, but work capacity is considerably reduced.

This view of respiratory control has far-reaching implications for human health and disease. Firstly, it provides a rational basis for different "set points" of metabolism, where "set point" refers to the prevailing basal condition ranging from a low-capacity, high efficiency state to a high-capacity, low efficiency state determined by coordinated multisite regulation. The need for different set points of oxidative phosphorylation can be readily seen from consideration of the balance between NADPH supply and NADH supply. NADPH is required for most biosynthetic and detoxication pathways; thus, conditions of growth and repair have much higher requirements for NADPH than do other conditions. The pentose phosphate pathway, which competes with glycolysis for utilization of glucose (55), produces much of the NADPH in cells. NADPH is also produced from oxidative decarboxylation of isocitrate and 2-oxoglutarate by enzymes that are $NADP^+$-specific (30) and by an energy-linked transhydrogenase that utilizes NADH as the reductant in the mitochondria (58). Because all of these major reactions that produce NADPH compete with provision of NADH for oxidative phosphorylation, mechanisms must be present to control the flux through these competing pathways. By controlling the amount of substrate oxidation directed toward NADH generation, the balance between substrate utilization for ATP production and NADPH supply is effectively regulated. The set point for oxidative phosphorylation must be variable to accomplish such shifts in metabolic needs of cells.

Changes in molecular machinery in response to altered work load, such as hypertrophy of muscle with exercise or tissue recovery following surgery, are well known to adjust the cells to a different work capacity and efficiency. Of considerable importance is the fact that coordinated multisite regulation can alter work capacity and energy efficiency without changes in constitutive molecular machinery. Under pathological conditions where detoxication and repair is critical, it may not be possible or even appropriate to therapeutically modulate this regulation of oxidative phosphorylation. On the other hand, supply of precursors for NADPH, such as glucose and citrate, may be beneficial to oxidative phosphorylation indirectly because of their use for NADPH production. Direct support of oxidative phosphorylation, bypassing the NAD^+-linked dehydrogenases, may also be affected by supply of succinate. Succinate is oxidized by succinate dehydrogenase and can, in principle, support oxidative phosphorylation and ATP production even though NAD^+-linked dehydrogenases are inhibited.

Until more is known about the mechanisms of regulation, therapy aimed at altering this response may not be very effective. On the other hand, variations in this regulation may underlie the contradictory results obtained with efforts to regulate oxidative metabolism. An example is the use of chloroacetate in an effort to stimulate pyruvate dehydrogenase activity and thereby stimulate energy production in septicemia and other conditions of apparent bioenergetic failure. Activation at a single site cannot activate the overall process.

A second implication of coordinated multisite regulation is that it provides a simple explanation for genetic differences in basal metabolic rate. Variations in expression of the machinery would be expected to alter capacity and efficiency of cell function because of the energy cost of maintaining more machinery. In addition, genetic differences in the regulatory components that result in altered steady-state concentrations of intermediates, increased membrane ion leak, or increased futile cycling would also affect the basal metabolic rate even without changes in the expression of molecular machinery.

Genetic differences can affect these responses and determine both the range of set point changes that are possible and the conditions under which these occur. Similarly, differences between sexes and age-related differences in expression and regulation may result in an altered set point and, hence, may account for differences in basal metabolic rate. A range of set points within a population provides adaptability of the population because it assures that some individuals have a high degree of metabolic preparedness for physiological challenges (high energy demands) and other individuals have a high energy efficiency to survive periods of starvation but have a relatively poor preparedness for physiological challenges.

A third area of importance for the concept of coordinated multisite regulation is that it emphasizes the need for exercise, in conjunction with proper nutrition, for optimal health. The metabolic set point of oxidative phosphorylation is variable and at least partially determined by the need to be prepared for altered work loads. In the absence of imposed work, the systems shift to a more highly efficient, low work-capacity level because of the lack of need to maintain the machinery to accommodate work. Thus, sedentary individuals are expected to be more energy efficient, i.e. have a lower basal metabolic rate and require fewer calories for weight maintenance. This is an extremely important issue because changes in calorie expenditure of only 1–2% per day can result in a 1 to 2 kg change in body weight per year. Changes in energy efficiency of 1 to 2% in forming Δp or using it to generate $[ATP]/([ADP]\cdot[P_i])$ would not be measurable with current techniques. Thus, improved methodologies will be needed to rigorously test whether coordinated, multisite regulation of oxidative phosphorylation controls the energy efficiency of nutrient oxidation and thereby contributes to differing tendencies of individuals toward obesity.

CONCLUSIONS

Contemporary studies of cellular energy metabolism in higher organisms indicate that primary regulation occurs independently of ATP, ADP, and inorganic phosphate concentration changes. Results indicate that regulation occurs at multiple sites and that coordinated regulation at these sites allows optimization of cell function through a range of metabolic set points. At one extreme, metabolic systems are at a high state of readiness to accommodate increased work loads, but this is relatively inefficient with regard to energy expenditure under basal conditions. At the other extreme, a low work capacity is associated with highly efficient energy utilization.

The sites of regulation include NAD^+-linked dehydrogenases, cytochrome oxidase, ATP synthase, adenine nucleotide transporter, and the mitochondrial phosphate carrier. Coordinated regulation at these sites has the ability to balance utilization of nutrients between NADPH supply for biosynthetic and detoxication functions and the need of ATP for these and other aspects of cell function. Coordinated regulation also provides the means to anticipate an increased work load and to adjust to increased work without decreases in $[ATP]/([ADP]\cdot[P_i])$. In addition, control at these sites allows mitochondrial function to be suppressed during anoxia to preserve osmotic stability and prolong the duration of anoxic tolerance.

The existence of this type of control has diverse implications for nutrition because it affects the strategies used for nutrient supply to individuals under various pathological and physiological states. It brings us to the conclusion that all calories are not equivalent with regard to cell and organ function and, hence, to human health. Numerous epidemiologic studies have arrived at this same conclusion: isocaloric supply of different types of fats, of different mixtures of fats, sugars, and protein, and of different types of proteins and carbohydrates are associated with different human health benefits and risks. Thus, the challenge before us is to identify the molecular factors involved in coordinated multisite regulation so that these factors can be nutritionally or therapeutically manipulated to improve human health. Recognition of this type of regulation directs our attention to certain precursors, such as citrate and succinate, which may be useful because their pathways of metabolism allow them to be used for NADPH supply or improved ATP production independently of the respiratory control state. Finally, recognition of coordinated multisite regulation reemphasizes the interplay between diet and exercise as determinants of health because it defines metabolic readiness in terms of metabolic conditioning.

Literature Cited

1. Andersson, B. S., Aw, T. Y., Jones, D. P. 1987. Mitochondrial transmembrane potential and pH gradient during anoxia. *Am. J. Physiol.* 252:C349–55
2. Anthony, G., Stroh, A., Lottspeich, F., Kadenbach, B. 1990. Different isozymes of cytochrome *c* oxidase are expressed in bovine smooth muscle and skeletal or heart muscle. *FEBS Lett.* 277:97–100
3. Aquila, H., Link, T. A., Klingenberg, M. 1987. Solute carriers involved in energy transfer of mitochondria form a homologous protein family. *FEBS Lett.* 212:1–9
4. Atkinson, D. E. 1970. Enzymes as control elements in metabolic regulation. In *The Enzymes*, ed. P. D. Boyer, 1:461–89. New York: Academic
5. Aw, T. Y., Andersson, B. S., Jones, D. P. 1987. Mitochondrial transmembrane ion distribution during anoxia. *Am. J. Physiol.* 252:C356–61
6. Aw, T. Y., Andersson, B. S., Jones, D.

P. 1987. Suppression of mitochondrial respiratory function following short-term anoxia. *Am. J. Physiol.* 252:C362–68
7. Aw, T. Y., Jones, D. P. 1989. Nutrient supply and mitochondrial function. *Annu. Rev. Nutr.* 9:229–51
8. Aw, T. Y., Jones, D. P. 1991. Intracellular respiration. In *The Lung: Scientific Foundations,* ed. R. G. Crystal, J. B. West, P. J. Barnes, N. S. Cherniack, E. R. Weibel, 2:1445–53. New York: Raven
9. Balaban, R. S. 1990. Regulation of oxidative phosphorylation in the mammalian cell. *Am. J. Physiol.* 258:C377–89
10. Becker, G. L., Fiskum, G., Lehninger, A. L. 1980. Regulation of free Ca^{2+} by liver mitochondria and endoplasmic reticulum. *J. Biol. Chem.* 255:9009–12
11. Brdiczka, D. 1990. Interaction of mitochondrial porin with cytosolic proteins. *Experientia* 46:161 67
12. Chance, B., Williams, C. M. 1956. The

respiratory chain and oxidative phosphorylation. *Adv. Enzymol.* 17:65–134

13. Crompton, M., Kessar, P., Al-Nassar, I. 1983. The α-adrenergic-mediated activation of the cardiac mitochondrial Ca^{2+} in vivo. *Biochem. J.* 216:333–42

14. Davis, J. N., Carlsson, A., MacMillan, V., Siesjo, B. K. 1973. Brain tryptophan hydroxylation: dependence on arterial oxygen tension. *Science* 182:72–74

15. De Nayer, P. 1987. Thyroid hormone action at the cellular level. *Horm. Res.* 26:48–57

16. Denton, R. M., McCormack, J. G. 1990. Ca^{2+} as a second messenger within mitochondria of the heart and other tissues. *Annu. Rev. Physiol.* 52:451–66

17. Erecinska, M., Wilson, D. F. 1982. Regulation of cellular energy metabolism. *J. Membr. Biol.* 70:1–14

18. Goldstein, S. A., Elwyn, D. H. 1989. The effects of injury and sepsis on fuel utilization. *Annu. Rev. Nutr.* 9:445–73

19. Groen, A. K., Wanders, R. J. A., Westerhoff, H. V., Van der Meer, R., Tager, J. M. 1982. Quantification of the contribution of various steps to the control of mitochondrial respiration. *J. Biol. Chem.* 257:2754–57

20. Heineman, F. W., Balaban, R. S. 1990. Control of mitochondrial respiration in the heart in vivo. *Annu. Rev. Physiol.* 52:523–42

21. Hoch, F. L. 1988. Lipids and thyroid hormones. *Prog. Lipid Res.* 27:199–270

22. Hochachka, P. W. 1988. Metabolic responses to reduced O_2 availability. In *Hypoxia: The Tolerable Limits,* ed, J. R. Sutton, C. S. Houston, G. Coates, pp. 41–47. Indianapolis: Benchmark

23. Huang, X. Q., Liedtke, A. J. 1989. Alterations in fatty acid oxidation in ischemic and reperfused myocardium. *Mol. Cell. Biochem.* 88:145–53

24. Jéquier, E., Acheson, K., Schutz, Y. 1987. Assessment of energy expenditure and fuel utilization in man. *Annu. Rev. Nutr.* 7:187–208

25. Jones, D. P. 1986. Intracellular diffusion gradients of O_2 and ATP. *Am. J. Physiol.* 250:C663–75

26. Jones, D. P. 1988. New concepts of the molecular pathogenesis arising from hypoxia. In *Oxidases and Related Redox Systems,* ed. T. E. King, H. S. Mason, M. Morrison, pp. 127–44. New York: Liss

27. Kacser, H., Burns, J. A. 1973. The control of flux. *Symp. Soc. Exp. Biol.* 27:65–104

28. Kadenbach, B., Stroh, A., Becker, A., Eckerskorn, C., Lottspeich, F. 1990.

Tissue- and species-specific expression of cytochrome *c* oxidase isozymes in vertebrates. *Biochim. Biophys. Acta* 1015:368–72

29. Kagawa, Y., Ohto, S. 1990. Regulation of mitochondrial ATP synthesis in mammalian cells by transcriptional control. *Am. J. Biochem.* 22:219–29

30. Kauffman, F. C., Evans, R. K., Thurman, R. G. 1977. Alterations in nicotinamide and adenine nucleotide systems during mixed-function oxidation of *p*-nitroanisole in perfused livers from normal and phenobarbital-treated rats. *Biochem. J.* 166:583–92

31. Krämer, R., Klingenberg, M. 1985. Structural and functional asymmetry of the ADP/ATP carrier from mitochondria. *Ann. NY Acad. Sci.* 456:289–90

32. McCormack, J. G., Denton, R. M. 1989. The role of Ca^{2+} ions in the regulation of intramitochondrial metabolism and energy production in rat heart. *Mol. Cell. Biochem.* 89:121–25

33. McCormack, J. G., Halestrap, A. P., Denton, R. M. 1990. Role of calcium ions in regulation of mammalian intramitochondrial metabolism. *Am. Phys. Soc.* 70:391–425

34. McMillin, J. B., Pauly, D. F. 1988. Control of mitochondrial respiration in muscle. *Mol. Cell. Biochem.* 81:121–29

35. Mills, E., Jobsis, F. F. 1972. Mitochondrial respiratory chain of carotid body and chemoreceptor response to changes in oxygen tension. *J. Neurophysiol.* 35:405–28

36. Mitchell, P. 1979. Keilin's respiratory chain concept and its chemiosmotic consequences. *Science* 206:1148–59

37. Mitchell, M. C., Herlong, H. F. 1986. Alcohol and Nutrition: Caloric value, bioenergetics, and relationship to liver damage. *Annu. Rev. Nutr.* 6:457–74

38. Moreno-Sanchez, R. 1985. Regulation of oxidative phosphorylation in mitochondria by external free Ca^{2+}-concentrations. *J. Biol. Chem.* 260:4028–34

39. Murphy, M. P. 1989. Slip and leak in mitochondrial oxidative phosphorylation. *Biochim. Biophys. Acta* 977:123–41

40. Ozawa, T., Tanaka, M., Suzuki, H., Nishikimi, M. 1987. Structure and function of mitochondria: Their organization and disorders. *Brain Dev.* 9:76–81

41. Pande, S. V., Gaswami, T., Parvin, R. 1984. Protective role of adenine nucleotide translocase in O_2-deficient hearts. *Am. J. Physiol.* 247:H25–H34

42. Park, Y., Devlin, T. M., Majde, J. A., Jones, D. P. 1991. Protective effect of

tricalciphor during mitochondrial failure. *Renal Fail.* In press.

43. Poehlmam, E. T., Horton, E. S. 1990. Regulation of energy expenditure in aging humans. *Annu. Rev. Nutr.* 10:255–75

44. Prosser, C. L. 1986. *Adaptational Biology: Molecules to Organisms.* New York: Wiley

45. Pullman, M. E., Monroy, G. C. 1963. A naturally occurring inhibitor of mitochondrial adenosine triphosphatase. *J. Biol. Chem.* 238:3862–69

46. Sapega, A. A., Sokolow, D. P., Graham, T. J., Chance, B. 1987. Phosphorus nuclear magnetic resonance: a noninvasive technique for the study of muscle bioenergetics during exercise. *Med. Sci. Sports Exerc.* 19:410–20

47. Savage, M. K., Jones, D. P., Reed, D. J. 1991. Calcium- and phosphate-dependent loading of glutathione by liver mitochondria. *Arch. Biochem. Biophys.* 290:51–56

48. Schultheiss, H. P., Klingenberg, M. 1985. Immunoelectrophoretic characterization of the ADP/ATP carrier from heart, kidney, and liver. *Arch. Biochem. Biophys.* 239:273–79

49. Schulz, L. O. 1987. Brown adipose tissue: Regulation of thermogenesis and implications for obesity. *Perspect. Pract.* 87:761–64

50. Senior, A. E. 1988. ATP synthesis by oxidative phosphorylation. *Am. Phys. Soc.* 68:177–231

51. Slater, E. C. 1987. The mechanism of the conservation of energy of biological oxidations. *Eur. J. Biochem.* 166:489–504

52. Snow, T. R., Kleinman, L. H., Lamanna, J. C., Wichsler, A. S., Jobsis, F. F. 1981. Response of cytochrome a,a$_3$ in the in situ canine heart to transient ischemic episodes. *Basic Res. Cardiol.* 76:289–304

53. Streffer, C. 1988. Aspects of metabolic change after hyperthermia. *Recent Result. Cancer Res.* 107:7–16

54. Tager, J. M., Wanders, R. J. A., Groen, A. K., Kunz, W., Bohnensack, R., et al. 1983. Control of mitochondrial respiration. *FEBS Lett.* 151:1–9

55. Tribble, D. L., Jones, D. P. 1990. Oxygen dependence of oxidative stress. Rate of NADPH supply for maintaining the GSH pool during hypoxia. *Biochem. Pharmacol.* 39:729–37

56. Wallace, D. C. 1987. Maternal genes: Mitochondrial diseases. *Birth Defects* 23:137–90

57. Waterlow, J. C. 1986. Metabolic adaptation to low intakes of energy and protein. *Annu. Rev. Nutr.* 6:495–526

58. Weigl, K., Sies, H. 1977. Drug oxidations dependent on cytochrome P-450 in isolated hepatocytes. *Eur. J. Biochem.* 77:401–8

59. Wendt-Gallitelli, M. F. 1986. Ca-pools involved in the regulation of cardiac contraction under positive inotropy. X-ray microanalysis on rapidly frozen ventricular muscle of guinea pig. *Basic Res. Cardiol.* 81:25–32

60. Westerhoff, H. V. 1989. Control, regulation and thermodynamics of free-energy transduction. *Biochimie* 71:877–86

61. Williams, R. S. 1986. Mitochondrial gene expression in mammalian striated muscle. *J. Biol. Chem.* 261:12390–94

62. Yamada, E. W., Shiffman, F. H., Huzel, N. J. 1980. Ca^{2+}-regulated release of an ATPase inhibitor protein from submitochondrial particles derived from skeletal muscles of the rat. *J. Biol. Chem.* 255:267–73

Annu. Rev. Nutr. 1992. 12:345–68

IRON-DEPENDENT REGULATION OF FERRITIN AND TRANSFERRIN RECEPTOR EXPRESSION BY THE IRON-RESPONSIVE ELEMENT BINDING PROTEIN

Elizabeth A. Leibold and Bing Guo

Human Molecular Biology and Genetics Program and Department of Medicine, University of Utah, Salt Lake City, Utah 84112

KEY WORDS: translational regulation, mRNA stability, RNA-binding proteins

CONTENTS

INTRODUCTION

Gene expression can be controlled at multiple levels including transcription, posttranscription, and posttranslation. In recent years, the mechanisms

345

0199-9885/92/0715-0345$02.00

regulating transcription have been studied intensely. Less is known about controls of posttranscriptional processes. Posttranscriptional control can be regulated by altering transcript processing, including capping, splicing, and polyadenylation, and by influencing the rate of transport into the cytoplasm. Once the mRNA is in the cytoplasm, the amount of protein synthesized depends on both the efficiency of translation and mRNA turnover. Translational control can be defined as a change in the efficiency of mRNA translation (37). This regulation can occur globally for all mRNAs, in order to increase total protein synthesis, or specifically for a particular mRNA. The regulation of mRNA turnover can be used to regulate the levels of specific mRNAs by altering the rates of degradation or stabilization (38, 65). In contrast to transcriptional regulation, which requires a delay due to RNA transcription and processing, posttranscriptional regulation occurs rapidly owing to the presence of preexisting mRNA in the cytoplasm. Thus, posttranscriptional regulation of specific genes is advantageous to cells that need to respond rapidly to biological stimuli.

Two examples of proteins regulated at the level of translation and mRNA stability are ferritin and transferrin receptor (TfR), respectively. Ferritin and TfR are the major proteins involved in the regulation of iron homeostasis (Figure 1). TfR is an integral membrane protein that brings iron into the cell via endocytosis of diferric transferrin. Following delivery to the endosome, iron is released and transferred to the iron-storage protein ferritin or to other cellular proteins requiring iron. The need for a posttranscriptional mechanism to regulate ferritin and TfR expression can be inferred by the toxicity of free iron in the cell. Although iron is an essential element and plays a role in many biological processes including oxygen and electron transfer, nitrogen fixation, and DNA synthesis, free iron is toxic to cells. This toxicity is due to the ability of free iron to form reactive hydroxyls that can cause peroxidation of lipid membranes and other cellular constituents. Most iron in cells is complexed. Ferritin is the molecule commonly used for sequestration. Cells regulate the concentration of free iron both by regulating its uptake via the TfR and by the sequestration of iron into ferritin.

Ferritin synthesis is increased by iron and is controlled primarily at the level of translation. Ferritin mRNAs are found in the cytoplasm in either inactive ribonucleoprotein (RNP) particles or in actively translating polysomes. The rate of ferritin protein synthesis is determined by the distribution of ferritin mRNAs between polysomes and RNP particles. The concentration of free iron in the cell regulates this distribution; high iron levels mobilize ferritin mRNAs onto polysomes for translation, whereas low iron levels cause ferritin mRNAs to be in RNP particles. In contrast, TfR synthesis is decreased by iron and is controlled at the level of mRNA stability. Cells respond to an increase

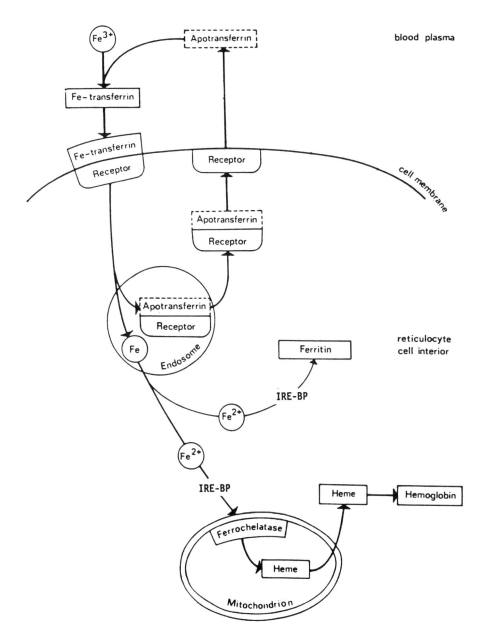

Figure 1   Scheme of iron uptake in the reticulocyte by the transferrin receptor and sequestration into ferritin. Reproduced with permission (68).

in iron concentration by altering the levels of TfR mRNA via an increase in the specific degradation of TfR mRNAs. The net result of ferritin and TfR regulation by iron is a decrease in the free iron levels in the cell, thereby preventing iron toxicity.

The coordinate regulation of ferritin and TfR is achieved through specific sequences within the mRNAs themselves. These sequences, termed iron-responsive elements (IREs), are located in the 5'- and 3'-untranslated regions of ferritin and TfR mRNAs, respectively. A cytoplasmic protein, the iron-responsive element binding protein (IRE-BP) modulates the posttranscriptional regulation of ferritin and TfR mRNAs through interaction with IREs in a process thought to be mediated by intracellular free iron. The IRE-BP has also been referred to as the ferritin repressor protein (FRP) and the iron regulatory factor (IRF).

Our purpose in this review is to discuss how the IRE-BPs regulate the posttranscriptional expression of ferritin and TfR. Several excellent reviews of the structure and the expression of ferritin (16, 25, 40, 57, 58, 82–85, 92) and TfR (40, 45, 84, 92) have been published in recent years. Thus, these topics are briefly covered here to provide the reader with sufficient background information.

STRUCTURE AND FUNCTION OF FERRITIN AND TRANSFERRIN RECEPTOR

In humans, approximately 25% of the total body iron stores is in the iron-storage protein ferritin and its derivative, hemosiderin (see 16, 58, 92). Hemosiderin is an aggregate of protein and iron found in lysosomes and is thought to result from from the degradation of ferritin. Ferritin is a ubiquitous protein found in vertebrates, invertebrates, plants, fungi, and bacteria. In vertebrates, the highest amount of ferritin is found in tissues that store iron, such as liver and spleen. Ferritin has a molecular weight of about 500,000. Crystallographic studies have shown that ferritin consists of a protein shell of 24 subunits assembled to form a cavity that can accommodate a core of up to 4500 iron atoms as ferric oxyhydroxide with variable amounts of phosphate (see 16, 82, 83, 85). The protein shell contains eight channels through which iron and small molecules can enter and exit. The ferritin shell consists of two types of subunits: heavy (H), 21,000 M_r, and light (L), 19,000 M_r. The primary amino acid sequences for ferritin H and L subunits have been obtained for a variety of organisms. Among mammals, the L subunits are 82–88% identical, whereas the H subunits are 95% identical (see 25, 57). *Rana castesbian* (bullfrog) ferritin contains three subunits (H, M, L) that are similar in size and sequence but differ in cell expression and regulation by iron (19, 21). To date, there is no evidence for the presence of tissue-specific H and L mRNAs in mammals.

Cells acquire iron via transferrin, a serum glycoprotein that binds to the TfR on the cell surface. TfR is a transmembrane protein; it is a dimer of two 90,000 dalton subunits linked by a disulfide bond. Each subunit has two N-asparagine-linked carbohydrates and binds one molecule of diferric transferrin with high affinity. After binding of diferric transferrin to TfR, the TfR-diferric transferrin complex is internalized by receptor-mediated endocytosis (Figure 1) (18, 39). At about pH 5.5, iron is dissociated from the TfR-transferrin complex, where it is transferred to ferritin and to other cellular proteins. The process by which iron is transferred from the endosome to other proteins is unknown. Apotransferrin bound to the TfR is recycled to the cell membrane, where it is dissociated from the receptor and is available for binding new iron.

THE IRON-RESPONSIVE ELEMENT

The cloning of the rat ferritin L gene (49) and the chicken ferritin H gene (81) led to the identification of the highly conserved IRE motif in their 5'-untranslated regions. This sequence has subsequently been identified in the 5'-untranslated regions of all vertebrate ferritin H and L sequences to date (Figure 2A). Functional IREs have been identified in other mRNAs involved in iron metabolism. Five IREs have been identified in the 3'-untranslated region of human (9, 42, 46, 55) and chicken (12, 42) transferrin receptor mRNAs. The ferritin and TfR IREs permit the coordinate regulation of ferritin and TfR synthesis (discussed below). One IRE has been identified in the 5'-untranslated region of the human erythroid 5-aminolevulinate synthase (erythroid ALAS) mRNA (15, 17, 22, 79) and one in the 5'-untranslated region of mitochondrial porcine aconitase mRNA (17). ALAS is the first enzyme in the heme biosynthetic pathway; aconitase is a mitochondrial iron-sulfur (Fe-S) enzyme in the Krebs cycle. IRE-like sequences have been reported in the mRNAs of *Drosophila melanogaster* maternal effect gene *toll* (17) and in *Azotobacter vinelandii* bacterioferritin (E. Stiefel, personal communication). The *toll* IRE does not appear to be functional (17). Whether the *A. vinelandii* IRE-like sequence functions as an IRE is not known.

Ferritin H and L IREs from human, rat, mouse, chicken, bullfrog, and the clawed toad *(Xenopus laevis)* are about 28–35 nucleotides in length and share approximately 90% homology with each other (Figure 2A). Although ferritin, TfR, mitochondrial aconitase, and erythroid ALAS IREs share limited sequence homology, secondary structure predictions suggest that all these IREs

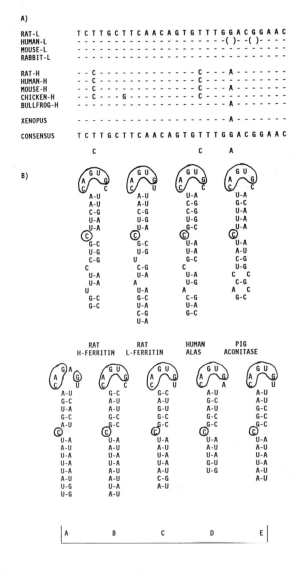

Figure 2 Sequence and structure of the iron-responsive element. (*A*) Sequences of ferritin L IREs from rat (49), human (78), mouse (25), and rabbit (25), and ferritin H IREs from rat (59), human (14), mouse (4, 47), chicken (81) bullfrog (19, 21), and *X. laevis* (54). A consensus sequence for the ferritin IRE is located below the sequences. Ambiguous sequences, ( ). (*B*) Computer-predicted IRE stem-loop structures. IRE stem-loop structures from the mRNAs for rat ferritin H (59) and L (2), human erythroid ALAS (15, 17, 22, 79), and TfR (9) and pig mitochondrial aconitase (17). Conserved nucleotides in the loop and the bulge cytosine in the stem are marked.

can be folded into a characteristic stem-loop structure (Figure 2*B*). This structure contains a six-membered loop and a base-paired stem with a bulge cytosine located five nucleotides away from the loop. The loop sequence is conserved in the IREs and has a consensus of C–A–G–U/A–G–U/C/A. The bulge cytosine in the stem and its location five nucleotides from the loop are conserved in all IREs. The sequences in the stem are conserved between ferritin H and L IREs but not between erythroid ALAS, aconitase, and TfR IREs. The negative free energy values for the IREs range between –20 and –30 kcal/mol (33, 96). As mentioned above, *A. vinelandii* was reported to contain an IRE-like sequence; however, the structure of this sequence is not similar to the consensus IRE. Possibly during evolution the prokaryotic IRE diverged from the eukaryotic IRE. The conformation of the IRE is important in binding to the IRE-BP and is discussed below.

STRUCTURE AND FUNCTION OF THE IRON-RESPONSIVE ELEMENT BINDING PROTEINS (IRE-BPs)

Conservation of IRE-BPs

IRE-BPs were first identified by RNA-protein gel shift and RNA-protein UV cross-linking assays (50). A ferritin IRE RNA was radiolabeled and incubated with a rat liver cytoplasmic protein extract, followed by resolution of the RNA-protein complexes on native polyacrylamide gels. UV-cross-linking of these complexes and analysis by SDS-polyacrylamide gel electrophoresis demonstrated that one of the complexes contained a protein of about 90,000 M_r. Subsequently, IRE-BPs have been identified in human (73), mouse (56), and rabbit (87) tissues and cells. IRE-BP binding activity has also been identified in protein extracts from *Xenopus,* fish, and *D. melanogaster* (72). No RNA-binding activity was found in extracts of budding or fission yeast or in bacteria using a human IRE (72). It is likely that either the IRE-BPs are not present in lower eukaryotes and prokaryotes or that IREs have diverged during evolution and no longer recognize mammalian IRE-BPs.

Purification, Cloning, and Characterization of IRE-BPs

IRE-BPs have been purified from human (74), rabbit (87), and rat (E. Leibold, unpublished data) livers and human placenta (60). Human placenta and liver IRE-BPs were purified by RNA-affinity chromatography, whereas rabbit liver and rat liver IRE-BPs were purified by conventional chromatography. The molecular weights determined by gel filtration and SDS polyacrylamide gels are similar for the IRE-BPs (90,000 100,000 daltons). Two IRE

BPs similar in size were purified from human placenta by RNA-affinity chromatography (60). Whether the placental proteins are proteolytic degradation products of one another, posttranslationally modified forms, or distinct proteins is not known.

IRE-BP cDNAs have been isolated from human (76) and rat liver (E. Leibold, unpublished data). The rat liver IRE-BP encodes a protein of 889 amino acids with a predicted molecular weight of 97,946. The amino terminus of rat liver IRE-BP was confirmed by direct sequence analysis and was determined to be 100 amino acids longer at the amino terminus than the reported human liver IRE-BP (76).

Comparison of the rat IRE-BP sequence with that of the human sequence demonstrated that they are 92% identical and 95% similar when conservative amino acid changes are considered. IRE-BPs share homology with mitochondrial pig (95) and yeast (28) aconitases ($\sim$ 32–36% identity and 53–56% similarity, respectively) (75). Aconitase is a Fe-S enzyme in the Krebs cycle that catalyzes the interconversion of citrate to isocitrate via the intermediate cis-aconitate. Mitochondrial aconitase exists in two states: an active state that contains a [4Fe–4S] cluster and an inactive state that contains a [3Fe-4S] cluster (5, 27, 70) (Figure 5). The highest degree of homology between the IRE-BP and aconitase was observed for amino acids, which form the active sites of aconitase, including three cysteines that are ligands for three iron-atoms. In mitochondrial aconitase, the fourth ligand on iron is a hydroxyl group that is protonated on addition of citrate (5). The structural similarities between the IRE-BP and mitochondrial aconitase suggest that the IRE-BP may have evolved from a common metal binding motif by the addition of a RNA-binding domain.

In addition to mitochondrial aconitase, a cytoplasmic form of aconitase has been characterized from pig heart (24) and bovine liver (C. Kennedy and H. Beinert, personal communication). The function of cytoplasmic aconitase is not known. The mitochondrial and cytoplasmic aconitases are similar in that they have a Fe-S cluster and aconitase activity, but they differ in their isoelectric points, stabilities, and molecular weights, which suggests that they are distinct proteins (24, 27). Interestingly, cytoplasmic aconitases and the human IRE-BP have similar molecular weights and have been localized to chromosome nine (36, 75, 80). Recent data indicate that bovine cytoplasmic aconitase binds the mitochondrial aconitase IRE, which would suggest that the IRE-BP and cytoplasmic aconitase are the same protein (C. Kennedy, H. Beinert, H. Zalkin, personal communication). What role aconitase activity plays in IRE-BP regulation is not clear. The amino acid sequence of cytoplasmic aconitase will be needed in order to obtain definitive proof of its identity.

IRON-DEPENDENT POSTTRANSCRIPTIONAL REGULATION BY IRE-BPs

Translational Regulation by IRE-BPs

The discovery of the IREs and the IRE-BPs has generated intense interest in the translational regulation of ferritin. For many years investigators have known that administration of iron to animals and cells resulted in an increase in ferritin synthesis (see 1, 25, 25a, 58, 71, 82, 83, 85, 90). This induction was not due to the production of new mRNA, since pretreatment of rats with actinomycin D, an inhibitor of RNA transcription, failed to suppress ferritin synthesis (13, 77, 94). It was demonstrated, however, that the increase in ferritin synthesis in rats treated with iron was due to an increase in the amount of ferritin mRNAs on polysomes (93). These experiments led Munro and co-workers (94) to postulate that a preexisting pool of ferritin mRNA that is present in the cytoplasm can be shifted by iron to actively translating polysomes.

The translational shift of ferritin mRNA from RNP particles to polysomes by iron was demonstrated in rat liver (1, 90) and human hepatoma cells (71) using ferritin H and L cDNAs to determine cytoplasmic location and levels of ferritin mRNA. Figure 3 shows an example of the redistribution of ferritin mRNA to polysomes after iron injection. Male rats were injected intraperitoneally with ferric ammonium citrate. Four hours later, livers were isolated and RNP particles and the polysomes were fractionated on sucrose gradients. Total RNA was isolated from each fraction and hybridized to radiolabeled ferritin H cDNA on Northern blots. After an injection of iron, liver ferritin H mRNAs are shifted from the inactive RNP pool (Figure 3*B*, lanes 1–5) to actively translating polysomes (Figure 3*B*, lanes 6–10). Ferritin L mRNA is shifted to a similar extent (data not shown). Approximately 75% of the ferritin H and L mRNAs shifted to polysomes within 30 min after an intraperitoneal injection of iron and showed a maximal shift by 4 hr (90). By 16 hr, ferritin mRNA returned to the RNP pool and was no longer translated at an increased rate. Metabolic labeling studies in rats demonstrated that ferritin synthesis in liver begins to increase within 1 hr, with maximal induction at 5 hr after iron injection (23, 58).

In contrast to the acute iron administration described above, a recent study showed that in female rats chronically fed a high iron diet, only ferritin L mRNA shifted to the polysomes, whereas ferritin H mRNA remained in the RNP pool (8). Other studies carried out with male rats fed a high iron diet showed that ferritin H and L mRNAs were unaffected by the high iron diet and did not shift to the polysomes (K. White, MIT dissertation, 1987). The

SALINE INJECTED

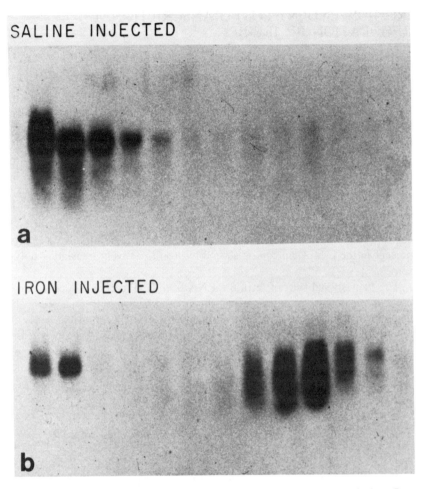

IRON INJECTED

Figure 3 Translational shift of ferritin mRNA from RNP particles to polysomes by iron. Rats were injected intraperitoneally with saline (*a*) or ferric ammonium citrate (*b*). Four hours later, livers were isolated and RNP particles and polysomes were fractionated on sucrose gradients. Total RNA was isolated from each fraction and hybridized to radiolabeled ferritin H cDNA on Northern blots. RNP fractions are in lanes 1–5; polysomal fractions are in lanes 8–12 (K. White, unpublished data).

final iron content of the livers differed in the two studies, suggesting that sex differences in iron metabolism may exist.

REGULATION BY IREs   The first studies to demonstrate a role for the IRE in the translational regulation of ferritin by iron were carried out using plasmids in which the 5'-untranslated region of rat ferritin L (2) or human ferritin H (34) genes was fused to heterologous reporter genes. When cells were transfected with these constructs and treated with iron, an increase in the activity or levels of the reporter proteins was observed. Deletions of the sequence containing the IRE abolished the response to iron (2, 34). A region of approximately 35 nucleotides containing the IRE was shown to be necessary and sufficient for iron-dependent translation regulation (11). Theil and co-workers (20) using a wheat germ in vitro translation lysate reported that a 70-nucleotide region in the 3'-untranslated region of bullfrog ferritin mRNA was involved in iron-mediated translational regulation of ferritin synthesis (20). This work was not substantiated by others, who have shown both in vitro (7) and in vivo (11) that the 3'-untranslated region does not contribute to the iron-dependent translational regulation of ferritin. The 3'-untranslated region may be required for translation, but may not be important for iron modulation.

Despite sequence differences in the stems of ferritin, erythroid ALAS, mitochondrial aconitase, and TfR IREs, all are predicted to form very similar, stem-loop structures, which suggests that the stem-loop structure may be important for IRE-BP binding. First, chemical and enzymatic nuclease footprinting experiments demonstrated that binding of the IRE-BP to the IRE was confined to the stem-loop (32). Second, mutant ferritin IRE RNAs, which contain base substitutions disrupting base-pairing of the stem, prevented IRE-BP binding (3, 48), whereas compensating substitutions, which restored base-pairing, restored IRE-BP binding (48). Third, deletion of a nucleotide in either the stem or the loop abolished protein binding to the IRE in vitro and iron-dependent regulation in vivo (10, 33, 73). These studies demonstrated that conformation of the IRE RNA is important for binding to the IRE-BP.

Because a particular RNA conformation appears to be important in IRE-BP binding, a computer search using an IRE motif was used to scan sequence databases to identify other IRE-containing mRNAs that may be regulated by the IRE-BPs (17). Two genes, *D. melanogaster toll* and porcine mitochondrial aconitase, contained IRE-like sequences consistent with the IRE consensus structure (Figure 2*B*). The *toll* IRE failed, however, to compete with either ferritin or erythroid ALAS IRE for binding to the IRE-BP (17). The predicted negative free energy of the *toll* IRE was lower than that of erythroid ALAS or ferritin IREs. These data suggested either that the *toll* IRE did not form a stable structure capable of binding the IRE-BP or that specific nucleotides not present in the *toll* sequence may be needed for IRE-BP binding. The IRE in mitochondrial aconitase mRNA appears to be functional, since it forms a RNA-protein complex when incubated with porcine lysates (C. Kennedy, H. Beinert, H. Zalkin, personal communication). These data suggested that mitochondrial aconitase may be translationally regulated by iron via the IRE-BPs.

POSITIONAL REQUIREMENTS OF THE IRE   The location of the IRE in ferritin mRNAs is important for translational regulation. Placement of either the B or C TfR IREs (Figure 2*B*) into the 5'-untranslated region of a reporter gene conferred iron-dependent regulation of translation similar to that observed for the ferritin mRNA (9). Thus, IREs located in the 5'-untranslated region of an mRNA regulate mRNA translation. Other studies demonstrated that translational repression of ferritin mRNA is dependent on the position of the IRE within the 5'-untranslated region (29). Ferritin IREs are typically located 30–39 nucleotides from the cap site of the mRNA (84), except in *X. laevis,* where the IRE is located 174 nucleotides from the cap site (54). When the IRE is moved 67 or more nucleotides downstream from the cap site, the IRE-BP/IRE complex does not interfere with translation (29). Secondary structure in the 5'-untranslated region, especially near the cap site, has been shown to limit accessibility of the cap to initiation factors and to inhibit translation (43, 44, 63). In contrast, secondary structure either in the coding region or in the 5'-untranslated region of mRNAs near the AUG initiation site is not as inhibitory (91). It is thought that the translational machinery can melt through RNA secondary structures during scanning. Thus, the data suggest that IRE-BPs repress translation either by physically blocking the cap binding protein complex from binding to the cap site or by interacting with one of the initiation factors. The location of the *X. laevis* IRE is difficult to reconcile with these data and requires further investigation.

TRANSLATIONAL REPRESSION BY IREs   Ferritin mRNAs were shown to be poorly translated in a rabbit reticulocyte lysate compared to translation in a wheat germ lysate, thus suggesting that a translational repressor specific for ferritin mRNA was present in the reticulocyte lysate (20, 86). The addition of purified IRE-BP to a wheat germ lysate programmed with mouse liver mRNA (86) or with a ferritin RNA synthesized in vitro (7, 86) repressed ferritin synthesis. This repression was relieved by the addition of excess ferritin IRE transcript, indicating that the IRE was required for regulation. These experiments showed a direct role for the IRE-BP in translational repression of ferritin synthesis.

As mentioned above, erythroid ALAS mRNA contains an IRE in its 5'-untranslated region (15, 17, 22, 79). ALAS is the first enzyme in heme biosynthesis and catalyzes the formation of 5-aminolevulinic acid by the condensation of glycine and succinyl CoA (see 22). ALAS activity is highest in hepatocytes and in erythroid cells, where heme is used in the synthesis of heme-containing proteins, such as cytochrome P450, and hemoglobin, respectively (see 22). ALAS is encoded by two separate genes: a ubiquitously expressed gene, termed the hepatic gene, and an erythroid-specific gene (69, 79). Only the erythroid ALAS mRNA contains an IRE. In hepatic tissues,

ALAS expression is feedback inhibited by heme. The repression of ALAS expression by heme is thought to be due to the transcriptional repression of the ALAS gene (80a) and/or to a decrease in the stability of the ALAS mRNA (31a). In contrast to hepatic tissues, the expression of ALAS in erythroid cells is stimulated in the presence of heme (27a). Evidence suggests that erythroid ALAS is regulated at the translational level by heme (26). The erythroid ALAS IRE is functional, as measured by binding to the IRE-BP in RNA-protein gel shift assays (15, 17), and is capable of regulating iron-dependent translation in transfected murine fibroblasts (17). Thus, the presence of an IRE in erythroid ALAS suggests that a translational mechanism is used to couple porphyrin synthesis with iron availability. Whether iron or heme is the direct inducer of erythroid ALAS synthesis via the IRE-BP remains unclear (discussed below).

Whether iron interacts with the IRE-BP as heme or in a nonheme chelatable form is a matter of controversy. Hemin has been shown to relieve the repression of ferritin translation by the IRE-BP in wheat germ lysates (51). Other forms of iron (Fe^{3+} plus H_2O_2, Fe^{3+}, and Fe^{2+} chelated with EDTA, Zn^{2+}-protoporphyrin IX, and protoporphyrin IX) did not relieve ferritin repression. Co^{3+}-protoporphyrin IX, however, was effective in relieving ferritin repression. Hemin was also shown to be cross-linked to a specific site on the IRE-BP in vitro (51a). Whether hemin is cross-linked to the IRE-BP in vivo is unknown. In other studies, heme inhibited binding of the IRE-BP to the IRE when measured by RNA-protein gel shift assays (31). Hemin appears to inhibit other nucleic acid protein interactions (31), which suggests that the inhibitory effect of hemin observed in vitro may not be specific for the IRE-BP/IRE interaction.

A clue to the nature of the iron moiety in IRE-BP comes from sequence analysis of the human IRE-BP cDNA (76). The homology between the amino acid sequences of the IRE-BP and mitochondrial aconitase, and specifically the conservation of the metal-binding sites (75), suggested that IRE-BPs may contain a Fe-S cluster. Furthermore the ability of cytoplasmic aconitase to bind IREs (C. Kennedy, H. Beinert, H. Zalkin, personal communication), in addition to the similarities in molecular weights (24, 27, 75) and chromosomal location (75, 80), suggested that cytoplasmic aconitase and the IRE-BP may be the same protein. To date, however, neither iron or heme has been reported in the IRE-BP.

Stabilization of TfR mRNA by IRE-BPs

TfR synthesis is regulated by intracellular iron and by the proliferative state of the cell. Cells treated with intracellular iron chelators increased TfR synthesis, whereas cells treated with iron salts or hemin decreased TfR synthesis (6,

52, 64, 67, 88, 89). The iron-dependent regulation of TfR is due to changes in TfR mRNA levels (66). The sequences that mediate TfR mRNA turnover are located in the 3'-untranslated region of TfR mRNAs (9, 10, 55, 62). The regulatory domain contains five IREs (A–E) (Figure 2B) that are predicted to be part of a larger RNA secondary structure (42, 55). The TfR IREs compete with ferritin IRE for the IRE-BP in RNA-protein gel shift assays (3, 42, 48, 56), thus suggesting that they have similar binding affinities. The TfR mRNA regulatory region was further delineated to a region of 250 nucleotides containing three of the five IREs (10). Deletion of a single nucleotide from each of the three IRE loops abolished RNA-binding activity in vitro and the iron-dependent regulation in mouse fibroblasts (10). Similar results were obtained by other investigators who showed that IRE-BP activity parallels TfR expression in phytohemagglutinin-stimulated lymphocytes and during the maturation of monocytes to macrophages (81a). Sequences other than the five IREs within the 250 nucleotide regulatory region were also shown to be important in regulation of TfR mRNA (10). These data suggest that the ability of the IRE-BP to interact with the TfR is important for regulation and that this interaction may protect the TfR mRNA from degradation by cellular RNases.

A Model for the Coordinate Regulation of Ferritin and TfR by IRE-BPs

A model for the coordinate regulation of ferritin and TfR expression by the IRE-BP has been postulated from data obtained by many laboratories (Figure 4). The function of an IRE is determined by its location within an mRNA. IREs located in the 5'-untranslated region of an mRNA mediate translational repression, whereas IREs located in the 3'- untranslated region mediate mRNA stabilization. The IRE-BP interacts with both IREs. When iron levels are low, IRE-BPs bind with high affinity to the IREs, repress ferritin synthesis, and stabilize TfR mRNA. These interactions result in decreased iron storage and increased iron uptake. When iron levels are high, IRE-BPs bind with low affinity to the IREs, increase ferritin synthesis, and destabilize TfR mRNAs. The result is increased iron storage and decreased iron uptake. The binding of the IRE-BP to the 5'-IRE may mediate translational repression by blocking the accessibility of the cap site to the cap-binding protein complex and the 43S pre-initiation complex. The binding of IRE-BP to 3'-IREs may stabilize TfR mRNAs either by forcing the TfR mRNA to adopt a conformation that makes it less susceptible to RNases or by protecting a specific RNAse cleavage site in the mRNA. The coordinate regulation of ferritin and TfR by the IRE-BPs allows cells to respond rapidly to incoming iron by regulating the amount of iron taken up by the cell via TfR and the amount sequestered by ferritin.

Control of Iron Uptake and Storage Within the Cell:
Responses to High and Low Iron Concentrations

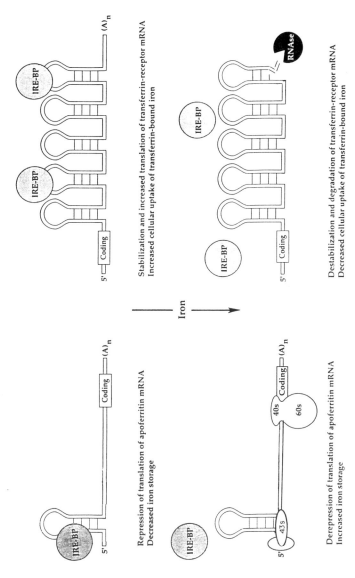

Figure 4 Proposed mechanism for the regulation of ferritin and TfR expression by the IRE-BP via the IREs. See text for detailed description. Reproduced with permission (41).

MECHANISM OF ACTIVATION OF IRE-BPs BY IRON

How do the IRE-BPs sense the iron status of the cell? The data indicate that iron regulates the activity of the IRE-BP. IRE-BP activity is increased in cells pretreated with the iron chelator desferrioxamine and is decreased in cells pretreated with iron (56, 73). The fact that induction was not blocked by actinomycin D, a transcription inhibitor, suggests that new mRNA was not necessary for increased IRE-BP activity. Cycloheximide, a translation inhibitor, delayed the increase in IRE-BP activity by 8 hr, suggesting a post-transcriptional mechanism for the activation of IRE-BP (56). IRE-BP activity was decreased in rats treated with iron over the course of 16 hr; no change in the levels of IRE-BP mRNA was noted (E. Leibold, unpublished data). These data indicate that the IRE-BP is regulated posttranscriptionally by iron.

To account for the differences observed in IRE-BP binding activities with iron and iron chelation treatments, investigators demonstrated that the IRE-BP has two distinct binding affinities for the IRE: a high affinity form (K_d = 20 pM) and a low affinity form (K_d = 3 nM) (30). In iron-depleted cells, which have increased RNA-binding activity, most of the IRE-BP is found in the high affinity form, whereas in iron-treated cells, which have decreased RNA-binding activity, very little of the IRE-BP is in the high affinity form. Thus, one effect of iron is to decrease the number of IRE-BPs in the high affinity form. The low affinity form in iron-treated cells can be activated in vitro by the addition of reducing agents, whereas no further activation of the IRE-BP can be obtained in iron-depleted cells (35). These data indicate that a functional sulfhydryl group in the IRE-BP can be reduced and oxidized, depending on the iron status of the cell. The interconversion of the IRE-BP from a high affinity form to a low affinity form would be a rapid way for IRE-BPs to respond to changes in cellular iron levels without the synthesis of new mRNA or protein.

One obvious model to explain IRE-BP activation by iron predicts that the IRE-BP is a regulatory protein that senses changes in intracellular iron concentration (75) (Figure 5A). The model is based on the interconversion of the Fe-S cluster of mitochondrial aconitase from the active [4Fe-4S] form to the inactive [3Fe-4S] form (5, 27). This model predicts that the IRE-BP is found in cells in either a [4Fe-4S] low RNA affinity form or a [3Fe-4S] high RNA affinity form. The proportion of the IRE-BP in these forms depends on the iron status of the cell. When iron levels are high, the IRE-BP is converted to the low RNA affinity form [4Fe-4S] by the addition of the fourth iron to the cluster. The fourth iron may prevent RNA binding either by sequestering a free cysteine or by stabilizing the IRE-BP structure, thereby rendering the RNA-binding site inaccessible. A structural comparison of the sequence of the active sites of mitochondrial aconitase with the human IRE-BP sequence

A)

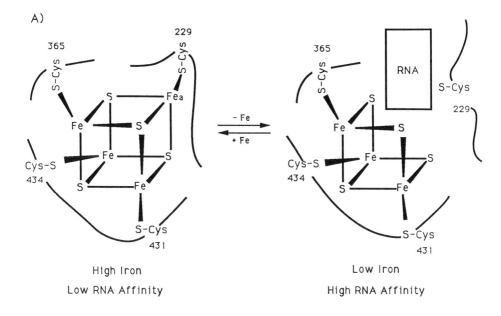

B)

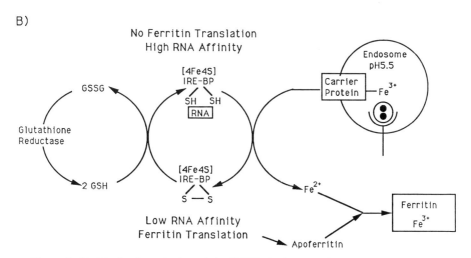

Figure 5 Models for the activation of the IRE-BP by iron. (*A*) Schematic model of the interconversion of the 4Fe and 3Fe cluster of IRE-BP based on the Fe-S cluster of aconitase (5). Fea is the labile fourth iron in aconitase. The numbers identifying the cysteine residues are from the human IRE-BP sequence (75). Adapted from Beinert (5). (*B*) Schematic model of IRE-BP activation by iron assuming the presence of an iron carrier protein. Transferrin receptor, ꓶ ; diferric transferrin, Ⓞ ; GSH, glutathione reduced; GSSG, glutathione oxidized. See text for detailed description.

showed that cysteine 229 in the IRE-BP is predicted to be positioned in the active cleft, where it could serve as the ligand for the fourth iron (75). When iron levels are low, the IRE-BP is converted to a high affinity form [3Fe-4S] containing a high affinity RNA-binding site. This RNA-binding site can be made accessible by a conformational change in the protein. Although this is an attractive model, we do not favor this model for the following reasons. First, cytoplasmic aconitase, in contrast to mitochondrial aconitase, contains four iron atoms when isolated, suggesting that the fourth iron is not labile (C. Kennedy and H. Beinert, personal communication). Second, cytoplasmic aconitase containing either no iron (apoenzyme), [3Fe-4S], or [4Fe-4S] binds the IRE with similar affinities (C. Kennedy, H. Beinert, and H. Zalkin, personal communication). Third, this model does not account for the in vitro activation of the IRE-BP by reducing agents.

An alternative model for IRE-BP activation by iron predicts that a putative iron carrier protein senses iron concentration and facilitates the conversion of the IRE-BP from a high RNA affinity form to a low RNA affinity form (Figure 5B). According to this model, the IRE-BP containing a [4Fe-4S] cluster exists in either a high RNA affinity form containing a pair of reduced thiolates or a low RNA affinity form containing an oxidized disulfide bond. In the reduced conformation, the sulfhydryl may interact directly with RNA or juxtaposed amino acids may form a high RNA affinity binding site. The model predicts that when iron enters the cell via receptor-mediated endocytosis, ferric irons dissociate from the TfR-transferrin complex and bind to the putative carrier protein. The Fe^{3+}-carrier protein complex interacts with the IRE-BP, creating a redox couple in which the oxidation of the thiols is coupled to the reduction of ferric ions on the carrier. The Fe-S cluster in the IRE-BP serves as an electron shuttle to the electrophilic Fe^{3+}-carrier. The net charge on the Fe-S cluster before and after the redox coupling remains constant. The oxidized form of IRE-BP is only a transient state and is rapidly reduced by glutathione to the high RNA affinity form. This model couples iron transport with translational regulation; the driving force is the endocytic iron uptake via the TfR. This model provides an explanation for data indicating that IRE-BP can be activated in vitro by reducing agents and is unaffected by iron (35). There is evidence that a oxidoreductase/iron transporter system is present in reticulocyte endocytic vesicles (61). This transporter system could function as the carrier protein in this model. Moreover, this model could account for the data showing that cytoplasmic aconitase containing no iron, [3Fe-4S], or [4Fe-4S] binds the IRE with similar affinities (C. Kennedy, H. Beinert, personal communication). However, this model is speculative: we do not have evidence to support the presence of an endocytic membrane carrier protein that interacts with the IRE-BP/IRE complex. Nonetheless, the presence of an iron carrier protein, which can sense the iron status of the cell, could couple IRE-BP activation with TfR-mediated iron uptake and translational regulation of ferritin synthesis.

RELATED FAMILIES OF IRE-BPs

Recent data indicate that there may be a family of related IRE-BPs. First, two IRE-BP species similar in size ($100,000$ M_r and $95,000$ M_r) have been isolated from human placenta by IRE affinity chromatography (60). These proteins have different isoelectric points, but similar IRE binding characteristics. Human liver contained only one of the IRE-BPs (60). Second, a cDNA isolated from a human T cell library has a predicted amino acid sequence 57% identical and 75% similar to IRE-BP (76). This cDNA was isolated using oligonucleotide probes derived from peptide sequences from the human IRE-BP. Third, RNA-protein gel shift assays showed that a second RNA-protein complex, in addition to the IRE-BP complex, was formed when a radiolabeled IRE was incubated with protein extracts from rat (50) and mouse (72). The second RNA-protein complex found in rats appears to be a specific complex, since it is resistant to RNase T1 digestion and heparin competition and can be purified by RNA affinity chromatography. The IRE-BP and the related IRE-BP in rat show differences, such as size and charge (B. Guo and E. Leibold, unpublished data). Whether the related IRE-BPs found in rat liver and human placenta are distinct proteins or the same protein altered by posttranslational modifications is unclear. The cloning and characterization of these proteins should allow experimental studies both in vivo and in vitro to determine their function.

CONCLUDING REMARKS

The past several years have witnessed important advances in our understanding of the iron-dependent translational regulation of ferritin expression by the IRE-BPs. In particular, the discovery of IREs in other mRNAs involved in iron metabolism has opened new pathways for research in iron metabolism and posttranscriptional regulation. Many unanswered questions remain. First, the homology of the IRE-BP with aconitase has provided clues to the mechanism by which IRE-BPs function. Does the IRE-BP act as a redox-sensitive regulatory molecule that modulates its activity, depending on the iron status of the cell? Or is IRE-BP activation facilitated by another protein that senses iron status and interacts with the IRE-BP? What are the important domains for RNA binding in the protein and what structural changes take place in the IRE-BP when RNA binds? Direct evidence is needed to demonstrate that the IRE-BP contains Fe-S and/or heme. Second, related IRE-BPs that are similar to the IRE-BPs have been identified. The function of these proteins is unknown. Are these proteins involved in iron homeostasis? Or if not, what are their functions? Third, the mechanism regulating the stabilization of TfR mRNAs by the IRE-BPs is not understood. How do IRE-BPs protect TfR mRNA from degradation? Where does this regulation occur in the cell? Are specific RNAses involved in this regulation? Finally, we do not know how iron is transferred from the endosome to the IRE-BP. Does the carrier protein physically couple endosomal transfer of iron to the IRE-BP/IRE complex? Over the next few years, as we seek and obtain answers to these questions, we will gain a better understanding of the regulation of iron metabolism as well as the mechanisms regulating posttranscriptional gene regulation.

ACKNOWLEDGMENTS

We are grateful to Dennis Winge, Liz Wyckoff, Alex Knisely, Charles Dameron, and Rich Ajioka for helpful suggestions on the text and stimulating discussions. We thank Claire Kennedy, Helmut Beinert, H. Zalkin, and Ed Stiefel for sharing data prior to publication. The work from the author's laboratory was supported by grants GM45201 and AM20630 from the NIH.

Literature Cited

1. Aziz, N., Munro, H. N. 1986. Both subunits of rat liver ferritin are regulated at a translational level by iron induction. *Nucleic Acids Res.* 14:915–27
2. Aziz, N., Munro, H. N. 1987. Iron regulates ferritin mRNA translation through a segment of its 5' untranslated region. *Proc. Natl. Acad. Sci. USA* 84:8478–82
3. Barton, H. A., Eisenstein, R. S., Bomford, A., Munro, H. N. 1990. Determinants of the interaction between the iron-responsive element binding protein and its binding site in rat L-ferritin RNA. *J. Biol. Chem.* 265:7000–15
4. Beaumont, C., Dugast, I., Renaudie, F., Souroujon, M., Grandchamp, B. 1989. Transcriptional regulation of ferritin H and L subunits in adult erythroid and liver cells from the mouse. *J. Biol. Chem.* 264:7498
5. Beinert, H., Kennedy, M. C. 1989. Engineering of protein bound iron-sulfur clusters. *Eur. J. Biochem.* 186:5–15
6. Bridges, K. R., Cudkowicz, A. 1984. Effect of iron chelators on the transferrin receptor in K562 cells. *J. Biol. Chem.* 259:12970–77
7. Brown, P. H., Daniels-McQueen, S., Walden, W. E., Patino, M. M., Gaffield, L., et al. 1989. Requirements for the translational repression of ferritin transcripts in wheat germ extracts by a 90-kDa protein from rabbit liver. *J. Biol. Chem.* 264:13383–86

8. Cairo, G., Tacchini, L., Schiaffonati, L., Rappocciolo, E., Ventura, E., et al. 1989. Translational regulation of ferritin synthesis in rat liver. *Biochem. J.* 264:925–28
9. Casey, J. L., Hentze, M. W., Koeller, D. M., Caughman, S. W., Rouault, T. A., et al. 1988. Iron-responsive elements: regulatory RNA sequences that control mRNA levels and translation. *Science* 240:924–28
10. Casey, J. L., Koeller, D. M., Ramin, V. C., Klausner, R. D., Harford, J. B. 1989. Iron regulation of transferrin receptor mRNA levels requires iron-responsive elements and a rapid turnover determinant in the 3' untranslated region. *EMBO J.* 8:3693–99
11. Caughman, S. W., Hentze, M. W., Rouault, T. A., Harford, J. B., Klausner, R. D. 1988. The iron-responsive element is the single element responsible for iron-dependent translational regulation of ferritin biosynthesis. *J. Biol. Chem.* 263:19048–52
12. Chan, L. N., Grammatikakis, N., Banks, J. M., Gerhardt, E. M. 1989. Isolation of cDNA and genomic clones for chicken transferrin receptor. *Nucleic Acids Res.* 17:3763–71
13. Chu, L. L. H., Fineberg, R. A. 1969. On the mechanism of iron-induced synthesis of apo-ferritin in HeLa cells. *J. Biol. Chem.* 244:3847–54

14. Costanzo, R., Colombo, M., Staempfli, S., Santoro, C., Marone, M. 1986. Structure of gene and pseudogenes of human apoferritin H. *Nucleic Acids Res.* 14:721–36

15. Cox, T. C., Bawden, M. J., Martin, A., May, B. M. 1991. Human erythroid 5-aminolevulinate synthase: promoter analysis and identification of an iron-responsive element in the mRNA. *EMBO J.* 10:1891–1902

16. Crichton, R. R. 1990. Proteins of iron storage and transport. *Adv. Protein Chem.* 40:281–363

17. Dandekar, T., Stripecke, R., Gray, N. K., Goosen, B., Constable, A., et al. 1991. Identification of a novel iron-responsive element in murine and human erythroid 5-aminolevulinic acid synthase mRNA. *EMBO J.* 10:1903–9

18. Dautry-Varsat, A., Ciechanover, A., Lodish, H. F. 1983. pH and the recycling of transferrin during receptor-mediated endocytosis. *Proc. Natl. Acad. Sci. USA* 80:2258–62

19. Dickey, L. F., Sreedharan, S., Theil, E. C., Didsbury, J. R., Wang, Y., et al. 1987. Differences in the regulation of messenger RNA for housekeeping and specialized-cell ferritin. *J. Biol. Chem.* 262:7901–7

20. Dickey, L. F., Wang, W. Y., Shull, G. E., Wortman, I. A., Theil, E. C. 1988. The importance of the 3'-untranslated region in the translational control of ferritin mRNA. *J. Biol. Chem.* 263:3071–74

21. Didsbury, J. R., Theil, E. C., Kaufman, R. E., Dickey, L. F. 1986. Multiple red cell ferritin mRNAs, which code for an abundant protein in the embryonic cell type, analyzed by cDNA sequence and by primer extension of the 5'-untranslated regions. *J. Biol. Chem.* 261:949–55

22. Dierks, P. 1990. Molecular biology of eukaryotic 5-aminolevulinate synthase. In *Biosynthesis of Heme and Chlorophylls*, ed. H. A. Hailey, pp. 201–33. New York: McGraw-Hill

23. Drysdale, J. W., Munro, H. N. 1966. Regulation of synthesis and turnover of ferritin in rat liver. *J. Biol. Chem.* 241:3630–37

24. Eanes, R. C., Kun, E. 1974. Inhibition of liver aconitase isozymes by (-)-erythro-fluorocitrate. *Mol. Pharmacol.* 10:130–39

25. Eisenstein, R. S., Bettany, A. J., Munro, H. N. 1990. Regulation of ferritin gene expression. *Adv. Inorg. Chem.* 8:91–138

25a. Eisenstein, R. S., García-Mayol, D., Pettingell, W., Munro, H. N. 1991.

Regulation of ferritin and heme oxygenase synthesis in rat fibroblasts by different forms of iron. *Proc. Natl. Acad. Sci. USA* 88:688–92

26. Elferink, C. J., Sassa, S., May, B. K. 1988. Regulation of 5-aminolevulinate synthase in mouse erythroleukemic cells is different from that in liver. *J. Biol. Chem.* 263:13012–16

27. Emptage, M. H. 1988. Metal clusters in proteins. *Am. Chem. Soc. Symp.* 372:343–71

27a. Fujita, H., Yamamoto, M., Yamagami, T., Hayashi, N., Sassa, S. 1991. Erythroleukemia differentiation: distinctive responses of the erythroid-specific and the nonspecific β-aminolevulinate synthase mRNA. *J. Biol. Chem.* 266:17494–17502

28. Ganglof, S. P., Marguet, D., Lauquin, G. 1990. Molecular cloning of the yeast mitochondrial aconitase gene (ACO1) and evidence of a synergistic regulation of expression by glucose plus glutamate. *Mol. Cell. Biol.* 7:3551–61

29. Goossen, B., Caughman, S. W., Harford, J. B., Klausner, R. D., Hentze, M. W. 1990. Translational repression by a complex between the iron-responsive element of ferritin mRNA and its specific cytoplasmic binding protein is position dependent *in vivo*. *EMBO J.* 9:4127–33

30. Haile, D. J., Hentze, M. W., Rouault, T. A., Harford, J. B., Klausner, R. D. 1989. Regulation of interaction of the iron-responsive element binding protein with iron-responsive RNA elements. *Mol. Cell. Biol.* 9:5055–61

31. Haile, D. J., Rouault, T. A., Harford, J. B., Klausner, R. D. 1990. The inhibition of the iron-responsive element RNA-protein interaction by heme does not mimic in vivo iron regulation. *J. Biol. Chem.* 265:12786–89

31a. Hamilton, J. W., Bement, W. J., Sinclair, P. R., Sinclair, J. F., Alcedo, J. A., Wetterhahn, K. E. 1991. Heme regulates hepatic 5-aminolevulinate synthase mRNA expression by decreasing mRNA half-life and not by altering its rate of transcription. *Arch. Biochem. Biophys.* 289:387–92

32. Harrell, C. M., McKenzie, A. R., Patino, M. M., Walden, W. E., Theil, E. C. 1991. Ferritin mRNA: interactions of iron regulatory element with translational regulator P-90 and the effect on base-paired flanking regions. *Proc. Natl. Acad. Sci. USA* 88:4166–70

33. Hentze, M. W., Caughman, S. W., Casey, J. L., Koeller, D. M., Rouault, T. A., et al. 1988. A model for the

structure and functions of iron-responsive elements. *Gene* 72:201–8

34. Hentze, M. W., Caughman, S. W., Rouault, T. A., Barriocanal, J. G., Dancis, A., et al. 1987. Identification of the iron-responsive element for the translational regulation of human ferritin mRNA. *Proc. Natl. Acad. Sci. USA* 238:1570–73

35. Hentze, M. W., Rouault, T. A., Harford, J. B., Klausner, R. D. 1989. Oxidation-reduction and the molecular mechanism of a regulatory RNA-protein interaction. *Science* 244:357–59

36. Hentze, M. W., Seuanez, H. N., O'Brien, S. J., Harford, J. B., Klausner, R. D. 1989. Chromosome localization of nucleic acid binding proteins by affinity mapping: assignment of IRE-binding protein to chromosome 9. *Nucleic Acids Res.* 17:6103–8

37. Hershey, J. W. B. 1991. Translational control in mammalian cells. *Annu. Rev. Biochem.* 60:717–55

38. Hunt, T. 1988. Controlling mRNA lifespan. *Nature* 334:567–68

39. Klausner, R. D., Ashwell, G., Van Renswoude, J., Harford, J. B., Bridges, K. R. 1983. Binding of apotransferrin to K562 cells: explanation of the transferrin cycle. *Proc. Natl. Acad. Sci. USA* 80:2263–66

40. Klausner, R. D., Harford, J. B. 1989. Cis-trans model for post-transcriptional gene regulation. *Science* 246:870–72

41. Knisely, A. 1992. Neonatal hemochromatosis. *Adv. Pediatr.* 39:In press

42. Koeller, D. M., Casey, J. L., Hentze, M. W., Gerhardt, E. M., Chan, L. L. 1989. A cytosolic protein binds to structural element within the iron regulatory region of the transferrin receptor mRNA. *Proc. Natl. Acad. Sci. USA* 86:3574–78

43. Kozak, M. 1986. Influences of mRNA secondary structure on initiation by eukaryote ribosomes. *Proc. Natl. Acad. Sci. USA* 83:2850–54

44. Kozak, M. 1989. The scanning model for translation: an update. *J. Cell Biol.* 108:229–41

45. Kuhn, L. C. 1989. The transferrin receptor: a key function in iron metabolism. *Schweiz. Med. Wochenschr.* 119:1319–26

46. Kuhn, L. C., McClelland, A., Ruddle, F. 1984. Gene transfer, expression and molecular cloning of the human transferrin receptor gene. *Cell* 37:95–103

47. Kwak, E. L., Torti, S., Torti, F. 1990. Murine ferritin heavy chain: isolation and characterization of a functional gene. *Gene* 94:255–61

48. Leibold, E. A., Laudano, A., Yu, Y. 1990. Structural requirements of iron-responsive elements for binding of the protein involved in both transferrin receptor and ferritin mRNA post-transcriptional control. *Nucleic Acids Res.* 18:1819–24

49. Leibold, E. A., Munro, H. N. 1987. Characterization and evolution of the expressed rat ferritin light subunit gene and its pseudogene family. *J. Biol. Chem.* 262:7335–41

50. Leibold, E. A., Munro, H. N. 1988. Cytoplasmic protein binds *in vitro* to a highly conserved sequence in the 5' untranslated region of ferritin heavy and light subunit mRNAs. *Proc. Natl. Acad. Sci. USA* 85:2171–75

51. Lin, J., Daniels-McQueen, S., Patino, M. M., Gaffield, L., Walden, W. E., et al. 1990. Derepression of ferritin messenger RNA translation by hemin in *in vitro*. *Science* 247:74–77

51a. Lin, J.-J., Patino, M. M., Gaffield, L., Walden, W. E., Smith, A., Thach, R. E. 1991. Crosslinking of hemin to a specific site on the 90-kDa ferritin repressor protein. *Proc. Natl. Acad. Sci. USA* 88:6068–71

52. Mattia, E. K., Rao, K., Shapiro, H. H., Sussman, H. H., Klausner, R. D. 1984. Biosynthetic regulation of the human transferrin receptor by desferrioxamine in K562 cells. *J. Biol. Chem.* 259:2689–92

53. Miyazaki, Y., Setoguchi, M., Higuchi, Y., Yoshida, S., Akizuki, S., Yamamoto, S. 1988. Nucleotide sequence of cDNA encoding the heavy subunit of mouse macrophage ferritin. *Nucleic Acids Res.* 16:373

54. Moskaitis, J. E., Pastori, R. L., Schoenberg, D. R. 1990. Sequence of *Xenopus laevis* ferritin mRNA. *Nucleic Acids Res.* 18:2184

55. Mullner, E. W., Kuhn, L. C. 1988. A stem-loop in the 3' untranslated region mediates iron-dependent regulation of transferrin receptor mRNA in the cytoplasm. *Cell* 53:815–25

56. Mullner, E. W., Neupert, B., Kuhn, L. C. 1989. A specific mRNA binding factor regulates the iron-dependent stability of cytoplasmic transferrin receptor mRNA. *Cell* 58:373–82

57. Munro, H. N. 1990. Iron regulation of ferritin gene expression. *J. Cell. Biochem.* 44:107–15

58. Munro, H. N., Linder, M. 1978. Ferritin: structure, biosynthesis, and role in iron metabolism. *Physiol. Rev.* 58:318–87

59. Murray, M. T., White, K., Munro, H.

N. 1987. Conservation of ferritin heavy subunit gene structure: implications for the regulation of ferritin gene expression. *Proc. Natl. Acad. Sci. USA* 84: 7438–42

60. Neupert, B., Thompson, N. A., Meyer, C., Kuhn, L. C. 1990. A high yield affinity purification method for specific RNA-binding proteins: isolation of the iron regulatory factor from human placenta. *Nucleic Acids Res.* 18:51–55

61. Nunez, M., Gaete, V., Watkins, J. A., Glass, J. 1990. Mobilization of iron from endocytic vesicles. *J. Biol. Chem.* 265:6688–92

62. Owen, D., Kuhn, L. C. 1987. Noncoding 3' sequences of the transferrin receptor gene are required for mRNA regulation by iron. 1987. *EMBO J.* 6:1287–93

63. Pelletier, J., Sonnenberg, N. 1985. Insertion mutagenesis to increase secondary structure within the 5' non-coding region of a eukaryotic mRNA. *Cell* 40:515–26

64. Pellicci, P. G., Tabillio, P., Thomopoulos, M., Titieux, M. 1982. Hemin regulates the expression of transferrin receptors in human hematopoietic cells lines. *FEBS Lett.* 145:350–54

65. Raghow, R. 1987. Regulation of messenger turnover in eukaryotes. *Trends Biochem.* 12:358–60

66. Rao, K., Harford, J. B. Rouault, T. A., McClelland, A., Ruddle, F. H., et al. 1986. Transcriptional regulation by iron of the gene for the transferrin receptor. *Mol. Cell. Biol.* 6:236–40

67. Rao, K. K., Shapiro, E., Mattia, E., Bridges, K., Klausner, R. 1985. Effects of alterations in cellular iron on biosynthesis of the transferrin receptor in K562 cells. *Mol. Cell. Biol.* 5:595–600

68. Rapoport, S. M. 1986. Uptake and fate of iron: the transferrin receptor. In *The Reticulocyte*, pp. 145–54. Boca Raton, Fla: CRC Press

69. Riddle, R. D., Yamamoto, M., Engel, J. D. 1989. Expression of 5-aminolevulinate synthase in avian cells: separate genes encode erythroid-specific and non-specific isozymes. *Proc. Natl. Acad. Sci. USA* 86:792–96

70. Robbins, A. H., Stout, C. D. 1989. The structure of aconitase. *Proteins: Struct. Function, Genet.* 5:289–312

71. Rogers, J., Munro, H. N. 1987. Translation of ferritin light and heavy subunit mRNAs is regulated by intracellular chelatable iron levels in rat hepatoma cells. *Proc. Natl. Acad. Sci. USA* 84:2277–81

72. Rothenberger, S., Mullner, E. L., Kuhn, L. 1990. The mRNA binding protein which controls ferritin and transferrin receptor expression is conserved during evolution. *Nucleic Acids Res.* 18: 1175–79

73. Rouault, T. A., Hentze, M. W., Caughman, S. W., Harford, J. B., Klausner, R. D. 1988. Binding of a cytosolic protein to the iron-responsive element of human ferritin messenger RNA. *Science* 241:1207–10

74. Rouault, T. A., Hentze, M. W., Haile, D. J., Harford, J. B., Klausner, R. D. 1989. The iron-responsive element binding protein: A method for the affinity purification of a regulatory RNA-binding protein. *Proc. Natl. Acad. USA* 86:5768–72

75. Rouault, T. A., Stout, C. D., Kaptain, S., Harford, J. B., Klausner, R. D. 1991. Structural relationship between an iron-regulated RNA-binding protein and aconitase: functional implications. *Cell* 64:881–83

76. Rouault, T. A., Tang, C. K., Kaptain, S., Burgess, W. H., Haile, D. J., et al. 1990. Cloning of the cDNA encoding an RNA regulatory protein—the human iron-responsive element-binding protein. *Proc. Natl. Acad. Sci. USA* 87: 7958–62

77. Saddi, R., Von Der Decken, A. 1965. The effect of iron administration on the incorporation of (^{14}C) leucine into ferritin by rat liver systems. *Biochim. Biophys. Acta* 111:124–33

78. Santoro, C., Marone, M., Ferrone, M., Costanzo, F., Columbo, M., et al. 1986. Cloning of the gene coding for human L apoferritin. *Nucleic Acids Res.* 14:2863–76

79. Schoenhaut, D. S., Curtis, P. J. 1989. Structure of mouse erythroid 5-aminolevulinate synthase gene and mapping of erythroid specific DNAse I hypersensitive sites. *Nucleic Acids Res.* 17:7013–28

80. Shows, T. B., Brown, J. A. 1977. Mapping AK-1, ACONs, and Ak3 to chromosome 9 in man employing an X/9 translocation and somatic cell hybrids. *Cytogenet. Cell Genet.* 19:26–37

80a. Srivastava, G., Borthwick, I. A., Maguire, D. J., Elferink, C. J., Bawden, M. J., et al. 1988. Regulation of 5-aminolevulinate synthase messenger-RNA in different rat tissues. *J. Biol. Chem.* 263:5202–9

81. Stevens, P. W., Dodgson, J. B., Engel, J. D. 1987. Structure of the chicken ferritin H-subunit gene. *Mol. Cell. Biol.* 7:1751–58

81a. Testa, U., Kühn, L., Petrini, M., Quaranta, M. T., Pelosi, E., Peschle, C.

1991. Differential regulation of iron regulatory element-binding protein(s) in cell extracts of activated lymphocytes versus monocytes-macrophages. *J. Biol. Chem.* 266:13925–30

82. Theil, E. C. 1983. Ferritin, structure, function and regulation. *Adv. Inorg. Chem.* 5:1–38
83. Theil, E. C. 1987. Ferritin: structure, gene regulation, and cellular function in animals, plants and microorganisms. *Annu. Rev. Biochem.* 56:289–314
84. Theil, E. C. 1990. Regulation of ferritin and transferrin receptor mRNAs. *J. Biol. Chem.* 265:4771–74
85. Theil, E. C. 1990. The ferritin family of iron-storage proteins. *Adv. Enzymol. Relat. Areas Mol. Biol.* 63:421–49
86. Walden, W. E., Daniels-McQueen, S., Brown, P. H., Gaffield, L., Russell, D., et al. 1988. Translational repression in eukaryotes: partial purification and characterization of a repressor of ferritin mRNA translation. *Proc. Natl. Acad. Sci. USA* 85:9503–7
87. Walden, W. E., Patino, M. M., Gaffield, L. 1989. Purification of a specific repressor of ferritin mRNA translation from rabbit liver. *J. Biol. Chem.* 264:13765–69
88. Ward, J. H., Jordan, I., Kushner, J. P., Kaplan, J. 1984. Heme regulation of HeLa cell transferrin receptor number. *J. Biol. Chem.* 259:13235–40
89. Ward, J. H., Kushner, J. P., Kaplan, J.

1982. Regulation of HeLa cell transferrin receptors. *J. Biol. Chem.* 257:10317–23
90. White, K., Munro, H. N. 1988. Induction of ferritin subunit synthesis by iron is regulated at both the transcriptional and translational levels. *J. Biol. Chem.* 263:8938–42
91. Wolin, S. L., Walter, P. 1988. Ribosome pausing and stacking during translation of a eukaryotic mRNA. *EMBO J.* 11:3561–69
92. Worwood, M. 1989. An overview of iron-metabolism at a molecular level. *J. Intern. Med.* 226:381–91
93. Zahringer, J., Baliga, B. S., Drake, R. L., Munro, H. N. 1977. Subcellular distribution of total-poly A containing RNA and ferritin mRNA in the cytoplasm of rat liver. *Biochem. Biophys. Res. Commun.* 65:583–90
94. Zahringer, J., Baliga, B. S., Munro, H. N. 1976. Novel translational control in regulation of ferritin synthesis by iron. *Proc. Natl. Acad. Sci. USA* 73:857–61
95. Zheng, L., Andrews, P. C., Hermodson, M. A., Dixon, J. E., Zalkin, H. 1990. Cloning and structural characterization of porcine heart aconitase. *J. Biol. Chem.* 265:2814–21
96. Zuker, M., Stiegler, P. 1981. Optimal computer folding of large RNA sequences using thermodynamics and auxiliary information. *Nucleic Acids Res.* 9:133–48

Annu. Rev. Nutr. 1992. 12:369–90

NUTRITIONAL ASPECTS OF COLLAGEN METABOLISM

Richard A. Berg

Department of Biochemistry, UMDNJ-Robert Wood Johnson Medical School, Piscataway, New Jersey 08854-5635

Janet S. Kerr

The Dupont/Merck Pharmaceutical Company, Experimental Station, Wilmington, Delaware 19880-0400

KEY WORDS: connective tissue, regulation, starvation, vitamins

CONTENTS

369

0199-9885/92/0715-0369$02.00

INTRODUCTION

Collagens are a family of proteins that constitute the major extracellular proteins found in the body. Most tissues and organs contain several types of collagen whose function is primarily structural [for reviews on collagen types, see (102)]. These types differ in both primary and higher order structure. The collagens polymerize extracellularly into a variety of structures that constitute the extracellular matrix [for reviews, see (151)]. Collagen associates both noncovalently and covalently with additional components of the extracellular matrix and with minerals to form specialized connective tissues including skin, tendon, and bone [see (49) for a review]. Collagen is remodeled continuously throughout growth and development. However, because it is extracellular, it must be depolymerized, degraded, and transported into cells to be hydrolyzed to amino acid residues for recycling into protein or for conversion to metabolites. Collagenous peptides enter the circulation and are transported either to the liver for breakdown to amino acids or to the kidney for excretion. Peptidases are present in both organs that cleave collagenous peptides. Since collagen is turned over continuously, a relatively constant amount of collagenous peptides containing the unique imino acid hydroxyproline is excreted every day and reflects the level of connective tissue metabolism and remodeling.

EXTRACELLULAR MATRIX

Collagen Types and Tissue Distribution

All collagens have two structural features in common: (a) collagens are trimers composed of three polypeptides, and (b) these trimers contain large amounts of the repeating amino acid sequence, X–Y–Gly, with a high content of proline in the X position and a high content of hydroxyproline in the Y position. This repeating amino acid sequence demands a helical conformation, and the three subunits of the protein fold into a triple-helical trimeric structure. The collagens are ubiquitous, and several types (discussed below) are found in most connective tissues. To date, at least fourteen collagen types have been identified [for review see (152)]. Types I, II, III, V, and XI are fibril-forming collagens and occur in almost all connective tissues; however, cartilage contains only types II and XI and not types I or III (152). Another group of collagens are recognized as *fibril-associated collagens with interrupted triple helices* (FACITs) (57). These collagens are types IX, XII, and XIV and are not fiber forming, but they are associated with fibers formed from collagens in the first group (152). For example, type IX is found on the surface of type II fibrils in cartilage (57, 154), and type XII, which shares similarity with type IX, may be associated with type I fibrils (36). Recently, another member of this group named type XIV has been identified based on homology to type IX and XII collagens (37). Types II, IX, X, and XI occur primarily in cartilage (152). Type X collagen is particularly restrictive in that it occurs exclusively in calcifying and hypertrophic cartilage during endochrondral bone formation (140).

In addition to the fibril-forming and FACIT collagens are those collagens that form specialized structures. Type IV forms a unique structure and is the major collagen of basement membrane (162). Type VII occurs in skin as anchoring fibrils of basement membranes in the dermal epidermal junction (19, 85). Type VIII, originally reported in endothelial cell cultures (135), is the major collagen in Desmemet's membrane (158), which separates the corneal endothelial cells from the stroma.

HETEROGENEITY OF COLLAGEN TYPES   Since each subunit or polypeptide is the product of a different gene, at least 25 different genes encode the polypeptides of the 14 collagen types. Most collagen polypeptides share some homology with polypeptides of another collagen type. Type I collagen occurs most abundantly as a heterotrimer consisting of two $\alpha1(I)$ polypeptides or subunits, and one $\alpha2(I)$ polypeptide or subunit. These two polypeptides exhibit 66% homology in the triple-helical domain (106). Type III collagen occurs as a homotrimer with three identical $\alpha1(III)$ subunits. The $\alpha1(II)$ subunit is 65% homologous to $\alpha1(I)$ and 58% homologous to $\alpha2(I)$ in the triple-helical regions of the polypeptide (106). Homotrimeric collagens are types II, III, VII, X, and XII. Type I collagen exists both as a homotrimer $\alpha1(I)$ and as a heterotrimer $\alpha1(I)\alpha2(I)$. Other collagens are heterotrimers containing two and even three distinct subunits. Types VI and IX contain three different subunits (149, 153).

Several collagen subunits occur in more than one form owing to alternative associations of subunits [for review, see (1)]. Type V collagen contains either two or three different subunits: $[\alpha1(V)]_3$; $[\alpha1(V)]_2\alpha2(V)$, or $\alpha1(V)\alpha2(V)\alpha3(V)$ (46). One of the subunits of type XI collagen, $\alpha3(XI)$, appears to be similar or identical to the $\alpha1(II)$ polypeptide of type II collagen (44). To date, five different polypeptides or subunits have been identified for type IV collagen (59, 70, 114). Although many different combinations of these subunits could exist, only a few such combinations have been isolated, the most common form being $[\alpha1(IV)]_2\alpha2(IV)$ (86).

VARIATION IN TYPES I AND III COLLAGEN GENE EXPRESSION   Since the collagen genes are developmentally regulated, a particular cell or tissue expresses only a subset of its collagen genes at any one time, depending on the cell's or tissue's state of development [for reviews, see (139)]. The regulation of gene expression is well demonstrated for types I and III (1).

Type I collagen occurs in greatest amounts in embryonic bone (109) and tendons (141) where it constitutes 65 and 30%, respectively, of the total protein. Type I collagen is also found in lung, skin, and muscle, although the amounts of type I found in these tissues are much less than in bone and tendon. Type I collagen accounts for 2.9% of total protein in embryonic sheep skin (declining to 1.6% at birth) and 3.7% of the protein in embryonic sheep lung (declining to 2.4% at birth) (150). Type III collagen is often found in association with type I collagen in most connective tissues (83). Type III occurs most abundantly in extensible tissues such as fetal skin and blood vessels. It accounts for a large part of the collagen of embryonic skin (126). Recently, some type III collagen and its mRNA have been found in the less extensible tissues of bone and tendon (84, 138).

REGULATION OF SYNTHESIS   Regulation of synthesis is achieved through a variety of modulators [for review, see (1)], including steroids (26, 27, 61, 127), drugs (87), interferon (131), growth factors (98, 128), cytokines (38, 40, 56, 78, 101, 125), and various metabolites including acetaldehyde (17, 69). Interferon has been used both to affect fibronectin synthesis (29) and to suppress collagen production in scleroderma fibroblasts in culture (79, 132); it may also have an effect on other fibrotic conditions (30) and wound healing (58).

Glucocorticoids decrease procollagen synthesis in fibroblast cell cultures as demonstrated by a decrease in total mRNA for type I procollagen. The decrease occurs in cellular, nuclear, cytoplasmic, and polysomal steady-state levels of mRNA for proα1 and proα2 polypeptides (27). More recent work by Cockayne & Cutroneo has shown that glucocorticoids may employ a mechanism that alters the level of binding of transcription factors to the proα2 collagen gene (26). Various tissues may respond differently; for example, estradiol was shown to increase mRNA levels for type I procollagen in osteoblastic cells of bone (42, 43).

Fibrosis that results from increased synthesis or decreased turnover of collagen is often the end result of tissue damage by drugs or other toxic agents (2, 25) or alcohol (4, 69). Carbon tetrachloride treatment of rats results in hepatic fibrosis (25) and the synthesis of a fibrogenic factor in liver (23).

Collagen Turnover

All tissues in the body contain collagen either serving as a structural element or performing a role in the pericellular matrix. Wherever collagen degradation has been measured, investigators have found that collagenous structures in general undergo continuous remodeling in turnover. As much as 3–5% of skin collagen may turn over per day (95). Early attempts to separate matrix collagen into pools destined for turnover at different rates were complicated

(95) and did not take into account either reutilization rates of labeled amino acids or the more recent finding that most secretory proteins are subject to a variable amount of intracellular turnover during their biosynthesis (7). Collagen was one of the first secretory proteins shown to be subject to intracellular turnover during its biosynthesis because of the ease with which degraded collagenous peptides containing hydroxyproline can be measured (12). Jackson & Heininger (72) used radioactive proline as a metabolic tracer and showed that collagen turned over at a rate of 2.6% per day and that proline was reused.

INTRACELLULAR ASSEMBLY AND TURNOVER   The biosynthesis of collagen occurs on membrane-bound polysomes and, simultaneously with its translation, collagen is transferred to the endoplasmic reticulum where it undergoes a number of posttranslational modification reactions that result in trimer formation (8). Like many secretory proteins, collagen is subject to a number of proofreading enzymatic reactions within the endoplasmic reticulum, and possibly the golgi apparatus, that cause abnormally modified collagen to undergo intracellular degradation to small peptides (7). These peptides are secreted from the cell through the default pathway leading from the endoplasm reticulum through the golgi to secretory vesicles and exocytosis. Proteins within the endoplasmic reticulum that bind to malfolded proteins such as immunoglobin heavy-chain binding protein (BIP) have been shown to bind to malfolded immunoglobulin heavy chains (60, 82) and may bind mutated collagen molecules (P. Byers, personal communication). Folding of proteins in vitro is a slow and inefficient process (55). The longer the time a protein exists in a malfolded, unstable state, the greater the chance for degradation or nonspecific aggregation, in vivo (133). Consequently, investigators predict that accessory factors will be identified as catalysts in the folding process. A heat shock glycoprotein, hsp47, was discovered because of its ability to bind both gelatin and native type I collagen (115). Its collagen-binding properties were found to decrease at pH 6.3 (134). Hsp47 was immunolocalized to the endoplasmic reticulum (134). Hsp47 possesses an RDEL sequence at its carboxyl terminal region (68) similar to the KDEL sequence of other ER resident proteins. Antibodies to hsp47 have shown that hsp47 and collagen are immunoprecipitated together (115a); possibly hsp47 may assist in collagen folding.

The enzymes responsible for the intracellular degradation of collagen either within the endoplasmic reticulum or within the golgi apparatus have not been identified (7). Although the exact site of intracellular degradation is unknown, evidence indicates that a fraction of collagen may be transferred to the lysosomal compartment for degradation by lysosomal enzymes (10). A variable fraction (10 40%) of newly synthesized collagen is subject to intracellular degradation, depending on the conditions (12). A required cofactor of collagen synthesis is ascorbic acid, which is essential for the hydroxylation of prolyl residues in newly synthesized collagen. Ascorbic acid deficiency in most systems results in underhydroxylated collagen that is unable to fold into a stable triple-helical conformation and is, therefore, subject to intracellular degradation (11). Genetic diseases that result in mutations within the triple-helical region of type I collagen result in collagen that is also incapable of assuming a triple helix and is subject to this pathway of intracellular degradation of newly synthesized collagen (157).

EXTRACELLULAR TURNOVER   Collagen present in the extracellular matrix, either in tissues such as skin, bone, or tendon or in organs such as the liver and kidneys, is subject to turnover and remodeling catalyzed primarily by metalloproteinases with an involvement of serine proteinases. A family of metalloproteinases has been shown to degrade extracellular matrix macromolecules and collagen in particular (100). This family of metalloproteinases share common features in that its members are synthesized and secreted from cells as precursors that are inactive and must be proteolytically activated either by serine proteinases (63) or other metalloproteinases [for review, see (9, 74)]. For example, the major type I collagen-degrading enzyme, commonly referred to as fibroblast collagenase, is activated by plasmin and superactivated by stromelysin (112, 146), another member of the metalloproteinase family. The major collagenase synthesized by neutrophils is activated by cathepsin G, a serine protease (20) that also activates stromelysin (123). The family of metalloproteinases that includes collagenase is also under cytokine regulation in that its members are induced by interleukin I (100). Remodeling of connective tissues within bone, tendon, and skin occurs as a result of either hormonal regulation or mechanical or electrical stimulation. For example, bone remodeling is altered as a result of mechanical tension, compression, and electrical field effects (105). Mechanical effects are also important for cartilage that depends on diffusion of nutrients for maintenance (121), since it has no vasculature.

AMINO ACID DEFICIENCY

Starvation

During advanced stages of starvation such as amino acid deficiency in which the animal goes into negative nitrogen balance, the two major sources of protein in the body are the muscle and connective tissues. The loss of bulk in these two organs may reflect the relative turnover of proteins within muscle and connective tissues; muscle tissue provides a ready source of amino acids as a result of intracellular catabolism of muscle proteins. Extracellular matrix, on the other hand, requires extracellular proteolysis for turnover. Since the turnover time for extracellular matrices is in general less than for muscle, the relative contribution of amino acids in the starved state is primarily from muscle, followed by connective tissues. Additional extracellular matrix components including fibronectin and proteoglycans are degraded during starvation (15, 21, 45, 71).

Starvation is a life-threatening condition. The responses of the organism are to protect the critical functions necessary for survival. Collagen is one of the proteins affected by fasting. Within 24 hr of fasting, collagen synthesis decreases to 50% of normal in the articular cartilage of guinea pigs, and this reduction continues to 8–12% of control levels after 96 hr (142). Although starvation may represent total fasting, most chronic conditions are not associated with such severe restriction. Other studies have investigated the response of collagen metabolism to partial, rather than complete, energy restriction. Significant reductions in collagen production have been reported in food-restricted rats as a function of both duration and degree of food deprivation. Even rats that gained weight decreased collagen production in articular cartilage. In contrast, noncollagen protein production was reduced only in articular cartilage from rats that lost weight, but it did not progressively decrease as the caloric intake continued to decrease (143). These data suggest that collagen production is sensitive to changes in food intake in both short-term diseases, such as in surgical patients and patients with broken bones, as well as in more long-term diseases such as osteoporosis. The data also suggest that malnutrition may have profound effects on collagen production (143). The level of urinary hydroxyproline was higher in fasting than in nonfasting deer (35) and may serve as a marker for nutritional balance. The effects of mechanics on bone calcium loss have been documented in postmenopausal women (116) who showed that walking affects bone density.

Changes in collagen production as well as in other connective tissues have also been reported in childen and animals that have undergone protein/calorie malnutrition. The amounts of collagen in skin biopsies taken from children with clinical protein energy malnutrition were lower than those of controls (148). During pregnancy, not only is the collagen content of bones changed but the composition of the bones changes in fetal rats whose dams were protein or energy deficient (107). The turnover of soluble collagen in skin is decreased in protein-malnourished rats, but the degradation and conversion into insoluble collagen occurs in the same proportions as in the rapidly growing rats (110). Whether the changes in collagen production result from changes in caloric intake (food deprivation or starvation) and/or protein malnutrition, evidence from many systems has shown that collagen production is a labile process, influenced by many factors. Certain diseases such as diabetes are accompanied by changes in nutritional status that result in undernourishment. Animals with streptozotocin-induced diabetes showed diminished growth with increased specific lung volume and weight. Collagen and elastin content were reduced (119) and there was decreased degradation of lung collagen (120). Undernourished, nondiabetic animals showed increased degradation of lung collagen (119).

Nutritional Emphysema

Several studies, including a recent review (88), have observed the effects of injury induced in the lung by prolonged nutritional alterations. The composition of the diet, age, and species of the animal determines the effects of nutritional interventions on lung structure and functions. The term "emphysema" may be misleading. Although the lung architecture is not normal, it is unclear whether or not there is actual tissue destruction. This is particularly true in models that use moderate to severe calorie restriction to study the effects of starvation on lung growth in juvenile rodents (62). In this model the lung connective tissue content is lower than in age-matched controls (89, 113, 137). Whether the decrease in lung collagen content results from the actual depletion of connective tissue or from lack of collagen production during lack of growth is not clear. The collagen and elastin may be depleted in meeting the nutritional requirements of the animals for survival (62, 136, 137). However, collagen and elastin levels are not lower than those found prior to the initiation of the nutritional intervention (81, 89). These findings suggest that in growing animals starvation prevents the normal accumulation of lung structural proteins rather than promoting their net loss.

In a model in which the animal is slowly growing or nongrowing, a net loss of connective tissue content, specifically collagen, has been observed. These findings are different from those cited above for the younger, growing rats in which similar food restriction results in less than normal connective tissue accumulation, but not the loss of these proteins (90). Thus, age as well as dietary restrictions determine the response of connective tissues, including collagen, to the specific insult.

CONTRIBUTION OF COLLAGEN TO NUTRITION

The contribution of connective tissue to overall body metabolism, specifically the contribution of collagen, is complicated [for a review, see (155)]. Body composition changes with aging: muscle mass and skeletal mass decrease with age while fat mass increases (28, 41). After menopause, both muscle mass and skeletal mass decrease (116); obesity may reduce bone mineral loss by increased weight bearing on the spine (3). Recent studies show that normally a significant basal level of collagen turnover occurs in lung and other tissues (95). This turnover takes place in the absence of either tissue remodeling or pathological processes. How turnover relates to the contribution of collagen to overall body metabolism or to the actual deposition of pathological amounts of the protein in tissues is unclear. Certainly, pathological conditions such as fibrosis lead to abnormal amount or types of collagen in many organs (see above).

One clinical manifestation of abnormal collagen turnover affected by malnutrition is inhibition of growth which, in turn, influences the well-being of the individual. An increased demand for collagen synthesis may occur in growing individuals or animals with acute injuries. Specific effects of malnutrition on collagen turnover may be age-dependent, and the young that are still growing may be more susceptible to such changes. In adolescents with Crohn's disease, collagen turnover responded to nutritional supplementation. Collagen was restored to several tissues of the body, suggesting that collagen synthesis had been decreased (111). In animal studies, weanling rats fed isocaloric diets containing varying amounts of casein experienced a weight gain related to the amount of protein in the diet, thus demonstrating that collagen metabolism is dependent on the animals' nutrition. (156). This finding holds true for both fetuses and dams. Once again, the type as well as the amount of collagen in an organ of a growing animal depends on the amount of protein and the caloric intake of the animal (77, 108, 159). Clearly, collagen is not the stable protein it was thought to be twenty years ago, but rather is a labile protein susceptible to various nutritional alterations. How specific alterations contribute to the overall metabolism of the individual remains to be determined. These contributions will be dependent upon tissue, age, the type of nutritional alteration, and whether they are of short-term or long-term duration.

ROLE OF VITAMINS IN COLLAGEN METABOLISM

Vitamin A

GENE INDUCTION   Vitamin A is known to have profound effects on epithelial cells. It has been used to cause the differentiation of F9 teratocytes, including induction of type IV collagen and laminin (144), and to cause the formation of cleft palates in newborn mice. All-*trans*-retinoic acid induces striking digit pattern duplications when locally applied to the developing chick limb bud (147). In other cases, retinoic acid has been shown to suppress differentiation of the epidermis. Retinoic acid can inhibit several differentiation-associated keratins (93). Excess vitamin A can induce metaplasia of both embryonic and adult keratinizing epithelia. In hamster cheek pouch epithelium given a topical application of vitamin A, the formation of keratin was inhibited and the epithelium lacked all features of keratinization (66). Chronic vitamin A intoxication leads to liver injury similar to that caused by alcoholic cirrhosis (5). Vitamin A deficiency has resulted in the synthesis of increased fibronectin in cultured hepatocytes. The addition of vitamin A reversed this effect, suggesting that vitamin A may have a direct effect on an extracellular matrix protein (91). There may be a relationship between retinol and fibrosis (14).

The work of Oikarinen et al (122) and Hein et al (65) indicated that various retinoids inhibited collagen production by human fibroblast cultures. Using concentrations of retinoids of 10^{-5} M or higher, they reported that all-*trans*- and 13-*cis*-retinoic acid inhibited collagen gene expression at the level of mRNA. This concentration is too high to suggest that inhibition occurs via retinoid-binding proteins (124) or the family of retinoid receptor proteins (16, 54); rather, it may occur through a different mechanism (see below). Several studies are consistent with retinoids inducing connective tissue proteins (94). Rat calvaria cultures transformed with sv40 have been induced with retinoic acid for type I procollagen and the noncollagen protein osteopontin (64).

EMBRYONIC DEVELOPMENT   Vitamin A has a dramatic effect on embryonic development; it can induce digit pattern duplications when applied locally to the developing chick limb bud (39). Thaller & Eichele have done extensive work on the identification and spatial distribution of retinoids and their effects in the developing chick limb bud (147) and have shown that retinoic acid may be a local chemical mediator of morphogenesis.

REGULATION OF COLLAGEN SYNTHESIS   Vitamin A has been reported to inhibit (51, 122) collagen synthesis in human dermal skin fibroblasts. Oikarinen et al (122) concluded that the reduction of collagen synthesis occurs at the level of mRNA production for the proα1(I) polypeptide chain. Geesin et al (51) reported that at high concentrations of retinoids both collagen synthesis and ascorbate-mediated lipid peroxidation were inhibited. These investigators concluded that the effect of vitamin A on collagen synthesis may be concentration dependent and that high, nonphysiological doses of retinoids affect ascorbate-induced lipid peroxidation, which in turn inhibits ascorbate-induced collagen synthesis (51).

Vitamin C

The dramatic effect of ascorbic acid on collagen synthesis was recognized, although not understood, as early as 1795 when the British Navy included citrus juice in the rations of its sailors to prevent scurvy. The measurement of vitamin C in human tissues continues as functional measures are proposed (73). The effect of ascorbate is not specifically limited to collagen but extends to glycosaminoglycan synthesis of proteoglycans (13). The importance of ascorbic acid in maintaining collagen production is supported by studies in which diabetic rats that have deficiencies in wound healing were shown to be deficient in ascorbic acid. When ascorbic acid was added to the diet, wound healing improved (103).

ANTIOXIDANT   In the early 1960s, ascorbic acid was shown to serve as a cofactor for the enzyme prolyl hydroxylase (126). This enzyme converts prolyl residues in collagen to hydroxyproline. The latter enables collagen to assume a triple-helical conformation at body temperature and to polymerize into functional collagen fibers. Prolyl hydroxylase is an iron-containing enzyme, and the iron must be in the ferrous state to form the catalytic complex. Ascorbic acid is believed to be the antioxidant that maintains the iron in the reduced ferrous state (92). Ascorbic acid may also serve as part of tissues' antioxidant defense by protecting collagen fibrils from damage caused by intensely energetic free radicals. Highly reactive free radicals produced either endogenously or by the metabolism of drugs or by radiation cause tissue damage. Vitamin C is a member of a group of micronutrients that may protect against tissue damage (99).

INDUCTION OF COLLAGEN GENES   Lipid peroxidation induces a wide range of cellular effects; it has been associated with alterations in second messenger pathways and cell proliferation. Chojkier et al (24) concluded that ascorbic acid induces lipid peroxidation and reactive aldehyde production in cultured human fibroblasts and that this process is necessary for stimulation of collagen synthesis. In their work, ascorbic acid and lipid peroxidation products stimulated equally the net production of collagen relative to noncollagen proteins, and the transcription of procollagen $\alpha 1(I) mRNA$ levels. Geesin et al (51, 52) reported that two different retinoids, at similar concentrations, inhibited both ascorbate-stimulated lipid peroxidation and collagen synthesis. They concluded that the ability of the retinoids to inhibit the oxidant effect of ascorbate, and not their receptor-mediated activity, is responsible for their effect on collagen synthesis.

In addition, ascorbic acid stimulates collagen synthesis by increasing the transcription rate of collagen genes and increasing the proportion of membrane-bound polysomes in collagen-synthesizing cells (80). Many investigators have noted that ascorbic acid stimulates increased transcription of collagen genes in different cultured cell types (24, 50, 97). Kao et al (80) reported that when tendon fibroblast cultures were exposed to ascorbic acid for more than six hours the fraction of membrane-bound polysomes increased significantly. Recently, Geesin et al (52, 53) provided evidence that linked ascorbate-induced lipid peroxidation and collagen synthesis. The alteration of cell membranes by lipid peroxidation may cause changes similar to those caused by growth factor treatment in which the cell responds to cell membrane-mediated signaling events. Geesin et al (53) also studied the effects of various growth factors on collagen synthesis by dermal fibroblasts; the two factors that significantly stimulated collagen synthesis, transforming growth factor-β (TGF-β) and fibroblast growth factor (FGF), were most sensitive to lipid peroxidation.

As discussed above, ascorbic acid was identified in the early 1960s as a cofactor for the enzyme prolyl hydroxylase in the hydroxylation of prolyl residues, an essential posttranslational modification of collagen. This reaction is essential for the formation and stabilization of the triple helical conformation and for polymerization of the trimers into functional collagen fibers.

Vitamin D

Recently, it has become apparent that vitamin D should be considered a steroid hormone. Its synthesis in the skin upon exposure to ultraviolet light obviates the dietary necessity for vitamin D. Vitamin D is metabolized in the liver to 25-hydroxyvitamin D_3, which is then metabolized in the kidney. The close regulation of renal production of 1,25-dihydroxyvitamin D_3 $[1,25(OH)_2D_3]$, the most potent of the naturally occurring derivatives of vitamin D, is suggestive of its hormonal nature. Additional support for this viewpoint is the presence of high affinity nuclear receptors for vitamin D in such tissues as the kidney and intestine. One of the target tissues for $1,25(OH)_2D_3$ is the intestine, which depends on this hormone for the absorption of dietary calcium [for a review of the effects of vitamin D on bone and intestinal cell differentiation, see (145)]. Vitamin D deficiency results in increased nonmineralized bone matrix, resulting in rickets in children and osteomalasia in adults. Evidence suggests an increased collagenase burden in the growth plate of rachitic rats (31). Vitamin D deficiency may be related to abnormal lung development (48).

Administration of $1,25(OH)_2D_3$ stimulates osteoclastic bone remodeling. It also promotes absorption of calcium and phosphorus across the intestinal epithelial cells. Exactly how mineralization is regulated is still not known. However, bone gla protein and matrix gla protein, which bind calcium, are involved in mineralization. Both are stimulated by $1,25(OH)_2D_3$ (47, 96, 161). Another bone protein, osteopontin, is also stimulated by $1,25(OH)_2D_3$ (160). Bone mineralization may have indirect effects on collagen metabolism, as a more rapid turnover is expected for less mineralized matrix. Specific effects of $1,25(OH)_2D_3$ on collagen synthesis are controversial (145).

The target tissues of vitamin D's endocrine action are the kidney, intestine, and bone (145). Recent evidence supports the interaction of $1,25(OH)_2D_3$ with the hematopoietic and immune systems. In hematopoietic cells, $1,25(OH)_2D_3$ promotes the differentiation of promonocytes into monocytes, then macrophages, and finally into osteoclasts, cells that mediate bone resorption. $1,25(OH)_2D_3$ stimulates increased production of osteoclast cells. Also, gamma interferon produced by activated T-lymphocytes stimulates activated macrophages to produce $25(OH)_2D_3$-1-hydroxylase, the enzyme essential for production of $1,25(OH)_2D_3$. Thus, the bone marrow contains a vitamin D paracrine system that produces and uses $1,25(OH)_2D_3$ (130).

The events leading to the calcification of the matrix of growth cartilage and/or the induction of new bone formation are poorly understood. Deposits of mineral occur extracellularly in a transforming matrix that contains collagens, glycoproteins, proteoglycans, and other proteins. Dietary intake of calcium, phosphate, vitamin D, protein, and copper is important in the process of bone mineralization. Vitamin D deficiency has been implicated in abnormalities of proteoglycans and in collagen metabolism in growing bones and lungs of rats (6, 48). Vitamin D metabolites are necessary for maximal stimulation of calcification of cartilage matrix (67). Although the exact mechanism by which vitamin D influences collagen metabolism remains unclear, apparently the vitamin is necessary for the normal production of this connective tissue. Not only is collagen influenced by 1,25-dihydroxy vitamin D_3, but noncollagen bone proteins such as osteocalcin are required for the assembly of bone and are induced by this vitamin (161).

Comparative studies indicate that in the growth plate cartilage from normal and vitamin D-phosphate deficient rats, different levels of collagenase are present. The growth plate cartilage from the vitamin D-deficient rats contains excess collagenase compared to the amount of tissue inhibitor of metalloproteinase (TIMP). As a result of the imbalance between the enzyme and the inhibitor, collagen is degraded. This process allows for thinning of the longitudinal septa and expansion of the hypertrophic cells in growth plates of rachitic rats. (31, 32). Thus, vitamin D deficiency can potentiate collagen breakdown. On the other hand, in rachitic chick epiphyseal growth cartilage, the hypertrophic chondrocytes are likely to be responsible for the increased levels of type X collagen that provide a maximum area of calcifiable matrix (129). All types of collagen may not be affected equally in deficiencies of vitamin D. In the transition from proliferation to hypertrophic cell zones in the growth plate, there is an increase in chondrocyte volume and production of type X collagen, but a decrease in overall collagen content, mediated by a proteinase-inhibitor imbalance.

MINERAL BALANCE AND BONE HOMEOSTASIS AS RELATED TO COLLAGEN METABOLISM

Calcitonin

Dietary calcium is the source of calcium for bone mineral (22). Calcitonin, a 32 amino acid peptide hormone secreted from thyroidal C-cells, is critical in maintaining calcium homeostasis because of its marked inhibition of osteoclastic bone resorption (117). Osteoclasts have been observed to shrink in size and decrease bone resorption within minutes of calcitonin application (34). Blood calcium concentration is the most important regulator of blood calcitonin secretion. As the calcium level rises, there is a proportional rise in calcitonin secretion (33). Because of its inhibition of osteoclastic bone resorption, calcitonin has been used successfully to treat diseases characterized by bone resorption and hypercalcemia. Calcitonin has been widely used in Paget's disease; it is also used to treat some cases of osteoporosis and the hypercalcemia of malignancy.

Many questions remain about the roles calcitonin plays in maintaining calcium homeostasis and skeletal integrity. To date, no skeletal disease has been attributed conclusively to calcitonin abnormalities (34). Do diseases of skeletal and calcium homeostasis result from primary and/or secondary calcitonin secretion abnormalities?

Parathyroid Hormone

Parathyroid hormone (PTH) regulates the levels of calcium and phosphate in the blood by affecting the activity of specific cells of bone and kidney. In turn, blood calcium is the prime regulator of PTH secretion; a rise in blood calcium levels results in a decrease in PTH secretion. Thus, a mutual regulatory interaction of PTH and calcium keeps the blood calcium level constant despite fluctuations in diet, bone metabolism, and renal function. In bone, PTH administration results in an increase in osteoclast number and function. Paradoxically, PTH effects this by binding to bone-forming osteoblasts, which then secrete factor(s) that stimulates osteoclasts to resorb bone (104). It may suppress the level of matrix proteins such as osteopontin (118).

PTH binds to specific G-protein-coupled receptors in bone and kidney, thereby activating adenylate cyclase and phospholipase C. Recently, the PTH receptor was cloned and characterized (76). Its striking homology with the calcitonin receptor (75) and its lack of homology with other G-protein-linked receptors indicate that receptors for calcium-regulating hormones are related and represent a new family of G-protein-coupled receptors. The effects of PTH on bone are complex and still poorly understood. PTH administration leads to an increase in osteoclast cell number and activity. Release of calcium is accompanied by an increase in phosphate release and the release of other bone matrix components, such as collagen (18).

Parathyroid hormone and vitamin D interact in a number of important ways. In the kidney, PTH activates the enzyme responsible for synthesizing $1,25(OH)_2D_3$, which is a potent inducer of calcium absorption in the intestine. Calcium absorption can increase from 10 to 70% in response to $1,25(OH)_2D_3$. This effect can operate synergistically with PTH to raise blood calcium levels. Bone is poorly mineralized in the absence of vitamin D metabolites, and PTH cannot mobilize calcium efficiently from poorly mineralized bone. Because vitamin D mediates the effect of PTH on blood calcium, it is used, in addition to oral calcium, to treat hypoparathyroidism.

CONCLUSION

The collagens are a family of proteins that constitute the major extracellular proteins of the body. They are ubiquitous proteins serving in either structural or pericellular matrix functions and are regulated during differentiation, growth, and remodeling of connective tissues. Collagen synthesis and degradation are altered by a large variety of metabolites, growth factors, hormones, cytokines, as well as mechanical stimuli. The regulation of collagen mass and specific types of collagen are related to the nutritional state of the animal and are affected by numerous disease processes because most disease processes involve changes in extracellular matrix either directly as in wound healing, inflammation, and fibrosis or indirectly as in starvation, diabetes, and abnormalities in mineral balance.

ACKNOWLEDGMENTS

The author's research has been supported in part by USPHS grant AM318389 from the NIH.

Literature Cited

1. Adams, S. L. 1992. Regulation of collagen gene expression. In *Extracellular Matrix: Chemistry, Biology, and Pathobiology,* ed. M. A. Zern, L. M. Reid. New York: Dekker. In press
2. Ala-Kokko, L., Pihlajaniemi, T., Myers, J. C., Kivirikko, K. I., Savolainen, E.-R. 1987. Gene expression of type I, III and IV collagens in hepatic fibrosis induced by dimethylnitrosamine in the rat. *Biochem. J.* 244:75–79
3. Aloia, J. F., McGowan, D. M., Vaswani, A. N., Ross, P., Cohn, S. H. 1991. Relationship of menopause to skeletal and muscle mass. *Am. J. Clin. Nutr.* 53:1378–83
4. Annoni, G., Weinber, F. R., Colombo, M., Czaja, M. J., Zern, M. A. 1990. Albumin and collagen gene regulation in alcohol- and virus-induced human liver disease. *Gastroenterology* 98:197–202
5. Baker, H., Ten-Hove, W., Kanagasundaram, N., Zaki, G., Leevy, C. B., et al. 1990. Excess vitamin A injures the liver. *J. Am. Coll. Nutr.* 9:503–9
6. Balmain, N., Cuisinier-Gleizes, P., Mathieu, H. 1982. Aspects ultrastructuraux du cartilage de croissance chez le rat rachitique par carence en vitamine D. *Biol. Cell* 44:37

7. Berg, R. A. 1986. Intracellular turnover of collagen. In *Regulation of Matrix Accumulation,* ed. R. P. Mecham, pp. 29–52. New York: Academic
8. Berg, R. A. 1991. General biologic processes of the lung. Synthesis of proteins. In *Lung: Scientific Foundation,* ed. R. G. Crystal, J. B. West, 1:13–24. New York: Raven
9. Berg, R. A., Capodici, C., D'Armiento, J. 1989. Collagenase activation. In *Therapeutic Control of Inflammatory Disease,* ed. A. J. Lewis, N. S. Doherty, N. R. Ackerman, pp. 84–91. New York: Elsevier
10. Berg, R. A., Schwartz, M. L., Rome, L. H., Crystal, R. G. 1984. Lysosomal function in the degradation of defective collagen in cultured lung fibroblasts. *Biochemistry* 23:2134–38
11. Berg, R. A., Steinman, B., Rennard, S. I., Crystal, R. G. 1984. Ascorbate deficiency results in decreased collagen production: Underhydroxylation of proline leads to increased intracellular degradation. *Arch. Biochem. Biophys.* 226:681–86
12. Bienkowski, R. A. 1984. Intracellular degradation of newly synthesized collagen. *Collagen Relat. Res.* 4:399–412

13. Bird, T. A., Schwartz, N. B., Peterkofsky, B. 1986. Mechanism for the decreased biosynthesis of cartilage proteoglycan in the scorbutic guinea pig. *J. Biol. Chem.* 261:11166–72
14. Blomhoff, R., Wake, K. 1991. Perisinusoidal stellate cells of the liver: important roles in retinol metabolism and fibrosis. *FASEB J.* 5:271–77
15. Bowersox, J. C., Scott, R. L. 1985. Plasma fibronectin metabolism during hemorrhagic shock and starvation. *J. Surg. Res.* 39:445–53
16. Brand, N., Petkovich, M., Krust, A., Chambon, P., de The, H., et al. 1988. Identification of a second human retinoic acid receptor. *Nature* 332:850–53
17. Brenner, D. A., Chojkier, M. 1987. Acetaldehyde increases collagen gene transcription in cultured human fibroblasts. *J. Biol. Chem.* 262:17690–95
18. Brignhurst, F. R. 1989. Calcium and phosphate distribution, turnover and metabolic interactions. In *Endocrinology*, ed. L. J. DeGroot, 2:805–43. New York: Grune & Stratton
19. Burgeson, R. E. 1988. New collagens, new concepts. *Annu. Rev. Cell Biol.* 4:551–77
20. Capodici, C., Muthukumaran, G., Amoruso, M. A., Berg, R. A. 1989. Activation of neutrophil collagenase by cathepsin G. *Inflammation* 13:245–58
21. Chadwick, S. J., Sim, A. J., Dudley, H. A. 1986. Changes in plasma fibronectin during acute nutritional deprivation in healthy human subjects. *Br. J. Nutr.* 55:7–12
22. Chan, G. M. 1991. Dietary calcium and bone mineral status of children and adolescents. *Am. J. Dis. Child.* 145: 631–34
23. Choe, I., Aycock, R. S., Raghow, R., Myers, J. C., Seyer, J. M., Kang, A. H. 1987. A hepatic fibrogenic factor stimulates the synthesis of types I, III, and V procollagens in cultured cells. *J. Biol. Chem.* 262:5408–13
24. Chojkier, M., Houglum, K., Solis-Herruzzo, J., Brenner, D. A. 1989. Stimulation of collagen gene expression by ascorbic acid in cultured human fibroblasts: a role for lipid peroxidation. *J. Biol. Chem.* 264:16957–62
25. Chojkier, M., Lyche, K. D., Filip, M. 1988. Increased production of collagen *in vivo* by hepatocytes and nonparenchymal cells in rats with carbon tetrachloride-induced hepatic fibrosis. *Hepatology* 8:808–14
26. Cockayne, D., Cutroneo, K. R. 1988. Glucocorticoid coordinate regulation of type I procollagen gene expression and

procollagen DNA-binding proteins in chick skin fibroblasts. *Biochemistry* 27:2736–45
27. Cockayne, D., Sterling, K. M. Jr., Shull, S., Mintz, K. P., Illeneye, S., Cutroneo, K. R. 1986. Glucocorticoids decrease the synthesis of Type I procollagen mRNAs. *Biochemistry* 25: 3202–9
28. Cohn, S. H., Vaswani, A., Zanzi, I., Aloia, J. F., Roginsky, M. S., Ellis, K. J. 1976. Changes in body composition with age measured by total body neutron activation. *Metabolism* 25:85–95
29. Czaja, M. J., Weiner, F. R., Eghbali, M., Giambrone, M.-A., Eghbali, M., Zern, M. A. 1987. Differential effects of γ interferon on collagen and fibronectin gene expression. *J. Biol. Chem.* 262: 3348–51
30. Czaja, M. J., Weiner, F. R., Takahaski, S., Giambrone, M.-A., van der Meide, P. H., et al. 1989. γ-Interferon treatment inhibits collagen deposition in murine schistosomiasis. *Hepatology* 12:795–800
31. Dean, D. D., Muniz, O. E., Berman, I., Pita, J. C., Carreno, M. R., et al. 1985. Localization of collagenase in the growth plate of rachitic rats. *J. Clin. Invest.* 76:716–22
32. Dean, D. D., Muniz, O. E., Woessner, J. F. Jr., Howell, D. S. 1990. Production of collagenase and tissue inhibitor of metalloproteinases (TIMP) by rat growth plates in culture. *Matrix* 10:320–60
33. Deftos, L. J. 1987. Calcitonin secretion in humans. In *Current Research on Calcium Regulating Hormones,* ed. C. W. Cooper, pp. 79–100. Austin, Tex: Univ. Texas Press
34. Deftos, L. J., Roos, B. 1989. Medullary thyroid carcinoma and calcitonin gene expression. In *Bone and Mineral Research,* ed. W. A. Peck, pp. 267–316. Amsterdam: Excerpta Media
35. DelGiudice, G. D., Seal, U. S., Mech, L. D. 1988. Response of urinary hydroxyproline to dietary protein and fasting in white-tailed deer. *J. Wildl. Dis.* 24:75–79
36. Dublet, B., Oh, S., Sugrue, S. P., Gordon, M. K., Gerecke, D. R., et al. 1989. The structure of avian type XII collagen. *J. Biol. Chem.* 264:13150–56
37. Dublet, B., van der Rest, M. 1991. Type XIV collagen, a new homotrimeric molecule extracted from bovine skin and tendon, with a triple helical disulfide-bonded domain homologous to type IX and type XII collagens. *J. Biol. Chem.* 266:6853–58

38. Duncan, M. R., Berman, B. 1989. Differential regulation of collagen, glycosaminoglycan, fibronectin and collagenase activity production in cultured human adult dermal fibroblasts by interleukin 1-α and β and tumor necrosis factor-α and β. *J. Invest. Dermatol.* 92:699–706

39. Eichele, G. 1986. Retinoids induce duplications in developing vertebrate limbs. *BioScience* 36:534–40

40. Elias, J. A., Freundlich, B., Adams, S. L., Rosenbloom, J. 1989. Regulation of human lung fibroblasts collagen production by recombinant interleukin 1, tumor necrosis factor and interferon γ. *Ann. NY Acad. Sci.* 580:233–44

41. Ellis, K. J., Yasumura, S., Vartsky, D., Vaswani, A. N., Cohn, S. H. 1982. Total body nitrogen in health and disease: Effects of age, weight, height and sex. *J. Lab. Clin. Med.* 99:917–26

42. Ernst, M., Heath, J. K., Rodan, G. A. 1989. Estradiol effects on proliferation, messenger ribonucleic acid for collagen and insulin-like growth factor-I, and parathyroid hormone-stimulated adenylate cyclase activity in osteoblastic cells from calvariae and long bone. *Endocrinology* 125:825–33

43. Ernst, M., Heath, J. K., Schmid, C., Foresch, R. E., Rodan, G. A. 1989. Evidence for a direct effect of estrogen on bone cells *in vitro*. *J. Steroid Biochem.* 34:279–84

44. Eyre, D. R., Wu, J. J. 1987. Type XI or 1α2α3α collagen. See Ref. 102, pp. 261–82

45. Fernandez Lemos, S. M., Kofoed, J. A., Caldarini, M., Venegas Agurre, M. I. 1989. Effect of total food deprivation on the proteoglycans of rat hyaline cartilage and bone. *Acta Physiol. Pharmacol. Lat. Am.* 39:273–80

46. Fessler, J. H., Fessler, L. I. 1987. Type V collagen. See Ref. 102. pp. 80–103

47. Fraser, J. D., Otawara, Y., Price, P. A. 1988. 1,25-Dihydroxyvitamin D stimulates the synthesis of matrix γ-carboxyglutamic acid protein by osteosarcoma cells. *J. Biol. Chem.* 263:911–16

48. Gaultier, C. L., Harf, A., Balmain, N., Cuisinier-Gleizes, P., Mathieu, H. 1984. Lung mechanics in rachitic rats. *Am. Rev. Respir. Dis.* 130:1108–10

49. Geesin, J. C., Berg, R. A. 1991. Biochemistry of skin, cartilage and bone. In *Clinical Use of Biomaterials in Facial Plastic Surgery*, ed. F. H. Silver, A. Glasgold, pp. 7–27. Boca Raton, Fla: CRC Press

50. Geesin, J, C , Darr, D., Kaufman, R.,

Murad, S., Pinnell, S. R. 1988. Ascorbic acid specifically increases type I and type III procollagen mRNA levels in human skin fibroblasts. *J. Invest. Dermatol.* 90:420–24

51. Geesin, J. C., Gordon, J. S., Berg, R. A. 1990. Retinoids affect collagen synthesis through inhibition of ascorbate-induced lipid peroxidation in cultured human dermal fibroblasts. *Arch. Biochem. Biophys.* 278:350–55

52. Geesin, J. C., Hendricks, L. J., Falkenstein, P. A., Gordon, J. S., Berg, R. A. 1991. Regulation of collagen synthesis by ascorbic acid: Characterization of the role of ascorbate-stimulated lipid peroxidation. *Arch. Biochem. Biophys.* 290:127–32

53. Geesin, J. C., Hendricks, L. J., Gordon, J. S., Berg, R. A. 1991. Modulation of collagen synthesis by growth factors: The role of ascorbate-stimulated lipid peroxidation. *Arch. Biochem. Biophys.* 289:6–11

54. Giguere, V., Ong, E. S., Segui, P., Evans, R. M. 1987. Identification of a receptor for the morphogen retinoic acid. *Nature* 330:624–29

55. Goldberg, M. E. 1985. The second translation of the genetic message: Protein and assembly. *Trends Biochem. Sci.* 10:388–91

56. Goldring, M. B., Krane, S. M. 1987. Modulation by recombinant interleukin 1 of synthesis of types I and III collagens and associated procollagen mRNA levels in cultured human cells. *J. Biol. Chem.* 262:16724–29

57. Gordon, M. K., Olsen, B. R. 1990. The contribution of collagenous proteins to tissue-specific matrix assemblies. *Curr. Opin. Cell Biol.* 2:833–38

58. Granstein, R. D., Deak, M. R., Jacques, S. L., Margolis, R. J., Flotte, T. J., et al. 1989. The systemic administration of γ interferon inhibits collagen synthesis and acute inflammation in a murine skin wounding model. *J. Invest. Dermatol.* 93:18–27

59. Gunwar, S., Saus, J., Noelken, M. E., Hudson, B. G. 1990. Glomerular basement membrane: identification of a fourth chain, α4, of type IV collagen. *J. Biol. Chem.* 265:5466–69

60. Haas, I. G., Wabl, M. 1983. Immunoglobulin heavy chains binding protein. *Nature* 306:387–89

61. Hamalainen, L., Oikarinen, J., Kivirikko, K. I. 1985. Synthesis and degradation of type I procollagen mRNAs in cultured human skin fibroblasts and the effect of cortisol. *J. Biol. Chem.* 260: 720–25

62. Harkema, J. R., Mauderly, J. L., Gregory, R. E., Pickrell, J. A. 1984. A comparison of starvation and elastase models of emphysema in the rat. *Am. Rev. Respir. Dis.* 129:584–91

63. He, C., Wilhelm, S. M., Pentland, A. P., Marmer, B. L., Grant, G. A., et al. 1989. Tissue cooperation in a proteolytic cascade activating human interstitial collagenase. *Proc. Natl. Acad. Sci. USA* 86:2632–36

64. Heath, J. K., Rodan, S. B., Yoon, K., Rodan, G. A. 1989. Rat calvarial cell lines immortalized with sv-40 large T antigen: Constitutive and retinoic acid-inducible expression of osteoblastic features. *Endocrinology* 124:3060–68

65. Hein, R., Mensing, H., Muller, P. K., Braun-Falco, O., Krieg, T. 1984. Effect of vitamin A and its derivatives on collagen production and chemotactic response of fibroblasts. *Br. J. Dermatol.* 111:37–44

66. Hill, M. W., Harris, R. R., Carron, C. P. 1982. A quantitative ultrastructural analysis of changes in hamster checkpouch epithelium treated with vitamin A. *Cell Tissue Res.* 226:541–54

67. Hinek, A., Poole, A. R. 1988. The influence of vitamin D metabolites on the calcification of cartilage matrix and the C-propeptide of type II collagen (chondrocalcin) *J. Bone Miner. Res.* 3:421–29

68. Hirayoshi, K., Kudo, H., Takechi, H., Nakai, A., Iwamatsu, A., et al. 1991. Hsp47: A tissue-specific, transformation-sensitive, collagen-binding heat shock protein of chick embryo fibroblasts. *Mol. Cell Biol.* 11:4036–44

69. Holt, K., Bennett, M., Chojkier, M. 1984. Acetaldehyde stimulates collagen and noncollagen protein production by human fibroblasts. *Hepatology* 4:843–48

70. Hostikka, S. L., Eddy R. L., Byers, M. G., Hoyhtya, M., Shows, T. B., Tryggvason, K. 1990. Identification of a distinct type IV collagen α chain with restricted kidney distribution and assignment of its gene to the locus of X chromosome-linked Alport syndrome. *Proc. Natl. Acad. Sci. USA* 87:1606–10

71. Howard, L., Dillon, B., Saba, T. M., Hofmann, S., Cho, E. 1984. Decreased plasma fibronectin during starvation in man. *J. Parenter. Enteral. Nutr.* 8:237–44

72. Jackson, S. H., Heininger, J. A. 1975. Proline recycling during collagen metabolism as determined by concurrent $^{18}O_2$- and ^{3}H-labeling. *Biochim. Biophys. Acta* 381:359–67

73. Jacob, R. A. 1990. Assessment of human vitamin C status. *J. Nutr.* 120:1480–85

74. Jeffrey, J. J. 1986. Biological regulation of collagenase. See Ref. 7, pp. 53–98

75. Jin, H. Y., Harris, T. L., Flannery, M. S., Aruffo, A., Kaji, E. H., et al. 1991. Expression cloning of an adenylate cyclase coupled calcitonin receptor. *Science* 254:1022–24

76. Juppner, H., Abdul-Badi, A., Freeman, M., Kong, X., Schipani, E., et al. 1991. A G protein-linked receptor for parathyroid hormone and parathyroid hormone-related peptide. *Science* 254:1024–26

77. Kaggwa, J. S. 1986. The effect of protein and energy deficiency on skin glycosaminoglycan levels in the rat. *Br. J. Nutr.* 56:329–39

78. Kahari, V.-M., Heino, J., Vuorio, E. 1987. Interleukin 1 increases collagen production and mRNA levels in cultured skin fibroblasts. *Biochim. Biophys. Acta* 929:142–47

79. Kahari, V.-M., Heino, J., Vuorio, T., Vuorio, E. 1988. Interferon γ and interferon α reduce excessive collagen synthesis and procollagen mRNA levels of scleroderma fibroblasts in culture. *Biochim. Biophys. Acta* 968:46–50

80. Kao, W. W.-Y., Flaks, J. G., Prockop, D. J. 1976. Primary and secondary effects of ascorbate on procollagen synthesis and protein synthesis by primary cultures of tendon fibroblasts. *Arch. Biochem. Biophys.* 173:638–48

81. Karlinsky, J. B., Goldstein, R. H., Ojserkis, B., Snider, G. L. 1986. Lung mechanics and connective tissue levels in starvation-induced emphysema in hamsters. *Am. J. Physiol.* 251:282–88

82. Kassenbrock, C. K., Garcia, P. D., Walter, P., Kelly, R. B. 1988. Heavy-chain binding protein recognizes aberrant polypeptides translocated *in vitro*. *Nature* 333:90–93

83. Keene, D. R., Sakai, L. Y., Bachinger, H. P., Burgeson, R. E. 1987. Type III collagen can be present on banded collagen fibrils regardless of fibril diameter. *J. Cell Biol.* 105:2393–2402

84. Keene, D. R., Sakai, L. Y., Burgeson, R. E. 1991. Human bone contains type III collagen, type VI collagen and fibrillin. *J. Histochem. Cytochem.* 38:59–69

85. Keene, D. R., Sakai, L. Y., Lunstrum, G. P., Morris, N. P., Burgeson, R. E. 1987. Type VII collagen forms an extended network of anchoring fibrils. *J. Cell Biol.* 104:611–21

86. Kefalides, N. A., ed. 1978. *Biology and*

Chemistry of Basement Membranes. New York: Academic

87. Kelley, J., Chrin, L., Shull, S., Rowe, D. W., Cutroneo, K. R. 1985. Bleomycin selectively elevates mRNA levels for procollagen and fibronectin following acute lung injury. *Biochem. Biophys. Res. Commun.* 131:836–43

88. Kerr, J. S. 1989. Lung injury induced by nutritional restriction. In *Handbook of Animal Model of Pulmonary Disease,* ed. J. O. Cantro, 2:159–67. Boca Raton, Fla: CRC Press

89. Kerr, J. S., Riley, D. J., Lanza-Jacoby, S., Berg, R. A., Spilker, H. C., et al. 1985. Nutritional emphysema in the rat. Influence of protein depletion and impaired lung growth. *Am. Rev. Respir. Dis.* 131:644–50

90. Kerr, J. S., Yu, S. Y., Riley, D. J. 1990. Strain specific respiratory air space enlargement in aged rats. *Exp. Gerontol.* 25:563–74

91. Kim, H.-Y., Wolf, G. 1987. Vitamin A deficiency alters genomic expression for fibronectin in liver and hepatocytes. *J. Biol. Chem.* 262:365–71

92. Kivirikko, K. I., Myllyla, R. 1982. Posttranslational enzymes in the biosynthesis of collagen: Intracellular enzymes. *Methods Enzymol.* 82:245–304

93. Kopan, R., Fuchs, E. 1989. The use of retinoic acid to probe the relation between hyperproliferation-associated keratins and cell proliferation in normal and malignant epidermal cells. *J. Cell Biol.* 109:295–307

94. Kopan, R., Traska, G., Fuchs, E. 1987. Retinoids as important regulators of terminal differentiation: Examining keratin expression in individual epidermal cells at various stages of keratinization. *J. Cell. Biol.* 105:427–40

95. Laurent, G. J. 1982. Dynamic state of collagen: Pathways of collagen degradation *in vivo* and their possible role in regulation of collagen mass. *Am. J. Phys.* 20:61–69

96. Lian, J. B., Carnes, D. L., Glimcher, M. J. 1987. Bone and serum concentrations of osteocalcin as a function of 1,25-dihydroxyvitamin D circulating levels in bone disorders in rats. *Endocrinology* 120:2123–30

97. Lyons, B. L., Schwarz, R. I. 1984. Ascorbate stimulation of PAT cells causes an increase in transcription rates and a decrease in degradation rates of procollagen mRNA. *Nucleic Acids Res.* 12:2569–79

98. Maachi, F., Heulin, M. H., el Farricha,

O., Belleville, F., Nabet, P. 1990. Respective roles of high and low molecular weight serum growth factors in the metabolism of proteoglycans and other cartilage macromolecules. *Reprod. Nutr. Dev.* 30:409–15

99. Machlin, L. J., Bendich, A. 1987. Free radical tissue damage: protective role of antioxidant nutrients. *FASEB J.* 1:441–45

100. Matrisian, L. M. 1990. Metalloproteinases and their inhibitors in matrix remodeling. *Trends Biochem. Sci.* 6:121–25

101. Mauviel, A., Teyton, L., Bhatnagar, R., Penfornis, H., Laurent, M., et al. 1988. Interleukin-1 α-modulated collagen gene expression in cultured synovial cells. *Biochem. J.* 252:247–55

102. Mayne, R., Burgeson, R. E., eds. 1987. *Structure and Function of Collagen Types.* New York: Academic

103. McLennan, S., Yue, D. K., Fisher, E., Capogreco, C., Heffernan, S., et al. 1988. Deficiency of ascorbic acid in experimental diabetes. *Diabetes* 37:359–61

104. McSheehy, P. M. J., Chambers, T. J. 1986. Osteoblastic cells mediate osteoclastic responsiveness to parathyroid hormone. *Endocrinology* 118:824–28

105. Meade, J. B. 1989. The adaptation of bone to mechanical stress: Experimentation and current concepts. In *Bone Mechanics,* ed. S. C. Cowin, pp. 211–51. Boca Raton, Fla: CRC Press

106. Miller, E. J. 1984. Chemistry of the collagens and their distribution. In *Extracellular Matrix Biochemistry,* ed. K. A. Piez, A. H. Reddi, pp. 41–81. New York: Elsevier

107. Miwa, T., Shoji, H., Solomonow, M., Yazdani, M., Nakamoto, T. 1987. Gestational protein-energy malnutrition affects the composition of developing skins of rat fetuses and their dams. *Br. J. Nutr.* 58:215–20

108. Miwa, T., Shoji, H., Solomonow, M., Yazdani, M., Nakamoto, T. 1989. The effect of prenatal protein-energy malnutrition on collagen metabolism in fetal bones. *Orthopedics* 12:973–77

109. Moen, R. C., Rowe, D. W., Palmiter, R. D. 1979. Regulation of procollagen synthesis during the development of chick embryo calvaria. *J. Biol. Chem.* 254:3526–30

110. Molner, J. A., Alpert, N. M., Wagner, D. A., Miyatani, S., Burke, F., Young, V. R. 1988. Synthesis and degradation of collagens in skin of healthy

and protein-malnourished rats *in vivo*, studied by $^{18}O_2$ labelling. *Biochem. J.* 250:71–76

111. Motil, K. J., Altchuler, S. I., Grand, R. J. 1985. Mineral balance during nutritional supplementation in adolescents with Crohn disease and growth failure. *J. Pediatr.* 1073:473–79

112. Murphy, G., Cockett, M. I., Stephens, P. E., Smith, B. J., Docherty, A. J. P. 1987. Stromelysin is an activator of procollagenase. *Biochem. J.* 248:265–68

113. Myers, B. A., Dubick, M. A., Gerreits, J., Rucker, R. B., Jackson, A. C., et al. 1983. Protein deficiency: effects on lung mechanics and the accumulation of collagen and elastin in rat lung. *J. Nutr.* 113:2308–15

114. Myers, J. C., Jones, T. A., Pohjolainen, E. R., Kadri, A. S., Goddard, A. D., et al. 1990. Molecular cloning of $\alpha5(IV)$ collagen and assignment of the gene to the region of the X chromosome containing the Alport syndrome locus. *Am. J. Hum. Genet.* 46:1024–33

115. Nagata, K., Yamada, K. M. 1986. Phosphorylation and transformation sensitivity of a major collagen-binding protein of fibroblasts. *J. Biol. Chem.* 261:7531–36

115a. Nakai, A., Hirayoshi, K., Nagata, K. 1990. Transformation of BALB/3T3 cells by simian virus 40 causes a decreased synthesis of a collagen-binding heat-shock protein (hsp47). *J. Biol. Chem.* 265:992–99

116. Nelson, M. E., Fisher, E. C., Dilmanian, F. A., Dallal, G. E., Evans, W. J. 1991. A 1-yr walking program and increased dietary calcium in postmenopausal women; effects on bone. *Am. J. Clin. Nutr.* 53:1304–11

117. Nicholson, G. C., Moseley, J. M., Sexton, P. M., Mendelsohn, F. A. O., Martin, T. J. 1986. Abundant calcitonin receptors in isolated rat osteoclasts. Biochemistry and autoradiographic characterization. *J. Clin. Invest.* 78:355–59

118. Noda, M., Rodan, G. A. 1989. Transcriptional regulation of osteopontin production in rat osteoblast-like cells by parathyroid hormone. *J. Cell Biol.* 108:713–18

119. Ofulue, A. F., Kida, K., Thurlbeck, W. M. 1988. Experimental diabetes and the lung. I. Changes in growth, morphometry, and biochemistry. *Am. Rev. Respir. Dis.* 137:162–66

120. Ofulue, A. F., Thurlbeck, W. M. 1988. Experimental diabetes and the lung. II. *In vivo* connective tissue metabolism. *Am. Rev. Respir. Dis.* 138:284–89

121. Ohara, B. P., Urban, J. P., Maroudas, A. 1990. Influence of cyclic loading on the nutrition of articular cartilage. *Ann. Rheum. Dis.* 49:536–39

122. Oikarinen, H., Oikarinen, A., Tan, E. M. L., Abergel, R. P., Meeker, C. A., et al. 1985. Modulation of procollagen gene expression by retinoids. *J. Clin. Invest.* 75:1545–53

123. Okada, Y., Nakanishi, I. 1989. Activation of matrix metalloproteinase 3 (stromelysin) and matrix metalloproteinase 2 ('gelatinase') by human neutrophil elastase and cathepsin G. *FEBS Lett.* 249:353–56

124. Omori, M., Chytil, F. 1982. Mechanism of vitamin A action gene expression in retinol-deficient rats. *J. Biol. Chem.* 257:14370–74

125. Phan, S. H., McGarry, B. M., Loeffler, K. M., Kunkel, S. L. 1988. Binding of leukotriene C_4 to rat lung fibroblasts and stimulation of collagen synthesis *in vitro*. *Biochemistry* 27:2846–53

126. Prockop, D. J., Berg, R. A., Kivirikko, K. I., Uitto, J. 1976. Intracellular steps in the biosynthesis of collagen. In *Biochemistry of Collagen*, ed. G. N. Ramachandran, A. H. Reddi, pp. 163–273. New York: Plenum

127. Raghow, R., Gossage, D., Kang, A. H. 1986. Pretranslational regulation of type I collagen, fibronectin and a 50-kilodalton noncollagenous extracellular protein by dexamethasone in rat fibroblasts. *J. Biol. Chem.* 261:4677–84

128. Raghow, R., Postlehwaite, A. E., Keski-Oja, J., Moses, H. L., Kang, A. H. 1987. Transforming growth factor β increases steady state levels of type I procollagen and fibronectin mRNAs posttranscriptionally in cultured human dermal fibroblasts. *J. Clin. Invest.* 79:1285–88

129. Reginato, A. M., Shapiro, I M., Lash, J. W., Jimenez, S. A. 1988. Type X collagen alterations in rachitic chick epiphyseal growth cartilage. *J. Biol. Chem.* 263:9938–45

130. Reighel, H., Koeffler, H. P., Norman, A. W. 1987. Synthesis *in vitro* of 1,25-dihydroxyvitamin D_3 and 24,25-dihydroxyvitamin D_3 by interferon-γ-stimulated normal human bone marrow and alveolar macrophages. *J. Biol. Chem.* 262:10931–37

131. Rosenbloom, J., Feldman, G., Freundlich, B., Jimenez, S. A. 1984. Transcriptional control of human diploid fibroblast collagen synthesis by γ in-

terferon. *Biochem. Biophys. Res. Commun.* 123:365–72

132. Rosenbloom, J., Feldman, G., Freundlich, B., Jimenez, S. A. 1986. Inhibition of excessive scleroderma fibroblast collagen production by recombinant γ interferon. *Arthritis Rheum.* 29:851–56

133. Rothman, J. E., Orci, L. 1990. Movement of proteins through the golgi stack: A molecular dissection of vesicular transport. *FASEB J.* 4:1460–68

134. Saga, S., Nagata, K., Chen, W.-T., Yamada, K. M. 1987. pH-dependent function, purification and intracellular location of a major collagen-binding glycoprotein. *J. Cell Biol.* 105:517–27

135. Sage, H., Bornstein, P. 1987. Type VIII Collagen. See Ref. 102, pp. 223–59

136. Sahebjami, H., MacGee, J. 1983. Changes in connective tissue composition of the lung in starvation and refeeding. *Am. Rev. Respir. Dis.* 128:644–47

137. Sahebjami, H., MacGee, J. 1985. Effects of starvation on lung mechanics and biochemistry in young and old rats. *J. Appl. Physiol.* 58:778–84

138. Sandberg, M. J., Makela, J., Multimaki, P., Vuorio, T., Vuorio, E. 1989. Construction of a human proα1(III) collagen cDNA clone and localization of type III collagen expression in human fetal tissues. *Matrix* 9:82–91

139. Sandell, L. J., Boyd, C. D. 1990. Conserved and divergent sequences and functional elements within collagen genes. In *Extracellular Matrix Genes*, ed. L. J. Sandell, C. D. Boyd, pp. 1–56. New York: Academic

140. Schmid, T. M., Linsenmayer, T. F. 1987. See Ref. 102, pp. 223–59

141. Schwarz, R. I., Bissell, M. J. 1977. Dependence of the differentiated state on the cellular environment: modulation of collagen synthesis in tendon cells. *Proc. Natl. Acad. Sci. USA* 74:4453–57

142. Spanheimer, R. G., Peterkofsky, J. B. 1985. A specific decrease in collagen synthesis in acutely fasted, vitamin C supplemented, guinea pigs. *J. Biol. Chem.* 260:3955–62

143. Spanheimer, R., Zlatev, T., Umpierrez, G., DiGirolamo, M. 1991. Collagen production in fasted and food-restricted rats: Response to duration and severity of food deprivation. *J. Nutr.* 121:518–24

144. Strickland, S., Smith, K. K., Marotti, K. R. 1980. Hormonal induction of differentiation in teratocarinoma stem-cells generation of parietal endoderm by retinoic acid and dibutyryl cAMP. *Cell* 21:347–55

145. Suda, T., Shinki, T., Takahashi, N. 1990. The role of vitamin D in bone and intestinal cell differentiation. *Annu. Rev. Nutr.* 10:195–211

146. Suzuki, K., Enghild, J. J., Morodomi, T., Salvesen, G., Nagase, H. 1990. Mechanisms of activation of tissue procollagenase by matrix metalloproteinase 3 (Stromelysin). *Biochemistry* 29:10261–70

147. Thaller, C., Eichele, G. 1987. Identification and spacial distribution of retinoids in the developing chick limb bud. *Nature* 327:625–28

148. Thavaraj, V., Sesikeran, B. 1989. Histopathological changes in skin of children with clinical protein energy malnutrition before and after recovery. *J. Trop. Pediatr.* 35:105–8

149. Timple, R., Engel, J. 1987. Type VI collagen. See Ref. 102, pp. 105–43

150. Tolstoshev, P., Haber, R., Trapnell, B. C., Crystal, R. G. 1981. Procollagen mRNA levels and activity and collagen synthesis during the fetal development of sheep lung, tendon and skin. *J. Biol. Chem.* 256:9672–79

151. Trelstad, R. L., ed. 1984. *The Role of Extracellular Matrix in Development.* New York: Liss

152. Van Der Rest, M., Garrone, R. 1991. Collagen family of proteins. *FASEB J.* 5:2814–23

153. van der Rest, M., Mayne, R. 1987. Type IX collagen. See Ref. 102, pp. 195–221

154. Vaughan, L., Mendler, M., Huber, S., Bruckner, P., Winterhalter, K. H., et al. 1988. D-Periodic distribution of collagen type IX along cartilage fibrils. *J. Cell Biol.* 106:991–97

155. Waterlow, J. C., Garlick, P. J., Millward, D. J. 1978. *Protein Turnover in Mammalian Tissues and in the Whole Body.* Amsterdam: North Holland

156. Wiley, E. R., McClain, P. E. 1988. The effects of graded levels of dietary protein on collagen metabolism in the skin of growing rats. *Nutr. Res.* 8(3):265–72

157. Willing, M. C., Cohn, D. H., Byers, P. H. 1990. Frameshift mutation near the 3' end of the Col1α(1) gene of type I collagen predicts an elongated pro-α-1(I) chain and results in osteogenesis imperfecta type I. *J. Clin. Invest.* 85:282–90

158. Yamaguchi, N., Benya, P. D., van der Rest, M., Ninomiya, Y. 1989. The cloning and sequencing of α1(VIII) collagen cDNAs demonstrate that type VIII collagen is a short chain collagen and con-

tains triple-helical and carboxyl-terminal nontriple-helical domains similar to those of type X collagen. *J. Biol. Chem.* 264:16022–29

159. Yazdani, M., Nakamoto, T. 1987. Gestational protein-energy malnutrition affects the composition of developing skins of rat fetuses and their dams. *Br. J. Nutr.* 58(2):215–20

160. Yoon, K., Buenaga, R., Rodan, G. A. 1987. Tissue developmental expression of rat osteopontin. *Biochem. Biophys. Res. Commun.* 148:1129–36

161. Yoon, K., Rutledge, S. J. C., Buenaga, R. F., Rodan, G. A. 1988. Characterization of the rat osteocalcin gene: Stimulation of promoter activity by 1,25-dihydroxyvitamin D_3. *Biochemistry* 27:8521–26

162. Yurchenco, P. D., Schittny, J. C. 1990. Molecular architecture of basement membranes. *FASEB J.* 4:1577–90

Annu. Rev. Nutr. 1992. 12:391–416

SERUM CHOLESTEROL AND CANCER RISK: An Epidemiologic Perspective

Stephen B. Kritchevsky

Department of Biostatistics and Epidemiology, University of Tennessee, Memphis, Tennessee 38163

David Kritchevsky

The Wistar Institute, 3601 Spruce Street, Philadelphia, Pennsylvania 19104-4286

KEY WORDS: HDL cholesterol, LDL cholesterol, epidemiology, randomized clinical trials

CONTENTS

INTRODUCTION

Although the possibility of a relationship between low serum cholesterol levels and the risk of cancer has been raised occasionally, it was not until

0199-9885/92/0715-0391$02.00

Pearce & Dayton reported an excess of cancer cases in a clinical trial of a lipid-lowering diet (75) that much interest was expressed. Their observations led to a review, of the then current work, by Ederer et al (25) who stated that if the findings of Pearce & Dayton were not considered, the remaining data suggested no danger from treatment for hypercholesterolemia; if the Pearce & Dayton data were included, a possible danger could be adduced. They concluded that there was no reason to question or suspend treatment of hypercholesterolemia but that further careful observation was warranted. In 1974, Rose et al (81) published an observational study of colon cancer and serum cholesterol in which, contrary to expectation, they found low serum cholesterol to be statistically associated with increased mortality from colon cancer.

By the early 1980s, the number of investigations reporting a serum cholesterol-cancer link was sufficient to warrant the convening of two meetings by the National Heart Lung and Blood Institute to discuss the association (unpublished). Investigations in this area were fairly evenly split between those finding a link between low serum cholesterol and elevated cancer risk and those not finding an association (28). There was also evidence that cancer prior to its diagnosis could itself lower serum cholesterol levels, so that the direction of causality between low serum cholesterol and cancer was in doubt (82).

Data from clinical trials and epidemiologic studies address two distinct but related questions: (*a*) In a given population, is having a serum cholesterol level in the lower end of the population distribution associated with increased cancer risk? (*b*) Does the reduction of serum cholesterol increase cancer risk? Despite the large number of studies in this area, a definitive evaluation of either of these questions has remained elusive. The great majority of the investigations were designed to study cardiovascular disease; thus methodologic issues relevant to cancer investigations frequently have not been addressed. Several deficiencies are frequently found: the lack of cancer incidence data, the lack of histologic verification of the cancer, the failure to exclude subjects with a history of cancer at baseline, the inadequacy of sample sizes for the detection of meaningful differences in cancer occurrence (especially for individual cancer sites), and the failure to consider potential confounders important to cancer investigations. The subject has raised sufficient concern to stimulate a number of reviews and many attempts to explain the findings. Two of the more thorough reviews have been those by McMichael et al (66, 67), who concluded that preclinical cancer leads to hypocholesterolemia and that low serum cholesterol may pose a risk of cancer, principally colon cancer. In this review, we examine the available literature and offer our perspective on this problem.

GROUND RULES

Part of the difficulty in arriving at a conclusion concerning the relationship between serum cholesterol and cancer is that very few studies in this area have approached the issue in exactly the same way. Though in all studies the measurement of serum cholesterol precedes the diagnosis of cancer, the analyses reflect a variety of assumptions concerning the natural history of cancer and the form of the statistical relationship between serum cholesterol and cancer. To facilitate comparisons across studies, we have made a number of decisions concerning the presentation and inclusion of data. Still, the diversity of approaches should be borne in mind when comparing study results.

Epidemiologic Studies Included

Tables 1–5 summarize the epidemiologic evidence relating to low serum cholesterol and increased cancer risk. In several instances, data from the same cohort at different follow-up times or from subpopulations within a cohort have been reported separately. Since these results are not independent, we have decided to include only one report from each study population. Data from the full cohorts are included in preference to data from subpopulations (18, 79, 102, 113, 118); cancer incidence data are preferred to mortality data (3, 20, 42, 115); and reports with longer follow-up periods are included in preference to those with shorter follow-up periods (4, 41, 76, 77, 83, 102). In three instances, single publications report the experience of multiple cohorts (24, 31, 48). Whenever sufficient information is provided, these cohorts are summarized separately. In one publication, an international collaborative group reported the combined experience of 11 separate cohorts (38). Since results from 7 of the 11 cohorts (accounting for 85% of the combined population) are reported elsewhere in detail and the remaining cohorts were either of modest size or had limited follow-up time, this report was not included in the summary tables. In this review we refer to "serum" cholesterol. It should be noted, however, that a number of studies examined plasma rather than serum cholesterol (3, 18, 39, 79, 82, 97, 100, 113, 118). Reports presenting data only in graphical form are not included in this review (28, 50, 57, 92).

Analysis

The studies presented in Tables 1–5 are shown in roughly descending order of the strength of evidence for an association between low serum cholesterol and cancer. Whenever possible, the comparison reported is the ratio of the risk of cancer development for those in the lowest fraction of the serum cholesterol

Table 1 The relative risk of cancer occurrence associated with low serum cholesterol (males)

Geographic location	Size of study population	Number of cases	Duration of follow-up (years)	Endpoint[a]	Most common site	Relative risk of cancer occurrence	Statistical significance ($p < 0.05$)	Ref.
North America	2,753	79	8.4	M	n.r.[b]	2.27	Yes	18
United States	5,125	459	10	I	Lung	2.2[c]	Yes	86
Honolulu	8,006	750	18	M	Lung	2.12, 1.25[c,d]	Yes, no[d]	102
Rural, Puerto Rico	2,585	52	8	M	n.r.	1.8, 2.2[e,f]	Yes, yes[f]	31
Japan	913	43	15	M	n.r.	2.01	Yes	48
Stockholm	—[g]	65	n.a.[h]	M	n.r.	1.8	No	32
Urban, Puerto Rico	6,208	127	8	M	n.r.	1.1, 1.7[e,f]	No, yes[e]	31
Scotland	7,000	630	12	I	n.r.	1.47[c]	Yes	39
Finland[i]	10,537	301	10	I	Prostate	1.3	No	51
18 US cities	361,662	2,989	8	M	Lung	1.3	No	91
N. Europe	2,322	132	15	M	n.r.	1.27	No	48
S. Europe	5,791	224	15	M	n.r.	1.19	No	48
Sweden	46,140	4,455	20	I	n.r.	1.04, 1.06[e,j]	No	110
N. California	73,671	3,218	9.9	I	Prostate	1.01	No	34
Netherlands	878	203	25	I	n.r.	—[k]	n.a.	56
United States	2,299	78	15	M	n.r.	0.98	No	48
E. Finland	2,745	65	7	I	n.r.	0.83	No	84
London	525[g]	267	n.a.	I	Skin	0.75	No	112

[a] M is total cancer mortality, I is total cancer incidence.
[b] n.r. is not reported.
[c] Compares low to high end of the cholesterol distribution.
[d] First and second entries are for 7–12 and ≥13 years of follow-up, respectively.
[e] Relative risk for a 40 mg/dL decrease in blood cholesterol.
[f] First and second entries are for those aged 45–54 and 55–64 years, respectively.
[g] Case-control study; 197 controls.
[h] n.a. is not applicable.
[i] Nonsmokers only.
[j] Range is for results at time intervals throughout the follow-up period.
[k] Inverted U-shaped relationship.

Table 2 The relative risk of cancer associated with low serum cholesterol in studies not fully accounting for a preclinical cancer effect (males)

Geographic location	Size of study population	Number of cases	Duration of follow-up (years)	Endpoint[a]	Most common site	Relative risk of cancer occurrence	Statistical significance ($p < 0.05$)	Ref.
Denmark	230	n.r.[b]	10	M	n.r.	3.75[c]	Yes	2
Malmö, Sweden	7,725	44	6	M	Stomach	2.67[c]	Yes	76, 77
Framingham	1,946	230	18	I	n.r.	1.66	Yes	115
Stockholm	3,486	156	14	M	Gastro-intest.	1.54	No	6
Evans Co., Ga.	1,250	74	14	I	n.r.	1.46[d]	No	42
Chicago	6,890	116	5	M	Lung	1.27	No	24
Israel	10,059	110	7	M	n.r.	1.25[e]	No	119
Chicago	1,899	78	17	M	Colo-rectal	1.09	No	24
Great Britain	7,690	226	9	M	n.r.	1.04	No	90
Chicago	1,233	99	18	M	Lung	0.72	No	24
Oslo, Norway	3,751	89	10	I	Stomach	0.64	No	114

[a] See Footnote a in Table 1.
[b] See Footnote b in Table 1.
[c] See Footnote c in Table 1.
[d] Compares below median to above median.
[e] Relative risk for a one standard deviation decrease in blood cholesterol.

Table 3 The relative risk of cancer occurrence associated with low serum cholesterol (females)

Geographic location	Size of study population	Number of cases	Duration of follow-up (years)	Endpoint[a]	Most common site	Relative risk of cancer occurrence	Statistical significance (p < 0.05)	Ref.
Studies accounting for a preclinical cancer effect								
Stockholm	—[b]	35	n.a.[c]	M	n.r.[d]	1.8	No	32
Finland[e]	14,783	549	10	I	Breast	1.2	No	51
N. California	86,464	4,259	10	I	Breast	1.14[f]	Yes	34
Scotland	8,262	554	12	I	n.r.	1.10	No	39
Sweden	46,570	4,580	20	I	n.r.	0.90–1.06[g,h]	No	110
United States	7,363	398	10	I	Breast	1.0[f]	No	86
E. Finland	4,221	78	7	I	n.r.	0.83	No	84
North America	2,476	65	8.4	M	n.r.	0.49	No	18
Studies not fully accounting for a preclinical cancer effect								
Evans Co., Ga.	1,370	53	14	I	n.r.	1.3[i]	No	42
Framingham	2,317	208	18	I	Breast	1.27	No	115
Stockholm	2,378	74	14	M	Gastro-intest.	0.97[f]	No	6
Chicago	5,750	55	5	M	Colo-rectal	0.40	No	24

[a] See Footnote a in Table 1.
[b] Case-control study; 196 controls.
[c] See Footnote h in Table 1.
[d] See Footnote b in Table 1.
[e] See Footnote i in Table 1.
[f] See Footnote c in Table 1.
[g] See Footnote e in Table 1.
[h] See Footnote j in Table 1.
[i] See Footnote d in Table 2.

Table 4 The relative risk of cancer occurrence associated with low serum cholesterol (combined sexes)

Geographic location	Size of study population	Number of cases	Duration of follow-up (years)	Endpoint[a]	Most common site	Relative risk of cancer occurrence	Statistical significance (p < 0.05)	Ref.
Europe[b]	822	18	3.1	M	n.r.[c]	3.9	Yes	99
United States	10,940	286	5	I	Lung	1.5	No	69
New Zealand	630	n.r.	17	M	n.r.	1.4[d]	n.r.	83
China	9,021	263	13	M	Lung	1.07[d]	No	14

[a] See Footnote a in Table 1.
[b] Preclinical period not considered in analysis.
[c] See Footnote b in Table 1.
[d] See Footnote e in Table 1.

Table 5 Site-specific associations between low serum cholesterol and cancer

Site	Sex	Size of study population	Number of cases	Duration of follow-up (years)	Endpoint[a]	Relative risk of cancer occurrence	Ref
Brain	Female	22,324	14	18	I	1.05[b]	5
	Male	17,718	32	18	M	0.68[c]	9
	Male	26,001	30	18	I	0.74[b]	5
Breast	Female	24,329	242	14	I	2.0[d]	11
	Female	46,570	1,182	20	I	1.04, 1.28[c,e]	11
	Female	14,783	95	10	I	1.14[f]	5
	Female	95,179	1,035	7.9	I	1.06[b]	3
Colon	Male	2,753	13	8.4	M	5.2	1
	Male	1,946	30	18	I	3.5	11
	Male	361,662	138	8	M	1.27[b]	9
	Male	8,006	80	12–14	I	Inv., No[g]	10
	Female	86,464	320	10	I	1.09[b]	3
	Male	73,671	278	10	I	0.90[b]	3
	Male	45,987	257	15	I	0.90	10
	Female	46,911	271	15	I	0.88	10
Lung	Male	5,791	44	15	M	2.54	4
	Male	913	5	15	M	2.08	4
	Male	10,537	34	10	I	1.85[f]	5
	Male	2,322	50	15	M	1.15	4
	Male	73,671	162	9.9	I	1.06[b]	3
	Male	2,299	29	15	M	1.05	4
	Male	361,662	437	8	M	1.04[b]	6

[a] See Footnote a in Table 1.
[b] See Footnote c in Table 1.
[c] See Footnote e in Table 1.
[d] For cases occurring under 51 years of age only.
[e] First entry is for all women; second is for women <50 years of age adjusting for beta-lipoprotein levels.
[f] See Footnote i in Table 1.
[g] Inverse relationship during follow-up years 5–9.9; no relationship after 10 years of follow-up.

distribution (usually fourths or fifths) compared to the risk of cancer for those in the rest of the distribution. When necessary, this was calculated from data supplied in the original report. Some reports compare the risk of cancer development for those in the lower end of the serum cholesterol distribution with the risk for those in the higher end. In other reports, statistical modeling was used to describe the risk relationship between serum cholesterol and cancer. In these instances the relative risk of cancer development associated with a 40-mg per deciliter (1.03 mmol per liter) decrement in serum cholesterol is presented. Results adjusted for multiple confounders were used when available. In some studies, the differences in mean serum cholesterol levels between noncases and cases were reported (1, 10, 53, 81, 82, 105). There is little basis on which to compare results from grouped data with results based on individual risk of cancer. Thus, studies reporting grouped differences are not included in the tables.

The Preclinical Cancer Effect

Substantial empirical evidence indicates that, in males, cancer can reduce low density lipoprotein cholesterol (LDL-C) prior to diagnosis, a process that McMichael et al (67) termed a preclinical cancer effect. Cancer patients, except those with breast cancer, have significantly lower serum cholesterol levels than do matched controls (2a, 73). In many epidemiologic studies, the relationship between serum cholesterol and increased cancer risk is strongest for those individuals whose cancer was diagnosed shortly after their cholesterol baseline measurement (10, 34, 38, 48, 51, 82, 91). The implication is that cancer, present but undiagnosed at baseline, was responsible for the lower serum cholesterol among at least some of those who went on to develop the disease. Finally, five studies have reported serial serum cholesterol measurements in individuals who eventually developed cancer. In each study, a decline in serum cholesterol was observed prior to diagnosis in at least some of those developing cancer (10, 55, 91, 98, 117). The effect appears to be manifest approximately two years prior to the diagnosis of clinical disease and four years prior to death, though these periods may be longer for slower growing tumors (55, 91, 98, 117). The mechanism by which cancer can lower serum cholesterol is unclear. However, Ueyama et al (110a) describe a gallbladder tumor cell line derived from a man whose serum cholesterol dropped prior to diagnosis. The culture medium from this cell line was found to increase LDL receptor activity in skin fibroblasts derived from normal individuals.

To avoid spurious associations between low cholesterol and cancer risk it is necessary to account for a preclinical cancer effect analytically. To do so, investigators commonly exclude from analysis individuals whose cancers developed in the first few years of follow-up (i.e. those cancers that are presumed to have been present but undiagnosed at the time of cholesterol measurement). Table 1 includes those studies of males that account for a preclinical cancer effect of at least two years prior to diagnosis or four years prior to death, and the estimates of the strength of association are from statistical analyses accounting for the effect. Table 2 summarizes studies of males that do not account for a sufficiently long preclinical cancer effect. Table 3 summarizes both types of studies for females. Table 4 describes studies that combined the sexes and includes one study that does not provide for a preclinical cancer effect.

Randomized Trials to Lower Serum Cholesterol

Table 6 presents the cancer experience of randomized trials of cholesterol-lowering interventions, both pharmacologic and dietary. Included are both primary and secondary prevention trials and trials in which cholesterol lowering was only one feature of the intervention. In several instances, both the in-trial and post-trial experience have been reported, and both are included in the table.

IS HAVING LOW SERUM CHOLESTEROL ASSOCIATED WITH INCREASED CANCER RISK?

As can be seen in Tables 1–4, the epidemiologic evidence relating low serum cholesterol and cancer risk is inconsistent: both direct and inverse relationships have been reported, and in one study, males in the middle third of the cholesterol distribution had the highest risk of cancer (56).

Taken as a whole, the literature supports a weak association between low serum cholesterol levels and increased occurrence of all cancers in males (Table 1). Among males, the median association across all studies is consistent with about a 30% increase in cancer risk for those with low serum cholesterol levels. The results range from a 227% increase in risk to a 25% decrease in risk. If there were no relationship at all, one would expect the median of the findings to be consistent with a relative risk of one (i.e. no association). Instead, 13 of the 18 study populations (72%) show a relative risk of cancer greater than 1.0. In those studies not fully accounting for a preclinical cancer effect (Table 2), the range of observed results is somewhat broader but the median finding is also consistent with about a 30% increase in cancer risk for those having low serum cholesterol levels.

In Table 1, a preponderance of the studies finding either little or no association between low serum cholesterol and cancer were studies of cancer incidence rather than cancer mortality. Within some study populations a stronger link is noted with mortality than with incidence. The Lipid Research Clinics Mortality Follow-up study found one of the strongest relationships between plasma cholesterol and cancer mortality (18). However, in two of the program's clinics, little or no relationship was found between plasma cholesterol and cancer incidence (113, 118). Similarly, Kagan et al (41) found a significant inverse relationship between serum cholesterol and cancer mortality through nine years of follow-up, but Stemmermann et al (103) found a relationship between serum cholesterol and cancer incidence only for cancer of the colon. On the other hand, Isles et al (39) found the association between plasma cholesterol levels and cancer mortality was weaker than that between plasma cholesterol and cancer incidence. Kark et al (42) found a relative risk of cancer incidence of 1.46 comparing those below and above the median of the serum cholesterol distribution. Looking at the same population, Davis et al (20) found a distinctly U-shaped relationship between cholesterol and mortality, with those in the middle tertile of the serum cholesterol distribution having a lower mortality rate than those in either the high or low tertile.

In females, the evidence for a low serum cholesterol-cancer association is weaker than in males (Table 3). In 11 of the 12 studies that included both males and females, the strength of the association between cancer and low serum cholesterol in females was weaker than or equal to the association seen in males. In one study (18), low plasma cholesterol was related to decreased cancer mortality among women but increased mortality among men. Only 1 of the 12 study populations found a statistically significant association in women (34). The median of the distribution of findings is consistent with a 5–10% increase in cancer risk for females with low serum cholesterol.

The difference in the strength of association by sex suggests an underlying role of hormonal or metabolic factors. To investigate this question, Kritchevsky (54) examined the role of body fat distribution in the association. Men tend to have more central fat than women both on an absolute and relative basis (59). Increased central adiposity has been associated with lower sex hormone-binding globulin levels in both sexes and with increased free testosterone in premenopausal women (26, 27, 101). In males participating in the National Health and Nutrition Survey Epidemiologic Follow-up Study (NHEFS) low serum cholesterol was associated with increased cancer incidence regardless of body fat distribution. In the females, however, low serum cholesterol was associated with increased cancer risk only among those with central adiposity. In females with peripheral adiposity, low serum cholesterol was associated with decreased cancer risk.

Though few studies have examined lipoprotein subfractions, the low serum cholesterol-cancer association appears to be attributable primarily to LDL-C (18, 100). The relationship between triglyceride levels and cancer has been inconsistent; several studies have reported very weak inverse or null associations (10, 18, 31, 76, 79, 100, 113). In general, no relationship between high density lipoprotein cholesterol (HDL-C) and cancer occurrence has been observed (18, 79, 116, 119). However, Keys (47) found increased HDL-C to be associated with increased cancer mortality.

It is noteworthy that, in males, all of the studies demonstrating a statistically significant association between low serum cholesterol and cancer examined populations that were either entirely or in part community based (2, 6, 18, 31, 48, 86, 102, 115). Cohorts consisting of employed males generally show little or no association. This finding suggests that factors associated with socioeconomic status play some role in the association. Indeed, the association between low serum cholesterol and cancer in Evans County white males was

Table 6 Cancer occurrence in trials of cholesterol reduction

Trial type	Cholesterol lowering intervention	Control	Treatment N	Control N	Treatment cases	Control cases	I/M[a]	Study duration (years)	Post-trial follow-up (years)	Ref.
Primary	Clofibrate	Olive oil	5331	5296	42	25	M	5.3	—	17
Primary[b]	Clofibrate	Olive oil	5331	5296	206	197	M	5.3	7.9	17
Primary	Diet[c]	None	604	628	5	8	M	5.0	—	36
Primary	Diet[d]	Usual care	6428	6438	81	69	M	6–8	—	71
Primary[b]	Diet[d]	Usual care	6428	6438	140	149	M	6–8	3.8	72
Primary	Cholestyramine and diet	Placebo and diet	1906	1900	16(57)	15(57)	M(I)	7.4	—	62
Primary	Diet, probucol, and clofibrate[e]	None	612	610	0	3	M	5.0	—	104
Primary[b]	Diet, probucol, and clofibrate[e]	None	612	610	13	21	M	5.0	10	104
Primary	Gemfibrozil and diet	Placebo and diet	2051	2030	11(31)	11(26)	M(I)	5.0	—	30
Mixed	Diet	Placebo	424	422	31(60)[c]	17(38)[f]	M(I)	8.3	—	75
Mixed[b]	Diet	Placebo	424	422	65	48	I	8.3	2	75
Mixed	Colestipol HCL	Placebo	548	546	2	2	M	1–3	—	23

Mixed	Diet	None	2197	2196	16	12	M	4.5	—	29
Mixed[g]	Diet	None	2344	2320	7	8	M	4.5	—	29
Secondary	Diet	None	229	229	4	4	I	5.0	—	60
Secondary[b]	Diet	None	206	206	7	5	M	5.0	6	61
Secondary	Niacin	Lactose	1119	2789	14	27	M	6.2	—	11
Secondary[b]	Niacin	Lactose	1119	2789	45	124	M	6.2	8.8	11
Secondary	Clofibrate	Lactose	1103	2789[h]	11	27	M	6.2	—	11
Secondary[b]	Clofibrate	Lactose	1103	2789	37	124	M	6.2	8.8	11
Secondary	Colestipol HCl, niacin, and diet	Placebo and diet	94	94	2	1	I	2.0	—	5
Secondary	Clofibrate, nicotinic acid, and diet	Diet	279	276	10	6	n.r.[i]	5.0	—	12

[a] Endpoint; I is cancer incidence, M is cancer mortality.
[b] Same study population as preceding entry.
[c] Treatment group also advised to quit smoking.
[d] Treatment also included blood pressure control through diet and drugs and a cigarette smoking intervention program.
[e] Probucol was used to treat type IIA hyperlipidemia and clofibrate was used to treat type IIB. Treatment also included drug treatment for blood pressure control and interventions design to encourage increased physical activity and smoking cessation.
[f] Deaths are from carcinomas only.
[g] Women only.
[h] Same control group as preceding entry.
[i] See Footnote b in Table 1.

much more pronounced in those of lower socioeconomic status than in those of higher socioeconomic status (42).

Site-Specific Relationships

Table 5 presents findings from investigations of several specific cancer sites. No indication of statistical significance is given because, in light of the small number of cases, lack of significance could be misinterpreted as a null result. In several studies of males, the strongest site-specific relationships have been found for colon cancer (18, 81, 103, 115). Generally, no relationship has been seen in females. In the Honolulu Heart Study, the relationship was particularly strong for men over age 55 at baseline (103). Several studies, however, have found no relationship (34, 37, 91), and Törnberg et al (109) found that males and females with both high serum cholesterol and high beta-lipoprotein levels were at elevated risk of colon cancer.

The location of the tumor in the bowel may be important. Stemmermann et al (103) found a relationship with low serum cholesterol levels primarily for tumors of the cecum and ascending colon. However, Sidney et al (94) found no association no matter which part of the bowel was examined.

The idea that low serum cholesterol is directly related to colon cancer risk is inconsistent with findings from studies of individuals with adenomatous polyps, the hypothesized colon cancer precursor lesion. In a case-control study, Mannes et al (64) found that subjects with high cholesterol levels were twice as likely to have polyps as those with low cholesterol levels. Two other case-control studies found no association between polyps and cholesterol levels (21, 73). Among other sites of the gastrointestinal tract, stomach cancer has been linked to low cholesterol levels while rectal cancer has been linked with elevated levels (107, 109).

Despite the attention focused on colon cancer, other sites have been linked more consistently with low serum cholesterol levels. Lung cancer has consistently, albeit weakly, been related to low cholesterol levels in males but not females. The strongest relationship was observed among three Southern European cohorts participating in the Seven Countries Study: those in the lowest quintile of the distribution were more than 2.5 times more likely to die from lung cancer (48). In the extended follow-up of the Honolulu Heart program, Stemmermann et al (102) reported that the mean baseline cholesterol level for subjects dying of lung cancer more than 12 years after the initial examination was 212.7 mg per deciliter (5.50 mmol per liter) compared to 218.4 mg per deciliter (5.64 mmol per liter) for the survivors.

Malignancies of the hematopoietic system have also been associated frequently with low cholesterol levels. In a study of over 360,000 males, participants in the lowest quintile were at over 80% increased risk (91). Knekt et al (51) found that among nonsmokers, males in the lowest quintile were five times more likely to develop either leukemia or lymphoma than were males with higher cholesterol levels. Nonsmoking women with low cholesterol were 1.7 times more likely to develop either leukemia or lymphoma than were nonsmoking women with higher cholesterol levels.

Among females, cervical cancer has been consistently related to low serum cholesterol levels. In a study of cancer incidence among health plan enrollees (34), the cervix was the only site for which a statistically significant relationship with low serum cholesterol was noted after excluding cases diagnosed during the first two years of follow-up. Schatzkin et al (87) also reported a strong association between cervical cancer and low serum cholesterol in the NHEFS. In an Australian case-control study, women with low plasma cholesterol levels ($<$154.4 mg per deciliter; $<$3.99 mmol per liter) were more than two times as likely to have *in situ* cervical cancer than were women with high cholesterol levels ($\geq$231.6 mg per deciliter; 5.98 mmol per liter) (8).

In two studies of breast cancer, tumors diagnosed in women below about 50 years of age were linked to low cholesterol levels, but cancer diagnosed at later ages was not (108, 111). Törnberg et al (108) found that in younger women, both higher serum beta-lipoprotein and lower serum cholesterol levels were associated with increased risk of breast cancer. This finding implies that low HDL-C was associated with increased cancer risk. As reviewed by Boyd & McGuire (7), a number of studies have reported depressed HDL-C levels in women with breast cancer. Several other studies have not found a relationship between total serum cholesterol and breast cancer, but none of the analyses were stratified by age of disease onset (35, 39, 51, 87).

Schatzkin et al (87) reported a strong relationship between low serum cholesterol and increased incidence of smoking-related cancers (sites: lung, mouth, larynx, esophagus, pancreas, bladder, cervix, and leukemia). The relationship was seen in both men and women and persisted throughout the follow-up period. The relationship was slightly stronger for nonsmokers than for smokers, a finding similar to that of Knekt et al (51) for all cancers. In males, Cowan et al (18) found that low total plasma cholesterol and LDL-C levels were only slightly more strongly related to mortality from smoking-related cancers (sites: larynx, lung, and bladder) than to mortality for all cancers. They found that females with low plasma LDL-C levels had elevated mortality from smoking-related tumors, contrary to the relationship seen for all cancers.

Distinct from many other sites, brain cancer has been linked to elevated cholesterol levels (see Table 5). Three studies have found higher cholesterol to be associated with increased occurrence of brain cancer, at least in males (1, 52, 97). Abramson & Kark (1) found that patients with primary brain tumors had serum cholesterol levels 22 mg per deciliter (0.57 mmol per liter) higher than controls. Two other studies have found high cholesterol to be associated with increased occurrence of malignant tumors of the nervous system (91, 112).

Explanations and Hypotheses

Several mechanisms have been offered to explain why low cholesterol levels might be associated with increased cancer risk.

1. CHOLESTEROL AND THE PHYSICAL CHEMISTRY OF THE CELL MEMBRANE   Researchers have speculated (74) that increased cell membrane fluidity, which might be associated with low serum cholesterol levels, may increase the likelihood of neoplastic transformation. However, Marenah et al (65) found that HDL-C and LDL-C levels were unrelated to cell membrane fluidity in monocytes from individuals with a wide range of serum cholesterol levels (124–387 mg per deciliter; 3.20–10.0 mmol per liter).

2. MEMBRANE CHOLESTEROL AS AN INFLUENCE ON TUMOR ANTI-GENICITY   Another suggestion is that the loss of membrane cholesterol may render tumor cells less antigenic (93), thereby allowing transformed cells to escape immune system surveillance. A link between low serum cholesterol and reduced tumor immunogenicity would explain why low serum cholesterol may be more strongly linked with cancer mortality than with cancer incidence. However, as a rule, solid tumors tend to contain more, not less, cholesterol than normal tissues (13).

3. ANTIMITOGENIC EFFECTS OF VERY LOW AND LOW DENSITY LIPO-PROTEINS   The presence of LDL can inhibit the activation of lymphocytes after exposure to a variety of mitogens and antigens (15, 19, 70), and the major apoproteins that constitute LDL (apo B and apo E) can inhibit lymphocyte activation in their purified forms (63). Ito et al (40) have also shown that LDL can inhibit cell proliferation in a number of cultured cell lines. Possibly, low levels of circulating lipoproteins are permissive of mitogenesis, or, conversely, high levels may exert an antipromoting effect.

4. LOW DENSITY LIPOPROTEINS AS INHIBITORS OF VIRALLY INDUCED CELL TRANSFORMATION   Chisari et al (16) found that physiologic concentrations of both LDL and VLDL suppress Epstein-Barr virus-induced immortalization of adult human B lymphocytes. Their findings suggest the possibility that hypocholesterolemia may play a role in other virally linked malignancies. Epidemiologic evidence points strongly to a viral etiology for cervical cancer (33). Cervical cancer is one of the cancers most strongly linked to low serum cholesterol levels in females.

5. GENETIC APOPROTEIN VARIANTS MAY BE LINKED TO INCREASED CAN-CER SUSCEPTIBILITY   Katan (45) has suggested that the apolipoprotein E (apo E) E-2 allele may explain the low serum cholesterol-cancer association. Apo E-2 is associated with lower total cholesterol levels in a number of populations, but it is not known whether the E-2 phenotype is associated with increased cancer risk. If there were a genetic explanation, one would expect to see an excess of cancer in families of probands with low cholesterol levels. Two studies have found that low levels of cholesterol were associated with increased cancer occurrence in mothers of probands (22, 120). Schrott et al (88) found that cancer was more commonly reported at death among the relatives of children with low serum cholesterol levels than among the relatives of children with high cholesterol levels. Contrary to these findings, Reed et al (80) found lower cancer family history scores in participants with low total cholesterol levels, but an excess of cancer mortality in the parents of males with higher HDL-C levels.

6. DIETARY DETERMINANTS OF SERUM CHOLESTEROL AND CANCER   If diet underlay the low cholesterol-cancer association, one would expect either increased polyunsaturated fat intake or lower saturated fat and dietary cholesterol intake among those developing cancer. In the NHEFS, despite having lower serum cholesterol levels, males developing cancer ingested a more hyper-cholesterolemic diet than those not developing cancer (54). The difference primarily reflected increased saturated fat and cholesterol intake; there was no difference in linoleic acid intake. Interestingly, Laskarzewski et al (58) found that the LDL-C level of progeny whose mothers had died of cancer was lower with increasing dietary cholesterol consumption than the level of those whose mothers had not died of cancer.

7. LEVELS OF FAT-SOLUBLE ANTIOXIDANTS OR VITAMINS AND CANCER   Vitamin E and several carotenoids are transported in the LDL particle, and their circulating levels are correlated with serum cholesterol (44, 96, 106). Low cholesterol levels, therefore, could also mean low levels of these sub-stances—especially in populations already at marginal levels. Low levels of beta-carotene have been linked fairly consistently to elevated risk of lung cancer, but the relationship between other carotenoids and cancer has not been well studied (68). Low levels of vitamin E have also been linked to increased risk of lung cancer, though less consistently than levels of beta-carotene (68). It may be, therefore, that the observed relationship between low serum cholesterol and cancer could, in fact, be due to a relationship between low levels of circulating vitamin E and/or carotenoids and not to any direct effect of serum cholesterol. Whether or not the association between low serum cholesterol and cancer is confounded by lipid-soluble vitamins and anti-

oxidants and cancer has not been determined. Kark et al (43) found that the low cholesterol-cancer association in the Evans County study was diminished after accounting for serum retinol levels. Subsequent study, however, has not supported an association between low serum vitamin A levels and increased cancer risk (68).

8. LOW SERUM CHOLESTEROL MIGHT REFLECT HIGHER BILE ACID FLUXES IN THE BOWEL   Subjects who eliminate dietary cholesterol more efficiently or those on drugs that enhance cholesterol excretion via the bile would have an increased flux of bile acids through the colon. Bile acids and their salts have been shown to be promoters of colon cancer in experimental animals. This explanation might be plausible for colon cancer, but low serum cholesterol levels are associated with other sites as well. The data have recently been reviewed by Broitman (9).

None of these explanations can explain why males should be preferentially affected. None of the explanations offered are mutually exclusive. The difficulty in understanding this relationship may be that the relationship between cholesterol and cancer depends on a number of different pathways for different tumor types.

DOES REDUCING SERUM CHOLESTEROL INCREASE CANCER RISK?

The epidemiologic evidence does not address the question of whether or not reducing serum cholesterol—as opposed to having low serum cholesterol—leads to an increase in cancer occurrence. The most direct evidence derives from randomized clinical trials of lipid-lowering interventions (see Table 6). Thirteen reports, including 14 experimental cholesterol-lowering interventions, have also provided data on cancer occurrence. With the exception of one investigation (29), the majority of study participants have been male.

A modest excess of cancer occurrence has been reported in trials of lipid-lowering interventions. In all, 252 cancers deaths (this number includes 6 incident cases from trials not reporting mortality) have been reported among 25,269 individuals randomized to the experimental interventions, and 208 cancer deaths (this number includes 5 incident cases) have been reported among the 25,774 individuals randomized to various control groups, for a 24% excess in the occurrence of cancer in the intervention groups ($p < 0.05$). The excess of cancer in treatment groups is seen to a greater or lesser extent in both dietary and pharmacologic interventions (23% and 26%, respectively) and in primary, secondary, and mixed prevention trials (18%, 20%, and 43%, respectively). Three trials provided both incidence and mortality data, but there is insufficient evidence to judge if patterns of incidence and mortality differ (30, 62, 75).

In four studies of pharmacologic interventions, both the placebo and treatment groups received dietary counseling for cholesterol reduction (5, 12, 30, 62). In other words, there was no real control group since serum cholesterol reduction was a feature of both treatment arms. When these four studies are dropped from consideration, a 26% excess in cancer occurrence is noted among the treatment groups.

The trials reviewed are of short duration relative to the carcinogenic process. Risk increases due to exposure to tumor initiators such as cigarette smoke are typically evident only after at least ten years from first exposure. If the lowering of serum cholesterol has an effect, it is on a later stage in the carcinogenic process, either promotion or progression. An important clue concerning the biologic process involved comes from the post-trial cancer experience of these trials.

The post-trial cancer experience has tended to be very different from the in-trial experience. During the WHO clofibrate trial, the annual cancer mortality rate was 1.9 per thousand in the clofibrate group and 1.2 per thousand in the placebo group. In the post-trial follow-up the rates were 2.61 and 2.78 per thousand, respectively (17). In a dietary intervention trial, Pearce & Dayton (75) reported an excess of 14 carcinoma deaths in the intervention group during the trial period and an excess of 3 deaths within one year of the end of the trial, but a deficit of 6 deaths in the second year after the trial. In both the Coronary Drug Project and the Multiple Risk Factor Intervention Trial, modest increases in cancer mortality in the experimental groups during the studies were followed by deficits in cancer mortality after the conclusion of the trials (11, 72).

The combination of an excess of cancers during the intervention period followed by a post-trial deficit suggests that serum cholesterol lowering provides an environment that somehow promotes tumor outgrowth. The reduction of circulating lipoprotein levels may lessen whatever antiproliferative effect they might exert. Alternatively, serum cholesterol lowering might accelerate the growth of tumors already established. Thus, after the trial stops and the pool of initiated tumors in the experimental group is depleted, the group experiences a lower post-trial rate of cancer occurrence than that of the control group.

Cholesterol is synthesized from acetate in a well-characterized pathway in which acetate is converted to beta-hydroxymethylglutarate (HMG), which under the influence of HGM-CoA reductase is converted to mevalonic acid. Mevalonic acid is metabolized to isoprenyl pyrophosphate. From this point, the pathway involves the combination of 2 isoprenoid units to give a geranyl derivative (10 carbon atoms), and then a farnesyl derivative (15 carbon atoms). Two farnesyl units combine to form squalene ($C_{30}H_{50}$), which is cyclyzed to lanosterol, which is eventually converted to cholesterol.

Until recently investigators thought that beyond HMG-CoA the synthetic process was committed to cholesterol, dolichol, and ubiquinone. It is now apparent that other vital metabolic pathways depend on products of cholesterol biosynthesis and may be affected if synthesis is affected. Siperstein et al (95) have shown that the inhibition of HMG-CoA activity blocks cellular DNA synthesis, which carries forward when exogenous mevalonate is provided. The p21[ras] growth protein is farnesylated as a posttranslational modification. Without this modification, the protein is inactive and cell growth is blocked (32a). Furthermore, some evidence suggests that the induction of *de novo* cholesterol synthesis may enhance cell growth. Kazanecki et al (46) found that reducing serum LDL by dietary manipulation stimulated cell proliferation in a number of tissues in weanling rats. In the context of carcinogenesis, Rao et al (78a) found that serum cholesterol reduction through either a high polyunsaturated fat diet or cholestyramine promoted 7,12-dimethylbenz[*a*]anthracene-induced mammary tumors in female rats.

It could be that by stimulating *de novo* cholesterol synthesis, lipid-lowering interventions cause an increase in the pool of substrate necessary to carry forward cell division. While this process may not affect tightly regulated normal cell populations, it may allow accelerated proliferation in transformed populations. Based on this model, clinical trials of HMG-CoA inhibitors would be expected to show a reduction of cancer occurrence in the treatment groups, since the amount of mevalonate and its metabolites would be reduced.

SUMMARY AND CONCLUSIONS

This review has examined the evidence surrounding two questions: (*a*) Is having low serum cholesterol associated with increased risk of cancer? (*b*) Does reducing serum cholesterol increase the occurrence of cancer? Some elevated risk of cancer for males with low serum cholesterol levels has been noted: the median of the studies examined is consistent with a 30% increased risk. The answer for females is less clear. The median of the studies examined suggests no more than a 5–10% increased risk associated with having low serum cholesterol. However, the risk seems to depend strongly on whether females have a central or peripheral body fat pattern (54). The cancers most consistently associated with low serum cholesterol levels are those of the colon and lung in males, the cervix and breast (but only for females under 50 years of age) in females, and leukemia in both sexes. In contrast, high cholesterol levels have been linked with an increase in brain cancer.

While immunologic, genetic, and dietary explanations have been offered to explain the association, it is difficult to support the idea that low serum cholesterol causes cancer in any direct manner. First, the findings themselves tend to be generally weak and somewhat inconsistent. Second, the strong influence of fat distribution in women suggests that a metabolic/hormonal basis underlies the association. One would not expect the results to differ by body fat pattern if the relationship were a causal one. Finally, if there were a direct causal role, one would expect populations with low serum cholesterol levels to have higher cancer rates. In China, counties with the lowest average plasma cholesterol levels have the lowest cancer rates (78). While this observation is open to a number of interpretations, it does not support the idea that low serum cholesterol is a tumor initiator.

In aggregate, the trials of lipid-lowering interventions reviewed here show an increase in cancer occurrence (primarily mortality) of approximately 24% in the cholesterol-lowered groups. However, the post-trial experience has shown a comparative deficit of cancer occurrence in the experimental groups. Recent evidence indicates that products in the cholesterol biosynthetic pathway affect DNA replication and cell proliferation. These findings suggests a mechanism by which cholesterol lowering might accelerate the development of tumors already initiated. The data that have been reviewed in no way suggest that treatment of hypercholesterolemia should not be pursued. They do suggest the presence of a relatively small subpopulation in whom reduction of plasma cholesterol may lead to increased occurrence of cancer. Vigorous efforts should be made to identify the susceptible population.

ACKNOWLEDGMENTS

The authors thank Diane Becton for her assistance with the tables and Nannette C. Gover for her editorial advice.

Literature Cited

1. Abramson, Z. H., Kark, J. D. 1985. Serum cholesterol and primary brain tumors: a case-control study. *Br. J. Cancer* 52:93–98
2. Agner, E., Hansen, P. F. 1983. Fasting serum cholesterol and triglycerides in a ten-year prospective study in old age. *Acta Med. Scand.* 214:33–41
2a. Alexopoulos, C. G., Blatsios, B., Avgerinos, A. 1987. Serum lipids and lipoprotein disorders in cancer patients. *Cancer* 60:3065–70
3. Anderson, K. M., Castelli, W. P., Levy, D. 1987. Cholesterol and mortality: 30 years of follow-up from the Framingham Study. *J. Am. Med. Assoc.* 257:2176–80
4. Beaglehole, R., Foulkes, M. A., Prior,

I. A. M., Eyles, E. F. 1980. Cholesterol and mortality in New Zealand Maoris. *Br. Med. J.* 280:285–87
5. Blankenhorn, D. H., Nessim, S. A., Johnson, R. L., Sanmarco, M. E., Azen, S. P., et al. 1987. Beneficial effects of combined colestipol-niacin therapy on coronary atherosclerosis and coronary venous bypass grafts. *J. Am. Med. Assoc.* 257:3233–40
6. Böttiger, L. E., Carlson, L. A. 1982. Risk factors for death for males and females. *Acta Med. Scand.* 211:437–42
7. Boyd, N. F., McGuire, V. 1990. Evidence of association between plasma high-density lipoprotein cholesterol and risk factors for breast cancer. *J. Natl. Cancer Inst.* 82:460–68

8. Brock, K. E., Hoover, R. N., Hensley, W. J. 1989. Plasma cholesterol and *in situ* cervical cancer: an Australian case-control study. *J. Clin. Epidemiol.* 42:87–89

9. Broitman, S. A. 1986. Cholesterol conundrums: The relationship between dietary and serum cholesterol in colon cancer. In *Dietary Fat and Cancer*, ed. C. Ip, D. F. Birt, A. E. Rogers, C. Mettlin, pp. 435–59. New York: Liss

10. Cambien, F., Ducimetiere, P., Richard, J. 1980. Total serum cholesterol and cancer mortality in a middle-aged male population. *Am. J. Epidemiol.* 112:388–94

11. Canner, P. L., Berge, K. G., Wenger, N. K., Stamler, J., Friedman, L., et al. 1986. Fifteen year mortality in Coronary Drug Project patients: Long-term benefit with niacin. *J. Am. Coll. Cardiol.* 8:1245–55

12. Carlson, L. A., Rosenhamer, G. 1988. Reduction of mortality in the Stockholm Ischaemic Heart Disease Secondary Prevention Study by combined treatment with clofibrate and nicotinic acid. *Acta Med. Scand.* 223:405–18

13. Chen, H. W., Kandutsch, A. A., Heiniger, H. J. 1978. The role of cholesterol in malignancy. *Prog. Exp. Tumor Res.* 22:275–316

14. Chen, Z., Peto, R., Collins, R., Mac-Mahon, S., Lu, J., et al. 1991. Serum cholesterol concentration and coronary heart disease in population with low cholesterol concentrations. *Br. Med. J.* 303:278–82

15. Chisari, F. V. 1977. Immunoregulatory properties of human plasma in very low density lipoproteins. *J. Immunol.* 119:2129–36

16. Chisari, F. V., Curtiss, L. K., Jensen, F. C. 1981. Physiologic concentrations of normal human plasma lipoproteins inhibit the immortalization of peripheral B lymphocytes by the Epstein-Barr virus. *J. Clin. Invest.* 68:329–36

17. Committee of the Principal Investigators. 1984. WHO cooperative trial on primary prevention of ischaemic heart disease with clofibrate to lower serum cholesterol: final mortality follow-up. *Lancet* 2:600–4

18. Cowan, L. D., O'Connell, D. L., Criqui, M. H., Barrett-Connor, E., Bush, T. L., et al. 1990. Cancer mortality and lipid and lipoprotein levels: the Lipid Research Clinics Program Mortality Follow-up Study. *Am. J. Epidemiol.* 131:468–82

19. Curtiss, L. K., Edgington, T. S. 1976. Regulatory serum lipoproteins: regulation of lymphocyte stimulation by a species of low density lipoprotein. *J. Immunol.* 116:1452–58

20. Davis, C. E., Knowles, M., Kark, J., Heyden, S., Hames, C. G., et al. 1983. Serum cholesterol levels and cancer mortality: Evans County twenty-year follow-up study. In *Dietary Fats and Health*, ed. E. G. Perkins, W. J. Visek, pp. 892–900. Champaign, Ill: Am. Oil Chem. Soc.

21. Demers, R. Y., Neale, A. V., Demers, P., Deighton, K., Scott, R. O., et al. 1988. Serum cholesterol and colorectal polyps. *J. Clin. Epidemiol.* 41:9–13

22. Deutscher, S. 1971. Levels of serum cholesterol among offspring of parents who died of cancer: a comparison with coronary heart disease. *J. Natl. Cancer Inst.* 46:217–24

23. Dorr, A. E., Gundersen, K., Schneider, J. C. Jr., Spencer, T. W., Martin, W. B. 1978. Colestipol hydrochloride in hypercholesterolemic patients—effect on serum cholesterol and mortality. *J. Chronic Dis.* 31:5–14

24. Dyer, A. R., Stamler, J., Paul, O., Shekelle, R. B., Shoenberger, J. A., et al. 1981. Serum cholesterol and risk of death from cancer and other causes in three Chicago epidemiological studies. *J. Chronic Dis.* 34:249–60

25. Ederer, F., Leren, P., Turpeinen, O., Frantz, I. D. Jr. 1971. Cancer among men on cholesterol-lowering diets: experience from five clinical trials. *Lancet* 2:203–6

26. Evans, D. J., Barth, J. H., Burke, C. W. 1988. Body fat topography in women with androgen excess. *Int. J. Obesity* 12:157–62

27. Evans, D. J., Hoffmann, R. G., Kalkhoff, R. K., Kissebah, A. H. 1983. Relationship of androgenic activity to body fat topography, fat cell morphology and metabolic aberrations in premenopausal women. *J. Clin. Endocrinol. Metab.* 57:304–10

28. Feinleib, M. 1982. Summary of a workshop on cholesterol and noncardiovascular disease mortality. *Prev. Med.* 11:360–67

29. Frantz, I. D. Jr., Dawson, E. A., Ashman, P. L., Gatewood, L. C., Bartsch, G. E., et al. 1989. Test of effect of lipid lowering by diet on cardiovascular risk: the Minnesota Coronary Survey. *Arteriosclerosis* 9:129–35

30. Frick, M. H., Elo, O., Haapa, K., Heinonen, O. P., Heinsalmi, P., et al. 1987. Helsinki Heart Study: primary-

prevention trial with gemfibrozil in middle-aged men with dyslipidemia. *New Engl. J. Med.* 317:1237–45

31. Garcia-Palmieri, M. R., Sorlie, P. D., Costas, R. Jr., Havlik, R. J. 1981. An apparent inverse relationship between serum cholesterol and cancer mortality in Puerto Rico. *Am. J. Epidemiol.* 114:29–40

32. Gerhardsson, M., Rosenqvist, U., Ahlbom, A., Carlson, L. A. 1986. Serum cholesterol and cancer: a retrospective case-control study. *Int. J. Epidemiol.* 15:155–59

32a. Goldstein, J. L., Brown, M. S. 1990. Regulation of the mevalonate pathway. *Nature* 343:425–30

33. Gusberg, S. B., Runowicz, C. D. 1991. Gynecologic cancers. In *Clinical Oncology*, ed. A. I. Holleb, D. J. Fink, G. P. Murphy, pp. 498–97. Atlanta, Ga: Am. Cancer Soc.

34. Hiatt, R. A., Fireman, B. H. 1986. Serum cholesterol and the incidence of cancer in a large cohort. *J. Chronic Dis.* 39:861–70

35. Hiatt, R. A., Friedman, G. D., Bawol, R. D., Ury, H. K. 1982. Breast cancer and serum cholesterol. *J. Natl. Cancer Inst.* 68:885–89

36. Hjermann, I., Byre, K. V., Holme, I., Leren, P. 1981. Effect of diet and smoking intervention on the incidence of coronary heart disease: report from the Oslo study group of a randomised trial in healthy men. *Lancet* 2:1303–13

37. Holtzman, E. J., Yaari, S., Goldbourt, U. 1987. Serum cholesterol and the risk of colorectal cancer. *New Engl. J. Med.* 317:114

38. International Collaborative Group. 1982. Circulating cholesterol levels and risk of death from cancer in men aged 40 to 69 years: experience of an International Collaborative Group. *J. Am. Med. Assoc.* 248:2853–59

39. Isles, C. G., Hole, D. J., Gillis, C. R., Hawthorne, V. N., Lever, A. F. 1989. Plasma cholesterol, coronary heart disease, and cancer in the Renfrew and Paisley survey. *Br. Med. J.* 298:920–24

40. Ito, F., Takii, Y., Suzuki, J., Masamune, Y. 1982. Reversible inhibition by human serum lipoproteins of cell proliferation. *J. Cell Physiol.* 113:1–7

41. Kagan, A., McGee, D. L., Yano, K., Rhoads, G. G., Nomura, A. 1981. Serum cholesterol and mortality in a Japanese-American population: the Honolulu Heart Program. *Am. J. Epidemiol.* 114:11 20

42. Kark, J. D., Smith, A. H., Hames, C.

G. 1980. The relationship of serum cholesterol to the incidence of cancer in Evans County, Georgia. *J. Chronic Dis.* 33:311–22

43. Kark, J. D., Smith, A. H., Hames, C. G. 1982. Serum retinol and the inverse relationship between serum cholesterol and cancer. *Br. Med. J.* 284:152–54

44. Kark, J. D., Smith, A. H., Switzer, B. R., Hames, C. G. 1981. Retinol, carotene, and the cancer/cholesterol association. *Lancet* 1:1371

45. Katan, M. B. 1986. Apolipoprotein E isoforms, serum cholesterol, and cancer. *Lancet* 1:507–8

46. Kazanecki, M. E., Melhem, M. F., Spichty, K. J., Kelly, R. H., Rao, K. N. 1989. Diet-induced reduction in serum lipoproteins stimulates cell proliferation in weanling rats. *Pharmacol. Res.* 21:533–47

47. Keys, A. 1980. Alpha lipoprotein (HDL) cholesterol in the serum and the risk of coronary heart disease and death. *Lancet* 2:603–6

48. Keys, A., Aravanis, C., Blackburn, H., Buzina, R., Dontas, A. S., et al. 1985. Serum cholesterol and cancer mortality in the Seven Countries Study. *Am. J. Epidemiol.* 121:870–83

49. Deleted in proof.

50. Klimov, A. N., Shestov, D. B. 1986. Ischemic heart disease in a Leningrad subpopulation: prevalence and follow-up studies. In *Atherosclerosis VII*, ed. N. H. Fidge, P. J. Nestel, pp. 45–49. New York: Elsevier Sci.

51. Knekt, P., Reunanen, A., Aromaa, A., Heliövaara, M., Hakulinen, T., et al. 1988. Serum cholesterol and risk of cancer in a cohort of 39,000 men and women. *J. Clin. Epidemiol.* 41:519–30

52. Knekt, P., Reunanen, A., Teppo, L. 1991. Serum cholesterol concentration and risk of primary brain tumours. *Br. Med. J.* 302:90

53. Kozarevic, D., McGee, D., Vojvodic, N., Gordon, T., Racic, Z., et al. 1981. Serum cholesterol and mortality: the Yugoslavia Cardiovascular Disease Study. *Am. J. Epidemiol.* 114:21–28

54. Kritchevsky, S. B. 1992. Dietary lipids and the low blood cholesterol-cancer association. *Am. J. Epidemiol.* In press

55. Kritchevsky, S. B., Wilcosky, T. C., Morris, D. L., Truong, K. N., Tyroler, H. A. 1991. Changes in plasma lipid and lipoprotein cholesterol and weight prior to the diagnosis of cancer. *Cancer Res.* 51:3198–3203

56. Kromhout, D., Bosschieter, E. B., Drijver, M., Coulander, C. 1988. Serum

cholesterol and 25-year incidence of and mortality from myocardial infarction and cancer: the Zutphen Study. *Arch. Intern. Med.* 148:1051–55

57. Lannerstad, O., Isacsson, S.-O., Lindell, S. 1979. Risk factors for premature death in men 56–60 years old. *Scand. J. Soc. Med.* 7:41–47

58. Laskarzewski, P., Khoury, P., Morrison, J. A., Kelly, K., Mellies, M., et al. 1982. Cancer, cholesterol and lipoprotein cholesterols. *Prev. Med.* 11:253–68

59. Leibel, R. L., Edens, N. K., Fried, S. K. 1989. Physiologic basis for the control of body fat distribution in humans. *Annu. Rev. Nutr.* 9:417–43

60. Leren, P. 1966. The effect of plasma cholesterol lowering diet in male survivors of myocardial infarction. *Acta Med. Scand. Suppl.* 466:1–92

61. Leren, P. 1970. The Oslo Diet-Heart Study: eleven year report. *Circulation* 42:935–42

62. Lipid Research Clinics Program. 1984. The Lipid Research Clinics Coronary Primary Prevention Trial Results: I. reduction in incidence of coronary heart disease. *J. Am. Med. Assoc.* 251:351–64

63. Macy, M., Okano, Y., Cardin, A. D., Avila, E. M., Harmony, J. A. K. 1983. Suppression of lymphocyte activation by plasma lipoproteins. *Cancer Res.* 43:2496s–2502s

64. Mannes, G. A., Maier, A., Thieme, C., Wiebecke, B., Paumgartner, G. 1986. Relation between the frequency of colorectal adenoma and the serum cholesterol level. *New Engl. J. Med.* 315:1634–38

65. Marenah, C. B., Lewis, B., Hassall, D., La Ville, A., Cortese, C., et al. 1983. Hypocholesterolaemia and noncardiovascular disease: metabolic studies on subjects with low plasma cholesterol concentrations. *Br. Med. J.* 286:1603–6

66. McMichael, A. J. 1991. Serum cholesterol and human cancer. In *Cancer and Nutrition*, ed. R. B. Alfin-Slater, D. Kritchevsky, pp. 141–58. New York: Plenum

67. McMichael, A. J., Jensen, O. M., Parkin, D. M., Zaridze, D. G. 1984. Dietary and endogenous cholesterol and human cancer. *Epidemiol. Rev.* 6:192–216

68. Merrill, A. H. Jr., Foltz, A. T., McCormick, D. B. 1991. Vitamins and cancer. See Ref. 66, pp. 261–320

69. Morris, D. L., Borhani, N. O., Fitzsimons, E., Hardy, R. J., Hawkins, C. M., et al. 1983. Serum cholesterol and

cancer in the Hypertension Detection and Follow-up Program. *Cancer* 52:1754–59

70. Morse, J. H., Witte, L. D., Goodman, D. S. 1977. Inhibition of lymphocyte proliferation stimulated by lectins and allogeneic cells by normal plasma lipoproteins. *J. Exp. Med* 146:1791–1803

71. Multiple Risk Factor Intervention Trial Research Group. 1982. Multiple Risk Factor Intervention Trial: risk factor changes and mortality results. *J. Am. Med. Assoc.* 248:1465–77

72. Multiple Risk Factor Intervention Trial Research Group. 1990. Mortality rates after 10.5 years for participants in the Multiple Risk Factor Intervention Trial: findings related to a priori hypotheses of the trial. *J. Am. Med. Assoc.* 263:1795–1801

73. Neugut, A. I., Johnsen, C. M., Fink, D. J. 1986. Serum cholesterol levels in adenomatous polyps and cancer of the colon: a case-control study. *J. Am. Med. Assoc.* 255:365–67

74. Oliver, M. F. 1981. Serum cholesterol—The knave of hearts or the joker. *Lancet* 2:1090–95

75. Pearce, M. L., Dayton, S. 1971. Incidence of cancer in men on a diet high in polyunsaturated fat. *Lancet* 1:464–67

76. Peterson, B., Trell, E. 1983. Premature mortality in middle-aged men: serum cholesterol as risk factor. *Klin. Wochenschr.* 63:795–801

77. Peterson, B., Trell, E., Sternby, N. H. 1981. Low cholesterol level as risk factor for noncoronary death in middle-aged men. *J. Am. Med. Assoc.* 245:2056–57

78. Peto, R., Borenham, J., Chen, J., Li, J., Campbell, T. C., et al. 1989. Plasma cholesterol, coronary heart disease, and cancer. *Br. Med. J.* 298:1249

78a. Rao, K. N., Melhem, M. F., Gabriel, H. F., Eskander, E. D., Kazenecki, M. E., et al. 1988. Lipid composition and de novo cholesterolgenesis in normal and neoplastic rat mammary tissues. *J. Natl. Cancer Inst.* 80:1248–53

79. Reed, D., Yano, K., Kagan, A. 1986. Lipids and lipoproteins as predictors of coronary heart disease, stroke and cancer in the Honolulu Heart Program. *Am. J. Med.* 80:871–78

80. Reed, T., Wagener, D. K., Donahue, R. P., Kuller, L. H. 1986. Family history of cancer related to cholesterol level in young adults. *Gen. Epidemiol.* 3:63–71

81. Rose, G., Blackburn, H., Keys, A., Taylor, H. L., Kannel, W. B., et al.

1974. Colon cancer and blood-cholesterol. *Lancet* 1:181–83
82. Rose, G., Shipley, M. J. 1980. Plasma lipids and mortality: a source of error. *Lancet* 1:523–26
83. Salmond, C. E., Beaglehole, R., Prior, I. A. M. 1985. Are low cholesterol values associated with excess mortality? *Br. Med. J.* 290:422–24
84. Salonen, J. T. 1982. Risk of cancer and death in relation to serum cholesterol: a longitudinal study in an Eastern Finnish population with high overall cholesterol level. *Am. J. Epidemiol.* 116:622–30
85. Deleted in proof
86. Schatzkin, A., Hoover, R. N., Taylor, P. R., Ziegler, R. G., Carter, C. L., et al. 1987. Serum cholesterol and cancer in the NHANES I Epidemiologic Followup Study. *Lancet* 2:298–301
87. Schatzkin, A., Hoover, R. N., Taylor, P. R., Ziegler, R. G., Carter, C. L., et al. 1988. Site-specific analysis of total serum cholesterol and incident cancer in the National Health and Nutrition Examination Survey I Epidemiologic Follow-up Study. *Cancer Res.* 48:452–58
88. Schrott, H. G., Clarke, W. R., Wiebe, D. A., Connor, W. E., Lauer, R. M. 1979. Increased coronary mortality in relatives of hypercholesterolemic school children: the Muscatine study. *Circulation* 59:320–26
89. Deleted in proof
90. Shaper, A. G., Phillips, A. N., Pocock, S. J. 1989. Plasma cholesterol, coronary heart disease, and cancer. *Br. Med. J.* 298:1381
91. Sherwin, R. W., Wentworth, D. N., Cutler, J. A., Hulley, S. B., Kuller, L. H., et al. 1987. Serum cholesterol levels and cancer mortality in 361 662 men screened for the Multiple Risk Factor Intervention Trial. *J. Am. Med. Assoc.* 257:943–48
92. Shestov, D. 1981. Results of four years of follow-up of USSR populations. In *USA-USSR 1st Lipoprotein Symp., Leningrad, May 26–27*, pp. 391–407. Bethesda, Md: Natl. Inst. Health
93. Shinitzky, M. 1984. Membrane fluidity in malignancy: adversative and recuperative. *Biochim. Biophys. Acta* 738:251–61
94. Sidney, S., Friedman, G. D., Hiatt, R. A. 1986. Serum cholesterol and large bowel cancer. *Am. J. Epidemiol.* 124:33–38
95. Siperstein, M. D., DoVale, H., Silber, J. R. 1987. Relationship of cholesterol to DNA synthesis in normal and cancerous cells. In *Drugs Affecting Lipid*

Metabolism, ed. R. Paoletti, D. Kritchevsky, W. L. Holmes, pp. 1–8. Berlin: Springer-Verlag
96. Smith, A. H., Hoggard, B. M. 1981. Retinol, carotene, and the cancer/cholesterol association. *Lancet* 1:1371–72
97. Smith, G. D., Shipley, M. J. 1989. Plasma cholesterol concentration and primary brain tumours. *Br. Med. J.* 299:26–27
98. Sorlie, P. D., Feinleib, M. 1982. The serum cholesterol-cancer relationship: an analysis of time trends in the Framingham Study. *J. Natl. Cancer Inst.* 69:989–96
99. Staessen, J., Amery, A., Birkenhäger, W., Bulpitt, C., Clement, D., et al. 1990. Is a high serum cholesterol level associated with longer survival in elderly hypertensives? *J. Hypertens.* 8:755–61
100. Stähelin, H. B., Rösel, F., Buess, E., Brubacher, G. 1984. Cancer, vitamins and plasma lipids: prospective Basel Study. *J. Natl. Cancer Inst.* 73:1463–68
101. Stefanick, M. L., Williams, P. T., Krauss, R. M., Terry, R. B., Vranizan, K. M., et al. 1987. Relationships of plasma estradiol, testosterone, and sex hormone-binding globulin with lipoproteins, apolipoproteins, and high density lipoprotein subfractions in men. *J. Clin. Endocrinol. Metab.* 64:723–43
102. Stemmermann, G. N., Chyou, P., Kagan, A., Nomura, A. M. Y., Yano, K. 1991. Serum cholesterol and mortality among Japanese-American men: the Honolulu (Hawaii) Heart Program. *Arch. Intern. Med.* 151:969–72
103. Stemmermann, G. N., Nomura, A. M. Y., Heilbrun, L. K., Pollack, E. S., Kagan, A. 1981. Serum cholesterol and colon cancer incidence in Hawaiian Japanese Men. *J. Natl. Cancer Inst.* 67:1179–82
104. Strandberg, T. E., Salomaa, V. V., Naukkarinen, V. A., Vanhanen, H. T., Sarna, S. J., et al. 1991. Long-term mortality after 5-year multifactorial primary prevention of cardiovascular diseases in middle-aged men. *J. Am. Med. Assoc.* 266:1225–29
105. Thomas, C. B., Duszynski, K. R., Shaffer, J. W. 1982. Cholesterol levels in young adulthood and subsequent cancer: a preliminary note. *Johns Hopkins Med. J.* 150:89–94
106. Thurnham, D. I. 1989. Lutein, cholesterol and risk of cancer. *Lancet* 2:441–42
107. Törnberg, S. A., Carstensen, J. M., Holm, L.-E. 1988. Risk of stomach can-

cer in association with serum cholesterol and beta-lipoprotein. *Acta Oncol.* 27:39–42

108. Törnberg, S. A., Holm, L.-E., Carstensen, J. M. 1988. Breast cancer risk in relation to serum cholesterol, serum beta-lipoprotein, height, weight, and blood pressure. *Acta Oncol.* 27:31–37

109. Törnberg, S. A., Holm, L.-E., Carstensen, J. M., Eklund, G.. A. 1986. Risks of cancer of the colon and rectum in relation to serum cholesterol and beta-lipoprotein. *New Engl. J. Med.* 315:1629–33

110. Törnberg, S. A., Holm, L.-E., Carstensen, J. M., Eklund, G. A. 1989. Cancer incidence and cancer mortality in relation to serum cholesterol. *J. Natl. Cancer Inst.* 81:1917–21

110a. Ueyama, Y., Matsuzawa, Y., Yamashita, S., Funahashi, T., Sakai, N., et al. 1990. Hypocholesterolaemic factor from gallbladder cancer cells. *Lancet* 336:707–9

111. Vatten, L. J., Foss, O. P. 1990. Total serum cholesterol and triglycerides and risk of breast cancer: a prospective study of 24,329 Norwegian women. *Cancer Res.* 50:2341–46

112. Wald, N. J., Thompson, S. G., Law, M. R., Densem, J. W., Bailey, A. 1989. Serum cholesterol and subsequent risk of cancer: results from the BUPA study. *Br. J. Cancer.* 59:936–38

113. Wallace, R. B., Rost, C., Burmeister, L. F., Pomrehn, P. R. 1982. Cancer incidence in humans: relationship to plasma lipids and relative weight. *J. Natl. Cancer Inst.* 68:915–18

114. Westlund, K., Nicolaysen, R. 1972. Ten-year mortality and morbidity related to serum cholesterol. *Scand. J. Clin. Lab. Invest.* 30(Suppl. 127):1–24

115. Williams, R. R., Sorlie, P. D., Feinleib, M., McNamara, P. M., Kannel, W. B., et al. 1981. Cancer incidence by levels of cholesterol. *J. Am. Med. Assoc.* 245:247–52

116. Wilson, P. W., Abbott, R. D., Castelli, W. P. 1988. High density lipoprotein cholesterol and mortality: The Framingham Heart Study. *Arteriosclerosis* 6:737–41

117. Winawer, S. J., Flehinger, B. J., Buchalter, J., Herbert, E., Shike, M. 1990. Declining serum cholesterol levels prior to diagnosis of colon cancer: a time-trend, case-control study. *J. Am. Med. Assoc.* 263:2083–85

118. Wingard, D. L., Criqui, M. H., Holdbrook, M. J., Barrett-Connor, E. 1984. Plasma cholesterol and cancer morbidity and mortality in an adult community. *J. Chronic Dis.* 37:401–6

119. Yaari, S., Goldbourt, U., Even-Zohar, S., Neufeld, H. N. 1981. Associations of serum high density lipoprotein and total cholesterol with total, cardiovascular, and cancer mortality in a 7-year prospective study of 10,000 men. *Lancet* 1:1011–15

120. Yamaguchi, N., Yamamura, J., Takahashi, K., Nakamura, R., Okubo, T. 1990. Familial history of cancer and dietary pattern, serum cholesterol, serum protein and blood hemoglobin. *Gan No Rinsho Japan.* n.v.:377–81. (In Japanese; from Medline abstr.)

Annu. Rev. Nutr. 1992. 12:417–41

LIPIDS IN HUMAN MILK AND INFANT FORMULAS

Robert G. Jensen, Ann M. Ferris, and Carol J. Lammi-Keefe

Department of Nutritional Sciences, University of Connecticut, Storrs, Connecticut 06269–4017

KEY WORDS: lipids, human milk, infant formulas, infant nutrition

CONTENTS

0199-9885/92/0715-0417$02.00

INTRODUCTION

Although the pace of research on lactation, human milk, and infant nutrition has slowed, work on milk lipids has continued because milk fatty acids can be altered by dietary as well as genetic factors. The lipids in human milk have been reviewed by Jensen through mid-1988 (51, 52). We have also compared the lipids in bovine and human milks (54). This review covers the period from mid-1988 through September 1991.

DETERMINATION OF MILK VOLUME

Test Weighing

In order to ascertain the quantities of nutrients conveyed to the infant in breast milk, the amount of milk must be determined. Estimation of 24-h milk intake is desirable, but often unattainable. Because of practical limitations, efforts have been made to lessen the problems. Arthur et al (3) measured milk intake of breast-fed infants by weighing the infant or the mother with a sensitive electronic balance. The method correlated highly ($r = 0.927$) with measurements done by direct test weighing and was validated in the field (43). Neville and colleagues (76, 79) reported that test weighing of the infant gives reliable estimates of breast milk intakes when an integrating electronic balance is employed and when done by motivated mothers in field conditions.

Deuterium Oxide Method

The deuterium oxide elimination method for measuring average daily milk intake has been validated (27). Deuterium oxide was given orally to the infants and isotope dilution was measured in the urine.

Doubly Labelled Water Method

The doubly labelled water method has made it possible to study the energy content of human milk without sampling the product (67, 68). Measured water output is equivalent to water (milk or formula) intake. Twenty-nine infants at 3 months of age consumed 892 ml of formula per day that had a metabolizable energy content of 66 kcal/dl. The calculated energy was 68 kcal/dl. At 6 weeks of age, the metabolizable energy content for formula-fed infants was 60 kcal/dl. The amounts for breast-fed infants were 53 kcal/dl at 6 weeks and 58 kcal/dl at 3 months. These data are lower than the accepted value of 70 kcal/dl (73). The authors suggested that expressed milk, either by hand or pump, contains more fat than the infant actually consumes. This occurs because hind milk always contains more fat than fore milk. Since the fat content of breast milk must be known to determine the quantities of fatty acids consumed by the infant, sampling before and after nursing to determine the fat content should be done. The fatty acid composition is not affected by amounts of fat expressed.

SAMPLING OF MILK FOR FAT CONTENT

Milk samples that represent the composition of milk consumed over 24 h should be obtained. Since it is often impractical to take samples from every nursing during 24 h, Jackson et al (46) evaluated the accuracy of predicting the 24-h fat content from daytime samples. Samples (0.5 ml) were hand expressed before and after nursing, the fat contents were determined by the Creamatocrit procedure (51), and the amounts were averaged (47). If only two random daytime samples were taken, the 24-h concentration could be predicted with 95% confidence limits of 7.0 g per liter or 21% of average amount. It was preferable to sample all daytime feeds, which reduced the 95% confidence limits to $+/-$ 3.3 g per liter (10%).

An automatic sampling shield was developed and used to obtain milk samples while the baby was nursing (48). The unit did not measure milk volume. A series of small samples (averaged total of 1.3 g) were taken while the baby suckled. The shield was well accepted by the babies, but mean milk intakes were 17% lower than without the shield. Furthermore, it was shown that the fat content increased about 2-fold and was nonlinear at the beginning and end.

DETERMINATION OF LIPID CONTENT

Earlier work is presented in Reference (51). Procedures most widely used are the Creamatocrit in which the height of a packed volume of milk fat globule is measured and calibrated against a method employing solvent extraction followed by gravimetric determination of lipid. Among more recent developments is an automated enzymatic method for measurement of fat in human milk (69). The method, which requires only 25 μl of milk, can be used in commercial sequential analyzers. Glycerol released from milk triacylglycerol (TG) is quantitated after enzymatic treatment. The enzymatic procedure was checked with the dry column procedure and gave similar results on milks with various fat contents (19).

A rapid procedure for quantitation of fat, protein, and lactose in human milk by infrared analysis has been reported (71). The procedure is used in the dairy industry. The equipment is expensive, hence not readily available, and must be recalibrated for human milk. It should be useful for selection of high protein milks by milk banks as was done by Michaelsen et al (71). Clark & Roche (18) have refined the gas-liquid chromatography (GLC) of total lipids in 100 μl of human milk. The amounts obtained were essentially equal ($r =$

0.99) to the quantities extracted by the dry column method. The fatty acid composition and lipid content of 100 μl of human milk can be determined simultaneously.

Collins et al (19) investigated the extraction, separation, and quantitation of human milk lipids by the dry column method. Milk (1 ml) was applied directly to the column (mainly Celite) and extracted. One milliliter of milk was separated into neutral and polar lipids, but there was some carryover of TG in the polar fraction. The completeness of recovery of TG, cholesterol, phospholipids, and vitamin E was comparable to that attained by the Folch extraction (51).

FACTORS AFFECTING TOTAL LIPID CONTENT

The increase in fat content that occurs during a nursing was confirmed (46, 48). A circadian rhythm was found in the milk from women in northern Thailand (47), but not in milk from women in the United States (65). The lack of an effect was attributed to individuality. The time elapsed between nursings influences the fat content. As the interval lengthened, the fat content of the subsequent nursing decreased (47). The time elapsed from birth or age postpartum influences the fat content. The amount of lipid increased from 3.98% at 12 weeks to 5.50% at 16 weeks; energy increased from 68.5 to 83.0 kcal/dl (25). The volume and nutrient composition of milk was determined during lactogenesis (1–8 days postpartum) and weaning (6 to 15 months postpartum) by Neville et al (78). Fat content and secretion rates (grams per day) increased during lactogenesis while both decreased after the initiation of weaning. Allen et al (2) examined the macronutrients and their daily secretion rates in the first year of lactation. Michaelsen and colleagues (70) observed a slight decrease in the fat content from 0 to 4 months postpartum, followed by an increase up to 17 months. The mean content was 3.9%. In contrast, the fat content in milk from California mothers increased only slightly from 3 months (3.62%) to 12 months (3.72%) while the volume consumed by the infant (grams per day) decreased from 811 to 514 ml (80). Allen et al (2) saw increases in fat content and amounts secreted from days 21 to 180 postpartum.

Few controlled investigations are available on changes that occur in the fat content of milk as a result of alterations in diet (51, 52). In general, the amounts of fat, protein, and lactose are little changed by variations in diet. Allen et al (2) reported that undernutrition can reduce the fat content by 25%, but they did not provide supporting data. Hachey et al (31) noted that the fat content of milk from 5 women was 2.5% on a low fat diet (5% fat, 80% carbohydrate) as compared to 3.3% on a high fat diet (40% fat, 45% carbohydrate). Nommsen et al (80) did not observe any relationship between milk fat content and maternal fat intake either in terms of total fat intake or percent of calorie intake. They did note a significant relation between milk fat content and maternal protein content in the later stages of lactation. Silber et al (88) fed diets high in carbohydrate (5% fat, 15% protein, 80% CHO) to 10 lactating mothers. After 5 days they found relative and absolute increases in the amounts of 10:0, 12:0, and 14:0 and decreases in 18:0, 18:1, and 18:2. The responses were similar in milk from mothers of term and preterm infants.

The effects of several diseases on milk lipids have been examined and reviewed (37). The milk from one diabetic patient contained 2.95% fat; control milks (n = 13) contained 4.53% fat at days 6–7 postpartum (7). However, the fat content in milk from diabetics later in lactation did not differ much from that of the controls (49, 64). The fat content of milk from two patients with cystic fibrosis was normal when the disease was mild but was reduced when it became severe (8, 87).

The milk from a patient with abetalipoproteinemia was low in TG (1.5%) at 2 weeks postpartum but was equivalent to normal concentrations later in lactation (103). Chylomicrons and VLDL, which are the major carriers of dietary fatty acids to the mammary gland, are completely absent in the syndrome.

The effect of the mother's level of adiposity has a positive influence on the level of fat in milk (51, 52). Nommsen et al (80) obtained a positive and significant correlation between milk lipid concentration and maternal ideal body weight at 6, 9, and 12 months postpartum. Volume of milk was not a factor, although Michaelsen et al (70) saw a higher fat content in mothers who produced large amounts of milk.

Parity, 4 or higher, reduced the fat content of milk (51, 52). Corroboration at different levels of parity was provided by Nommsen et al (80) and Michaelsen et al (70). Seasonal effects may be related to parity and region in the populations studied. However, few data are available.

Lipogenesis is the ultimate major factor controlling the fat content in human milk. Unfortunately very little information is available on this aspect. Neville (77) noted that "the regulation of the total fat content of human milk is less well understood, in part because the lipid content of secreted milk is extremely difficult to measure accurately." The fat in milk is derived from fat in the blood or by synthesis in the mammary gland. The sources of fat in the blood are the diet via the gut, the liver, and adipose tissue. Dietary fatty acids are transported to the mammary gland in chylomicrons from the intestine and VLDL from the liver while adipose tissue fatty acids travel bound to albumin. In the mammary gland, fatty acid synthesis is limited to 10:0–14:0 using glucose as the source of carbon. Diet and adipose are sources of 16:0 and longer fatty acids.

Peak amounts of dietary fatty acids appear in milk about 8–10 h after consumption with an exponential decay from 48 to 72 h (32). About 10% of the total acids, primarily 10:0 to 14:0, are synthesized in the mammary gland, 60% are derived from tissue synthesis and adipose tissue, and 30% come from the diet. In this study (32), the percentages of kcal in the diet were protein, 17; carbohydrate, 56; and fat, 27. Later, Hachey et al (31) determined the effects of low and high fat diets on the synthesis of endogenous fatty acids by the mammary gland. Some of these data are presented above in the section on diet. The relative amounts of dietary fat and carbohydrate influence the amounts of 10:0–14:0 synthesized in the mammary gland. Very little production of 16:0 and 18:0 occurred in the high fat diet, but the amounts were increased 6-fold in the low fat diet. One women had a substantial decrease in body fat after changing from a low to a high fat diet, but no changes were observed in the fat content or amounts of 10:0–14:0 in her milk. In continuations of these studies, Silber et al (88) confirmed the influence of dietary carbohydrate. Emken et al (23) showed that milk TG, phospholipids, and cholesteryl esters were synthesized from the same fatty acid pool, but the influence of dietary fatty acids was greater for TG. Hachey et al (30) reported that on the low-high fat regimens mentioned above, consumption of the low fat diet produced milk with a fat content of 2.2% while consumption of the high fat diet produced milk with a fat content of 3.28%. The amounts of fat produced per day per breast, however, were almost the same: 8.0 g and 7.74 g. Milk fat content was inversely correlated with body water flux, i.e. volume adjusted for fat content.

LIPID CLASSES

Introduction

Information on the general composition of human and bovine milk (54) and several infant formulas is given in Table 1. The data were obtained from the manufacturers of these products and from Reference 96. The lipids in human milk are mostly TGs (98 + %), cholesterol (0.4%), phospholipids (1.3%), and traces of other lipids (51, 52, 54). Small amounts (<5%) of 3-chloropropanediol in the alkyldiacylglycerols and monoalkylglycerols have been detected in bovine and human milk lipids (63). The lipid composition of milk from mothers with diabetes (64) and cystic fibrosis (8) has been reported. Several papers on milk sterols have appeared (10, 19, 25, 56, 65).

Triacylglycerols

Some papers have been published on the structure of TG in human milk. Dotson et al (22) separated and tentatively identified TGs by high performance liquid chromatograph (HPLC). The major TGs were composed of 16:0, 18:1, and 18:2 regardless of variations in dietary fat. The kinds and amounts of major TGs are listed in Table 2. The identifications in Table 2 are

Table 1   General composition (%) of human and bovine milks and some infant formulas

	Milk[a]		Formulas[b]				
Component	Human (36 days)	Bovine	1	2	3	4	5
Protein	1.0	3.4	1.52	1.5	1.5	1.6	1.5
Casein, % protein	40	82	40	82	40	0	40
Fat	3.9	3.3	3.8	3.6	3.6	3.3	3.4
Lactose	6.8	4.8	7.0	7.2	7.2	7.2	7.6
Kilocalories/dl	72	75	68	68	68	65	67

[a] Adapted from Reference 54.
[b] Adapted from Reference 96, and personal correspondence. 1 = Enfamil, Mead Johnson; 2 = Similac, Ross; 3 = SMA, Wyeth; 4 = Good Start, Carnation, Reconstituted; protein is hydrolyzed whey; contains lactose and maltodextrin; 5 = NAN, Nestle. All are for healthy term infants and the proteins in all except 4 are bovine milk whey and casein.

carbon numbers only and do not indicate position of the fatty acids. For example, $16:0-18:1-18:1$ represents $16:0-18:1-18:1$, $18:1-18:1-16:0$, and $18:1-16:0-18:1$. However, we know that most of the $16:0$ is located at sn-2, so there will be more $18:1-16:0-18:1$ (51, 52). The 8 TGs in Table 2 totaled 89.5%, but 19 more were detected. Milk TGs were separated by carbon number with capillary GLC in the search for chloropropanediol diesters (63), and TGs were analyzed in milk from a patient with abetalipoproleinemia (103). The amounts of lower molecular weight TGs were greater than normal in these milks.

Phospholipids

Phospholipids originate from the plasma membrane of the secreting cell. The major classes are (wt%): phosphatidyl choline, 30.0; phosphatidyl ethanolamine, 28.0; and sphingomyelin, 32 (51, 52). Smaller amounts of phosphatidyl serine and inositol, along with cerebrosides and gangliosides, have been found. The gangliosides bind enterotoxins. Milk from cystic fibrosis patients contained more sphingomyelin and less phosphatidylethanolamine than normal milk (8, 36).

van Beusekom et al (98) found a positive relationship between the $6:0-14:0$ contents of total milk lipids and total phospholipids. Incorporation of $6:0-14:0$ into phospholipids was favored over polyunsaturated fatty acids (PUFAs) in the mammary gland. Diets rich in CHO increased secretion of $6:0-14:0$ into phospholipids at the expense of PUFAs.

Sterols

Earlier research has established that the sterol content of human milk ranges from 10 to 20 mg/dl with cholesterol as the major component. Most of the

Table 2   Major triacylglycerols in human milk[a]

TG	Area %	TG	Area %
14:0–14:0–12:0[b]	2.62	16:0–12:0–18:1	4.84
18:2–18:2–18:1	3.27	12:0–16:0–16:0	14.69
18:1–14:0–14:0	4.70	12:0–18:0–18:1	6.21
16:0–14:0–14:0	4.47	18:1–18:1–18:1	3.17
18:2–18:2–16:0	8.84	16:0–16:0–18:1	11.96
18:2–16:0–12:0	6.07	18:0–18:0–18:1	7.38
12:0–18:1–18:1	1.99	18:0–18:0–18:1	3.44

[a] Adapted from Reference 22.
[b] 14:0–14:0–12:0 represents the isomeric TGs 14:0–14:0–12:0, 12:0–14:0–14:0, and 14:0–12:0–14:0.

cholesterol is located in the milk fat globule membrane and the amount is not affected by diet (51, 52). Cholesterol is not present in formulas unless animal fats are added. Kallio et al (56) analyzed milk for cholesterol and its precursors by GLC and mass spectrometry. At 2 months postpartum values (μg/dl) were: cholesterol, 15800; desmosterol, 1509; squalene, 386; lanosterol, 94; methosterol, 48; dimethylsterol, 45; and lathosterol, 45. Other sterols detected in small and unreported amounts were cholestanol, campesterol, and sitosterol. Data are also given for milks at 6 and 9 months postpartum.

Lammi-Keefe et al (65) detected a diurnal pattern in the cholesterol content of milk—ranging from 8.75 mg/dl at 0600 h to 11.2 mg/dl at 2200 h. Clark & Hundrieser (17) found that the mean total cholesterol content of 25 milk samples was 13.5 mg/dl and was significantly correlated with the lipid content. The cholesteryl ester content was about 20%. Collins et al (19) found 10.58 mg/dl total cholesterol in 12 samples. The composition of the esters is given in (17). Emken et al (23) determined that the influence of dietary fatty acids on cholesteryl esters occurred at 8 to 10 h after consumption, the same as for TG and phospholipids. Boersma et al (10) measured cholesterol in milk from women in St. Lucia as follows: 36.0 mg/dl at 0 to 4 days, 19.7 mg/dl at 5 to 9 days, and 19.0 mg/dl at 10 to 30 days postpartum.

FATTY ACIDS

Introduction

Most of the earlier data listed in Table 3 were obtained by GLC instruments equipped with packed columns (94). These columns are incapable of the resolution attainable with wide-bore capillary columns of suitable length (at least 15 m, preferably 30 m), coated with polar stationary phases and utilizing temperature programming. Recently, newer stationary phases have been used to separate *trans* isomers and PUFAs. Details of operation are in the papers

Table 3 Saturated fatty acid (wt%) in human milk lipids

Fatty acid	Reference number[a]							
	94	60	85	10	22	22	86	86
		$n = 15$	$n = 23$	$n = 12$	$n = 30$	CV[b]		±SEM[b]
4:0	0.19	—	—	—	—	—	—	—
6:0	0.15	—	—	0.07	—	—	—	—
8:0	0.46	—	—	0.37	0.02	0.04	0.03	—
10:0	1.03	0.71	0.92	2.39	0.06	0.05	1.4	—
11:0	—	—	—	—	—	—	—	—
12:0	4.40	4.41	6.99	12.32	4.34	0.11	6.2	—
13:0	0.06	0.05	—	—	0.05	0.14	—	—
i-14:0[c]	0.04	—	—	—	—	—	—	—
14:0	6.27	6.73	8.80	11.78	4.65	0.07	7.6	—
a-15:0[c]	0.21	—	—	—	—	—	—	—
15:0	0.43	0.46	—	—	0.34	0.09	—	—
i-16:0	0.17	—	—	—	—	—	—	—
16:0	22.00	21.83	14.10	23.61	19.25	0.03	20.5	0.70
a-17:0	0.23	—	—	—	—	—	—	—
17:0	0.58	0.57	—	—	0.43	0.07	—	—
i-18:0	0.11	—	—	—	—	—	—	—
18:0	8.06	8.15	3.94	5.83	7.97	0.06	9.0	0.46
19:0	—	0.04	—	—	0.10	0.07	—	—
20:0	0.44	0.22	0.47	0.24	0.21	0.10	0.3	0.02
21:0	0.13	0.35	—	—	—	—	—	—
22:0	0.12	0.09	0.30	0.12	0.12	0.06	—	—
23:0	—	—	—	—	0.03	0.02	—	—
24:0	0.25	—	—	0.07	0.22	0.14	0.5	0.01
26:0	—	—	—	—	0.03	0.31	—	—
Total saturates	45.33	43.61	34.90	56.8	37.82	—	38.6	0.72

[a] Normalized data from 15 papers (94). Mature milks from German donors (60), Gambian mothers (85), St. Lucian mothers (10), mothers from Illinois (22) and Florida (86).
[b] CV, coefficient of variation; SEM, standard error of mean.
[c] i is iso a, anteiso.

quoted below. In addition, we urge investigators to determine the fat content of milk and formulas, so that the actual amounts of fatty acids conveyed to the infant can be calculated, and to present their data as weight percent (grams per 100 g fatty acid) and as weight of fatty acid per deciliter of milk (51, 52).

Fatty acids are converted to methyl esters prior to analysis by GLC to increase volatility and efficiency of separation. Methanolysis of the extracted fat is usually done by reactions with sodium hydroxide-methanol or boron-trifluoride methanol. Lepage & Roy (66) observed slightly better recovery of human milk fatty acids with a direct transesterification procedure they developed. However, Bitman & Wood (9) did not find any differences in 35 milk fatty acid profiles when they compared direct transesterification with methanolysis by boron trifluoride-methanol. Many investigators add anti-oxidants to the solvents to prevent possible loss of PUFA by oxidation (99).

When analyzing human milk lipids for fatty acids, we recommend that investigators (*a*) use columns that will resolve *trans* isomers and PUFA, (*b*) determine and report the total lipid content, and (*c*) if possible, obtain information on maternal diets. Analysts may want to investigate the bracketing procedure used by van der Steege and co-workers (99). Internal standards were employed to quantitate the fatty acids. Data on within series and series-to-series precision and biological variation are given in this study.

Classes of Fatty Acids

Normalized older values from 15 studies are listed in Tables 3–5 (94). Most of these data were obtained by GLC analyses with packed columns and in some cases with prior separation of the fatty acid classes. The resolution and level of detection were not as good as observed in the later studies using capillary columns and electronic integrators (10, 22, 60, 85, 86, 98, 99). The data from an earlier study on Floridian milks (86) are given for comparison and also to exemplify the remarkable absence of a large data base on the fatty acid profiles of human milk obtained with the latest methods.

A few other recent publications report on changes due to age postpartum

Table 4 Monounsaturated fatty acid (wt%) in human milk lipids

Fatty acid	Reference number[a]							
	94	60	85	10	22	22	86	86
		n = 15	*n* = 23	*n* = 12	*n* = 30	CV		±SEM
13:1	—	—	—	—	0.03	0.10	—	—
c- 14:1n5[b]	0.41	0.29	0.23	0.29	0.31	0.20	—	—
t- 14:n5[b]	0.07	0.19	—	—	—	—	—	—
15:1	0.11	—	—	—	0.09	0.09	—	—
c- 16:1n7	3.29	2.68	0.66	3.55	2.58	0.07	—	—
t- 16:1n7	0.36	0.46	—	—	—	—	—	—
17:1	0.37	0.32	—	—	0.33	0.07	—	—
c- 18:1n7	—	—	—	3.64	—	—	—	—
c- 18:1n9	31.30	34.31[b]	47.0	22.63	33.23	0.04	37.6	0.75
t- 18:1n9	2.67	3.12	—	—	4.72	0.07	—	—
20:1n11	—	—	—	—	0.17	0.20	—	—
20:1n9	0.67	—	0.83	0.42	0.38	0.11	0.9	0.7
21:1n9	—	—	—	—	0.01	0.29	—	—
22:1n9	0.08	0.08	0.22	—	0.07	0.05	0.1	0.02
24:1n9	0.12	—	0.05	0.04	0.03	0.07	—	—
Total monoenes	39.45	42.38	48.80	30.6	41.95	—	38.5	—

[a] See Table 3.
[b] c is *cis;* t, *trans.*

Table 5 Polyunsaturated fatty acid (wt%) in human milk lipids

Fatty acid	Reference number[a]							
	94	60	85	10	22	22	86	86
n6 series		$n = 65$	$n = 23$	$n = 12$	$n = 30$	CV	—	±SEM
18:2cc	10.85	10.76	13.0	9.57	15.55	0.05	15.8	0.61
18:2tt	0.46	0.14	—	—	—	—	—	—
18:2ct	0.69	0.14	—	—	—	—	—	—
18:2tc	—	0.07	—	—	—	—	—	—
18:3	0.25	0.16	—	0.09	0.18	0.08	—	—
19:2	—	—	—	—	0.51	0.05	—	—
20:2	0.27	0.34	0.83	0.31	0.38	0.04	0.4	0.03
20:3	0.32	0.26	0.21	0.42[b]	0.46	0.05	0.4	0.03
20:4	0.46	0.36	0.31	0.58	0.53	0.06	0.6	0.03
22:2	0.11	0.11	—	—	0.05	0.06	—	—
22:4	0.09	0.08	0.08	0.15	0.06	0.07	0.2	0.02
22:5	0.09	—	0.30	0.07	—	—	0.1	0.02
Total n6	13.59	12.26	14.7	11.2	17.72	—	17.4	0.62
n3 series								
18:3	1.03	0.81	0.84	0.62	1.11	0.08	0.8	0.09
20:3	—	0.06	—	—	0.03	0.13	—	—
20:4	0.09	—	—	—	—	—	—	—
20:5	0.12	0.04	—	0.07	0.07	0.12	0.1	0.03
22:3	—	—	—	—	0.13	0.05	—	—
22:5	0.19	0.17	0.20	0.16	—	—	0.1	0.01
22:6	0.25	0.22	0.39	0.56	0.16	0.08	0.1	0.01
Total n3	1.68	1.38	1.6	1.4	1.5	—	1.1	0.09
Total PUFA	15.27	13.64	16.3	15.6	19.22	—	17.4	0.62
Total n6/n3	8.10	9.23	9.19	8.0	11.8	—	18.8	1.2

[a] See Table 3.
[b] 0.05% of 20:3n3 not listed.

(10), diet and parity of rural Gambian mothers (85), dietary n3 fatty acids in milk from Inuit and Canadian women (44), effects of vegetarian and non-vegetarian diets (93), and fatty acids in milk from mothers in Tanzania, Curacao, and Surinam (72).

Diabetes increased the quantities of 10:0–14:0 in milk (49). Linoleic acid and its derivatives were low in milks from patients with cystic fibrosis, but other PUFAs were elevated (8). The contents of long-chain fatty acids in the milk from the women with abetalipoproteinemia were much lower than normal (103).

Human milk lipids contain more fatty acids than those listed in Tables 3–5, a total of 185 at the last count (54). Among these is 9,11–18:2ct, a conjugated fatty acid that is anticarcinogenic and a potent antioxidant (29, 54). We found an average content of 0.186% in 20 samples (55). The quantities found are dependent on the amounts of dietary bovine milk lipids and oxidized food lipids consumed. This compound is an example of an important nonnutritive role of milk lipids.

Human milk contains prostaglandins in very small quantities (51, 52). The prostaglandin content in milk from mothers of term babies was low and comparable to that of milk from mothers of preterm babies (75). Prostaglandins E2 and F2 were stable in human milk and gastric fluid. The compounds may protect and maintain the integrity of the intestinal epithelial cells in the developing infant (5).

MILK FAT GLOBULE EMULSION AND MEMBRANES

The Emulsion

Lipids in human milk are dispersed as emulsified globules composed mostly of nonpolar TG, cholesteryl esters, and other lipids in the core (51, 52, 54). Bipolar substances, proteins, phospholipids, and cholesterol from the plasma membrane of the secreting cell envelop the globule as it is extruded, forming an emulsion-stabilizing membrane. The globules have an average diameter of 4 μm, with a surface area of 1.4 m/g of fat and numbering about 1.1×10^{10} ml. There are almost no data on factors affecting the size and numbers of globules.

The Globule Membrane

The membranes are derived from apical plasma membrane of mammary epithelial cells (11, 51, 52, 54, 57). A hypothetical model (57) includes an inner layer of phospholipids with fatty acids oriented into the TG core (cholesterol will also be here). A second layer contains phospholipids, proteins, and glycolipids. Human globule membranes are coated with an array of glycoprotein filaments whose functions are unknown but may enhance digestion and release of TG fatty acids by binding lipases. The filaments are removed by heating the milk. Some of the globules have a cytoplasmic crescent (42, 84). Although quantitatively negligible, the crescents could contain trace elements, enzymes, hormones, and growth factors of importance to the infant.

FAT-SOLUBLE VITAMINS

Introduction

Since the fat-soluble vitamins A, D, E, and K are required for growth and development, the belief that human milk is the ideal food for infants has focused attention on their amounts and availability (51, 52 74a). Analyses of these vitamins, once arduous, have been eased by HPLC and other developments. Some of the analyses can be done quickly and reliably.

Retinoids and Carotenoids

The quantities of vitamin A (retinol) range from 40 to 70 μg/dl and of carotenoids from 20 to 40 μg/dl in well-nourished women in Europe and the United States (74a). Most of the retinol is esterified and the main carotenoid is beta-carotene. The bioavailability of carotenoids varies, so retinol equivalents (RE) are used. One RE is defined as 1 μg of all-*trans* retinol, 6 μg of all *trans* beta-carotene, or 12 μg of other provitamin A carotenoids. Vitamin A deficiency, which can lead to xerophthalmia, is rare in the US but is a major problem elsewhere affecting mostly children. The current RDA for infants 0 to 0.5 yr is 375 μg RE (74a).

Recent data are presented in Table 6 (58, 82, 83). All of these analyses were done by HPLC. Patton et al (83) observed a marked (10-fold) decrease in carotenoids and retinoids as lactation progresses. Patton et al (83) reported a drop in carotenoid contents 3 to 126 h after parturition. Lycopene, alpha-carotene, and lutein were detected in mature milks (58), and beta-cryptoxanthin and possibly zeaxanthin in colostrum (83).

Vitamin D

Earlier papers on vitamin D in human milk revealed a content of 0.63 to 1.25 μg per liter (51, 52, 74a). The Recommended Dietary Allowance (RDA) for infants 0.0–0.5 months postpartum is 7.5 μg and for their mothers is 10 μg. Since the content in human milk does not provide 7.5 μg, infants not exposed to sunlight may be at risk for vitamin D deficiency. Lack of the vitamin causes inadequate bone mineralization. However, Atkinson et al (4) found 80 IU per liter (2 μg) for preterm milk and 60 IU (1.5 μg) for term milk (Table 6). These quantities are higher than those quoted above. On the other hand, the quantities determined in Finnish fore milks ranged from 0.35 to 3.1 μg per liter depending on the season (1). The amounts were greater and the seasonal differences not so pronounced in hind milks. Supplementation of mothers with 25 or 50 μg of vitamin D significantly increased the amounts in February and April. Theoretically, the calculated antirachitic activity of milk in winter should have been increased by supplementation (50 μg) to levels of unsupplemented mothers in September, but responses were variable.

In metabolic studies, a maternal vegetarian diet was associated with increased serum 1,25-dihydroxyvitamin D during lactation (92), and unsupplemented breast-fed infants (n = 22) in Madison, Wisconsin, had no evidence of vitamin D deficiency during the first 6 months of life (28). The vitamin-D supplemented term infant fed human milk or cow milk or soy-based formula regulates mineral metabolism normally (41). The RDAs for

Table 6 Recent data on fat-soluble vitamins in human milk

Vitamin and milk type			Reference
Retinoids and carotenoids μg/dl			
Mature milk			
Retinol	52, 57		58, 82
Beta-carotene	23		51, 52
Colostrum			
Carotenoids, parity 1	114 ± 1.32		83
parity 2–3	218 ± 1.96	66 beta-carotene	83
D μg per liter[a]			
Preterm milk	2.0		4
Term milk	1.5		4
Winter foremilk	0.35		1
Summer foremilk	3.1		1
E mg/dl			
Preterm milk, 3 days	1.45		39[b]
Preterm milk, 36 days	0.29		39[b]
Term milk 3 days	1.14		39[b]
Term milk 36 days	0.28		39[b]
Mature milk	0.34		19
Colostrum, 0–4 days	2.2		10[c]
Transitional milk, 5–9 days	1.4		10
Mature milk, 10–30 days	0.8		10
K μg per liter			
Colostrum, 30–81 h	3.39[d]	Range: 2.3 to 7.6	12–14
Mature milk, 6 months	2.87[d]	Range: 2.1 to 9.28	12–14

[a] As activity of 1 μg of cholecalciferol, which is 40 IU.
[b] As alpha-tocopherol equivalents.
[c] Alpha-tocopherol.
[d] Difference between colostrum and mature milks is not significant.

maternal and infantile uptakes of vitamin D seem to be more than adequate (74a), but the large differences between the low amounts in breast milk and the RDAs suggest that breast-fed infants are at risk for vitamin D deficiency and should be supplemented. Preterm infants and babies who receive prolonged breast-feeding and insufficient exposure to sunlight are at higher risk and need greater supplementation.

Vitamin E

Vitamin E refers to a group of tocopherols, alpha, beta, gamma, and delta, that differ in biopotency (74a). Natural α-tocopherol is prefixed by RRR- (formerly d-). One milligram of this isomer is the RRR-tocopherol equivalent

(TE). The other forms have reduced biopotency. The compounds are antioxidants that slow peroxidation of PUFA in membrane phospholipids by trapping free radicals. Cellular damage and neurological symptoms can be prevented by adequate levels of tocopherols in the diet. Deficiencies can occur in premature, very low birth weight infants and in subjects who do not absorb fat properly. Infants may need 0.4 mg of tocopherol per gram of dietary PUFA, although a fixed ratio has not been established.

The concentrations of alpha-tocopherol in milk range from 3.0 to 4.5 mg per liter. These analyses were done by HPLC. The alpha, beta, gamma, and delta isomers have been resolved. The RDA (mg TE) is 3 for infants 0 to 0.5 yr and 12 for lactating women 0 to 0.54 yr (74a). Premature infants may need an oral supplementation of 17 mg per day (74a).

In a recent study, the vitamin E contents (medians and ranges) of milks at day 3 and 36 postpartum (mg TE)/dl were 1.45 (0.64–6.4) and 0.29 (0.19–0.86) preterm and 1.14 (0.63–4.21) and 0.28 (0.19–0.86) term (39). Collins et al (19) detected 0.34 mg of alpha-tocopherol per deciliter in mature fresh milk and 0.33 mg in milk stored for 2 weeks at −70°C. Total lipids were extracted by the dry column method (19). The data from (39) indicate that the needs of the preterm infant may not be met by breast-feeding. Haug et al (39) observed a decrease in the ratios of alpha- to gamma-tocopherol of 10:1 to 4:1 during the first 2 weeks of lactation. The ratio remained constant for 36 weeks. They did not detect delta-tocopherol nor was an increase in milk vitamin E seen when the mothers were given 50 mg per day for a week. Collins et al (19) noted high correlations between alpha-tocopherol and TG or cholesterol but not between alpha-tocopherol and phospholipid. Boersma et al (10) found 0.8 mg/dl of alpha-tocopherol in mature milks from women in St. Lucia. The postpartum decrease was 22 mg TE/dl at 0 to 4 days and 14 at 8 days postpartum. Alpha-tocopherol in human milk (mg/dl) was 0.6 in Curacao, 0.5 in Dominica, and 0.5 in Belize.

Vitamin K

HPLC has shown that human milk contains about 2 μg per liter of vitamin K (51, 52, 74a). The vitamin K denotes a group of compounds containing the 2-methyl-1,4-naphthoquinone moiety. Phylloquinone is the plant form of the vitamin, has a phytyl group at position 3, and is the most prevalent homolog in milk. The RDA for infants, 0 to 0.5 yr, is 5 μg per day, well above the amount in breast milk (74a). Newborn term infants in the US, are given 0.5 to 1 mg of vitamin K by intramuscular injection; preterm infants receive at least 1 mg. The RDA for lactating mothers is 65 μg per day. The vitamin is required for the biosynthesis of prothrombin and other blood clotting factors (74a). In vitamin K deficiency, abnormal proteins are formed. These can be detected in serum by a sensitive assay. Exclusively breast-fed infants are at risk for hemorrhagic disease of the newborn, hence the supplementation above. Synthesis of vitamin K (menaquinones) by microorganisms is minimal, and in newborns and preterm infants liver stores are low (12, 74a).

Most of the recent analyses of vitamin K in milk have been done by Canfield and associates (12–14). These data are shown in Table 6. Canfield & Hopkinson (12) reviewed the older data. The HPLC assay was described (13). With this method Canfield et al (13) found 2.94 ± 1.94 μg per liter in pooled milk and 3.15 ± 2.87 in individual milks. Later, Canfield et al (14) reported the quantities in colostrum (30–81 h) and milk in Table 6. Postpartum differences were not significant. The amounts of vitamin K in milk were not predicted by dietary intakes of vegetables or fat. The vitamin was located in the fat core of the globule and thus was not associated with the membrane. Canfield et al stated that the amounts of vitamin K are inadequate to meet the recommended intakes of infants <6 months of age. Human milk does not contain adequate vitamin K to prevent hemorrhagic disorders in neonates.

INFANT FORMULAS

The composition of the lipids in formulas is uniform because of legislation and the desire to imitate human milk (38, 91, 96). The formulas contain 3.3 to 3.8% fat (Table 1) and the total lipid is composed of 98 + % TG, 0.03 to 0.1% sterols, and about 0.2% phospholipids. The sterols are phytosterols unless animal fats, usually butter or destearinated tallow, are added. The sterols and phospholipids will include those that were in the oils and lecithin added as an emulsifier. All analyzed formulas contained phosphatidylcholine, and many contained sphingomyelin (105). The fat globules in these formulas are uniformly 0.3 μm in diameter, which results in a surface area 48.6 m^2 per deciliter based on a 3.6% fat content. This is about ten times more globule surface area than in human milk (51, 52).

The fatty acids in formulas represent the profiles in the added fats and oils. These are given as classes of fatty acids in Table 7. Additional information for European formulas is in Reference 59. The formula must contain at least 2.7 en% 18:2n6, and no formula can have less than 0.3 g per 100 kcal of this acid (91). Most formulas contain 18:3n3, but none contain any but trace amounts of 20:4:n6, 20:5n3, and 22:6n3 (59, 86, 91). Since 20:4n6, 20:5n3, and 22:6n3 in as yet unknown quantities (in addition to 18:2n6 and 18:3n3) appear to be needed for optimal growth and development, efforts are underway to determine the effects of adding these acids to formulas (91). Carroll (15) recommended that n6 PUFA should not exceed 20% of total fatty acids; 18:3n3 no more than 3%; and 20:5n3 + 22:6n3, no more than 1%. Total n3 acids should not exceed en% in standard formulas. Uauy (97) also made recommendations to be discussed later. If partially hydrogenated oils or butter

Table 7 Fatty acid class composition (%) of formulas[a]

| Fatty acid class | Formula[b] | | | | |
	Enfamil	Similac	SMA	Good Start	NAN
Saturates	58.2	45.7	44.2	44.6	50.1
Monounsaturates	14.2	17.2	41.3	33.2	36.3
Polyunsaturates	27.6	37.1	14.5	22.2	13.6
P/S ratio	0.47	0.81	0.33	0.50	0.27

[a] Compiled by authors from personal communications.
[b] Enfamil, Mead Johnson—coconut and soy oils; Similac, Ross—coconut and soy oils; SMA, Wyeth-Ayerst—oleo, high oleic, coconut, and soy oils; Good Start, Carnation—palm olein, high oleic, and coconut oils; NAN, Nestle—butter and corn oils.

are used in formulas, they will contain varying amounts of *trans* fatty acids. There are no reports of these in US formulas, but European products contained 0.2 to 4.6% (59). Carroll (15) recommended that *trans* fatty acids not be used, but if they are the quantity should be no more than 6% of total acids or 3 en%. The gangliosides, specifically GM1, in human and bovine milks and in infant formula bind enterotoxins and are part of the host defense systems (62).

NUTRITIONAL ASPECTS

Introduction

The components in milk are an interrelated system in which compartmentation is one of the factors controlling the flow of nutrients and metabolites to the breast-fed infant (53). These compartments and their interactions help control the temporal sequence of events leading to intestinal absorption of nutrients. The other factors controlling nutrient flow are the amount of milk, the surface area and topography of the fat globules, activity of relevant enzymes, and the status of the absorptive cells in the small intestine. A description of the physiological basis of infant feeding (104) and a discussion of lipids as an energy source for infants are available (21).

Absorption of Dietary Lipids

Few articles have been published in this area since the last reviews (51, 52). Milk and formula TGs are hydrolyzed in sequence, first by gastric lipase in the stomach (34, 35) and then by pancreatic colipase (90) and milk bile salt-stimulated lipase (BSSL) (40) in the small intestine. Formulas and bovine milk do not contain BSSL, which is synthesized in the human mammary gland and apparently enhances absorption of milk lipids. The enzyme is destroyed by heating at 56°C for 30 min and is nonspecific, also hydrolyzing retinyl and cholesteryl esters. Some of the fatty acids released and monoacylglycerols (MGs) formed during lipolysis of dietary fats in the stomach and small intestine are highly microbicidal, constituting one of the infant's many host defense systems (34, 40).

Milk fat globules must first be altered by gastric lipase before the core TGs can be digested by pancreatic colipase in the small intestine (6). Predigestion by gastric lipase compensates for the relatively low concentrations of pancreatic colipase and bile salts in infants as compared to children and adults (6). Most of the activity designated as lingual in earlier papers is of gastric origin in the human.

Calcium soaps of fatty acids are highly insoluble in water and may be excreted. Milk contains ionized calcium, and insoluble soaps can be formed in the gut. Absorption of calcium and fatty acids may be reduced in the small intestine. Jandacek (50) observed solubilization of calcium soaps of long-chain fatty acids by liquid fatty acids. The solubilities of Ca-16:0, Ca-12:0, and Ca-18:1 were 15.6, 22.8, and 53.3 wt% in 18:1 at 40°C. The solubility of Ca-18:1 in a bile salt micellar system was enhanced by 18:1. Solubilization may explain the high bioavailability of some Ca soaps.

The belief that the preponderance of 16:0 in the *sn*-2 position of milk TG is responsible for the almost complete absorption of human milk fat is based on two studies (26, 95). It was assumed that much of the 16:0 is present as the 2 MG after digestion. However, the experimental subjects in either study did not resemble the milk-infant dyad. Mixtures of oils were fed to rats by Tomarelli et al (95). Rats, unlike humans, do not have much gastric as compared to lingual lipase (35). Native and randomized lards were given to infants (26), but as a formula and presumably homogenized into the mixture, resulting in a large increase in globular surface area. In native lard, most of the *sn* 2 position contains 16:0, similar to the fatty acid in milk TG. The BSSL of human milk was not present in either investigation. Also the composition and topography of the globule membrane differs from that of milk. The complete digestion of milk TG (6) by the lipolytic sequence described above eliminates the relationship between TG structure and absorption because 2-monopalmitoylglycerol is not present in human milk, since all acylglycerols are digested.

Formulas containing different amounts of 16:0 at the *sn*-2 position of TGs did not affect absorption in preterm infants (100, 101). When preterm infants (26.5 to 37.5 weeks) were fed Almiron AB or a modified lard Almiron, neither the fat absorption coefficients, 69.6 and 58.6%, nor the energy absorption coefficients, 80.3 and 75.4%, were significantly different (100). Ninety three percent of the 16:0 was present at the *sn*-2 position of the TGs in the formula containing lard. A similar study (101) analyzed fecal, plasma, and erythrocyte lipids. The carbon numbers of the formula fecal TGs were also determined. Most (97–98%) of the fecal lipids from the infants fed either formula were free fatty acids (FFAs), whose compositions closely resembled those of the formula. The long-chain fatty acid profiles of plasma and erythrocyte membranes were also correlated with the formula fatty acids. The authors suggested that the fecal FFAs were the result of rapid transit of formula TG through the small intestine, reduced lipase activity in the small intestine, and lipolysis (microbial and residual pancreatic lipases?) and absence of absorptive cells in the colon. No differences in overall absorption of fat were observed, but TGs with 16:0 esterified at *sn*-2, similar to those of human milk, resulted in higher absorption of fatty acids. Arachidonic acid (20:4n6) and 20 other carbon unsaturates were preferentially absorbed, whereas some 16:0, 18:2n6, and 18:3n3 were not absorbed (101).

While all of the lipases described above hydrolyze medium-chain triglycerides (MCT, 8:0 and 10:0) more rapidly than long-chain fatty acid TG (34, 35, 94), absorption rates were equivalent (37). Substantial quantities of 8:0 and lesser amounts of 10:0 were apparently absorbed through the stomach wall. The authors (37) doubted that large amounts of MCT (40–50% of total fat) would improve fat absorption in preterm infants. They recommended a maximum of 10–15%. They noted that MCT acids act primarily as sources of energy and are not used in membranes, etc.

The Host Defense Effects of Lipids

PRODUCTS OF LIPOLYSIS   Protozoa, bacteria, and viruses are destroyed in vitro by lipolysis products: primarily 12:0, 18:2, and their MGs (40, 51, 52). In vivo, these compounds would quickly become available in the stomach and intestines of infants and reduce the colonization of microorganisms at these sites. Recent data (45) revealed inactivation of several enveloped viruses including HIV by human milk stored at 4°C for several days. Inactivation was caused by FFA released by serum lipoprotein lipase, which apparently leaks into milk. Dissolution of viruses also occurred when they were exposed to stomach contents after feeding of milk. The FFA were also produced by gastric lipase.

Enterotoxins from *Vibrio cholerae* and *Escherichia coli* are inhibited by a monosialoganglioside GM1 found in human and bovine milks and in a formula at levels of 12, 1.2, and <1.0 μg per liter respectively (51, 52). There are no new data, but a brief review is available (62).

Requirements for Polyunsaturated Fatty Acids

The requirements of humans for 18:2n6, the original essential fatty acid (EFA), are well established (51, 52, 74). The minimum amount for infants is set at 3 en% (240 mg/dl) by the American Academy of Pediatrics (20).

Deficiency of EFA was established by abnormal triene-tetrane $(20:3n9/20:4n6)$ ratios in plasma lipids or erythrocyte phospholipids. An abnormal ratio is greater than 0.4. Uauy (97) recommended 500 to 700 mg of $18:2n6$ per kilogram of infant weight per day, with a maximum of 12 en% (1.5 kg per day). Recent data on EFA deficiency in premature infants resulted in a recommendation that the average amount of $18:2n6$ required to achieve normality was 1.19 g/kg per day (24). In this study, 67% of the premature infants had low levels of plasma $18:2n6$. The quantities recommended above are higher than those suggested by Uauy (97) in order to overcome the deficiency in premature infants. Human milk and infant formulas contain more than enough $18:2n6$ for the infant's needs. The $18:2n6$ content of milk ranges from 8 to 16% or higher depending on the maternal diet. The amounts of $18:3n3$ in milk range from 0.3 to 1.1% and are responsive to maternal diet. Milk fat contains about 2% of the elongation-desaturation products of $18:2n6$ and $18:3n3$ and may also be essential.

The necessity for inclusion of n3 PUFA in the human diet has been established (74, 89). The requirements are based on inference from the fatty acid profiles in milk (16) of neonates for these fatty acids and the acids in infant tissues, both postmortem (16) and in erythrocyte phospholipids (89, 91). Uauy made the following suggestion. (a) Total n6 and n3 acids should be set at 4–5 en% of total energy for infants, with a maximum of 12 en%. These energy levels represent 600 to 800 mg/kg per day. (b) The intake of $18:2n6$ should be 500 to 700 mg/kg per day and the total supply of n3 fatty acids 70 to 150 mg/kg daily (97). Since the activity of neonatal desaturases and elongases may be suboptimal, half of the fatty acids should be provided as LCPUFA, 20 and 22C derivatives. (c) Formulas for preterm infants should provide 35–75 mg of LCPUFA/kg per day as DHA or EPA + DHA. Ideally, the ratio of n-6 to n-3 PUFA should be maintained within a range of 5/1 to 15/1. Clandinin et al (16) have suggested the following physiological intakes of PUFA (percent of total acids): 1% of 20 and $22:n6$, 0.7% of 20 and $22:n3$, 12% of $18:2n6$, and 0.9% of $18:3n3$.

Trans Fatty Acids

The presence of *trans* fatty acids in milk continues to attract attention (51, 52), yet only recently have the contents been regularly reported. Koletzko et al (60; Tables 4, 5) found 4.4% (by weight) *trans* fatty acid with seven isomers. All of these acids are derived primarily from partially hydrogenated oils. Some are derived from bovine milk lipids. The acids were detected in the plasma lipids of 30 mother-infant pairs (61). The amounts of all fatty acids that are present are given in this study, e.g. maternal plasma contained 7.75% n6 LCPUFA and 2.71% n3 LCPUFA; the values for cord blood were 15.62 and 3.82%.

SUMMARY AND CONCLUSIONS

About 50 metabolically important fatty acids can be identified in human milk. The extent of absorption of milk fatty acids varies considerably from infant to infant, particularly in pre-term infants, and requires more study. Human milk provides sufficient vitamins A and E for the term infant, but supplementation with vitamins D and K may be necessary. More research is needed on the amounts of the fat-soluble vitamins in human milk, the efficiency of transfer from mother to infant, the reasons for variation in different women, and the consequences to breast-fed infants of inadequate intake of vitamins D and K.

Breast milk contains the PUFA needed by term infants who are able to synthesize the long-chain PUFA soon after birth. Pre-term infants fed formulae need supplementation with n3 and n6 long-chain PUFA, since formulas currently do not contain these acids. More work is needed to determine the requirements for n3 and n6 fatty acids, expressed as weights per kilogram.

A larger data base using improved analytical procedures to study the nature and content of lipids in human milk is needed. The impact of maternal genetics and diet on fatty acids in milk should be studied, as well as the effect of maternal diet on eicosanoids secreted by the mammary gland. Information on the structure and function of the milk fat globule and its membrane is needed. Little is known about the effect of milk banking on milk lipids. The reader of this review will no doubt find other gaps in our knowledge of the lipid composition and nutritional value of milk that require additional investigation.

ACKNOWLEDGMENTS

Some of the research reported herein was supported in part by federal funds made available through provisions of the Hatch Act and by NICHHD Contracts NO1-HD-2817 and 6-2917. Scientific Contribution No. 1429, Storrs Agricultural Experiment Station, University of Connecticut, Storrs, CT, 06269.

Literature Cited

1. Ala-Houhala, M., Koskinen, T., Parviainen, M. T., Visakurpi, J. K. 1988. 25-Hydroxyvitamin D and vitamin D in human milk: effects of supplementation and season. *Am. J. Clin. Nutr.* 48:1057–60

2. Allen, J. C., Keller, R. P., Neville, M. C., Archer, P. 1991. Studies in human lactation: milk composition and secretion rates of macronutrients in the first year of lactation. *Am. J. Clin. Nutr.* 54:69–80

3. Arthur, P. G., Hartman, P. E., Smith, M. 1987. Measurement of the milk intake of breast-fed infants. *J. Pediatr. Gastroenterol. Nutr.* 6:758–63

4. Atkinson, S. A., Reinhardt, T. A., Hollis, B. W. 1987. Vitamin D activity in maternal plasma and milk in relation to gestational stage of delivery. *Nutr. Res.* 7:1005–11

5. Bedrick, A. D., Britton, J. R., Johnson, S., Koldovsky, O. 1989. Prostaglandin

stability in human milk and infant gastric fluid. *Biol. Neonate* 56:192–97

6. Bernback, S., Blackberg, L., Hernell, O. 1990. The complete digestion of human milk triacylglycerol requires gastric lipase, pancreatic colipse-dependent lipase, and bile salt-stimulated lipase. *J. Clin. Invest.* 85:1221–26

7. Bitman, J., Hamosh, M., Lutes, V., Neville, M. C., Seacat, J., Wood, D. L. 1989. Milk composition and volume during the onset of lactation in a diabetic mother. *Am. J. Clin. Nutr.* 50:1364–69

8. Bitman, J., Hamosh, M., Wood, D. L., Freed, L. M., Hamosh, P. 1987. Lipid composition of milk from mothers with cystic fibrosis. *Pediatrics* 80:927–32

9. Bitman, J., Wood, D. L. 1987. Comparison of direct transesterification of fatty acids with procedures applied to extracts of human and cow milk fat. *J. Am. Oil Chem. Soc.* 64:637 (Abstr. 74)

10. Boersma, E. R., Offringa, P. J., Muskiet, F. A. J., Chase, W. M. 1991. Vitamin E, lipid fractions, and fatty acid composition of colostrum, transitional milk, and mature milk: an international collaborative study. *Am. J. Clin. Nutr.* 53:1197–1204

11. Buchheim, W., Welsch, U., Patton, S. 1988. Electron microscopy and carbohydrate histochemistry of the human milk fat globule membrane. In *Biology of Human Milk*, ed. L. A. Hanson, pp. 27–44. New York: Raven

12. Canfield, L. A., Hopkinson, J. M. 1989. State of the art: Vitamin K in human milk. *J. Pediatr. Gastroenterol. Nutr.* 8:430–41

13. Canfield, L. M., Hopkinson, J. M., Lima, A. F., Martin, G. S., Burr, J., et. al. 1990. Quantitation of vitamin K in human milk. *Lipids* 25:406–11

14. Canfield, L. M., Hopkinson, J. M., Lima, A. F., Silva, B., Garza, C. 1991. Vitamin K in colostrum and mature human milk over the lactation period—a cross-sectional study. *Am. J. Clin. Nutr.* 53:730–35

15. Carroll, K. 1989. Upper limits of nutrients in infant formulas: polyunsaturated fatty acids and *trans* fatty acids. *J. Nutr.* 119:1810–13

16. Clandinin, M. T., Chappell, J. E., Aerde, E. E. 1989. Requirements of newborn infants for long chain polyunsaturated fatty acids. *Acta Paediatr. Scand. Suppl.* 351:63–71

17. Clark, R. M., Hundrieser, K. E. 1989. Changes in cholesteryl esters of human milk with total lipid. *J. Pediatr. Gastroenterol. Nutr.* 9:347–50

18. Clark, R. M., Roche, M. E. 1990. Gas chromatographic procedure for measuring total lipid in breast milk. *J. Pediatr Gastroenterol. Nutr.* 10:271–72

19. Collins, S. E., Jackson, M. B., Lammi-Keefe, C. J., Jensen, R. G. 1989. The simultaneous separation and quantitation of human milk lipids. *Lipids* 24:746–49

20. Committee on Nutrition, American Academy of Pediatrics. 1985. Fats and fatty acids. In *Pediatric Nutrition Handbook*, ed, G. B. Forbes, C. W. Woodruff, 11:105–10. Elk Grove Village, Il: Am. Acad. Pediatr.

21. Denn, D., Schmidt-Sommerfeld, E. 1989. Lipids as an energy source for the fetus and newborn infant. In *Textbook of Gastroenterology in Infancy*, ed. E. Lebenthal, pp. 293–310. New York: Raven. 2nd ed.

22. Dotson, K. D., Jerrell, J. P., Picciano, M. F., Perkins, E. G. 1992. High performance liquid chromatography of human milk triacylglycerols and gas chromatography of component fatty acids. *Lipids* 27:In press

23. Emken, E. A., Adlof, R. O., Hachey, D. L., Garza, C., Thomas, M. R., Brown-Booth, L. 1989. Incorporation of deuterium-labeled fatty acids into human milk, plasma, and lipoprotein phospholipids and cholesteryl esters. *J. Lipid Res.* 30:395–402

24. Farrell, P. M., Gutcher, G. R., Palter, M., DeMets, D. 1988. Essential fatty acid deficiency in premature infants. *Am. J. Clin. Nutr.* 48:220–29

25. Ferris, A. M., Dotts, M. A., Clark, R. M., Ezrin, M., Jensen, R. G. 1988. Macronutrients in human milk at 2, 12, and 16 weeks postpartum. *J. Am. Diet. Assoc.* 88:694–97

26. Filer, L. J., Mattson, F. H., Fomon, S. J. 1969. Triglyceride configuration and fat absorption by the human infant. *J. Nutr.* 99:293–98

27. Fjeld, C. R., Brown, K. H., Schoeller, D. A. 1988. Validation of the deuterium oxide method for measuring average daily intake in infants. *Am. J. Clin. Nutr.* 46:671–79

28. Greer, F. R., Marshall, S. 1989. Bone mineral content, serum vitamin D metabolite concentrations, and ultraviolet B light exposure in infants fed human milk with and without vitamin D-2 supplements. *J. Pediatr.* 114:204–12

29. Ha, Y. L., Storkson, J., Pariza, M. W. 1990. Inhibition of benzo(α)-pyrene-induced mouse forestomach neoplasia by conjugated dienoic derivatives of

linoleic acid. *Cancer Res.* 50:1097–1101

30. Hachey, D. L., Motil, D. J., Wong, W. W., Garza, C., Klein, P. D. 1989. Milk production and fat concentration are related to whole body water flux during human lactation. *FASEB J.* 3:A454 (Abstr. 1321)

31. Hachey, D. L., Silber, G. H., Wong, W. W., Garza, C. 1989. Human lactation II. Endogenous fatty acid synthesis by the mammary gland. *Pediatr. Res.* 25:63–68

32. Hachey, D. L., Thomas, M. R., Emken, E. A., Garza, C., Brown-Booth, L., et al. 1987. Human lactation: maternal transfer of dietary triglycerides labeled with stable isotopes. *J. Lipid Res.* 28:1185–92

33. Hamosh, M. 1988. Deleted in proof

34. Hamosh, M. 1988. Fat needs for term and preterm infants. In *Nutrition During Infancy*, ed. R. G. Tsang, B. L. Nichols, pp. 133–59. Philadelphia: Hanley & Belfus

35. Hamosh, M. 1991. *Lingual and Gastric Lipases: Their Role in Fat Digestion.* Boca Raton, Fla: CRC Press

36. Hamosh, M., Bitman, J. 1992. Human milk in disease: Lipid composition. *Lipids* 27:In press

37. Hamosh, M., Mehta, N. R., Fink, C. S., Coleman, J., Hamosh, P. 1991. Fat absorption in premature infants: Medium-chain triglycerides and long-chain triglycerides are absorbed at similar rates. *J. Pediatr. Gastroenterol. Nutr.* 13:143–49

38. Hansen, J. W., Cook, D. A., Cordano, A. 1988. Human milk substitutes. See Ref. 34, pp. 378–98

39. Haug, M., Laubach, C., Burke, M., Harzer, G. 1987. Vitamin E in human milk from mothers of preterm and term infants. *J. Pediatr. Gastroenterol. Nutr.* 6:605–9

40. Hernell, O., Blackberg, L., Bernback, S. 1989. Milk lipases and *in vivo* lipolysis. In *Protein and Non-protein Nitrogen in Human Milk,* ed. S. A. Atkinson, B. Lonnerdal, 16:221–33. Boca Raton, Fla: CRC Press

41. Hillman, L. S., Chow, W., Salmons, S. S., Weaver, E., Erickson, M., Hansen, J. 1988. Vitamin D metabolism, mineral homeostasis, and bone mineralization in term infants fed human milk, cow milk-based formula, or soy-based formula. *J. Pediatr.* 112:864–74

42. Huston, G. E., Patton, S. 1990. Factors related to the formation of cytoplasmic crescents on milk fat globules. *J. Dairy Sci.* 73:2061–66

43. Imong, S. M., Jackson, D. A., Woolridge, M. W., Wongsawasdii, L., Ruckphacophunt, S., et al. 1988. Indirect test weighing: A new method for measuring overnight breast milk intakes in the field. *J. Pediatr. Gastroenterol. Nutr.* 7:699–706

44. Innis, S. M., Kuhnlein, H. V. 1988. Long-chain n-3 fatty acids in breast milk of Inuit women consuming traditional foods. *Early Hum. Dev.* 18:185–89

45. Isaacs, C. E., Thormar, H. 1990. Human milk lipids inactivate enveloped viruses. In *Breastfeeding, Nutrition, Infection, and Infant Growth in Developed and Emerging Countries,* ed. S. A. Atkinson, L. A. Hanson, R. K. Chandra, pp. 161–74. St. Johns, Newfoundland, Canada: Arts Biomed.

46. Jackson, D. A., Imong, S. M., Silprasert, A., Preunglumpoo, S., Leelapat, P., et. al. 1988. Estimation of 24 h breast-milk fat concentration and fat intake in rural northern Thailand. *Br. J. Nutr.* 59:365–71

47. Jackson, D. A., Imong, S. M., Silprasert, A., Ruckphaopunt, S., Woolridge, S., et. al. 1988. Circadian variation in fat concentration of breast-milk in a rural northern Thai village. *Br. J. Nutr.* 59:349–63

48. Jackson, D. A., Woolridge, M. W., Imong, S. M., McLeod, C. N., Yutabootr, Y., et. al. 1987. The automatic sampling shield: a device for sampling suckled breast milk. *Early Hum. Dev.* 15:295–306

49. Jackson, M. B., Lammi-Keefe, C. J., Ferris, A. M., Jensen, R. G. 1988. Total lipid and medium chain fatty acid contents of milk from diabetic and control women. *FASEB J.* 2:A653 (Abstr. 2069)

50. Jandacek, R. J. 1991. The solubilization of calcium soaps by fatty acids. *Lipids* 26:250–53

51. Jensen, R. G. 1989. *The Lipids of Human Milk.* Boca Raton, Fla: CRC Press

52. Jensen, R. G. 1989. Lipids in human milk-composition and fat soluble vitamins. See Ref. 21, pp. 157–208

53. Jensen, R. G., Ferris, A. M., Lammi-Keefe, C. J., Henderson, R. A. 1988. Human milk as a carrier of messages to the nursing infant. *Nutr. Today.* Nov/Dec:20–25

54. Jensen, R. G., Ferris, A. M., Lammi-Keefe, C. J., Henderson, R. A. 1990. Lipids of bovine and human milks: A comparison. *J. Dairy Sci.* 73:223–40

55. Jensen, R. G., Henderson, R. A., Ferris, A. M., Lammi-Keefe, C. J. 1990. Conjugated linoleic acid in human milk

with possible anticarcinogenecity. *FASEB J*. A914 (Abstr. 3760)

56. Kallio, M. J. T., Siimes, M. A., Perheentupa, J., Salmenpera, L., Miettenen, T. A. 1989. Cholesterol and its precursors in human milk during prolonged exclusive breast-feeding. *Am. J. Clin. Nutr*. 50:782–85

57. Kanno, C. 1990. Secretory membranes of the lactating mammary gland. *Protoplasma* 159:184–208

58. Kim, Y., English, C., Reich, P., Gerber, L. E., Simpson, K. L. 1990. Vitamin A and carotenoids in human milk. *J. Agric. Food Chem*. 38:1930–33

59. Koletzko, B., Bremer, H. J. 1989. Fat content and fatty acid composition of infant formulas. *Acta Paediatr. Scand*. 78:513–21

60. Koletzko, B., Mrotzek, M., Bremer, H. J. 1989. Fatty and acid composition of mature human milk in Germany. *Am. J. Clin. Nutr*. 47:954–59

61. Koletzko, B., Muller, J. 1990. *Cis* and *trans* isomeric fatty acids in plasma lipids of newborn infants and their mothers. *Biol. Neonate* 57:172–78

62. Kolsto-Otnaess, A-B. 1989. Nonimmunoglobulin components in human milk-candidates for prophylaxis against infantile infections. See Ref. 40, pp. 211–20

63. Kuksis, A., Marai, L., Cerbulis, J., Farrell, H. M. Jr. 1986. Comparative study of the molecular species of chloropropandeiol diesters and triacylglycerols in milk fat. *Lipids* 22:183–90

64. Lammi-Keefe, C. J., Ahn, H. S., Jackson, M. B., Jensen, R. G. 1989. Total cholesterol in breast milk from insulin dependent diabetic women (IDDM) and controls (C). *FASEB J*. 3:A769 (Abstr. 3153)

65. Lammi-Keefe, C. J., Ferris, A. M., Jensen, R. G. 1990. Changes in human milk at 0600, 1000, 1400, 1800, and 2200 h. *J. Pediatr. Gastroenterol. Nutr*. 11:83–88

66. Lepage, G., Roy, C. C. 1984. Improved recovery of fatty acids through direct transesterification without prior extraction or purification. *J. Lipid Res*. 25:1391–96

67. Lucas, A., Davies, P. S. W., Phil, M. 1990. Physiologic energy content of human milk. See Ref. 45, pp. 337–57

68. Lucas, A., Ewing, G., Roberts, S. B., Coward, W. A. 1987. How much energy does the breastfed infant consume and expend? *Br. Med. J*. 295:75–77

69. Lucas, A., Hudson, G. J., Simpson, P., Cole, T. J., Baker, B. A. 1987. An automated enzymic micromethod for the measurement of fat in human milk. *J. Dairy Res*. 54:487–92

70. Michaelsen, K. F., Pedersen, S. B., Skafte, L., Jaeger, P., Pettersen, B. 1988. Infrared analysis for determining macronutrients in human milk. *J. Pediatr. Gastroenterol. Nutr*. 7:229–35

71. Michaelsen, K. F., Skafte, L., Badsberg, J. H., Jorgensen, M. 1990. Variation in macronutrients in human bank milk: Influencing factors and implications for human milk banking. *J. Pediatr. Gastroenterol. Nutr*. 11:229–39

72. Muskiet, F. A. J., Hutter, N. H., Martini, I. A., Jonxis, J. H. P., Offrins, P. J., Boersma, E. R. 1986. Comparison of the fatty acid composition of human milk from mothers in Tanzania, Curacao, and Surinam. *Hum. Nutr. Clin. Nutr*. 41C:149–59

73. National Research Council. 1989. *Recommended Dietary Allowances*, pp. 24–38. Washington, DC: Natl. Acad. Press. 385 pp. 10th ed.

74. National Research Council. 1989. See Ref. 73, pp. 44–51

74a. National Research Council. 1989. See Ref. 73, pp. 78–114

75. Neu, J., Chi-Ying, W-W., Measel, C. P., Gimotty, P. 1988. Prostaglandins in human milk. *Am. J. Clin. Nutr*. 47:649–52

76. Neville, M. C. 1987. The measurement of milk transfer from mother to breastfeeding infant. *J. Pediatr. Gastroenterol. Nutr*. 6:659–62

77. Neville, M. C. 1989. Regulation of milk fat synthesis. *J. Pediatr. Gastroenterol. Nutr*. 8:426–29

78. Neville, M. C., Allen, J. C., Archer, P. C., Casey, C. E., Seacat, J., et al. 1991. Studies in human lactation: milk volume and nutrient composition during weaning and lactogenesis. *Am. J. Clin. Nutr*. 54:81–92

79. Neville, M. C., Keller, R., Seacat, J., Lutes, V. 1988. Studies in human lactation: Milk volumes in lactating women during the onset of lactation and full lactation. *Am. J. Clin. Nutr*. 48:1375–86

80. Nommsen, L. A., Lovelady, C. A., Heinig, M. J., Lonnerdal, B., Dewey, K. G. 1991. Determinants of energy, protein, lipid, and lactose concentrations in human milk during the first 12 mo of lactation: the Darling study. *Am. J. Clin. Nutr*. 53:457–65

81. Deleted in proof

82. Ollilainen, V., Heinonen, M., Linkola, E., Varo, P., Koivistoinen, P. 1989.

Carotenoids and retinoids in Finnish foods: Dairy products and eggs. *J. Dairy Sci.* 72:2257–65

83. Patton, S., Canfield, L. M., Huston, G. E., Ferris, A. M., Jensen, R. G. 1990. Carotenoids of human colostrum. *Lipids* 25:159–65

84. Patton, S., Huston, G. E. 1988. Incidence and characteristics of cell pieces on human milk fat globules. *Biochim. Biophys. Acta* 965:146–53

85. Prentice, M., Landing, M. A. J., Drury, P. J., Dewit, O., Crawford, M. A. 1989. Breast-milk fatty acids of rural Gambian mothers: Effects of diet and maternal parity. *J. Pediatr. Gastroenterol. Nutr.* 8:486–90

86. Putnam, J. C., Carlson, S. E., DeVoe, P. W., Barness, L. A. 1982. The effect of variations in dietary fatty acids on the erythrocyte phosphatidylcholine and phosphatidylethanolamine in human infants. *Am. J. Clin. Nutr.* 36:106–14

87. Shiffman, M. L., Seale, T. W., Flux, M., Rennert, O. R., Swender, P. T. 1989. Breast-milk composition in women with cystic fibrosis: report of two cases and a review of the literature. *Am. J. Clin. Nutr.* 49:612–17

88. Silber, G. H., Hachey, D. L., Schanler, R. J., Garza, C. 1988. Manipulation of maternal diet to alter fatty acid composition of human milk intended for premature infants. *Am. J. Clin. Nutr.* 47:810–14

89. Simopoulos, A. P. 1991. Omega-3 fatty acids in health and disease and in growth and development. *Am. J. Clin. Nutr.* 54:438–63

90. Small, D. M. 1991. The effects of glyceride structure on absorption and metabolism. *Annu. Rev. Nutr.* 11:413–34

91. Smith, L., Darling, P., Roy, C. C. 1989. Pitfalls in the design and manufacture of infant formulas. See Ref. 21, pp. 435–48

92. Specker, B. L., Tsang, R. C., Ho, M., Miller, D. 1987. Effect of vegetarian diet in serum 1,25-dihydroxyvitamin D concentrations during lactation. *Obstet. Gynecol.* 70:870–74

93. Specker, B. L., Wey, H. E., Miller, D. 1987. Differences in fatty acid composition of human milk in vegetarian and nonvegetarian women: Long-term effect of diet. *J. Pediatr. Gastroenterol. Nutr.* 6:764–68

94. Tomarelli, R. M. 1988. Suitable fat formations for infant feeding. In *Dietary Fat Requirements in Health and Development*, ed. J. Beare-Rogers, pp. 1–27. Champaign, Il: Am. Oil Chem. Soc.

95. Tomarelli, R. M., Meyer, B. J., Weaber, J. R., Bernhart, F. W. 1968. Effect of positional distribution on the absorption of fatty acids of human milk and infant formulas. *J. Nutr.* 95:583–90

96. Tsang, R. C., Nichols, B. L. 1988. Nutrient content of infant formulas. See Ref. 34, pp. 418–24

97. Uauy, R. 1990. Are w-3 fatty acids required for normal eye and brain development in the human? *J. Pediatr. Gastroenterol. Nutr.* 11:296–300

98. van Beusekom, C. M., Martini, I. A., Rutgers, H. M., Boersma, E. R., Muskiet, F. A. J. 1990. A carbohydrate-rich diet not only leads to incorporation of medium-chain fatty acids (6:0–14:0) in milk triglycerides but also in each milk phospholipid subclass. *Am. J. Clin. Nutr.* 52:326–34

99. van der Steege, G., Muskiet, F. A. J., Martini, I. 1987. Simultaneous quantification of total medium- and long-chain fatty acids in human milk by capillary gas chromatography with split injection. *J. Chromatogr.* 415:1–11

100. Verkade, H. J., Van Asselt, W. A., Vonk, R. J., Bijleveld, C. M. A., Fernandes, J., et al. 1989. Fat absorption in preterm infants: The effect of lard and antibiotics. *Eur. J. Pediatr.* 149: 126–29

101. Verkade, H. J., Hoving, E. B., Muskiet, F. A. J., Martini, I. A., Jansen, G., et al. 1991. Fat absorption in neonates: comparison of long-chain fatty acid and triglyceride compositions of formula, feces, and blood. *Am. J. Clin. Nutr.* 53:643–51

102. Deleted in proof

103. Wang, C-S., Illingsworth, D. R. 1987. Lipid composition and lipolytic activities in milk from a patient with homozygous familial hypobetalipoproteinemia. *Am. J. Clin. Nutr.* 44:730–36

104. WHO. 1989. Infant feeding: The physiological basis. *Bull. WHO* 67(Suppl.):1–108

105. Ziesel, S. H., Char, D., Sheard, N. F. 1986. Choline, phosphatidylcholine in human and bovine milk and infant formulas. *J. Nutr.* 116:50–58

Annu. Rev. Nutr. 1992. 12:443–71

REGULATION OF GENE EXPRESSION BY VITAMIN A: The Role of Nuclear Retinoic Acid Receptors

Martin Petkovich

Cancer Research Laboratories, Botterell Hall, Queen's University, Kingston, Ontario, Canada K7L 3N6

KEY WORDS: steroid hormone receptors, transcription factors, responsive elements, retinoids, development

CONTENTS

PERSPECTIVES

Determination of the genetic interactions responsible for definition of the body plan during embryogenesis remains one of the most interesting problems in developmental biology. Different species employ rather diverse strategies

443

0199-9885/92/0715-0443$02.00

in establishing ultimate body plan or pattern. One basic mechanism found in many phylogenetically unrelated invertebrates such as echinoderm and nematode (28) is asymmetric cell division involving canonical patterns of cell cleavage during blastula formation; cleavage planes formed during cell division define spatial domains restricted to specific cell fates. Although some of the differentiated cell types in these organisms depend on cell-cell interactions (48), certain cell lineages appear to be established by the inheritance of regionally sequestered maternal factors (28).

In another strategy, which is the predominant mechanism in higher vertebrates, the establishment of axial polarity and the patterning of cells depend on local signals distributed in the form of gradients that derive from restricted regions of the embryo (93). In addition, identity of cell lineages, and the ultimate patterns of differentiation in their progeny, depend qualitatively on contact with other cells, either directly or via signal molecules. In vertebrates, early cell lineages arising from blastomeres are for the most part indeterminate (28, 73, 115). Lineage indeterminacy suggests that the establishment of cell fate requires position-dependent cell interactions that may begin as early as cleavage (28, 32). However, positional information is required not only during the very early stages of development but as long as cells remain multipotent, even throughout adulthood. Consequently, numerous signal molecules have been identified, such as retinoids, fibroblast growth factors (FGFs), and transforming growth factors (TGFβs), which play important roles in the determination of cell fate both in the embryo and in the adult (2, 20, 32, 92).

Retinoids and Development

A major breakthrough in the study of signal molecules and pattern formation in vertebrates has come from recognition that certain metabolites of vitamin A,[1] retinoic acid (RA) in particular, are able to re-specify positional cues in the chick limb bud in a dose-dependent manner (41). More recently it has been shown that retinoic acid may also play an essential role in the establishment of the primary embryonic axes (1a, 40, 42, 129, 149).[2] The effectiveness of RA as a potent inducer of cell differentiation is not limited to embryonic tissues; the epithelium, connective tissues, including bone and cartilage, as well as hematopoetic tissues are some of the better known RA targets in the adult (99, 107, 113, 133, 134, 158, and references therein). The pleiotropy of RA activity has supported the hypothesis that RA may function

---

[1]A comprehensive review on retinoid metabolism can be found elsewhere in this volume.

[2]A complete discussion of the physiological effects of RA is beyond the scope of this review; several excellent reviews have been published recently on the effects of retinoids on vertebrate development (12–14, 41, 63, 72, 138).

through several different pathways. However, the identification of RA nuclear receptors, which belong to the steroid hormone receptor superfamily (50, 106), suggests that RA may effect cellular responses mainly through the regulation of gene transcription (43, 56). Can the ability of RA to regulate gene expression through these receptors alone account for the diversity of RA responses? In this review, an attempt has been made to summarize what is known about retinoic acid receptors and to describe how they might directly or indirectly interact with other cellular components to generate diversity in cellular responsiveness to RA.

RETINOIC ACID RECEPTORS

Nuclear Receptor Structure and Function

At present, more than 30 members of the nuclear receptor family have been identified (43, 56). Specific ligands for many of these receptors include the steroid hormones estrogen, androgen, mineralocorticoid, glucocorticoid, vitamin D_3, and progesterone (43, 56). However, the discoveries that the actions of thyroid hormone and retinoic acid are also mediated through such nuclear receptors indicated that this class of transcription factors has evolved to recognize many structurally unrelated ligands (43, 50, 56, 106). Indeed, given the growing number of orphan receptors (nuclear receptors for which ligands have not yet been identified) and the present lack of obvious prospective ligands, it is likely that the molecular nature of activating ligands will further diversify. Despite the structural unrelatedness of ligands, nuclear receptors are remarkably well conserved (43, 56). Little variation is found in the linear arrangement of their functional elements, which typically comprise six domains (43, 52, 56), denoted A–F in Figure 1. Also the degree of amino acid identity between physiologically unrelated receptors is striking, particularly in the DNA- and ligand-binding regions. Although each nuclear receptor exhibits peculiarities distinguishing it from others in the family, the conservation of basic structure suggests conservation of basic function as well (43, 45, 47, 56, 77). Consequently, structure/function relationships for one subclass of receptors can in many cases be inferentially applied to another, but always with due caution. In many respects (described below), the retinoic acid receptors resemble other steroid receptors; however, they also display features unique to this subclass.

The DNA-Binding Domain

The DNA-binding domain, region C, is the hallmark of the nuclear receptor family. Of the 66–68 amino acids that make up this domain, 19 residues are invariant, including 8 cysteine residues involved in chelation of 2 Zn^{2+} ions, thus forming the base of the two so-called Zn-binding fingers (60, 86). X-ray

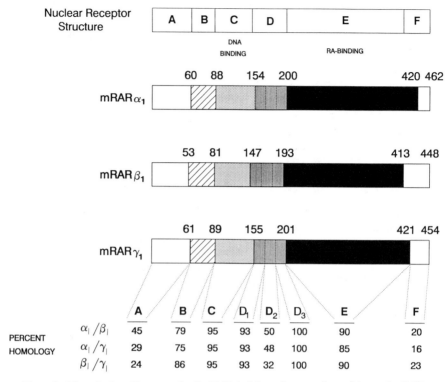

Figure 1 The retinoic acid receptor family (RARs). Schematic comparison of the murine RAR proteins. The basic linear organization of nuclear receptor functional domains is shown at the *top* of the figure (see text). The primary amino acid sequences (numbered above the protein structure) have been aligned on the basis of sequence identity, the percentage of which is indicated in the table at the *bottom* of the figure. The DNA- and hormone-binding domains are indicated. The subdomains of region D in the RARs are indicated by the *hatched* lines (see text for details).

crystallographic studies of the glucocorticoid receptor DNA-binding domain indicate that unlike other finger-containing proteins such as TFIIIA, whose fingers act as independent units, each contributing to DNA binding, the zinc fingers of nuclear receptors fold together as part of a larger, more integrated globular domain (86). These studies further indicate that the most N-terminal finger (C1) of the glucocorticoid receptor (GR) DNA-binding domain adopts a configuration allowing its close apposition to key nucleotide residues in the cognate hormone responsive element (HRE; discussed below) that are important for binding selectivity (86 and references therein). The second finger (C2) may stabilize this binding through nonspecific interactions. Crystallographic analysis of the GR, and NMR studies on the estrogen receptor (ER) and GR DNA-binding regions, confirm earlier studies showing (*a*) that swapping the C1 finger between the ER and GR was sufficient to trade specificities for the estrogen and glucocorticoid responsive elements and (*b*) that three amino acid residues in the C1 finger were implicated in this specificity (26, 77, 87, 144).

Retinoic acid receptors (RARs) were first isolated by virtue of their sequence similarity to other nuclear receptors in the DNA-binding region (50, 106). Subsequently, three types of retinoic acid receptors (RARs), RAR-α (50, 106), RAR-β (11), and RAR-γ (67, 75), have been isolated from several species of vertebrates. When compared with other members of the nuclear receptor family, the retinoic acid receptors exhibit a similar structure showing the highest degree of homology in the DNA (C) binding region. For example the 66 amino acid RAR-α region C (see Figure 1) is 62%, 58%, and 45% homologous to the corresponding domains of the thyroid hormone receptor (c-ErbA), ER, and GR, respectively. When compared with each other, the RARs are almost completely identical in region C (see Figure 1). Figure 2 shows the hypothetical structure of the human RAR-α region C as it might be

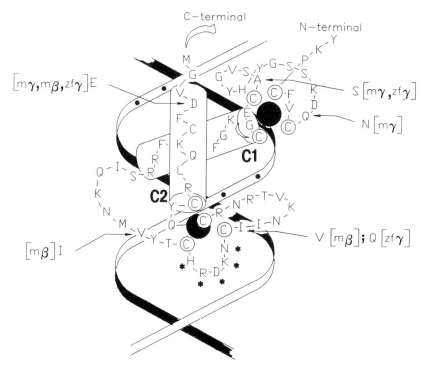

Figure 2 Schematic representation of the mouse RAR-α DNA-binding domain bound to DNA. Hypothetical modelling of the DNA-binding domain based on the resolution of the glucocorticoid receptor DNA-binding domain (see text for references). The amino- and carboxy-terminal ends (residues 88 and 153, respectively) are indicated as are the two "zinc-finger" domains C1 and C2 (see text). Zinc ions are depicted by solid discs. Residues marked by asterisks correspond to the D-box referred to in the text. Arrows indicate amino acid residue substitutions found in mouse RAR-β and -γ (mβ and mγ respectively) and zebrafish RAR-γ; zfγ.

folded together to contact DNA. The few differences in amino acid sequence between RAR α, β, and γ are noted in the diagram. In addition, the conservation between region C of zebrafish RAR-γ and mouse RARs is also noted. This comparison illustrates that this sequence has remained essentially unchanged for over 450 million years.

Responsive Elements

DNA sequences required for the action of particular classes of nuclear receptors have been identified. For example, several natural glucocorticoid responsive elements (GREs) contain an imperfect inverted repeat of the conserved hexanucleotide TGTTCT, separated by a spacer of three nonconserved nucleotides (26, 87, 144, and references therein). Receptors for androgens, mineralocorticoids, and progestins, each of which on the basis of amino acid sequence homology have been classified as belonging to the GR subfamily, also appear to share specificity for this target site motif (26, 87, 90, 144). The estrogen responsive element (ERE), though similar in structure to the GRE, contains the conserved hexanucleotide TGACCT (56). Most of the other members of the nuclear receptor family, including the RARs, belong to the estrogen receptor/thyroid hormone receptor (ER/TR) subfamily (also on the basis of amino acid homology) and display varying degrees of cross-recognition of palindromic responsive elements containing canonical or degenerate forms of the ER-type half-site TGACCT (90). However, as first noted for the RARs, the spacing between ER-type response element half-sites as well as their relative orientation appears to be important for the selective discrimination of target responsive elements by members of the ER/TR subfamily (26, 145). For example, whereas palindromic thyroid hormone responsive elements (TREs) (e.g. TGACCGGTCA) confer retinoic acid responsiveness to heterologous promoters, direct repeat TREs do not (26, 145). Recent studies indicate that the presence of 3, 4, or 5 spacer nucleotides between direct repeats of the TGACCT (or inverted AGGTCA) half-site will determine the specificity of response to vitamin D_3, T_3, and RA, respectively (26, 145). However, response element configuration alone is insufficient to specify receptor-target gene interactions. How do GR target sites for the GREs select for response to the GR when progesterone receptors (PR) and androgen receptors (AR) appear to recognize the target element equally well? Even more puzzling, do RA responsive elements (RAREs) exist that are specific for RARα, β, and γ subtypes?

RAREs and RARs

Minor differences between the DNA-binding domains of the three RARs appear to be conserved between species (e.g. see Figures 1 and 2). Could these minor changes target subtle differences in RAREs? Co-transfection of the RAR subtypes α, β, and γ with various RARE promoter constructs has thus far failed to reveal selectivity for one subtype over another (116, 136). Unfortunately, co-transfection experiments in which receptor and target copy numbers greatly exceed physiological levels might create conditions that favor nonspecific interactions between receptor and target and thus obscure the more subtle protein-DNA and/or protein-protein contacts required for specificity. In this regard, it is interesting to note that in transgenic mice harboring a construct containing the RARE from the RAR-β promoter and a "neutral" promoter (hsp 68) spliced to a β-galactosidase *(lacZ)* encoding reporter gene, the resulting pattern of *lacZ* expression closely resembles that of the RAR-β gene (116), even though in transient transfection experiments the same RARE was recognized by all three RARs (116, 136). Since RAR-α gene expression is ubiquitous throughout most developmental stages, and RAR-γ is expressed in regions of the embryo different from RAR-β, this particular RARE apparently displays a preference for the latter (116, 119, 130). However, these results may also indicate the requirement for the coexpression of factors other than RARs in tissues in which the RARE is active (see below).

Although the expression of many genes appears to be regulated by RA (e.g. 21, 23, 29, 35, 57, 65, 68, 78, 84, 85, 107, 133, 137, 142, 146), relatively few RA-responsive elements (RAREs) have been characterized. One of the first naturally occurring RAREs to be characterized was found in the RAR-β promoter (23, 64). In both the human (23) and mouse (64) forms of this promoter, which exhibit extensive homology with each other, the RARE is a direct repeat of the motif GTTCA separated by a gap of 6 nucleotides. Related RAREs have been identified in other genes, including the human alcohol/dehydrogenase promoter (145), the mouse complement factor H gene promoter (97, 145), the mRAR-α2 isoform promoter (81), and the mouse cellular retinol-binding protein type-I (mCRBP1) (132). For the RAR-β2 RARE, mutagenesis has demonstrated that both repeats are essential for RAR binding (98, 122). Moreover, the 6 nucleotide spacing is optimal, although a spacing of 3 nucleotides was also shown to be effective both in binding RAR and in conferring RA responsiveness to a heterologous promoter (122).

Although the kinetics of receptor binding to response elements has not been thoroughly investigated, within receptor subclasses some degeneracy in response elements clearly is permissible. As mentioned above, the ER will not bind detectably to a GRE but will bind to certain degenerate forms of the ERE. Similarly, the GR does not bind detectably to EREs. In some instances an ERE half-site (TGACCT) has been shown capable of confering estrogen inducibility to a promoter. However, if the formation of an ER dimer is required for estrogen-inducible transactivation, and one ER monomer binds to one half-site, how is this possible? Perhaps a receptor dimer can still functionally bind to a half-site. In this case, one ER monomer would make specific contacts with the half-site while the second would make nonspecific contacts with adjacent nucleotides. Results from crystallographic studies of the GR suggest that this may indeed be the case (86). These studies show that the DNA-binding domain of the GR is capable of making contacts with non-GRE motifs as long as there is a stabilizing interaction with another monomer bound to a half-site (86). Thus if this kind of interaction is widespread, isolated half-sites that confer hormone responsiveness may in fact be binding receptor dimers. Furthermore, receptors may functionally interact with "poor" or "nonconsensus" responsive elements provided that this interaction is stabilized by "strong" or "specific" protein-protein contacts with other proteins bound to DNA. Although this latter possibility has yet to be demonstrated experimentally, some evidence suggests that certain nuclear receptors interact specifically with other proteins bound to DNA (see below).

The laminin B1 (LamB1) promoter contains several degenerate forms of isolated ERE half sites that have been shown to be important for the RA responsiveness of this promoter (147, 148). The LamB1 RAREs appear not to be selective for a particular RAR subtype in transient transfection experiments (147, 148). Do RARs bind to these sites as monomers, homodimers, or heterodimers? Furthermore, since the most efficient RARE is in the form of a direct repeat of half-sites rather than inverted, as would be the case for the ER or GR, how might RAR monomers be oriented with respect to each other? Since evidence suggests that other nuclear receptors may form heterodimers with RARs (Ref. 54 and see below), it will be interesting to see how RAR-RAR or RAR-nuclear receptor contacts differ from those of ER and GR homodimers.

The Ligand-Binding Domain

The ligand-binding region of nuclear receptors, characterized by the moderate degree of conservation it displays with other members of the receptor family, fulfils numerous functions. For those receptors known to bind specific ligands, mutational analyses have unequivocally demonstrated the importance of this region for ligand binding (43, 56, 77). Other functions ascribed to this region, such as receptor dimerization and transcription activation, have been shown to be ligand dependent (44, 45, 58, 151). The molecular nature of the ligand dependency of these functions has not yet been defined. For receptors such as the PR, GR, and ER, one model suggests that the binding of ligand dissociates the receptor from the protein-inactivating function of the heat-shock protein HSP 90, thus permitting receptor dimerization, high affinity DNA binding, and transcriptional activation to occur (92, 108–110). However, recent studies using in vitro transcription systems indicate that hormone-dependent transcription activation can occur in the absence of heat shock proteins (4). Furthermore, there is no evidence that nuclear receptors such as RAR, TRs, and vitamin D_3 receptor (VDR) form complexes with Hsp 90 (25, 107). The ligand binding domains of receptors for structurally unrelated ligands share significant regions of homology (43, 56), Perhaps the key to the mechanism of ligand activation is formed in the amino acid residues that are conserved among these different receptors.

The retinoic acid receptors exhibit almost complete conservation in their retinoic acid binding domains (see Figure 1). Of the 220 amino acids that make up this domain, all but 35 residues are identical when RAR-α, -β, and -γ are compared with each other. In addition, most of these subtype specific residues are also conserved when these receptors are compared with their respective counterparts in other species. Therefore, it seems likely that these amino acid differences are indicative of important functional differences between the three RAR subtypes.

The ligand-binding region (E) of the RARs may, like other nuclear receptors, be involved in several functions: (a) ligand binding, (b) receptor dimerization, and (c) transactivation.

LIGAND BINDING   The RAR retinoic acid binding domain has been defined as a 220 amino acid region in the C-terminal half of these receptors. This region was originally designated as the RA-binding region because of its relative homology to ligand-binding domains of other steroid receptors (92, 106). Investigators have now shown that when the C-terminal portions of RAR-α and RAR-β, extending from the beginning of region D to the C-termini, are expressed in *Escherichia coli,* they bind retinoic acid apparently as efficiently as do whole receptors (16, 24, 136, 156).

When human (hRAR-α) and mouse (mRAR-β) were compared with respect to their abilities to activate reporter gene transcription, approximately 5–10 fold higher concentrations of RA were required to achieve the same level of activation with RAR-α as was achieved with RAR-β (11). This study provided the first evidence that the abilities of the three receptors to bind retinoic acid could be different. With the identification of RAR-γ, transcription activation studies suggested that, of the three receptors, RAR-γ might have the highest affinity for RA (51). The notion that these differences might be exploited to identify synthetic subtype-specific retinoid analogs has led to the identification of several such compounds exhibiting subtype preference (3, 16, 61, 113, 141). One example is the retinobenzoic acid compound Am80 (24, 139), which activates RAR-α with a higher apparent affinity than it does RAR-β, the converse of observations with RA. Interestingly, Am80, which binds cellular retinoic acid binding protein (CRABP) with a much lower affinity than does RA, mimics exactly the effects of RA in causing digit duplications in chick wing bud experiments (139).

Tagged onto the C-terminal end of the ligand-binding domain is region F. Little is known about the function, if any, of this region of the RARs. Although it is relatively well conserved between man and mouse, little homology is seen in more divergent species such as *Xenopus* (42), newt (111), and zebra fish (J. White and M. Petkovich, unpublished).

RAR DIMERS AND HETERODIMERS   Receptor dimerization has been shown to occur for many receptors (44, 45, 80, 86). Crystallographic and NMR analyses of the ER and GR indicate that dimerization occurs at least in part through interactions between DNA-binding domains (86). Several structure/ function studies have also indicated an important role for the ligand-binding domain in the dimerization process. The identification of a conserved motif in ligand-binding domains of various nuclear receptors including the TR and RARs, which is reminiscent of the α-helical "zipper" motifs (47, 79) that mediate homo- and heterodimer formation between the transcription factors c-jun, C/EBP, DBP, and c-myc, indicated a possible point of interaction between receptors (44–46, 62, and references therein).

The formation of homodimers is essential for the function of many transcription-regulating proteins. This appears to be the case for several nuclear receptors including the ER and GR (44, 76, 135, and references therein). The existence of multiple genes encoding RARs, each of which through alternate exon usage can give rise to several different isoforms (see below), raises the possibility that RAR homo/hetero dimer formation may be an important regulatory mechanism. Co-transfection studies indicate that the RAR-$\gamma2$ isoform can inhibit the activity of RAR-β (66). Whether this antagonism is the result of heterodimer formation or due to competition for limiting transcription factors is not clear.

Cross-linking studies have shown that the TR and RAR-α may form complexes with several distinct nuclear proteins (co-regulators) that are expressed in a cell-dependent fashion (27, 46, 54, 55). These interactions are apparently mediated by protein-protein interactions occurring through the ligand-binding domain "zipper-like" motif (44–46, 54, 103). Although the identities of these co-regulators are under investigation in several laboratories, at least one of them corresponds to another nuclear receptor, TR (54). This relationship between RAR and TR has been quite revealing in terms of the oncogenic activity of the v-erbA TR derivate and is discussed in further detail below. Nuclear receptors other than TR are likely to form complexes with RARs. Most recently, a novel class of retinoid receptors (RXRs) has been characterized (88). Interestingly, RXRs have been shown to form heterodimers with RARs, thereby enhancing manyfold RAR binding to RAREs (see below). Furthermore, transcription activation studies indicate that certain intermediary factors functionally, and perhaps physically, link the nuclear receptors to the transcriptional machinery (see below). The nature of these intermediary factors remains to be determined.

TRANSCRIPTION ACTIVATION FUNCTIONS   Several classes of domains are capable of mediating transcriptional activation; many of them are acidic in nature, such as the yeast transcription factor Gal4 and the herpes simplex virus protein VP16 (151, 152, 154, and references therein). Studies with hormone receptors have shown that regions important for transcription activation can be found both in the N-terminal A/B domains as well as in the ligand-binding domain (region E) (140, 143, 151). Interestingly, the A/B region activation domain can function independently of ligand, whereas the C-terminal activation domain appears to be hormone dependent (140, 151). Neither of these domains in the estrogen receptor contains stretches of acidic amino acids whereas both comparable domains in the glucocorticoid receptor (τ1, located in the A/B domain, and τ2, in the N-terminal regiona of the glucocorticoid-binding domain) are acidic in content.

By expressing portions of these receptors either alone or as chimeras with other transcription factors, researchers have demonstrated that the two separate transactivation domains in steroid receptors can function independently by interacting with different, specific intermediary factors (140, 151, and references therein). These intermediary factors are currently uncharacterized.

N-TERMINAL REGION ISOFORMS   Several classes of steroid receptors including those for progesterone, retinoic acid, and thyroid hormone exhibit alternative exon usage in the most N-terminal or A/B region of receptors (71, 75, 140, 143, 155). The PR exists in two forms, which differ in their A domains and exhibit both cell-type and promoter specificity in their ability to activate reporter gene transcription (143). Possibly, each of the two different N-terminal alternate regions of the PR interacts with different intermediary factors that exhibit cell-specific patterns of expression (140). This possibility is particularly interesting given that several alternate A regions for each of the three RAR subtypes (α, β, and γ) as well as for TR-β have been characterizcd. Expression of these alternative A regions, at least for the RAR and PR, appears to be regulated independently by different promoters in a tissue-specific manner (81, 159).

At least seven different isoforms have been identified for mouse RAR-α, three for RAR-β, and seven for RAR-γ (51, 71, 159). For most of these isoforms, regions B through F (Figure 1) are constant (10). The functional importance of the A region for RARs is not know. However, sequence comparisons with homologous isoforms found in species other than mouse, chicken (100, 131), man (75), newt (111), *Xenopus* (42), and zebra fish (J. White and M. Petkovich, unpublished) reveal a high degree of conservation. Although the tissue-specific distribution of most of these isoforms is not presently known, several of the isoforms have been expressed in a temporally and spatially restricted manner. For example, whereas the RAR-α1 isoform is ubiquitously expressed, RAR-α2 appears to be specifically expressed in intestine and lung with only trace amounts of transcript detectable in other tissues (81). Similarly, RAR-β2 is found in kidney, heart, and skeletal muscle, whereas the RAR-β1 and -β3 isoforms are expressed more abundantly in brain, lung, and skin. RAR-γ1 is the predominant RAR-γ isoform expressed in skin, and both RAR-γ1 and -γ2 isoforms are expressed during embryogenesis; the RAR-γ2 isoform is expressed more abundantly during early stages of development (71). Interestingly, certain isoforms of RAR-α and -β appear to be regulated by different promoters (81, 159). The expression of RAR-α2, for example, could be induced by retinoic acid in embryonal carcinoma cells, whereas RAR-α1 appears constitutively expressed (81). In addition, each different isoform mRNA contains a different 5'-stretch of noncoding mRNA that may affect its stability or translation efficiency in a tissue-specific manner (74).

Sequence comparisons between RAR-α1, RAR-β1, and RAR-γ1 or between RAR-α2, RAR-β2, and RAR-γ2 reveal significant homologies between these different subtype isoforms. These homologies indicate that the related subtypes may have functional similarities (17). The functional differences between A regions remain to be determined, yet the tissue-specific distribution, independent regulation of different isoforms, and evolutionary sequence conservation strongly suggest that specific differences do exist (17).

Hinge Region: Region D

The region joining the ligand- and DNA-binding domains (region D) has gained little attention, because its sole purpose for the ER and GR appears to be to serve as a structural joint between the two functional domains (43, 56). The RARs, however, exhibit an unusual degree of subtype-specific sequence conservation in this region (158). On the basis of homology, this region can be subdivided into three separate regions: D_1, D_2, and D_3 (158) (Figure 1). The most N-terminal subdivision is conserved 14 out of 15 residues between RAR-α, -β, and -γ (Figure 1) and contains a sequence resembling the nuclear localization signal for the large T antigen (106) while the most C-terminal region, D_3, is completely conserved (7/7 residues). The central region D_2 is less than 50% conserved when RAR-α, -β, and -γ are compared with each other but is almost entirely conserved in cross species comparisons (17). For example the hRAR-γ D$_2$ is identical in 17 out of 24 residues with respect to its zebra fish counterpart (J. White and M. Petkovich, unpublished). This degree of conservation indicates that the D region plays a specific functional role that possibly is unique to this class of receptors. The importance of this region remains to be determined.

RAR Phosphorylation

Several classes of nuclear receptors including the PR, GR, ER, VDR, and TR are posttranslationally modified by phosphorylation (31, 69, 114). Recently, RAR-γ1 has also been shown to be phosphorylated (114). Although the phosphorylated amino acid residues were not identified, regions A/B and D were shown to be targets of phosphorylation (114). Both regions A/B and D of RAR-γ contain several kinase recognition motifs for serine phosphorylation (see Table II in Ref. 114). Although the phosphorylation of PR, GR, and VDR could be increased by the presence of their respective ligands, this was apparently not the case with RAR-γ (31, 69, 114). What function phosphorylation serves for nuclear receptor function has yet to be determined; however, D region phosphorylation may be important in nuclear localization. Furthermore, amino-terminal phosphorylation would not be inconsistent with the possibility that phosphorylation may modulate the efficiency of transcription activation (114).

Developmental Expression of RARs

The three genes encoding RARs are differentially expressed during mouse development. Several comprehensive studies comparing the temporal and spatial distribution of RAR, CRBP, and CRABP mRNA transcripts have been carried out in mouse (15, 38, 39, 94, 119, 120) and chicken (123, 131), and the reader is directed to these studies for detailed accounts of expression patterns. These studies show that RAR expression is widespread throughout various developing tissues and organs but is topographically restricted in a subtype-specific manner (38, 39, 119, 120). RAR-α in general is rather ubiquitously expressed. However, as previously discussed, at least one iso-form of RAR-α exhibits a more tissue-specific pattern of expression (81). Tracheal epithelium has long been suspected as a prime target for the effects of RA. The expression of RAR-β in trachea, among other epithelial tissues, implies that this subtype has a particular role in mediating retinoid effects. Skin is an important RA target tissue and RAR-γ appears to be the pre-dominantly expressed RAR in this organ (102, 158). Similarly, the expression of RAR-γ in chondrogenic regions is consistent with previous studies describing effects of retinoids on bone growth and development (107, 120). This is particularly noteworthy in the developing vertebrate limb where RAR-γ expression appears to be restricted to precartilagenous condensations that will eventually form the bone models (38, 120).

Regulating RAR Function: Interactions with Other Gene Products

OTHER TRANSCRIPTION FACTORS   Several lines of evidence indicate that membrane and nuclear receptor signalling pathways converge on common regulatory elements and can exert opposite effects on gene transcription. For example, it has been shown that a site in the human osteocalcin promoter can be recognized both by the VDR and RAR as well as by the AP1 site binding proteins c-jun and c-fos. Whereas this responsive element can mediate induction of osteocalcin gene transcription by RA and vitamin D_3, the coexpression of c-jun and c-fos suppresses basal transcription of this gene as well as induction by the two vitamins (125). Whether these opposing effects of the receptors and fos/jun complex are the result of direct protein-protein interactions or of competition for the common DNA motif is not clear. Conversely, studies have shown that the collagenase promoter is positively regulated by the AP-1 site, which binds the fos/jun complex, and that this positive regulation can be blocked by the GR (37, 124, 157). DNA binding and protein cross-linking experiments indicate that a direct interaction between the GR and either fos or jun is responsible for the mutual corepression (157).

Interestingly, the promoter of interstitial collagenase and a synthetic AP-1-dependent promoter were repressed both by RAR-α and the thyroid hormone receptor (c-ErbAα) in a ligand-dependent manner. The mechanism of this repression is like that of the GR and is also thought to be through a specific receptor-mediated decrease in activity of c-fos/c-jun, which positively regulate these promoters (33). Without the AP-1 site, RAR-α and c-ErbA-α do not affect promoter activity (33, 99). Although there is no evidence as yet that RAR-α and c-Erb-α are directly interacting with c-fos and/or c-jun proteins, the v-erbA oncoprotein (126), in a dominant negative fashion, abrogates the RAR-α mediated repression of AP-1 activity. Possibly, v-erbA competes with RAR-α and c-ErbA for receptor-binding sites, but due to specific mutations in its C-terminal region, v-erbA is unable to inactivate or repress AP-1 activity. Such a mechanism may account for the oncogenic activity of v-erbA. Since c-ErbAα has been shown to form heterodimers with RAR-α (53, 54), a similar heterodimer formed between v-erbA and RARα might be inactive (33, 53, 54).

Although the exact nature of receptor-fos/jun interactions is not known, such interactions may be essential for the reciprocal control of cell proliferation and differentiation. Such interactions may also provide a basis for understanding some of the mechanisms underlying the hormonal responsiveness of tumors.

CELLULAR RETINOIC ACID BINDING PROTEINS   In addition to the RARs, small intracellular retinoic acid binding proteins (CRABPs) have also been identified (5, 49, 127, 128, 134, 153). CRABPs belong to a multigene family of proteins that includes the retinol binding proteins CRBPI and II (5, 49) as well the myelin protein P2 (91) and fatty acid binding protein (FABP) (91). These proteins, which are approximately 15 kd in size, do not bear an obvious similarity to any known transcription factors (128), nor is there any similarity in the structure of these proteins with the RAR ligand-binding domain; yet they bind RA selectively and with high affinity. The function of these proteins is unknown; however, the lack of expression of these proteins in certain RA-responsive target tissues and cells suggests that CRABPs are not directly involved in mediating RA effects (15, 134). Recent studies have shown that CRABP-I, when stably transfected into F9 teratocarcinoma stem cell lines, can inhibit RA-induced differentiation into extraembryonic endoderm (9). This finding supports the notion that CRABPs may limit the amount of RA available to the nuclear receptors, thus attenuating RA effects on gene regulation (87). This model is consistent with the observation that sites of expression of CRABPs, such as in the developing central nervous system, craniofacial area, and limbs, are targets for retinoid teratogenicity (6, 15, 112).

RETINOID X RECEPTORS

In addition to the RARs, a second family of nuclear receptors, the retinoid X receptors (RXRs), appear to be involved in mediating cellular responses to retinoids (59, 88, 118). Their structures are shown in Figure 3. RXR-α was identified by low stringency cross-hybridization with the RAR-α DNA-binding domain (88) while RXR-β (H-2R11BP) was identified as a transcription factor binding to the region II enhancer of major histocompatability complex (MHC) class I genes (59). A closely related receptor has recently been identified in *Drosophila melanogaster;* however, this RXR-like receptor neither binds retinoids nor activates gene expression in a RA-dependent manner (104). Although RARs and RXRs differ substantially in primary structure (see Figure 3), RXRs respond specifically to RA (88, 89). The degree of homology between RXR-α and RXR-β in both the DNA- and ligand-binding domains is greater than 85% (Figure 3). RXR-α requires relatively higher concentrations of RA for activation than do the RARs (88).

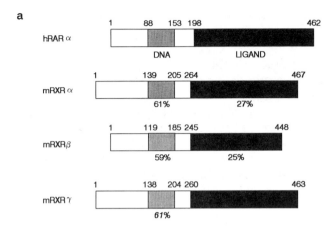

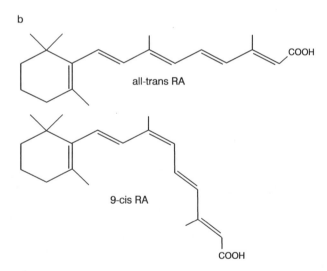

Figure 3 (*a*) The retinoid-X-receptor (RXR) family. Schematic comparison of the amino acid sequence homology between RAR-α and mouse RXR-α, RXR-β, and RXR-γ. Protein residues are indicated above the protein structures. Only the DNA-binding and ligand-binding domains of RXRs exhibit significant amino acid identity with RAR-α (shown by the *shaded* and *solid* regions respectively). Sequence comparisons between the three RXRs indicate almost complete conservation of DNA- and ligand-binding regions (>90%) (see reference 80a). (*b*) Structural representations of all-*trans (top)* and 9-*cis (bottom)* isomers of retinoic acid.

In co-transfection experiments, well-characterized retinoid analogues (e.g. TTNBP) were much less effective at stimulating a transcriptional response with RXR-α than with RAR-α (88). These results suggest that RXR-α might be specific for a retinoid other than RA (7, 88). Recently, 9-*cis* isomer of RA has been identified as a likely candidate for the RXR ligand (Figure 8*b*) (62a). This naturally occurring retinoid binds to RXRs with higher affinity than does all-*trans* RA, the metabolite from which it is derived (62a, 81a).

RXR-α and RXR-β exhibit unique tissue-specific patterns of expression distinct from those seen for the RARs (59, 88, 118). Interestingly, the tissue distribution of the chicken RXR-α (cRXR-α) overlaps extensively with that for CRABP (118) in the developing peripheral nervous system; however, it is not clear whether RXR and CRABP functions are in any way integrated (118). Comparatively high levels of RXR-α expression have been observed in both chicken and rat liver (88). This high level of expression in the liver may suggest a regulatory role for RXR-α in retinoid metabolism and transport (88). Note that RXR-α is also expressed in the intestinal villi (118), where it may be involved in regulating the genes involved in retinol transport (118, and see below). Less is presently known about the expression of RXR-β; however, Northern blot analysis indicates that unlike RXR-α, RXR-β is not highly expressed in the liver and may be specific for the central nervous system (59, 88). A third RXR, RXR-γ has recently been identified (80a); however, little is presently known about its tissue-specific expression.

Although the RXRs, like RARs, can recognize several ERE and thyroid hormone responsive element (TRE) related elements, at least two responsive elements that confer differential regulation by RARs and RXRs have been identified (89, 117). One of these elements is located in the upstream region of the cellular retinol-binding protein-type II (CRBPII), which is thought to be important in the movement of vitamin A across the intestinal lumen (89). This element is unusual because it contains five nearly perfect tandem repeats of the sequence AGGTCA that are separated by single spacer nucleotides (89). Furthermore, although both RAR and RXR can bind to this responsive element, only RXR can activate reporter gene expression in co-transfection experiments (89). Co-expression of RAR with RXR in these experiments resulted in almost complete inhibition of transcription activation (89). A second RXR-specific responsive element was recently identified in the apolipoprotein A1 (apoA1) gene. ApoA1 encodes a plasma protein involved in lipid transport and it is synthesized in both the liver and intestine (117). Site A in this promoter (-214 to -192) can bind in vitro to RXR-α but only weakly to RAR-α and RAR-β. Accordingly, it can confer a RA response to a reporter gene when co-transfected with RXR-α but not with RAR-α and -β (117).

RXRs Interact with Several Classes of Nuclear Receptors

In a remarkable twist of scientific fate, researchers seeking the identity of a factor essential for the efficient binding of RARs to RAREs purified and identified what turned out to be an RXRβ (80a, 157a). Thus, heterodimers of RARs and RXRs appear to be more efficient at binding to RAREs than do homodimers of these receptors. This linkage between two receptor subclasses unifies two RA signal transduction pathways; different RAR/RXR combinations may exhibit unique specificities in target gene activation. A better picture of RXR/RAR overlap, in terms of their domains of tissue specific expression, will tell us much about the particular roles of individual RXR/RAR complexes. Even more surprising is the degree of promiscuity of RXRs; RXRs interact not only with RARs but also with thyroid hormone receptors and vitamin D_3 receptors (73a, 80a, 157a, 160). These interactions indicate that RXRs may play a more central role in nuclear receptor signalling pathways (73a). The extent to which RXRs interact with other members of the superfamily of nuclear receptors remains to be determined.

RAR-α AND PROMYELOCYTIC LEUKEMIA

The assignment of the RAR-α locus to chromosome 17 (see Table 1) led to speculation that the RAR-α might be involved in the pathogenesis of acute promyelocytic leukemia (APL) (8, 18, 19, 22, 34, 82). This hypothesis was substantiated by the fact that APL patients could be effectively treated by RA (18, 19, 22, 82, 95). Several groups have shown that the chromosome 15;17 translocation t(15;17), which has become essentially pathonomic for APL, fuses most of the coding portion of the RAR-α, from breakpoints clustered within the second intron of the gene, to the promyelocytic leukemia locus (PML) on chromosome 15 (1, 30, 35, 36, 83). PML, first identified by characterization of the translocation locus on chromosome 15, also appears to encode a transcription factor (36, 70, 105). The translocation gives rise to a novel protein comprising most of RAR-α (regions B through F) fused in frame and downstream of a large portion of PML (36, 70). Expression of the RAR-α PML chimera in co-transfection studies indicates that it may act as a dominant negative regulation of promyelocyte differentiation (36). However, depending on the exact point of fusion which can vary from one APL patient to another (36, 70, 105), certain PML-RAR fusions can function as transcription activators with greater or lesser efficiencies than those of wild-type RAR-α. Although it is presently unclear whether the PML-RAR-α protein is causally related to the pathogenesis of APL, the mechanism of PML-RARα function may be consistent with that proposed for v-erb A (see above) (33). As in the case where v-erb A expression appears to abrogate the normal inhibitory actions of RAR-α with the AP1 element (see above), PML-RAR-α may prove to be ineffective in suppressing AP1-mediated transcription activation on those target genes where this function is necessary for the normal differentiation of promyelocytes (33).

CONCLUSIONS

Modulation of Retinoid Signal Transduction

The recent identification of a number of interacting components of the retinoid signal transduction system supports the view that the cellular response to RA depends on numerous factors. Below are listed some of the critical factors that may determine the biological activity of RA in a given tissue.

RETINOID METABOLISM   Relatively little is known about the physiological regulation of retinoid concentrations (7). Whether or not RA concentrations are modulated at the level of the target tissue is of particular interest (7). In recent years, there has been considerable interest in the role of RA in chick limb pattern formation (41, 138). When RA is locally applied in the form of a RA-soaked bead to the anterior margin of the wing bud, a minor duplication of digit pattern develops, exactly mimicking the pattern resulting from grafts from the zone of polarizing activity (ZPA) located in the posterior margin of the limb bud (41, 138). Although RA may be a morphogen in the limb bud that provides graded spatial clues in the form of a concentration gradient emanating either from the ZPA or from RA-soaked beads, recent studies suggest that RA may act to convert cells into ZPA-like regions that release a morphogen other than RA (101, 150). In either case, the regulation of local levels of RA is likely to be important. Perhaps the graded distribution of CRABP(S) (41, 138), which appears to be opposite in polarity to the RA gradient presumed to exist in the limb bud, may be important to ensure that only one specified region of the limb forms a ZPA. In any case, modulation of RA concentration in the limb bud results in a very discrete change in tissue response. Thus, the maintenance of normal tissue responses may require factors such as CRBPs, CRABPs as well as retinoid-metabolizing enzymes (7, 138). The discovery that 9-*cis* RA is a more potent ligand for RXRs than all-*trans* RA indicates that retinoid isomerization may be a pivotal point for regulating target tissue responses to retinoids (62a, 81a). There are, of course, other metabolites of RA, and it will be important to determine if any of these other metabolites are receptor selective (141).

RARs   The three subtypes of retinoic acid receptors are RAR-α, RAR-β, and RAR-γ (11, 50, 67, 75, 106). These receptors, by regulating target gene expression in a RA-dependent manner, play a pivotal role in the cellular response to RA. Although these nuclear receptors are highly similar in structure, the minor differences between them are well conserved throughout evolution, suggesting that each RAR may serve different functions. This observation is supported by the finding that the efficiency of transactivation by RA and other synthetic retinoids appears to be dependent on the RAR subtype mediating the effect (see above). Moreover, the different subtypes are differentially expressed in retinoid target tissues. Little is known about the functional significance of the RAR isoforms that differ in their A regions (see Figure 1). These different A regions (see above), by interacting specifically with transcription intermediary factors (TIFs), may play a role in determining promoter and cell selective transactivation, as appears to be the case for other nuclear receptors (143, 151, 154). Thus, the range of possible responses to RA may vary greatly depending on (a) which isoforms are present and (b) what intermediary factors (TIFs) are available in a given cell-type. The possibility that RARs may form homo- and heterodimers would further increase the possible RAR configurations on target gene RAREs.

RXRs   The discovery of a second subclass of RA-activated nuclear receptors, the RXRs, has added a further level of control in regulating the cellular response to RA. In this parallel RA signal transduction pathway are also three receptor subtypes: RXR-α, RXR-β, and RXR-γ (80a, 157a). Because of the comparatively high levels of RA required to activate RXRs, some studies have suggested that other retinoids may specifically act through these receptors (88), and indeed 9-*cis* RA appears to be the likely candidate (62a, 81a). The finding that RXRs interact with RARs to enhance their specific activities indicates that the effect of RA on the expression of a particular target gene will depend on (i) which combinations of RAR/RXRs are present and (ii) whether all-*trans* RA, 9-*cis* RA, or both isomers are present in a given cell.

RAREs   RARs have been shown to activate gene transcription through a number of different types of responsive elements including EREs, TREs, RAREs, as well as ERE half-sites (see above), consisting of direct repeats, inverted repeats, and isolated half-sites of the motif G(G/T)TCA. How effectively do these sites promote the RA activation of gene expression by binding RARs? The affinity of the RARs for these sites will likely affect the fold induction of the target gene. How does the orientation of half-sites (direct, or inverted) affect the ability of RAR subtypes and isoforms to bind and activate transcription from these response elements? Do different combinations of RAR/RXR complexes exhibit different response element selectivities? The answer to this latter question appears to be yes. Consistent with this notion are the results of in vitro DNA binding experiments using different response elements and, indirectly, the results of co-transfection experiments (73a, 80a, 157a, 160). The answers to these fundamental questions will become clearer as additional RAREs become characterized.

OTHER TRANSCRIPTION FACTORS   The possibility that RARs form heterodimers with c-ErbA suggests that associations similar to that observed for RARs and RXRs may also exist between RARs and other nuclear receptors. RAR mutants such as PML-RAR-α observed in acute promyelocytic leukemias may also exert deleterious effects on RA response through heterodimer formation (33, 43, 54). Conceivably, such interactions would result not only in different levels of transcription activation but could also affect target gene specificity (53, 54, 80, 103).

Finally, the contextual location of an RARE in a promoter may significantly influence whether or not a particular RA target gene is expressed. For example, the pattern of expression of a β-galactosidase marker gene under the control of the RAR-$\beta2$ promoter in transgenic mice suggests that the mere presence of RA is insufficient to activate the RARE-bearing promoter (94). The activity of a RARE, nestled among other *cis*-acting elements, may depend on the presence or absence of the corresponding *trans*-acting factors that bind to them. The interactions between RARs and AP1-binding factors are particularly interesting: they may be competitive, as in the case of the osteocalcin promoter, or direct, as in the regulation of the collagenase promoter (see above). These interactions link two separate signal transduction systems. It is not difficult to see how this link is necessary for the orchestration of cell growth and differentiation during the highly complex process of development.

SUMMARY

Over the past five years, a wealth of information has accumulated concerning the molecular mechanisms mediating RA effects on gene expression. The molecular cloning of the 3 retinoic acid receptors (RARs), of their 16 or so different isoforms, and of the 3 retinoid X receptors (RXRs) as well as the identification of at least 2 different active isomers of RA (all *trans*- and 9-*cis*-RA) and of several different CRABPs and CRBPs now provide the essential tools to explain the pleiotropy that has become associated with RA effects. In the years to come, a concentrated effort to delineate the complex interactions between the various components of the retinoid signal transduction system should shed light on the mechanisms underlying pattern formation during vertebrate development and point to new ways in which retinoids can be exploited therapeutically.

ACKNOWLEDGMENTS

I thank J. White for preparing the figures, P. Greer and R. Deeley for helpful comments, and S. Tirreli and M. McCallum for typing the manuscript. This review is dedicated to Pierre Chambon and the retinoic acid receptor group at the L.G.M.E. in Strasbourg, in particular, N. Brand, A. Krust, A. Zelent, C. Mendelsohn, and P. Kastener, for creating an exciting environment for postdoctoral studies.

Literature Cited

1. Alcalay, M., Zangrilli, D., Pandolfi, P. P., Longo, L., Mencarelli, A., et al. 1991. Translocation breakpoint of acute promyelocytic leukemia lies within the retinoic acid receptor alpha locus. *Proc. Natl. Acad. Sci. USA* 88:1977–81

1a. Altaba, A. R. I. 1991. Vertebrate development: an emerging synthesis. *Trends Genet.* 7:276–80

2. Altaba, A. R., Melton, D. A. 1989. Interaction between peptide growth factors and homeobox genes in the establishment of antero-posterior polarity in frog embryos. *Nature* 341:33–38

3. Astrom, A., Pettersson, U., Krust, A., Chambon, P., Voorhees, J. J. 1990. Retinoic acid and synthetic analogs differentially activate retinoic acid receptor dependent transcription. *Biochem. Biophys. Res. Commun.* 173:339–45

4. Bagchi, M. K., Tsai, S. Y., Tsai, M., O'Malley, B. W. 1991. Progesterone enhances target gene transcription by receptor free of heat shock proteins hsp90, hsp56 and hsp70. *Mol. Cell. Biol.* 11:4998–5004

5. Bailey, J. S., Siu, C. H. 1990. Purification of cellular retinoic acid-binding proteins types I and II from neonatal rat pups. *Methods Enzymol.* 189:356–63

6. Balling, R. 1991. CRABP and the teratogenic effects of retinoids. *Trends Genet.* 7:35–36

7. Blomhoff, R., Green, M. H., Berg, T., Norum, K. R. 1990. Transport and storage of vitamin A. *Science* 250:399–403

8. Borrow, J., Goddard, A. D., Sheer, D., Solomon, E. 1990. Molecular analysis of acute promyelocytic leukemia breakpoint cluster region of chromosome 17. *Science* 249:1577–80

9. Boylan, J. F., Gudas, L. J. 1991. Overexpression of the cellular retinoic acid binding protein-I (CRABP-1) results in a reduction in differentiation-specific gene expression in F9 teratocarcinoma cells. *J. Cell Biol.* 112:965–79

10. Brand, N. J., Petkovich, M., Chambon, P. 1990. Characterization of a functional promoter for the human retinoic acid receptor-alpha (hRAR-alpha). *Nucleic Acids Res.* 18:6799–6806

11. Brand, N., Petkovich, M., Krust, A., Chambon, P., de The, H., et al. 1988. Identification of a second human retinoic acid receptor. *Nature* 332:850–53

12. Brockes, J. 1991. We may not have a morphogen. *Nature* 350:15

13. Brockes, J. P. 1989. Retinoids, homeobox genes, and limb morphogenesis. *Neuron* 2:1285–94

14. Brockes, J. P. 1990. Retinoic acid and limb regeneration. *J. Cell Sci. Suppl.* 13:191–98

15. Busch, C., Sakena, P., Funa, K., Nordlinder, H., Eriksson, U. 1990. Tissue distribution of cellular retinol-binding protein and cellular retinoic acid-binding protein: Use of monospecific anti-bodies for immunohistochemical and cRNA for in situ localization. *Method Enzymol.* 189:315–24

16. Cavey, M. T., Martin, B., Carlavan, I., Shroot, B. 1990. In vitro binding of retinoids to the nuclear retinoic acid receptor alpha. *Anal. Biochem.* 186:19–23

17. Chambon, P., Zelent, A., Petkovich, M., Mendelsohn, C., Leroy, P., et al. 1991. The family of retinoic acid nuclear receptors. In *Retinoids. 10 Years On*, ed. J-H. Saurat, pp. 10–27. Basel, Switzerland: Karger

18. Chang, K. S., Trujillo, J. M., Ogura, T., Castiglione, C. M., Kidd, K. K., et al. 1991. Rearrangement of the retinoic acid receptor gene in acute promyelocytic leukemia. *Leukemia* 5:200–4

19. Chen, Z., Chen, S. J., Tong, J. H., Zhu, Y. J., Huang, M. E., et al. 1991. The retinoic acid alpha receptor gene is frequently disrupted in its 5' part in Chinese patients with acute promyelocytic leukemia. *Leukemia* 4:288–92

20. Cho, K. W. Y., De Robertis, E. M. 1990. Differential activation of *Xenopus* homeobox genes by mesoderm-inducing growth factors and retinoic acid. *Genes Dev.* 4:1910–16

21. Clarke, C. L., Graham, J., Roman, S. D., Sutherland, R. D. 1991. Direct transcriptional regulation of the progesterone receptor by retinoic acid diminishes progestin responsiveness in the breast cancer cell line T-47D. *J. Biol. Chem.* 266:18969–75

22. Cleary, M. L. 1991. Oncogenic conversion of transcription factors by chromosomal translocations. *Cell* 66:619–22

23. Clifford, J. L., Petkovich, M., Chambon, P., Lotan, R. 1990. Modulation by retinoids of mRNA levels for nuclear retinoic acid receptors in murine melanoma cells. *Mol. Endocrinol.* 4:1546–55

24. Crettaz, M., Baron, A., Siegenthaler, G., Hunziker, W. 1990. Ligand specificities of recombinant retinoic acid receptors RAR alpha and RAR beta. *Biochem. J.* 272:391–97

25. Dalman, F. C., Sturzenbecker, L. J., Levin, A. A., Lucas, D. A., Perdew, G. H., et al. 1991. Retinoic acid receptor belongs to a subclass of nuclear receptors that do not form "docking" complexes with hsp90. *Biochemistry* 30:5605–8

26. Danielsen, M., Hinck, L., Ringold, G. M. 1989. Two amino acids within the knuckle of the first zinc finger specify DNA response element activation by the glucocorticoid receptor. *Cell* 57:1131–38

27. Darling, D. S., Beebe, J. S., Burnside, J., Winslowk, E. R., Chin, W. W. 1991. 3,5,3'-Triiodothyronine (T3) receptor-auxiliary protein (TRAP) binds DNA and forms heterodimers with the T3 receptor. *Mol. Endocrinol.* 5:73–84

28. Davidson, E. H. 1990. How embryos work: a comparative view of diverse modes of cell fate specification. *Development* 108:365–89

29. de Groot, R. P., Pals, C., Kruijer, W. 1991. Transcriptional control of c-jun by retinoic acid. *Nucleic Acids Res.* 19:1585–91

30. Dejean, A., de The, H. 1990. Hepatitis B virus as an insertional mutagene in a human hepatocellular carcinoma. *Mol. Biol. Med.* 7:213–22

31. Denner, L. A., Weigel, N. L., Maxwell, B. L., Schrader, W. T., O'Malley, B. W. 1990. Regulation of progesterone receptor-mediated transcription by phosphorylation. *Science* 250:1740–43

32. de Pablo, F., Roth, J. 1990. Endocrinization of the early embryo: an emerging role for hormones and hormone-like factors. *Trends Biochem. Sci.* 15:339–42

33. Desbois, C., Aubert, D., Legrand, C., Pain, B., Samarut, J. 1991. A novel mechanism of action for v-ErbA: Abrogation of the inactivation of transcription factor AP-1 by retinoic acid and thyroid hormone receptors. *Cell* 67:731–40

34. de The, H., Chomienne, C., Lanotte, M., Degos, L., Dejean, A. 1990. The t(15;17) translocation of acute promyelocytic leukemia fuses the retinoic acid receptor alpha gene to a novel transcribed locus. *Nature* 347:558–61

35. de The, H., del Mar Vivanco-Ruiz, M., Tiollais, P., Stunnenberg, H., Dejean, A. 1990. Identification of a retinoic acid responsive element in the retinoic acid receptor beta gene. *Nature* 343:177–80

36. de The, H., Lavau, C., Marchio, A., Chomienne, C., Degos, L., et al. 1991. The PML-RAR alpha fusion mRNA generated by the t(15;17) translocation in acute promyelocytic leukemia encodes a functionally altered RAR. *Cell* 66:675–84

37. Diamond, M. I., Miner, J. I., Yoshinga, S. K., Yamamoto, K. R. 1990. Transcription factor interactions: selectors of positive or negative regulation from a single DNA element. *Science* 249:1266–72

38. Dolle, P., Ruberte, E., Kastner, P., Petkovich, M., Stoner, C. M., et al. 1989. Differential expression of gene encoding alpha, beta, and gamma retinoic acid receptors and CRABP in the developing limbs of the mouse. *Nature* 342:702–5

39. Dolle, P., Ruberte, E., Leroy, P., Morriss-Kay, G., Chambon, P. 1990. Retinoic acid receptors and cellular retinoid binding proteins. I. A systematic study of their differential pattern of transcription during mouse organogenesis. *Development* 110:1133–51

40. Durston, A. J., Timmermans, J. P. M., Hage, W. J., Hendriks, H. F. J., de Vries, N. J., et al. 1989. Retinoic acid causes an anteroposterior transformation in the developing central nervous system. *Nature* 340:140–44

41. Eichele, G. 1989. Retinoids and vertebrate pattern formation. *Trends Genet.* 5:246–51

42. Ellinger-Ziegelbauer, H., Dreyer, C. 1991. A retinoic acid receptor expressed in the early development of *Xenopus laevis*. *Genes Dev.* 5:94–104

43. Evans, R. M. 1988. The steroid and thyroid hormone receptor superfamily. *Science* 240:889–95

44. Fawell, S. E., Lees, J. A., White, R., Parker, M. G. 1990. Characterization and colocalization of steroid binding and dimerization activities in the mouse estrogen receptor. *Cell* 60:953–62

45. Forman, B. M., Herbert, H. S. 1991. Interactions among a subfamily of nuclear hormone receptors: The regulatory zipper model. *Mol. Endocrinol.* 4:1293–1301

46. Forman, B. M., Yang, C. R., Au, M., Casanova, J., Ghysdael, J., et al. 1989. A domain containing leucine-zipper-like motifs mediate novel in vivo interactions between the thyroid hormone and retinoic acid receptors. *Mol. Endocrinol.* 3:1610–26

47. Frankel, A. D., Kim, P. S. 1991. Modular structure of transcription factors: Implications for gene regulation. *Cell* 65:717–19

48. Gerhart, J. 1989. The primacy of cell interactions in development. *Trends Genet.* 5:233–36

49. Giguere, V., Lyn, S., Yip, P., Siu, C-H., Amin, S. 1990. Molecular cloning of cDNA encoding a second cellular retinoic acid-binding protein. *Proc. Natl. Acad. Sci. USA* 87:6233–37

50. Giguere, V., Ong, E. S., Segui, P., Evans, R. M. 1987. Identification of a receptor for the morphogen retinoic acid. *Nature* 330:624–29

51. Giguere, V., Shago, M., Zirngibl, R., Tate, P., Rossant, J., et al. 1990. Identification of a novel isoform of the retinoic acid receptor gamma expressed in the mouse embryo. *Mol. Cell. Biol.* 10:2335–40

52. Giguere, V., Yang, N., Segui, P., Evans, R. M. 1988. Identification of a new class of steroid hormone receptors. *Nature* 331:91–94

53. Glass, C. K., Devary, O. V., Rosenfeld, M. G. 1990. Multiple cell type-specific proteins differentially regulate target sequence recognition by the alpha retinoic acid receptor. *Cell* 63:729–38

54. Glass, C. K., Lipkin, S. K., Devary, O. V., Rosenfeld, M. G. 1989. Positive and negative regulation of gene transcription by a retinoic acid-thyroid hormone receptor heterodimer. *Cell* 59:697–708

55. Graupner, G., Willia, K. N., Tzukerman, M., Zhang, X., Pfahl, M. 1989. Dual regulatory role for thyroid-hormone receptors allows control of retinoic-acid receptor activity. *Nature* 340:653–56

56. Green, S., Chambon, P. 1988. Nuclear receptors enhance our understanding of transcriptional regulation. *Trends Genet.* 4:309–14

57. Gudas, L. J. 1990. Assays for expression of genes regulated by retinoic acid in murine teratocarcinoma cell lines. *Methods Enzymol.* 190:131–40

58. Guiochon-Mantel, A., Lescop, P., Christin-Maitre, S., Loosfelt, H., Perrot-Applanat, M., et al. 1991. Nucleocytoplasmic shuttling of the progesterone receptor. *EMBO J.* 10:3851–59

59. Hamada, K., Gleason, S. L., Levi, B. Z., Hirschfeld, S., Appella, E., et al. 1991. H-2RIIBP, a member of the nuclear hormone receptor superfamily that binds both the regulatory element of major histocompatibility class I genes and the estrogen response element. *Proc. Natl. Acad. Sci. USA* 86:8289–93

60. Hard, T., Kellenbach, E., Boelens, R., Maler, B. A., Dahlman, K., et al. 1990. Solution structure of the glucocorticoid receptor DNA-binding domain. *Science* 249:157–60

61. Hashimoto, Y., Kagechika, H., Shudo, K. 1990. Expression of retinoic acid receptor genes and the ligand-binding selectivity of retinoic acid receptors (RAR's). *Biochem. Biophys. Res. Commun.* 166:1300–7

62. He, X., Rosenfeld, M. G. 1991. Mechanisms of complex transcriptional regulation: Implications for brain development. *Neuron* 7:183–96

62a. Heyman, R., Mangelsdort, P. J., Dyck, J. A., Stein, R. B., Eichele, G. et al. 1992. 9-cis retinoic acid is a high affinity ligand for the retinoid x receptor. *Cell* 68:397–406

63. Hoffman, M. 1990. The embryo takes its vitamins. *Science* 250:372–73

64. Hoffmann, B., Lehmann, J. M., Zhang, X-K., Hermann, T., Husmann, M., et al. 1990. A retinoic acid receptor-specific element controls the retinoic acid receptor beta promoter. *Mol. Endocrinol.* 4:1727–36

65. Hudson, L. G., Santon, J. B., Glass, C. K., Gill, G. N. 1990. Ligand-activated thyroid hormone and retinoic acid receptors inhibit growth factor receptor promoter expression. *Cell* 62:1165–75

66. Husmann, M., Lehmann, J., Hoffmann, B., Hermann, T., Tzukerman, M., et al. 1991. Antagonism between retinoic acid receptors. *Mol. Cell. Biol.* 11:4097–4103

67. Ishikawa, T., Umesono, K., Mangelsdorf, D. J., Aburatani, H., Stanger, B. Z., et al. 1990. A functional retinoic acid receptor encoded by the gene on human chromosome 12. *Mol. Endocrinol.* 4:837–44

68. Izpisua-Belmonte, J-C., Tickle, C., Dolle, P., Wolpert, L., Duboule, D. 1991. Expression of the homeobox Hox-4 genes and the specification of position in chick wing development. *Nature* 350:585–89

69. Jones, B., Jurutka, P., Haussler, C., Haussler, M., Whitfield, K. 1991. Vitamin D receptor phosphorylation in transfected ROS 17/2.8 cells is localized to the N-terminal region of the hormone-binding domain. *Mol. Endocrinol.* 5:1137–46

70. Kakizuka, A., Miller, W. H. Jr., Umesono, K., Warrell, R. P., et al. 1991. Chromosomal translocation t(15;17) in human acute promyelocytic leukemia fuses RAR alpha with a novel putative transcription factor, PML. *Cell* 66:663–74

71. Kastner, P., Krust, A., Mendelsohn, C., Garnier, J. M., Zelent, A., et al. 1990. Murine isoforms of retinoic acid receptor gamma with specific patterns of expression. *Proc. Natl. Acad. Sci. USA* 87:2700–4

72. Kim, K. H., Griswold, M. D. 1990. The regulation of retinoic acid receptor mRNA levels during spermatogenesis. *Mol. Endocrinol.* 4:1679–88

73. Kimmel, C. B., Warga, R. M., Schilling, T. F. 1990. Origin and organization of the Zebrafish fate map. *Development* 108:581–94

73a. Kliewer, S. A., Umesono, K., Mangelsdorf, D., Evans, R. M. 1992. Retinoid X receptor interacts with nuclear receptors in retinoic acid, thyroid hormone and vitamin D₃ signalling. *Nature* 355:446–49

74. Kozak, M. 1991. An analysis of vertebrate mRNA sequences: Intimations of translational control. *J. Cell Biol.* 115:887–903

75. Krust, A., Kastner, P., Petkovich, M., Zelent, A., Chambon, P. 1989. A third human retinoic acid receptor, hRAR-alpha. *Proc. Natl. Acad. Sci. USA* 86:5210–14

76. Kumar, V., Chambon, P. 1988. The estrogen receptor binds tightly to its responsive element as a ligand-induced homodimer. *Cell* 55:145–56

77. Kumar, V., Green, S., Stack, G., Berry, M., Jin, J. R., Chambon, P. 1987. Functional domains of the human estrogen receptor. *Cell* 51:941–51

78. Kuo, C. J., Mendel, D. B., Hansen, L. P., Crabtree, G. R. 1991. Independent regulation of HNF-1 alpha and HNF-1 beta by retinoic acid in F9 teratocarcinoma cells. *EMBO J.* 10:2231–36

79. Landschultz, W. H., Johnson, P. F., McKnight, S. L. 1988. The Leucine zipper: a hypothetical structure common to a new class of DNA binding proteins. *Science* 240:1759–64

80. Lazar, M. A., Berrodin, T. J., Harding, H. P. 1991. Differential DNA binding by monomeric, homodimeric and potentially heteromeric forms of the thyroid hormone receptor. *Mol. Cell. Biol.* 11:5005–15

80a. Leid, M., Kautner, P., Lyons, R., Nakshatri, H., Saunders, M., et al. 1992. Purification, cloning, and RXR identity of the HeLa cell factor with which RAR or TR heterodimerizes to bind target sequences efficiently. *Cell* 68:377–95

81. Leroy, P., Krust, A., Zelent, A., Mendelsohn, C., Garnier, J-M., et al. 1991. Multiple isoforms of the mouse retinoic acid receptor alpha are generated by alternative splicing and differential induction by retinoic acid. *EMBO J.* 10:59–69

81a. Levin, A., Sturzenbecker, L., Kazmer, S., Bosakowski, T., Huselton, C., et al. 1992. 9-cis Retinoic acid stereoisomer binds and activates the nuclear receptor RXRα. *Nature* 355:359–61

82. Longo, L., Donti, E., Mencarelli, A., Avanti, G., Pegoraro, L., et al. 1990. Mapping of chromosome 17 breakpoints in acute myeloid leukemias. *Oncogene* 5:1557–63

83. Longo, L., Pandolfi, P. P., Biondi, A., Rambaldi, A., Mencarelli, A., et al. 1990. Rearrangements and aberrant expression of the retinoic acid receptor alpha gene in acute promyelocytic leukemias. *J. Exp. Med.* 172:1571–75

84. Lucas, P. C., Forman, B. M., Samuels, H. H., Granner, D. K. 1991. Specificity of a retinoic acid response element in the phosphoenolpyruvate carboxykinase gene promoter: Consequences of both retinoic acid and thyroid hormone receptor binding. *Mol. Cell. Biol.* 11:5164–70

85. Lucas, P. C., O'Brien, R. M., Mitchell, J. A., Davis, C. M., Imai, E., et al. 1991. A retinoic acid response element is a part of a pleiotropic domain in the phosphoenolpyruvate carboxykinase gene. *Proc. Natl. Acad. Sci. USA* 88:2184–88

86. Luisi, B. F., Xu, W. X., Otwinowski, Z., Freedman, L. P., Yamamoto, K. R., et al. 1991. Crystallographic analysis of the interaction of the glucocorticoid receptor with DNA. *Nature* 352:497–505

87. Mader, S., Kumar, V., de Verneuil, H., Chambon, P. 1989. Three amino acids of the oestrogen receptor are essential to its ability to distinguish an oestrogen from a glucocorticoid responsive element. *Nature* 338:271–74

88. Mangelsdorf, D. J., Ong, E. S., Dyck, J. A., Evans, R. M. 1990. Nuclear receptor that identifies a novel retinoic

acid response pathway. *Nature* 345:224–29

89. Mangelsdorf, D. J., Umesono, K., Kliewer, S. A., Borgmeyer, U., Ong, E. S., et al. 1991. A direct repeat in the cellular retinol-binding protein type II gene confers differential regulation by RXR and RAR. *Cell* 66:555–61

90. Martinez, E., Givel, F., Wahli, W. 1991. A common ancestor DNA motif for invertebrate and vertebrate hormone response elements. *EMBO J.* 10:263–68

91. Matareses, V., Buelt, M. K., Chinander, L. L., Bernlohr, D. A. 1990. Purification of adipocyte lipid-binding protein from human and murine cells. *Methods Enzymol.* 189:363–69

92. McDonnell, D. P., Nawaz, Z., O'Malley, B. W. 1991. In situ distinction between steroid receptor binding and transactivation at a target gene. *Mol. Cell. Biol.* 11:4350–55

93. Melton, D. A. 1991. Pattern formation during animal development. *Science* 252:234–41

94. Mendelsohn, C., Ruberte, E., LeMeur, M., Morriss-Kay, G., Chambon, P. 1991. Developmental analysis of the retinoic acid-inducible RAR-b2 promotor in transgenic animals. *Development* 113:723–34

95. Miller, W. H. Jr., Warrell, R. P. Jr., Frankel, S. R. A., Jakubowski, J. L., Gabrilove, J. M., et al. 1990. Novel retinoic acid receptor-a transcripts in acute promyelocytic leukemia responsive to all-trans-retinoic acid. *J. Natl. Cancer Inst.* 82:1932–33

96. Mornon, J. P., Bissery, V., Gaboriaud, C., Thomas, A., Ojasoo, T., et al. 1989. Hydrophobic cluster analysis (HCA) of the hormone-binding domain of receptor proteins. *J. Steroid Biochem.* 34:355–61

97. Munoz-Canoves, P., Vik, D. P., Tack, B. F. 1990. Mapping of a retinoic acid-responsive element in the promoter region of the complement factor H gene. *J. Biol. Chem.* 265:20065–68

98. Naar, A. M., Boutin, J., Lipkin, S. M., Yu, V. C., Holloway, J. M., et al. 1991. The orientation and spacing of core DNA-binding motifs dictate selective transcriptional responses to three nuclear receptors. *Cell* 65:1267–79

99. Nicholson, R. C., Mader, S., Nagpal, S., Leid, M., Rochette-Egly, C., et al. 1990. Negative regulation of the rat stromelysin gene promoter by retinoic acid is mediated by an AP1 binding site. *EMBO J.* 9:4443–54

100. Nohno, T., Muto, K., Noji, S., Saito, T., Taniguchi, S. 1991. Isoforms of

retinoic acid receptor beta expressed in the chick embro. *Biochim. Biophys. Acta* 1089:273–75

101. Noji, S., Nohno, T., Koyama, E., Muto, K., Ohyama, K., et al. 1991. Retinoic acid induces polarizing activity but is unlikely to be a morphogen in the chick limb bud. *Nature* 350:83–86

102. Noji, S., Yamaai, T., Koyama, E., Nohno, T., Fujimoto, W., et al. 1989. Expression of retinoic acid receptor gene in keratinizing front of skin. *FEBS Lett.* 259:86–90

103. O'Donnell, A. L., Rosen, E. D., Darling, D. S., Koenig, R. J. 1991. Thyroid hormone receptor mutations that interfere with transcriptional activation also interfere with receptor interaction with a nuclear protein. *Mol. Endocrinol.* 5:94–99

104. Oro, A. E., McKeown, M., Evans, R. M. 1990. Relationship between the product of the ultraspiracle locus and the vertebrate retinoid X receptor. *Nature* 347:298–301

105. Pandolfi, P. P., Grignani, F., Alcalay, M., Mencarelli, A., Biondi, A., et al. 1991. Structure and origin of the acute promyelocytic leukemia myl/RAR alpha cDNA and characterization of its retinoid-binding and transactivation properties. *Oncogene* 6:1285–92

106. Petkovich, M., Brand, N. J., Krust, A., Chambon, P. 1987. A human retinoic acid receptor which belongs to the family of nuclear receptors. *Nature* 330:444–50

107. Petkovich, P. M., Heersche, J. N. M., Tinker, D. O., Jones, G. 1984. Retinoic acid stimulates 1,25-dihydroxyvitamin D3 binding in rat osteosarcoma cells. *J. Biol. Chem.* 259:8274–80

108. Picard, D., Khursheed, B., Garabedian, M. J., Fortin, M. G., Lindquist, S., et al. 1990. Reduced levels of hsp90 compromise steroid receptor action in vivo. *Nature* 348:166–68

109. Picard, D., Salser, S. J., Yamamoto, K. R. 1988. A movable and regulable inactivation function within the steroid binding domain of the gucocorticoid receptor. *Cell* 54:1073–80

110. Pratt, W. B., Jolly, D. J., Pratt, D. V., Hollenberg, S. M., Giguere, V., et al. 1988. A region in the steroid binding domain determines formation of the non-DNA-binding, 9 S glucocorticoid receptor complex. *J. Biol. Chem.* 263:267–73

111. Ragsdale, C. W., Petkovich, M., Gates, P. B., Chambon, P., Brockes, J. P. 1989. Identification of a novel retinoic acid receptor in regenerative tissues of the newt. *Nature* 341:654–57

112. Redfern, C. P. F., Daly, A. K. 1990. Cellular retinoic acid-binding protein from neonatal rat skin: purification and analysis. *Methods Enzymol.* 189:307–14

113. Redfern, C. P. F., Daly, A. K., Latham, J. A. E., Todd, C. 1990. The biological activity of retinoids in melanoma cells: Induction of expression of retinoic acid receptor beta by retinoic acid in S91 melanoma cells. *FEBS Lett.* 273:19–22

114. Rochette-Egly, C., Lutz, Y., Saunders, M., Scheuer, I., Gaub, M-P., et al. 1991. Retinoic acid receptor gamma: Specific immunodetection and phosphorylation. *J. Cell Biol.* 115:535–45

115. Rossant, J., Joyner, A. L. 1989. Towards a molecular genetic analysis of mammalian development. *Trends Genet.* 5:277–83

116. Rossant, J., Zirngibl, R., Cado, D., Shago, M., Giguere, V. 1991. Expression of a retinoic acid response element-hsplacZ transgene defines specific domains of transcriptional activity during mouse embryogenesis. *Genes Dev.* 5:1333–44

117. Rottman, J. N., Widom, R. L., Madal-Ginard, B., Mahdavi, V., Karathanasis, S. K. 1991. A retinoic acid-responsive element in the apolipoprotein AI gene distinguishes between two different retinoic acid response pathways. *Mol. Cell. Biol.* 11:3814–20

118. Rowe, A., Eager, N. S. C., Brickell, P. M. 1991. A member of the RXR nuclear receptor family is expressed in neural-crest-derived cells of the developing chick peripheral nervous system. *Development* 111:771–78

119. Ruberte, E., Dolle, P., Chambon, P., Morriss-Kay, G. 1991. Retinoic acid receptors and cellular retinoid binding proteins. II. Their differential pattern of transcription during early morphogenesis in mouse embryos. *Development* 111:45–60

120. Ruberte, E., Dolle, P., Krust, A., Zelent, A., Morriss-Kay, G., et al. 1990. Specific spatial and temporal distribution of retinoic acid receptor gamma transcripts during mouse embryogenesis. *Development* 108:213–22

121. Deleted in proof.

122. Ruiz, M. d. M. V., Bugge, T. H., Hirschmann, P., Stunnenberg, H. G. 1991. Functional characterization of a natural retinoic acid responsive element. *EMBO J.* 10:3829–38

123. Sani, B. P., Singh, R. K., Reddy, L. G., Gaub, M-P. 1990. Isolation, partial purification and characterization of nu-

clear retinoic acid receptors from chick skin. *Arch. Biochem. Biophys.* 283:107–13

124. Schule, R., Rangarajan, P., Kliewer, S., Ransone, L. J., Bolado, J., et al. 1990. Functional antagonism between oncoprotein c-jun and the glucocorticoid receptor. *Cell* 62:1217–26

125. Schule, R., Umesono, K., Mangelsdorf, D. J., Bolado, J., Pike, J. W., et al. 1991. Jun-fos and receptors for vitamin A and D recognize a common response element in the human osteocalcin gene. *Cell* 61:497–504

126. Selmi, S., Samuels, H. H. 1991. Thyroid hormone receptor/and v-erbA. *J. Biol. Chem.* 266:11589–93

127. Shubeita, H. E., Sambrook, J. F., McCormick, A. M. 1987. Molecular cloning and analysis of functional cDNA and genomic clones encoding bovine cellular retinoic acid-binding protein. *Proc. Natl. Acad. Sci. USA* 84:5645–49

128. Siegenthaler, G. 1990. Gel electrophoresis of cellular retinoic acid-binding protein, cellular retinol-binding protein and serum retinol-binding protein. *Methods Enzymol.* 189:299–307

129. Sive, H. L., Cheng, P. F. 1991. Retinoic acid perturbs the expression of Xhox lab genes and alters mesodermal determination in *Xenopus laevis. Genes Dev.* 5:1321–32

130. Smith, J. C. 1989. Induction and early amphibian development. *Curr. Opin. Cell Biol.* 1:1061–70

131. Smith, S. M., Eichele, G. 1991. Temporal and regional differences in the expression pattern of distinct retinoic acid receptor-beta transcripts in the chick embryo. *Development* 111:245–52

132. Smith, W. C., Nakshatri, H., Leroy, P., Rees, J., Chambon, P. 1991. A retinoic acid response element is present in the mouse cellular retinol binding protein I (mCRBP1) promoter. *EMBO J.* 10:2223–30

133. Stellmach, V., Leask, A., Fuchs, E. 1991. Retinoid-mediated transcriptional regulation of keratin genes in human epidermal and squamous cell carcinoma cells. *Proc. Natl. Acad. Sci. USA* 88:4582–86

134. Stoner, C. M., Gudas, L. J. 1989. Mouse cellular retinoic acid binding protein: Cloning, complementary DNA sequence, and messenger RNA expression during the retinoic acid-induced differentiation of F9 wild type and RA-3-10 mutant teratocarcinoma cells. *Cancer Res.* 49:1497–1504

135. Struhl, K. 1991. Mechanisms for diversity in gene expression patterns. *Neuron* 7:177–81

136. Sucov, H. M., Murakami, K. K., Evans, R. M. 1990. Characterization of an autoregulated response element in the mouse retinoic receptor type beta gene. *Proc. Natl. Acad. Sci. USA* 87:5392–96

137. Suva, L. J., Ernst, M., Rodan, G. A. 1991. Retinoic acid increases zif268 early gene expression in rat preosteoblastic cells. *Mol. Cell. Biol.* 11:2503–10

138. Tabin, C. J. 1991. Retinoids, homeoboxes, and growth factors: Toward molecular models for limb development. *Cell* 66:199–217

139. Tamara, K., Kagechika, H., Hashimoto, Y., Shudo, K., Ohsugi, K., et al. 1990. Synthetic retinoids, retinobenzoic acids, Am80, Am580 and Ch55 regulate morphogenesis in chick limb bud. *Cell Differ. Dev.* 32:17–26

140. Tasset, D., Tora, L., Fromental, C., Scheer, E., Chambon, P. 1990. Distinct classes of transcriptional activating domains function by different mechanisms. *Cell* 62:117–87

141. Thaller, C., Eichele, G. 1990. Isolation of 3,4-didehydroretinoic acid, a novel morphogenic signal in the chick wing bud. *Nature* 345:815–19

142. Tomomura, M., Kodomatsu, K., Nakamoto, M., Muramatsu, H., Kondoh, H., et al. 1990. A retinoic acid responsive gene, MK, produces a secreted protein with heparin binding activity. *Biochem. Biophys. Res. Commun.* 171:603–9

143. Tora, L., Gronemeyer, H., Turcott, E. B., Gaub, M. P., Chambon, P. 1988. The N-terminal region of the chicken progesterone receptor specifies target gene activation. *Nature* 333:185–88

144. Umesono, K., Evans, R. M. 1989. Determinants of target gene specificity for steroid/thyroid hormone receptors. *Cell* 57:1139–46

145. Umesono, K., Murakami, K. K., Thompson, C. C., Evans, R. M. 1991. Direct repeats as selective response elements for the thyroid hormone, retinoic acid, and vitamin D3 receptors. *Cell* 65:1255–66

146. Urios, P., Duprez, D., Le Caer, J-P., Courtois, Y., Vigny, M., et al. 1991. Molecular cloning of RI-HB, A heparin binding protein regulated by retinoic acid. *Biochem. Biophys. Res. Commun.* 175:617–24

147. Vasios, G., Mader, S., S., Gold, J. D., Leid, M., Lutz, Y., et al. 1991. The late retinoic acid induction of laminin B1 gene transcription involves RAR binding to the responsive element. *EMBO J.* 10:1149–58

148. Vasios, G. W., Gold, J. D., Petkovich, M., Chambon, P., Gudas, L. J. 1989. A retinoic acid-responsive element is present in the 5' flanking region of the laminin B1 gene. *Proc. Natl. Acad. Sci. USA* 86:9099–9103

149. Wagner, M., Thaller, C., Jessell, T., Eichele, G. 1990. Polarizing activity and retinoid synthesis in the floor plate of the neural tube. *Nature* 345:819–22

150. Wanek, N., Gardiner, D. M., Muneoka, K., Bryant, S. V. 1991. Conversion by retinoic acid of anterior cells into ZPA cells in the chick wing bud. *Nature* 350:81–83

151. Webster, N. J. G., Green, S., Jin, J. R., Chambon, P. 1988. The hormone-binding domains of the estrogen and glucocorticoid receptors contain an inducible transcription activation function. *Cell* 54:199–207

152. Webster, N. J. G., Green, S., Jin, J. R., Hollis, M., Chambon, P. 1988. The yeast UAS is a transcriptional enhancer in human HeLa cells. *Cell* 52:169–78

153. Wei, L-N., Tsao, J-L., Chu, Y-S., Jeannotte, L., Nguyen-Huu, M. C. 1990. Molecular cloning and transcriptional mapping of the mouse cellular retinoic acid-binding protein gene. *DNA Cell Biol.* 9:471–78

154. White, J. H., Brou, C., Wu, J., Burton, N., Egly, J-M., et al. 1991. Evidence for a factor required for transcriptional stimulation by the chimeric acidic activator GAL-VP16 in HeLa cell extracts. *Proc. Natl. Acad. Sci. USA* 88:7674–78

155. Wood, W. M., Ocran, K. W., Gordon, D. F., Ridgway, E. C. 1991. Isolation and characterization of mouse complementary DNA's encoding alpha and beta thyroid hormone receptors from thyrotrope cells: The mouse pituitary-specific beta-2 isoform differs at the amino terminus from the corresponding species from rat pituitary tumor cells. *Mol. Endocrinol.* 5:1049–61

156. Yang, N., Schule, R., Mangelsdorf, D. J., Evans, R. M. 1991. Characterization of DNA binding and retinoic acid binding properties of retinoic acid receptor. *Proc. Natl. Acad. Sci. USA* 88:3559–63

157. Yang-Yen, H-F., Chambard, J-C., Sun, Y-L., Smeal, T., Schmidt, T. J., et al. 1990. Transcriptional interference between c-jun and the glucocorticoid receptor: Mutual inhibition of DNA binding due to direct protein-protein interaction. *Cell* 62:1205–15

157a. Yu, V. C., Delsert, C., Andersen, B.,

Holloway, J. M., Derary, O., et al. 1991. RXRβ: A coregulator that enhances binding of retinoic acid, thyroid hormone, and vitamin D receptors to their cognate response elements. *Cell* 67:1251–66

158. Zelent, A., Krust, A., Petkovich, M., Kastner, P., Chambon, P. 1989. Cloning of murine alpha and beta retinoic acid receptors and a novel receptor gamma predominantly expressed in skin. *Nature* 339:714–17

159. Zelent, A., Mendelsohn, C., Kastner, P., Krust, A., Garnier, J-M., et al. 1991. Differentially expressed isoforms of the mouse retinoic acid receptor beta are generated by usage of two promoters and alternate splicing. *EMBO J.* 10:71–81

160. Zhang, X., Hoffman, B., Tran, P., Graupner, G., Pfahl, M. 1992. Retinoid X receptor is an auxiliary protein for thyroid hormone receptors and retinoic acid receptors. *Nature* 355:441–46

Annu. Rev. Nutr. 1992. 12:473–87

FAT SUBSTITUTES: A Regulatory Perspective[1]

J. E. Vanderveen and W. H. Glinsmann

Division of Nutrition, Center for Food Safety and Applied Nutrition, US Food and Drug Administration, 200 C Street SW, Washington, DC 20204

KEY WORDS: fat substitutes, food safety, regulation of dietary fats

CONTENTS

INTRODUCTION

The term *fat substitute* implies that a substance, when used as a replacement for the traditional fat contained in a food, has certain desirable physical or organoleptic properties of the fat that it replaces while lacking undesirable properties of this fat. Usually, a fat substitute is not a fat; however, the term also has been used to refer to those lipids that, because of their structure or elevated melting point, are not digestible or only partially digestible when consumed by humans.

---

[1]The US government has the right to retain a nonexclusive, royalty-free license in and to any copyright covering this paper.

Fats contribute favorably to the texture, flavor, and appearance of foods. Brief descriptions of the effects of fat on these three properties follow. A full description of these organoleptic attributes and their importance to perceived food quality as indicated by consumer acceptance would require substantial discussion and is not the purpose of this article; several reviews serve that need (8–10, 19). In-depth knowledge of the desirable properties of fats is necessary, however, for the successful development and use of fat substitutes (19). Fats differ substantially with regard to their effects on the organoleptic properties of foods, but the success of a fat substitute in providing desired functional properties largely dictates its potential use.

The amount of fat present in a food and the physical properties of that fat may determine many of the characteristics of the food. For example, the amount of fat in a food and the melting point of the fat will greatly affect the texture of the food. The texture of a food, in turn, helps determine "mouth feel" and other organoleptic characteristics that contribute to food acceptance. Fat content also affects the structure and color of a food. Finally, fats are important in determining the flavor or aromatic characteristics of some foods, because many flavor or aromatic components are fat-soluble. The effects of some flavor components may also be modified by the type and quantity of fat in the food.

The major impetus for the development of fat substitutes is that the nutritional properties of fat contribute to excessive energy intake and the development of disease. Not only do fats contain more than twice as many calories as other macronutrients in foods, but the dietary intake of total fat and specific fatty acids and lipids, such as cholesterol, is now considered a risk factor for some degenerative diseases. The Surgeon General's Report on Nutrition and Health (15), the National Academy of Sciences' Report on Diet and Health (13), and the National Cholesterol Education Program's Report of the Expert Panel on Population Strategies for Blood Cholesterol Reduction (12) review these relationships and provide a comprehensive consensus on diet-disease relationships. The development of fat substitutes is in part a market response to the perceived health benefits of lower consumption of fat (11).

In theory, replacement of fat with fat substitutes that reduce calorie content and lower exposure to specific lipids that augment the risk of degenerative diseases should contribute to the nutritional quality of the food supply. In practical terms, the validity of this hypothesis depends on (a) the physical and biochemical properties of the fat substitutes, (b) the extent to which such fat substitutes replace fat in the diets of those population segments for which a reduction in fat intake would be beneficial while not affecting the fat intake of those segments of the population that may not require reduced fat intake, and (c) the extent to which their use does not result in adverse effects such as nutritional deficiencies, excesses, or imbalances.

Also, when considering the potential health effects of fat substitutes, it is important to note that specific fats (i.e. fatty acids) have specific metabolic effects that vary according to the ratio of specific dietary fats, genetic predisposition, developmental stage, or disease condition. These metabolic effects influence hepatic cholesterol metabolism and the many biological functions that are altered by eicosanoid production. Cell membrane function, which is altered by lipid components, may also be affected. Therefore, the potential influence of fat substitutes on dietary lipid intake and any relationship to risk of developing chronic diseases is not easily assessed.

Historically, dietary fats and oils have been considered a primary source of energy without regard to the health effects of their specific complement of fatty acids and sterols. Economic considerations related to availability, cost, and technical function in a food product largely determined the use of a fat or oil. Many processing techniques, such as hydrogenation and restructuring of lipids, have been developed to use fats and oils from a wide variety of sources for many technical and functional effects. The wide application of fat-processing technology coupled with the use of "and/or" labeling has provided flexibility in marketing similar products with different ingredient sources and fat composition. And/or labeling refers to the regulatory provision in ingredient labeling that allows a manufacturer to cite the use of alternate fats without requiring a label change.

Traditional safety concerns have been largely limited to standard toxic endpoints, such as myocarditis associated with the high erucic acid content in rapeseed oils. In some cases, standards of identity, which defined the composition of products, governed the specific lipid composition of items such as butter, ice cream, and certain cheeses. When alternative fat sources were substituted in standardized products, they were labeled pejoratively as imitation products. If a standardized product was manufactured with a substitute fat (e.g. ice cream with partially hydrogenated soybean oil), then the product was considered adulterated and it was prevented from being marketed. In response to the recommendations of the White House Conference on Food, Nutrition, and Health (18) the Food and Drug Administration (FDA) published regulations that permitted these foods to be called substitutes provided such foods were not nutritionally inferior to the traditional food (17). The regulatory requirements for nutritional equivalency are that the substitute food must have equal or greater amounts of all nutrients that are contained in the traditional food at levels of two percent or more of the US RDA.

In recent years the marketing of fats has changed dramatically. So-called "heart-healthy" fats and the potential health benefits of diets with altered lipid composition are an active area for research.

The FDA is now engaged in extensive rule-making to modify food labeling according to the provisions of the Nutrition Labeling and Education Act of 1990 (NLEA) (14). Rule-making, as authorized by the Procedures Act, provides for regulations to implement the requirements of a law. The procedures involve publishing a proposal in the *Federal Register* for the purpose of receiving comment by individuals, industry, or other interested groups; reviewing comments and considering their merits in fulfilling the intent of the law; and publishing a final regulation, which has the force of law. Under special circumstances, any person who will be adversely affected by a final regulation may file objections with the Secretary of the Department of Health and Human Services and request a public hearing. Such action, under some circumstances, can cause a regulation to be stayed until final action on such objections is taken, including the holding of a hearing by an administrative law judge. Under these procedures, proposals for fat labeling were issued by the FDA in November 1991. These proposals, which would allow health claims for fats, are discussed later in this chapter.

SAFETY REQUIREMENTS FOR FOODS

The Federal Food, Drug, and Cosmetic (FD&C) Act (2) and its implementing regulations require that foods be inherently safe to be offered to the public for sale. The 1938 FD&C Act prohibits traffic in food that is injurious to health and prohibits the addition of poisons to food.

The 1958 Food Additive Amendment to the FD&C Act established a petition procedure for premarket approval of food additives. The amendment stipulates that a food is deemed to be adulterated if it contains any unsafe food additive and specifies that an unapproved food additive is unsafe. The amendment also describes the type of data necessary to evaluate the safety of a food additive, and it provides criteria for determining the safety and the suitability of the food additive for approval. Of particular note is the need to consider:

(a) The probable consumption of the additive and any substance formed in the food because of the use of the additive; (b) the cumulative effect of such additive in the diet . . ., taking into account any chemically or pharmacologically related substance or substances in such diet; and (c) safety factors which in the opinion of experts qualified by scientific training and experience to evaluate the safety of food additives are generally recognized as appropriate for the use of animal experimentation data.

The amendment also contains the Delaney Clause, which prohibits the approval of any food additive that is found to induce cancer when ingested by humans or animals. It provides administrative procedures for premarket evaluation and approval of a food additive and for judicial review of agency decisions. These procedures require that all critical data supporting an agency action are made publically available.

The Food Additive Amendment establishes that a food additive must be shown to be safe before it can be added to food; it stipulates that a food additive must achieve some physical or functional effect in food; and it establishes criteria for relative safety rather than for absolute safety and directs the agency to set tolerance levels for use of a food additive. Notably absent from criteria is any mention of a potential health benefit from an additive in food.

Confusion in marketing products with a novel composition may occur when safety is determined by independent scientific experts outside the FDA. The 1958 Food Additive Amendment to the FD&C Act establishes, as part of the definition of a food additive, a category for substances that are generally recognized as safe (GRAS) for their intended use in food. General recognition of an additive's safety is accorded by "experts qualified by scientific training and experience to evaluate its safety." The general recognition of safety may be founded either on "experience based on common use in food" before the amendment was introduced in 1958 or on "scientific procedures," which are essentially the same as the FDA uses in approving a new food additive. New fat substitutes that are made by processing of common food components require a determination of the GRAS status of their new uses. To establish a food use as GRAS, the FDA must carefully review previous uses and data publically available in the scientific literature. An independent GRAS determination cannot be based on data that is not publically available and cannot be supported by usage and experimental data when each alone would be inadequate to document safety of use. When an independent GRAS determination is made, the full responsibility for the safety of the food product resides with the manufacturer. The FDA may challenge such a determination or may affirm GRAS status using rule-making procedures.

Approving new fat substitutes and meeting the requirements of the food additive amendment may involve complex decisions. Often it is difficult to estimate the probable consumption and the cumulative effect on the diet that new fat substitutes and related substances may have. Application of a significant safety factor extrapolated from animal data may not be possible. Approaches used to prove the safety of fat substitutes may differ from those used for most food ingredients because such substances potentially may constitute a large portion of the diet of an individual. In some diets 40% of the calories are from fat. Such diets consist of more than 20% fat on a weight basis. If half of the fat in such a diet were to be replaced by a fat substitute, then the diet would consist of more than 10% of the fat substitute by weight. At such high consumption, it would be impossible to achieve a large safety factor by conducting only traditional toxicology studies with animals.

A large safety factor has been the standard for most new food additive approvals. Thus, to establish use with a lesser safety factor, one would need to establish safety with data from clinical trials in which human subjects consume levels of the fat substitute at and preferably above the highest expected exposure level to provide a margin of safety. Data from such studies must confirm and be consistent with safety data derived from animal studies. When inconsistencies occur, the data from the human study must demonstrate a high level of safety when compared with data from animal studies.

When using data from animal studies, one must consider whether an animal model is appropriate for determining health effects that are relevant to humans. Many traditional animal models are not interpretable in terms of human health effects causally related to altered gastrointestinal physiology or nutrient availability because of significant differences between animal and human dietary tolerances, nutrient requirements, absorption mechanisms, or effects on metabolism. The use of animal models to estimate adverse health effects of novel fat substitutes in human subpopulations with a potentially increased risk typically has not been possible because appropriate models have not been identified.

The FD&C Act [Section 409(i)] provides the FDA with authority to issue regulations to allow human investigational testing of unapproved food additives, but, to date, such regulations have not been proposed or promulgated. The development of regulations in this area could facilitate the evaluation of novel macronutrient replacement products by providing a uniform approach to human clinical testing for food additive safety evaluations.

New issues that may arise concerning the safety of use of fat substitutes include the theoretical long-term influences of novel macro food ingredients on nutritional status or on gastrointestinal physiology. A number of these issues may not be practical to investigate before marketing a product. Often, no surrogate measures can predict long-term effects on health outcome. Thus, in some cases it may be desirable to consider some form of postmarket surveillance to confirm the safety of long-term use. However, surveillance would be difficult to implement for food additives unless their use was very limited and surveillance was actively pursued. Also, the endpoints for safety evaluation would need to be anticipated and relatively discrete. Complex effects, such as those caused by multiple agents or those that result from interactions between dietary ingredients over time, are less likely to be uncovered by a postmarket surveillance system, particularly one that is passive.

Exposure Assessment

An important step in assuring the safety of new food ingredients, such as fat substitutes, is the calculation of the level of exposure to an individual if the product is approved for use. This step is frequently referred to as exposure assessment. A very simplified approach to exposure assessment is to estimate the amount of a new product that could be placed in the food supply for specific purposes and to divide that amount by the exposed population. This method of estimating exposure is unacceptable for safety assessments, as it results in unrealistically low individual exposures. It is necessary to use an exposure assessment model that is based on food consumption data from national probability surveys. These surveys indicate both the frequency of consumption and the serving size of individual foods.

The exposure assessment model used by the FDA and other regulatory agencies permits estimation of the fraction of the population that would consume more than a given amount of the product under evaluation. Usually, regulatory agencies require that safety assessments be based on exposure assessments made for the 90th or 95th percentile level of consumption by individuals who use the product.

When estimated exposure levels prove to be higher than safety data would support, they can be lowered by limiting product content or the categories of food in which the product may be used. In such cases, approval of the product may be contingent on the manufacturer agreeing to provide actual exposure data obtained through postmarket surveillance. Postmarket exposure surveillance, as opposed to safety surveillance, can be effectively conducted.

For fat substitutes that are reduced in caloric content, one must also consider some expanded use for caloric compensation. In addition, where such ingredients will have preferential markets and will be selected for high use levels (e.g. for weight reduction programs or for individuals requiring diets low in saturated fat to lower the risk of coronary heart disease), such use must be taken into account when evaluating safety. In this regard, exposure assessments for fat substitutes cannot rely solely on exposures to products that they are replacing.

Approaches to Safety Testing

To assess the safety of new food ingredients such as fat substitutes, one must develop tests that can demonstrate that the substance is safe. Procedures considered appropriate for testing new products are outlined in the FDA monograph "Toxicological Principles for the Safety Assessment of Direct Food Additives and Color Additives Used in Food," referred to as the Redbook (4). It is not our purpose here to discuss the detailed data requirements for approval of new ingredients. In the case of fat substitutes, however, discussion of some unusual nutritional considerations is appropriate.

As indicated above, a fat substitute could become a significant portion of the diet. If the substance is poorly digested or is not digested at all, the absorption of essential nutrients by an individual may be decreased if the substance (*a*) absorbs the nutrients on its surface (in the case of a fiber or fiber-like substance), (*b*) dissolves the nutrients (in the case of substances with lipid-like characteristics), or (*c*) alters the function of the digestive tract (e.g. alters transit times or microbial ecology). Another nutrition concern with nondigestible fat substitutes is the effect on the total consumption of nutrients. Consideration must be given to the replacement of those nutrients associated with fats, such as essential fatty acids and fat-soluble vitamins. If nutrient interferences can occur, the FDA must have data to indicate that they are trivial in terms of public health or that they can be accurately estimated and safely corrected by fortifying products that contain the substitute.

There could also be an increase in the consumption of nutrients, such as proteins and some carbohydrates, that may be present in digestible form as components of fat substitutes. In this regard, current levels of dietary protein in the United States are high, and further increases are not recommended (13).

The potential effect of fat substitutes on the microflora in the gastrointestinal tract must also be considered. Some of these microflora are associated with the synthesis of nutrients, such as vitamin K, biotin, and volatile fatty acids. Alterations could also result in altered pathogenicity or long-term effects on bowel health.

General health concerns and populations with potentially increased risk must be considered in the approval process for new food ingredients when safety is predicted on nonabsorption. Consider, for example, a new ingredient that is not absorbed in healthy animals, does not affect the absorption or synthesis of essential nutrients by intestinal bacteria, does not have any toxic effect on the gastrointestinal tract, but does accumulate in body tissues or shows toxic effects when it is injected into animals. It may then be necessary to demonstrate that the ingredient would not cross the gastrointestinal tract wall in the event that the epithelial tissues were compromised by a disease or injury that is likely to occur in a subpopulation.

Ingredients that are unmodified during transit through the gastrointestinal tract may cause or contribute to the formation of intestinal blockage. Substances that have a tendency to clump when changes in hydration or pH occur are of special concern.

The laxative effects of ingredients that are not digested may also be of concern. The laxative effects of nondigested substances among individuals may vary. Some individuals have a low tolerance for nondigested substances, which results in frequent defecation or anal leakage. In either situation, a reduction in absorption of nutrients may occur. Generally, most adults can recognize an association between the occurrence of the laxation and the consumption of the nondigested ingredient, and they avoid such products. However, some adults and many children cannot make the association. Under such circumstances, a limitation on the maximum exposure or special labeling may be necessary.

The safety evaluation of food additives is based on a projected effect on the general population. Specific effects in subpopulations are usually addressed through the provision of appropriate label information that can (but usually does not) consist of warning statements. In the case of novel fat substitutes, it may be difficult to predict the effect on subpopulations. Additional clinical trials may be necessary to define safety of use in subgroups at increased risk.

Safety and Nutritional Benefits of Prominent Fat Substitutes

In assessing the safety and long-term health effects of fat substitutes, one usually considers the origin of the substance and the degree of processing. In reality, each product must be evaluated on the basis of data on its chemistry, metabolism, and stability during processing and storage rather than on its origin. For discussion, however, it is convenient to group fat substitutes into categories based on their origins: (*a*) fat substitutes derived from traditional food sources, (*b*) fat substitutes produced through chemical synthesis, and (*c*) fat substitutes derived from novel food sources.

FAT SUBSTITUTES FROM TRADITIONAL FOOD SOURCES

The fat substitutes derived from traditional food sources are primarily carbohydrates, proteins, or combinations of both. Gums such as alginates, guar, carrageenan, and xanthan are widely used in substitute cream toppings and certain candies. Proteins such as gelatin or whey have often been used to provide stability and improve mouth feel. Although these gums have been used safely for some time in limited amounts, assessment of expanded use of these substances has in many cases not been made nor has a recent exposure assessment been made for consumers of these products. In addition to gums, several water-soluble bulking agents derived from hemicellulose and other soluble fibers may be used to replace some of the fat that is traditionally added to foods. Although many of these substances have been affirmed as GRAS for functional uses, they have not been assessed for safety of expanded use for the purpose of replacing macronutrients, such as fat. Responsibility for the safety of increased use of these products as fat substitutes rests with the manufacturers. The FDA will take action against such expanded uses when it has reason to believe that such uses are unsafe; however, expanded uses may occur without the agency's knowledge.

A number of ingredients have been derived from various sources of starches, such as potatoes, tapioca, and corn (1, 8, 9). These ingredients are derived through acid or enzymatic hydrolysis and consist of low molecular weight starches, dextrins, and maltodextrins that are readily digestible. These ingredients have been used in a variety of foods, including luncheon meats, salad dressings, frozen desserts, table spreads, dips, baked goods, and confections. Frequently, a specific ingredient can be used successfully in only a limited range of foods. High water-holding capacity and film-forming capacities enable these products to replace a substantial amount of fat in foods. These carbohydrates are generally considered safe and can be used at moderate levels in the diet.

More recently, proteins have been microparticulated under carefully controlled conditions to form particles smaller than 2 μm. These protein particles can bind with water at the rate of one part protein to two parts water by weight, thus yielding a product with one third of the caloric content of protein by weight. This hydrated product has excellent mouth feel and is particularly useful for frozen or refrigerated products, but it breaks down with thermal processing and thus cannot be used for frying or baking. Because the proteins are not altered chemically, they are digested and metabolized like ordinary protein. The FDA determined that use of the microparticulated proteins was GRAS for use in reduced calorie frozen desserts (6). By limiting the use of these products to certain categories of food, the increase in protein added to the diet of the 90th percentile consumer can be kept within a prudent range.

FAT SUBSTITUTES PRODUCED BY CHEMICAL SYNTHESIS

Fat substitutes that are derived through chemical synthesis are designed to closely mimic the properties of traditional fats, except that they are partially or totally undigested. The processes associated with the synthesis of these fat substitutes range from routine hydrogenation to more complex synthesis involving esterification and condensation. Polydextrose, a polymer made from glucose, is known as a reduced calorie, partially digestible bulking agent but also has been promoted as a fat substitute. Polydextrose is made by a high-temperature polymerization process that provides a number of glycosidic bonds in a 1–6 linkage, similar to the linkage that occurs in dietary fiber. Polydextrose was approved by the FDA in 1981 for use in products in eight food categories (10). Use of polydextrose in some of the products in these eight categories results in substantial reductions in fat. The categories include confections, frostings, salad dressings, and frozen dairy desserts. For these products and other approved uses, there are no limits on the amount of polydextrose that may be used. If a single serving of food contains more than 15 g of polydextrose, however, a statement on the label must be included to advise consumers that a laxative effect may be caused by excessive consumption of the product.

Data submitted to the FDA on the metabolism of polydextrose indicate that only one fourth of the product is absorbed by humans. Therefore, the agency has authorized the use of the value of 1 calorie per gram for calorie calculations in conjunction with nutrition labeling and nutrition claims. Perhaps the most important aspect of polydextrose from a nutrition perspective is the apparent lack of any effect on the absorption or metabolism of essential nutrients. The basic carbohydrate structure of polydextrose and its solubility in aqueous media reduce the probability that nutrients will be bound to the surface of the molecule.

Many fat substitutes derived through chemical synthesis are altered fats. For example, several attempts have been made to modify triglycerides to prevent or decrease the hydrolytic action of human lipase. One approach is to form a triglyceride composed of long-chain saturated fatty acids. Such triglycerides can be formed either by hydrogenating triglycerides containing long-chain unsaturated fats or through transesterification, with long-chain saturated fatty acids as the starting material. The resulting fats have high melting points, and they consequently pass through the gastrointestinal system mostly undigested (16). When significant amounts of high melting point fats are consumed, individuals may experience severe gastrointestinal disturbances and frequent defecation with nonformed stools. Another approach is to modify the ester linkage of a triglyceride, in a manner which would block the normal action of lipases because of steric hindrance. Whether the reduced absorption of either high melting point triglycerides or sterically modified triglycerides promotes reduced absorption of fat-soluble vitamins or other nutrients is not known.

The most intense research on the development of fat substitutes during the last two decades has been with carbohydrate fatty acid esters. These compounds are easy to synthesize by transesterification, with chemically active fatty acid forms, such as anhydrides, or by other organic chemistry techniques. By using a variety of fatty acids and carbohydrates, one can produce products with a wide range of physical and biological properties. The most investigated carbohydrate fatty acid esters for use as fat substitutes are derivatized monosaccharides and disaccharides.

The Procter and Gamble Company has petitioned the FDA for a food additive approval of an octa-fatty acid ester of sucrose, called sucrose polyester (SPE) (7). Early animal research showed that SPE was not hydrolyzed and there was no evidence of absorption, but early human testing revealed a problem of anal leakage. Extensive research on this product has improved its functional properties and eliminated anal leakage.

Extensive testing in animals and humans showed, however, that consumption of SPE lowered the absorption of some fat-soluble vitamins from the diet. Furthermore, at least for vitamin E, prolonged use of SPE has a negative effect on the body stores (7). Research has shown that vitamin supplements can be effective in maintaining animal body stores when the level of SPE consumption is maintained within certain limits. Data from clinical studies involving humans are currently being obtained and evaluated.

Another concern with SPE has been the absorption of lipophilic drugs. Preliminary data have been obtained through clinical studies on the most commonly used lipophilic drugs. At planned levels of use, no reductions in serum drug levels were observed. Should SPE be approved, however, drug manufacturers may have to perform tests on a broad range of current and future drugs to assure that absorption will not affect treatment.

The concern about substances such as SPE crossing the intestinal mucosa when compromised by disease or injury is being addressed. Researchers have shown that SPE injected into animals was accumulated primarily in the liver and was excreted slowly into the bile (7).

Other concepts for the development of fat substitutes through chemical synthesis have been proposed, including esters of polycarboxylic acids and the conversion of ester linkages of triglycerides to ethers. Although limited animal research has been reported for such fat substitutes, the approval of these products for human use is probably years away.

FATS DERIVED FROM NOVEL SOURCES

There has long been an interest in naturally occurring lipids that are partially or completely nondigested. Considerable research has been conducted on paraffins and mineral oils. In addition to promoting frequent defecations and anal leakage, these substances may be unsafe. Evidence indicates that consumption of mineral oils causes a decrease in absorption of fat-soluble vitamins because these vitamins dissolve in such oils. Furthermore, there is some evidence that mineral oils may be absorbed in small amounts and accumulate in tissues.

Oils from novel plant sources have also generated interest as potential fat substitutes. One such fat is jojoba oil. Jojoba oil is poorly digested and therefore may lower caloric intake, but little is known about its toxicological effects. Early indications are that its functional qualities in foods are marginally useful (9).

TOTAL DIET PERSPECTIVE

The FDA must maintain the safety and nutritional adequacy of the food supply when approving new food additives or affirming uses of ingredients as GRAS. Part of the strategy for carrying out this mission must be to maintain a total diet perspective. In an environment in which multiple new fat substitutes are being reviewed, the FDA must consider the cumulative effects that the substitutes may have. The combination of such substances may have significant additive or interactive effects on nutrient bioavailability, gastrointestinal tract disturbances, or general health. The existence of reduced or calorie-free carbohydrate substitutes must also be considered. Theoretically, the approval of nondigestible fat and carbohydrate substitutes could result in the production of a number of nonnutrient foods, which, if incorporated into an individual's diet, could have serious health consequences (5).

To make informed decisions, the FDA needs data that are sometimes difficult to obtain. To protect proprietary information, the agency frequently cannot require manufacturers to account for other products also being reviewed for approval. The alternatives are to obtain needed data through in-house research or to restrict initial approval to levels that are consistent with all existing data.

LABELING OF FOOD CONTAINING FAT SUBSTITUTES

With the enactment of the Nutrition Labeling and Education Act (NLEA) of 1990 (14), the labeling of all foods is subject to new requirements. Nutrition labeling will be required on virtually all packaged foods except those that are not a meaningful source of nutrients and those produced by small businesses. The NLEA requires that nutrition labels include total calories and the amounts of total fat, saturated fat, protein, carbohydrates, and fiber. The new law also requires that all foods have ingredient statements and that claims for nutrient content and claims for diet-disease relationships be defined by regulation.

In response to the new law, the FDA has proposed to define nutrient content claims such as fat free, low fat, and reduced fat. Proposals have also been published to permit health claims for a relationship between the level of dietary fats and cardiovascular disease, as well as for a relationship between the level of dietary fats and cancer.

Although labeling regulations do not require a specific listing of fat substitutes or the nutrition panel, the amounts of fat substitutes added to foods will affect labeling information. Fat substitutes, like all other ingredients, must be listed in order of predominance in the list of ingredients. The number of calories listed in the nutrition label must equal the digestible calories contained in a serving of food. In the case of a nondigestible ingredient, the manufacturer must have data to demonstrate the level of digestible calories (e.g. polydextrose has 1 calorie per gram).

The amounts of nondigestible lipid substances would not be included in total fats, just as nondigestible carbohydrates are not included in total carbohydrates. Most nondigestible carbohydrates will, however, be included as part of the total dietary fiber. In this regard, the physiological effects of these substitutes are not known. The use of fat substitutes may qualify a food for the label of fat free, low fat, or reduced fat, provided that the product complies with established definitions. Similarly, reductions in total fat or saturated fat made possible by the use of fat substitutes may qualify a product to be labeled with an appropriate health claim.

SUMMARY

Fat substitutes, in theory, may provide special health benefits to certain population segments. The most probable benefits are a reduction in total fat intake and a subsequent reduction in intake of calories from fat. Whether individuals who consume high intakes of fat substitutes that are partially or totally nondigestible also benefit from lower calorie intake on a long-term basis is unknown. It is likely that many individuals will compensate by increasing total food intake to maintain calorie intake.

Consumption of fat substitutes presents nutrition problems. Those fat substitutes that are partially or totally nondigested may reduce the bioavailability of other nutrients. Similarly, fat substitutes may have adverse effects on normal gastrointestinal tract function or intestinal tract flora.

Unlike other functional food additives, fat substitutes can make up a significant portion of the total diet. For this reason, traditional safety factors cannot be applied. Consequently, more reliance on data from clinical studies involving human subjects and requirements for postmarket surveillance will be necessary as part of the approval process.

ACKNOWLEDGMENT

The assistance of Keith B. Vanderveen in preparing this manuscript is greatly appreciated.

Literature Cited

1. Duxbury, D. D., Meihold, N. M. 1991. Special report: Diet, nutrition, and health. *Food Process.* 52:58–72
2. Federal Food, Drug, and Cosmetic Act as amended. 1989. Washington, DC: US Gov. Print. Off.
3. Food and Drug Administration. 1981. Polydextrose: Food additives permitted for direct additions to food for human consumption. *Fed. Regist.* 46:30080–81
4. Food and Drug Administration. 1982. *Toxicological Principles for the Safety Assessment of Direct Food Additives and Color Additives used in Food.* Washington, DC: Bureau of Foods
5. Food and Drug Administration. 1985. Report of the nonnutrient foods task force. Washington, DC: Cent. Food Safety Appl. Nutr. Unpublished report
6. Food and Drug Administration. 1990. Direct food substance affirmed as GRAS: Microparticulated protein. *Fed. Regist.* 55:6391
7. Jandacek, R. J. 1991. Developing a fat substitute. *ChemTech* 21:398–402
8. La Barge, R. G. 1988. The search for a low-calorie oil. *Food Technol.* 42(1):84–90
9. La Barge, R. G. 1991. Other low-calorie ingredients: Fat and oil substitutes. In *Alternative Sweeteners,* ed. L. O. Nabors, R. C. Gielari, pp. 423–50. New York: Dekker. 2nd ed.
10. Moppett, F. K. 1991. Polydextrose. See Ref. 9, pp. 401–21
11. Morrison, R. M. 1990. Food research and policy: National food review. *US Dep. Agric.* 32:24–30
12. National Cholesterol Education Program. 1990. Report of the expert panel on population strategies for blood cholesterol reduction. *US Dep. Health Serv. NIH Publ. 90-3046*
13. National Research Council, Committee on Diet and Health. 1989. *Diet and Health: Implications for Reducing Chronic Disease Risk.* Washington, DC: Natl. Acad. Press
14. US Code 301. November 8, 1990. Nutrition labeling and education act. *Public Law 101-535*

15. US Department of Health and Human Services. 1988. The Surgeon General's report on nutrition and health. *US Dep. Health Hum. Serv. Publ. 88. 50211*

16. Vanderveen, J. E., Heidelbaugh, N. D., O'Hara, M. J. 1966. Study of man during a 56-day exposure to an oxygen-helium atmosphere at 258 mm. Hg total pressure. IX. Nutritional evaluation of feeding bite-size foods. *Aerospace Med.* 37:591–94

17. Vanderveen, J. E. 1987. Nutritional equivalency from a regulatory perspective. *Food Technol.* 41(2):131–32

18. White House Conference. 1987. *White House Conference on Food, Nutrition, and Health, Final Report.* Washington, DC: US Gov. Print. Off.

19. Yackel, W. C. 1992. Application of starch-based fat replacers. *Proc. Symp. Starch Funct., Inst. Food Technol. Annu. Meet., Dallas. Food Technol.* In press

SUBJECT INDEX

A

Abetalipoproteinemia
  human milk and, 421, 423, 427
Acetaldehyde
  collagen synthesis and, 372
Acetate
  dietary fiber and, 27, 29
Acetazolamide
  placental nutrient transport and, 195
Acetic acid
  dietary fiber and, 28
Acetyl CoA
  cobalamin deficiency and, 63-64
Acetyl CoA carboxylase
  adipose tissue and, 213
N-Acetyl-L-cysteine
  food processing and, 127, 131-32
N-Acetylglutamate
  urea cycle and, 83
Acetyl LDL
  vitamin A transport and, 48
Acid phosphatase
  keratomalacia and, 7
Acitretin
  retinoids and, 170
Aconitase
  iron-responsive element binding proteins and, 352, 355, 357, 360-62
  iron-responsive elements and, 349-51
Actin
  placental nutrient transport and, 184
Actinic keratosis
  retinoids and, 167, 174
Actinomycin D
  ferritin and, 353
  iron-responsive element binding proteins and, 360
Acyl CoA
  cellular energy metabolism and, 336
Acyl CoA-binding protein
  adipose tissue and, 213, 217
Acyl CoA:retinol acyltransferase
  esterification of retinol in enterocytes and, 39, 50, 52
Acyl CoA synthetase
  adipose tissue and, 218

Adaptation
  to nutritional stress, 4-6
Adenine
  cobalamin deficiency and, 71, 74
Adenocarcinomas
  retinoids and, 168
Adenopathy
  diffuse
    toxic oil syndrome and, 247
Adenosine
  cobalamin deficiency and, 74
  placental nutrient transport and, 193
Adenosine diphosphate (ADP)
  cellular energy metabolism and, 331-36, 338, 340-41
Adenosine triphosphate (ATP)
  cellular energy metabolism and, 331-41
  placental nutrient transport and, 194-95
Adenosylcobalamin-dependent methylmalonyl CoA mutase
  cobalamin neuropathy and, 63-67
Adenosylhomocysteine
  cobalamin neuropathy and, 70-71
  hyperhomocyst(e)inemia and, 281
Adenosylmethionine
  cobalamin neuropathy and, 67-70, 74
  hyperhomocyst(e)inemia and, 281-82
Adenylate cyclase
  parathyroid hormone and, 382
Adenyl cyclase
  placental nutrient transport and, 184
ADH3 gene
  retinoic acid synthesis and, 51
Adipocyte lipid-binding protein
  adipose tissue and, 213, 217-18, 221-23
Adipose tissue
  brown, 207-8, 214-15, 223-26
  cancer risk and, 401, 410-11
  cell differentiation in vivo and, 215-24

chylomicron remmant retinyl esters and, 40
  human milk and, 421-22
  lactational capacity and, 103-4
  research trends and, 225-26
  secretory cells and, 224-25
  space flight and, 260
  white, 207-15, 225-26
Adipsin
  adipose tissue and, 213, 218, 221-22, 224, 226
Adrenocorticotropic hormones
  tryptophan and, 237
α-Adrenoreceptors
  adipose tissue and, 213, 217
β-Adrenoreceptors
  adipose tissue and, 213-15, 217, 223
  placental nutrient transport and, 184
Adults
  osteomalasia in, 380
Aflatoxin B₁
  retinoids and, 163
  structure of, 163
Africa
  albinism and, 154
  human milk composition and, 427
  lactose intolerance and, 13
Age
  urea cycle and, 82
Alanine
  food processing and, 121, 132
  placental nutrient transport and, 186-87
Albinisim
  β-carotene and, 154
Albumin
  human milk and, 421
  hyperhomocyst(c)inemia and, 281
  marasmus and, 5
Alcohol
  gallstones and, 317
Alcohol dehydrogenases
  retinoic acid synthesis and, 51
Alditol acetates
  dietary fiber and, 23
Alginates
  dietary fiber and, 25

placental nutrient transport
and, 192
Monoacylglycerols
human milk and, 434
Monoalkylglycerols
human milk and, 422
Monoamine oxidase
tryptophan and, 238
Monobutyrin
adipose tissue and, 210, 226
Monocarboxylates
placental nutrient transport
and, 189-90
Monoglyceride lipase
adipose tissue and, 210
2-Monopalmitoylglycerol
human milk and, 434
Monosaccharides
fat substitutes and, 483
placental nutrient transport
and, 185-86
Mono unsaturated fatty acids
human milk and, 426
infant formulas and, 433
Mortality
cancer and, 394-408
cobalamin deficiency and, 60,
62
hyperhomocyst(e)inemia and,
295
Motion sickness
space
microgravity and, 260, 267
Motretinide
cancer, 166
Mouth cancer
cholesterol and, 405
Mucilages
dietary fiber and, 22
Mucin
gallstones and, 303-5, 311,
312, 318
Mucopolysaccharides
dietary fiber and, 27
Muscle
cobalamin neuropathy and,
67
collagen and, 372
eosinophilia–myalgia syn-
drome and, 247
space flight and, 260
toxic oil syndrome and, 248
Mutagenicity
food processing and, 124
Myelin sheath
cobalamin neuropathy and,
60, 62, 64-66, 68, 70-
73, 75-76
eosinophilia–myalgia syn-
drome and, 245
Myelodysplastic syndromes
retinoids and, 162, 168,
173

Myocardial infarction
hyperhomocyst(e)inemia and,
288, 290-91

N

Na+
placental nutrient transport
and, 186-92, 195-99
space flight and, 268, 270-71
NAD(P)
cellular energy metabolism
and, 332-36, 338-41
α-Naphthyl acetate hydrolase
adipose tissue and, 210
Nausea
space motion sickness and,
267
toxic oil syndrome and, 247
Neck cancer
retinoids and, 162, 167
Neurocognitive symptoms
eosinophilia–myalgia syn-
drome and, 246
Neurolathyrism
in India, 11-12
Neuromuscular coordination
space flight and, 263
Neuropathy
cobalamin deficiency and, 60-
76
eosinophilia–myalgia syn-
drome and, 245, 247,
250
toxic oil syndrome and, 248
Neurotoxin
eosinophilia–myalgia syn-
drome and, 245-46, 250
Neutrophils
retinoic acid and, 164
New Zealand
cancer risk and, 397
Niacin
cancer risk and, 403
pellagra and, 11
tryptophan and, 237-38
Nicotinamide
pellagra and, 11
tryptophan and, 238
Nicotinamide adenine di-
nucleotide (NAD)
cellular energy metabolism
and, 332-35, 337-41
pellagra and, 11
tryptophan and, 237-38
Nicotinic acid
cancer risk and, 403
pellagra and, 10-11
placental nutrient transport
and, 191
tryptophan and, 237-38
Nitrate reductase
dietary fiber and, 29

Nitric oxide
urea cycle and, 82, 95
N-[4-(5-Nitro-2-furyl)-2-
thiazolylformamide
(FANFT)
retinoids and, 163, 171, 174
Nitrogen
food processing and, 124
placental nutrient transport
and, 188
space flight and, 263, 268-69
urea cycle and, 85
Nitroreductase
dietary fiber and, 29
Nitrosamines
cancer and, 163, 169-71
food processing and, 125
structure of, 163
vitamin C and, 144, 147
vitamin E and, 144
Nitroso compounds
food processing and, 131
vitamin C and, 142
Nitrosourea
cancer and, 163
Nitrous oxide (N2O)
cobalamin neuropathy and,
60-62, 66-71, 75
Nonsteroidal antiinflammatory
drugs (NSAIDS)
eosinophilia–myalgia syn-
drome and, 247
North America
cancer risk and, 394, 396
eosinophilia–myalgia syn-
drome and, 236
human milk composition and,
427
N system
placental nutrient transport
and, 187
Nuclear magnetic resonance
(NMR)
mitochondrial respiration in
heart and, 331
Nucleic acids
oxidative damage and, 140,
142
Nucleophiles
food processing and, 131
Nucleosides
placental nutrient transport
and, 193
5'-Nucleotidase
placental nutrient transport
and, 184
Numbness
cobalamin neuropathy and, 60
Nutritional stress
adaptation to, 4-6
Nuts
antioxidant properties and,
146

CUMULATIVE INDEXES

CONTRIBUTING AUTHORS, VOLUMES 8–12

515

CHAPTER TITLES, VOLUMES 8–12

ANNUAL REVIEWS INC.

a nonprofit scientific publisher
4139 El Camino Way
P. O. Box 10139
Palo Alto, CA 94303-0897 • USA

ORDER FORM
ORDER TOLL FREE **1-800-523-8635** (except California)
FAX: 415-855-9815

Annual Reviews Inc. publications may be ordered directly from our office; through booksellers and subscription agents, worldwide; and through participating professional societies.

Prices are subject to change without notice. ARI Federal I.D. #94-1156476

- **Individuals:** Prepayment required on new accounts by check or money order (in U.S. dollars, check drawn on U.S. bank) or charge to MasterCard, VISA, or American Express.

- **Institutional Buyers:** Please include purchase order.

- **Students: $10.00 discount** from retail price, per volume. Prepayment required. Proof of student status must be provided. (Photocopy of Student I.D. is acceptable.) Student must be a degree candidate at an accredited institution. Order direct from Annual Reviews. Orders received through bookstores and institutions requesting student rates will be returned.

- **Professional Society Members:** Societies who have a contractual arrangement with Annual Reviews offer our books at reduced rates to members. Contact your society for information.

- **California orders** must add applicable sales tax.

- **CANADIAN ORDERS:** We must now collect 7% General Sales Tax on orders shipped to Canada. Canadian orders will not be accepted unless this tax has been added. Tax Registration # R 121 449-029. **Note:** Effective 1-1-92 Canadian prices increase from USA level to "other countries" level. See below.

- **Telephone orders,** paid by credit card, welcomed. Call Toll Free **1-800-523-8635** (except in California). California customers use 1-415-493-4400 (not toll free). M-F, 8:00 am - 4:00 pm, Pacific Time. Students ordering by telephone must supply (by FAX or mail) proof of student status if proof from current academic year is not on file at Annual Reviews. Purchase orders from universities require written confirmation before shipment.

- **FAX: 415-855-9815   Telex: 910-290-0275**

- **Postage paid by Annual Reviews** (4th class bookrate). UPS domestic ground service (except to AK and HI) available at $2.00 extra per book. UPS air service or Airmail also available at cost. UPS requires street address. P.O. Box, APO, FPO, not acceptable.

- **Regular Orders:** Please list below the volumes you wish to order by volume number.

- **Standing Orders:** New volume in the series will be sent to you automatically each year upon publication. Cancellation may be made at any time. Please indicate volume number to begin standing order.

- **Prepublication Orders:** Volumes not yet published will be shipped in month and year indicated.

- **We do not ship on approval.**

ANNUAL REVIEWS SERIES *Volumes not listed are no longer in print*		Prices, postpaid, per volume		Regular Order Please send Volume(s):	Standing Order Begin with Volume:
		Until 12-31-91 USA & Canada / elsewhere	After 1-1-92 USA / other countries (incl. Canada)		
Annual Review of ANTHROPOLOGY					
Vols. 1-16	(1972-1987)	$33.00/$38.00 ⎤			
Vols. 17-18	(1988-1989)	$37.00/$42.00 ⎬ $41.00/$46.00			
Vols. 19-20	(1990-1991)	$41.00/$46.00 ⎦			
Vol. 21	(avail. Oct. 1992)	$44.00/$49.00	$44.00/$49.00	Vol(s)._____	Vol._____
Annual Review of ASTRONOMY AND ASTROPHYSICS					
Vols. 1, 5-14,	(1963, 1967-1976)				
16-20	(1978-1982)	$33.00/$38.00 ⎤			
Vols. 21-27	(1983-1989)	$49.00/$54.00 ⎬ $53.00/$58.00			
Vols. 28-29	(1990-1991)	$53.00/$58.00 ⎦			
Vol. 30	(avail. Sept. 1992)	$57.00/$62.00	$57.00/$62.00	Vol(s)._____	Vol._____
Annual Review of BIOCHEMISTRY					
Vols. 30-34, 36-56	(1961-1965, 1967-1987)	$35.00/$40.00 ⎤			
Vols. 57-58	(1988-1989)	$37.00/$42.00 ⎬ $41.00/$47.00			
Vols. 59-60	(1990-1991)	$41.00/$47.00 ⎦			
Vol. 61	(avail. July 1992)	$46.00/$52.00	$46.00/$52.00	Vol(s)._____	Vol._____